International Gas Union Research Conference (IGRC 2020)

AA005352

Muscat, Oman
24 – 26 February 2020

ISBN: 978-1-7138-1245-6

Printed from e-media with permission by:

Curran Associates, Inc.
57 Morehouse Lane
Red Hook, NY 12571

Some format issues inherent in the e-media version may also appear in this print version.

Printed with permission by Curran Associates, Inc. (2020)

International Gas Union (IGU)
IGU Secretariat
Att: Gas Natural Fenosa
Placa de Gas, 1
Building B 3rd floor
08003 Barcelona
Spain

Phone: +34 93 412 97 89
Fax: +34 93 402 54 26

secretariat@igu-gasnatural.com

Additional copies of this publication are available from:

Curran Associates, Inc.
57 Morehouse Lane
Red Hook, NY 12571 USA
Phone: 845-758-0400
Fax: 845-758-2633
Email: curran@proceedings.com
Web: www.proceedings.com

TABLE OF CONTENTS

SOLUTIONS FOR OPENING A DOOR; GAS FOR TRANSPORTATION 1
I. Nadhairi

THE DEVELOPMENT OF CRYOGENIC MILLING SYSTEM FOR OUTSOURCEING BUSINESS IN LNG TERMINAL 6
T. Mori

'HYDROGEN AS A FUEL FOR HEATING PROCESSES' 34
S. Gersen

ENGIE LAB CRIGEN: RESEARCH AND INNOVATION CENTER TO ACCERLATE TEH ZERO CARBON TRANSITION HYDROGEN 44
F. Legrand

TECHNOLOGY IN THE LNG CHAIN – AN UNFINISHED TRAIL OF INNOVATION 50
P. Hagyard

ADVANCED BOG MANAGEMENT IN LNG TRANSFER OPERATIONS 59
R. Tawk

COMBINING ARTIFICIAL INTELLIGENCE AND RESERVOIR SIMULATION FOR OPTIMAL CONTROL OF UNDERGROUND GAS STORAGE 73
J. Stopa

ACCURACY OF FISCAL METERING SYSTEM GAS NETWORK 84
F. Raisi

RESEARCH ON THE INJECTION OF NEW GASES INTO STORAGES 89
G. Caumette

STOCHASTIC APPROACH TO MINIMIZING MAINTENANCE COST INCORPORATING ATTITUDE TOWARDS RISK 94
M. Al-Kindi, N. Al-Azri

CONTRIBUTING TO A KNOWLEDGE BASED ECONOMY 103
S. Riyami

IMPACT OF CO2 CONTENT ON THE PERFORMANCE OF BIOGAS/H2 INVERSE DIFFUSION FLAMES 108
F. Quintino, E. Fernandes

ACHIEVING WELLBORE INTEGRITY IN HIGH H2S AND CO2 ENVIRONMENT IN OMAN 117
H. Yousef

QATARGAS FLARE REDUCTION JOURNEY 124
H. Al-Mohannadi

SOLUTIONS FOR THE FUTURE, TECHNOLOGY FOR FUTURE 131
S. Torun

CH4 EMISSIONS FROM END USE 147
J. Schweitzer

QATARGAS FLARE REDUCTION PROJECT 155
A. Benhamidcha

FIT FOR THE FUTURE: MAJOR PROJECTS AT OMAN LNG 165
H. Lawati

MODULARIZATION FOR LNG PLANTS 172
Y. Tamimi

NOVEL ACTIVATED CARBONS FOR CO2/CH4 ADSORPTION 177
K. Hanashi

VISUALIZATION / IOT OF INDUSTRIAL EQUIPMENT 188
M. Yanaka

ABSORPTION OF CO2 IN BLENDED AQUEOUS SOLUTION OF 2-AMINO-2-HYDROXYMETHYL-1, 3-PROPANEDIOL (AHPD) AND PIPERAZINE (PZ): MODELING OF CO2 SOLUBILITY DATA USING ARTIFICIAL NEURAL NETWORK APPROACH 204
G. Murshid

USE OF „ RENEWABLE HEAT“ TO INCREASE „INTELLIGENT 3-GENERATIONTECHNOLOGY´S“ EFFICIENCY 219
I. Discantiny

AUTONOMOUS INTEGRITY RECOGNITION (AIR) 227
H. Tholen

INTEGRATION OF GAS AND ELECTRICITY WITH RESIDENTIAL FUEL CELL 233
K. Ito

EVALUATING THE RESILIENCE OF INTEGRATED ENERGY SYSTEMS USING DYNAMIC SIMULATION 243
C. Bode, A. Senkel, G. Schmitz

CONVERGENCE BETWEEN ELECTRICITY & FUEL-GAS FOR ENERGY TRANSITION TO ZERO CARBON FOOTPRINT 250
P. Buchet

INVESTIGATION OF BIOMETHANE TRACE ELEMENTS COMPOSITIONS: DATA SCIENCE STUDY TO IMPROVE THE KNOWLEDGE 262
A. Lovat

DEVELOPMENT OF RFID FOR LOCALIZATION OF PE GAS BURIED PIPES 271
M. Hardy

GRTGAZ R&D ROADMAPS ON RENEWABLE NATURAL GAS AND PRESENTATION OF EXPERIMENTAL PROJECTS 278
D. Ballestas, C. Tastard, A. Louvat

THE FUTURE ROLE OF GAS IN A CARBON CONSTRAINED WORLD 288
J. Knijp

CONTROL VALVE MAINTENANCE DIGITALIZATION 304
A. Mahrouqi

TENSILE PERFORMANCE EVALUATION OF PVC PIPES FOR STEAM INJECTION WELLS 314
Z. Qamar, T. Pervez, F. Al-Jahwari

BUSINESS CONTINUITY PLAN (BCP) BY UTILIZATION OF MICRO GAS ENGINE COMBINED HEAT AND POWER (CHP) SYSTEM 342
K. Yasuda

CHALLENGES OF SUBSEA PIPELINES SLEEPERS AND CWC AGEING ISSUESSHARING OF QATARGAS RECENT EXPERIENCE WITH THE INDUSTRY 350
S. Khalaf

HEALTH MONITORING OF WELDED PIPELINES WITH MECHANICAL WAVES AND FUZZY INFERENCE SYSTEMS 358
M. Mohammadzaheri

TECHNO-ECONOMIC ASSESSMENT OF CO2 MONETIZATION TOWARDS METHANOL PRODUCTION 366
L. Sammarraie

CONTRIBUTION OF NATURAL GAS AS MID- AND LONG-TERM COUNTERMEASURE FOR GLOBAL WARMING AND THE ROAD TO THE CARBON NEUTRALITY 371
T. Nagata, F. Sano, K. Akimoto

SYNTHETIC NATURAL GAS (SNG) PRODUCTION: OPTIMAL CONFIGURATION AND SIGNIFICANT VARIABLES 400
D. Devaraj, E. Syron, P. Donnellan

CO2 RECYCLING: CO2 FIXATION INTO METHANE USING RUTHENIUM BASE CATALYSTS 408
N. Berahim, A. Seman, M. Noh

HODLNG 420
J. Finidori

DIGITALIZATION IS TRANSFORMING OUR BUSINESS 428
K. Venkatesan, T. Beerthuis

ENERGY DETERMINATION WHEN RENEWABLE GASES ARE INJECTED INTO CITY GRIDS 435
S. Rickelt, Y. Nordlund

IMPACT OF NEW GASES ON GAS INFRASTRUCTURE 444
D. Lanzi

ARTIFICIAL INTELLIGENCE MAINTENANCE TAB 452
J. Mahyijari

VARIABLE CYCLING FOR MOLECULAR SIEVE DEHYDRATION 458
H. Ismaili

ADVANCEMENTS IN LNG TECHNOLOGY 467
F. Legrand

DEVELOPMENT OF SOLID OXIDE FUEL CELL SYSTEM FOR COMMERCIAL USE AT TOHO GAS AND DENSO CORPORATION 475
Y. Sasamata

PROGRESS OF SOFC RESIDENTIAL CHP SYSTEM, ENE-FARM TYPE S IN OSAKA GAS 483
T. Nakao, S. Inoue, M. Suzuki, S. Iwata

CONTEXT AND OBJECTIVES: MICRO-CHP IN FRANCE, A QUITE COMPLEX SITUATION 497
P. Millin

Author Index

نعبر المحيطات لضمان المستقبل
Crossing Oceans to Secure the Future

Ship propulsion in the last 70 years

→ From wind power to gas fuel

OMAN SHIPPING COMPANY S.A.O.C

1 **Clipper *Sovereign of the Seas***

The clipper ship built in 1852 was a sailing vessel setting the 1854 world record for fastest sailing ship at 22 knots

2 ***SS Agamemnon* long-distance steamship**

Built in 1865 to compete with tea clippers *Agamemnon* could steam at 10 knots consuming only 20 tons of coal a day (half of previous steamships) thanks to the compound steam engine

3 ***Selandia* efficient diesel-driven ship**

Selandia was the largest and most advanced diesel-driven ship at the time of her maiden voyage from Aalborg to London in January 1912 (12 knots)

4 **CMA CGM choose LNG for its biggest ships**

CMA CGM decides to equip its future nine 22,000+ teu ultra-large container ships (UCLS) to be delivered in 2020 with engines using liquefied natural gas

Crossing Oceans to Secure the Future

Environmental sustainability

Roles and Regulations

OMAN SHIPPING COMPANY S.A.O.C

2018 — EU MRV
First mandatory reporting period starts 1 Jan.

2019 — IMO DCS
First mandatory reporting period starts 1 Jan.

2020 — IMO Sox
Global 0.5% Sulphur cap enters into force on 1 Jan.
IMO EEDI
Phase 2 enters into force on 1 Jan (up to 20% reduction).

2021 — IMO NOx
North Sea & Baltic Sea ECA enters into force on 1 Jan (tire III).

2023 — IMO GHG
Adoption of revised strategy (cop 23)

2025 — IMO EEDI
Phase 3 enters into force on 1 Jan (up to 30% reduction)

2030 — IMO GHG
Reduction of CO_2 emissions per tone-mile by at least 40%, aiming for 70% by 2050.

2050 — IMO GHG
Annual emission reduction at least 50% compared to 2008.

Crossing Oceans to Secure the Future

Available options to comply with IMO 2020

Use of LNG as a Marine Fuel to comply with IMO 2020?

OMAN SHIPPING COMPANY S.A.O.C

	VLSFO	Scrubber	LNG

Explanation	0.1% S – 0.5% S Very Low Sulphur Fuel Oil.	New/Retrofit	Reduction SOx, NOx & PM emissions Liquefied natural gas.
Benefits	- Minimize operational difficulty. - Does not require large investments. - Seen as a short-term option for compliance with Sulphur emission regulations.	- Relatively cheaper. - Shorter payback period. - The cost of installing a scrubber is higher for retrofit system than for the new system.	- Proven technology for large container vessels cruise ships & bulk carriers. - Readily available, clean. - Cost advantage to marine gasoil. - Viable option in the medium & long term.
Risks	- Uncertainties about 0.5% compatibility & stability. - Possible quality & availability problems.	- Limited operating experience. - Risk of future regulations on carbon emissions. - Compliance concerns for open loop scrubber. - Reduces space in the ships.	- Inappropriate for most ships. - Infrastructure for storing & refueling LNG. - Requires longer terms planning (after 2025). - LNG tanks occupy huge space in the vessels.

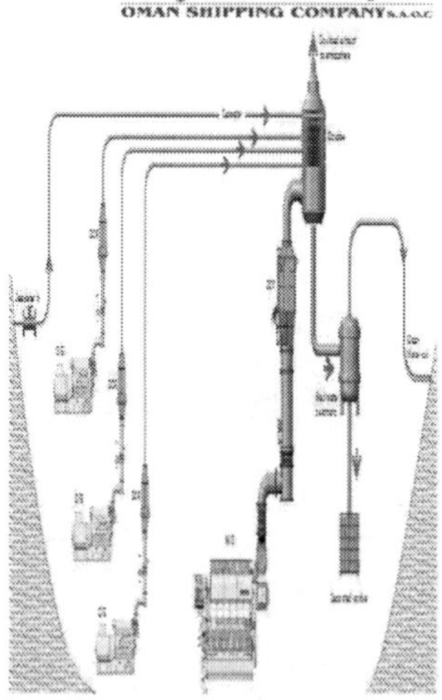

Crossing Oceans to Secure the Future

2020 IMO global Sulphur Cap

Where do we stand?

→ Uptake of solutions so far

Today	Scrubber	LNG capable
Fleet installed	1,739 ships (1.8%) 128.7m GT (9.2%)	530 ships (0.5%) 40.6m GT (2.9%)
Fleet incl. pending installation	2,846 ships (2.9%) 219.1m GT (15.7%)	
Orderbook (newbuilds)	578 ships (15.6%) 46.7m GT (35.4%)	348 ships (9.4%) 24.0m GT (18.2%)

Source: Clarksons Research (SIW 15 Nov 2019)

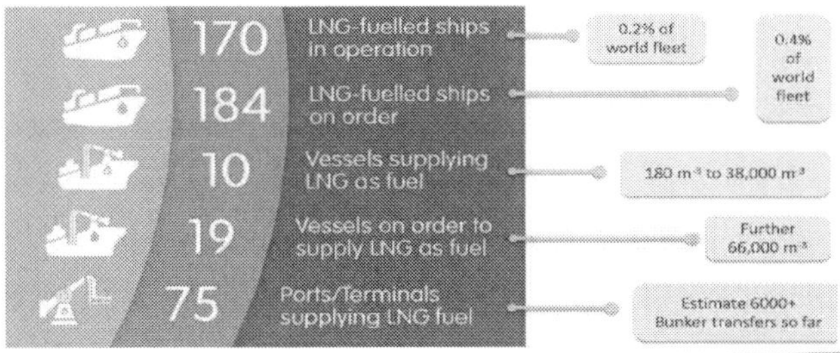

Ref: DNV GL / SGMF members database Sept 2019

→ Outlook

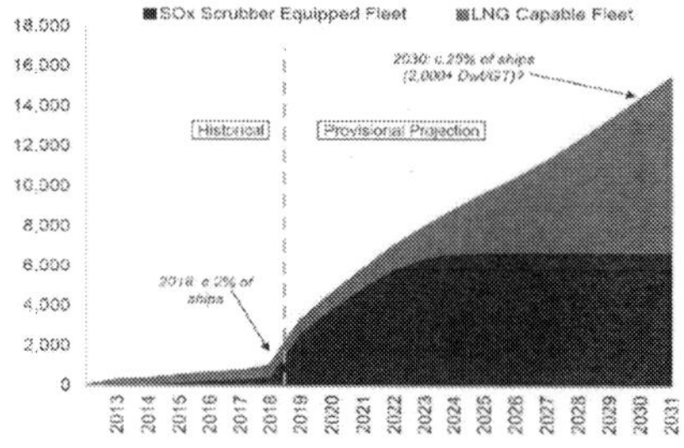

Source: Clarksons Research (SFC Sep 2019)

OMAN SHIPPING COMPANY S.A.O.C

Crossing Oceans to Secure the Future

Alternative fuels outlook

Target 2050

OMAN SHIPPING COMPANY S.A.O.C

→ LNG as fuel going forward

- *Only mature solution directly compliant with all current regulations*
- *LNG is cheaper than IFO380 (based on equivalent energy)*
- *Rapidly expanding bunkering infrastructure*
- *Long term viability when switching to biogas or synthetic methane/substitute natural gas (SNG)*
- *Combined cycle diesel/fuel cell could reduce CO2 emissions by up to 80% using carbon capture*

→ Other alternative fuels

- LPG & methanol/ethanol technically feasible, but not carbon neutral while lacking bunkering infrastructure
- Growing availability of biofuels, but high demand competition (other industries) & sustainability issues
- Hydrogen/ammonia carbon-free, but in R&D phase
- Ammonia has higher potential for shipping due to easier storage & handling (even if toxic & caustic)

1,036 teu feeder container ship WESAmelie was retrofitted to dual fuel (DF) in 2017
Drop in fuel (approx. 17%), saving 56 t of CO_2 emissions per round trip

Terntank has ordered a pair of 15,000 dwt DF-Hybrid chemical tankers. The ships can run on LNG and liquid biogas (LBG)

Crossing Oceans to Secure the Future

Compliant Fuel Choices - Today… and Tomorrow What are the options?

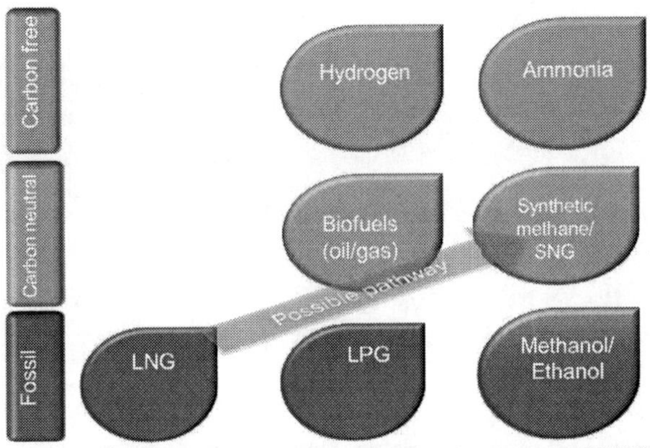

Key considerations
Maturity & availability of technology
Specific energy (weight) & density (volume)
Safety considerations (flammability, toxicity)
Regulatory framework
Global availability of fuel (terminal network)
Availability of bunkering facilities
Sustainability (ESG/CSR aspects)
CAPEX
OPEX
Flexibility for future adaptation

Crossing Oceans to Secure the Future

Closing note

- LNG is a fully compliant IMO 2020 fuel – Reduction: Sox; Nox; GHG
- Upstream CH4 losses can be reduced
- Further on board CH4 slip improvement is needed
- Use of Bio-LNG and Synthetic LNG would further improve emission

LNG is Safe to Use and becoming readily available

- Marine Propulsion Engines can readily burn Methane
- Most ship types now have a Gas Fueled variant in service or on order
- We have a good but still developing regulatory framework

Crossing Oceans to Secure the Future

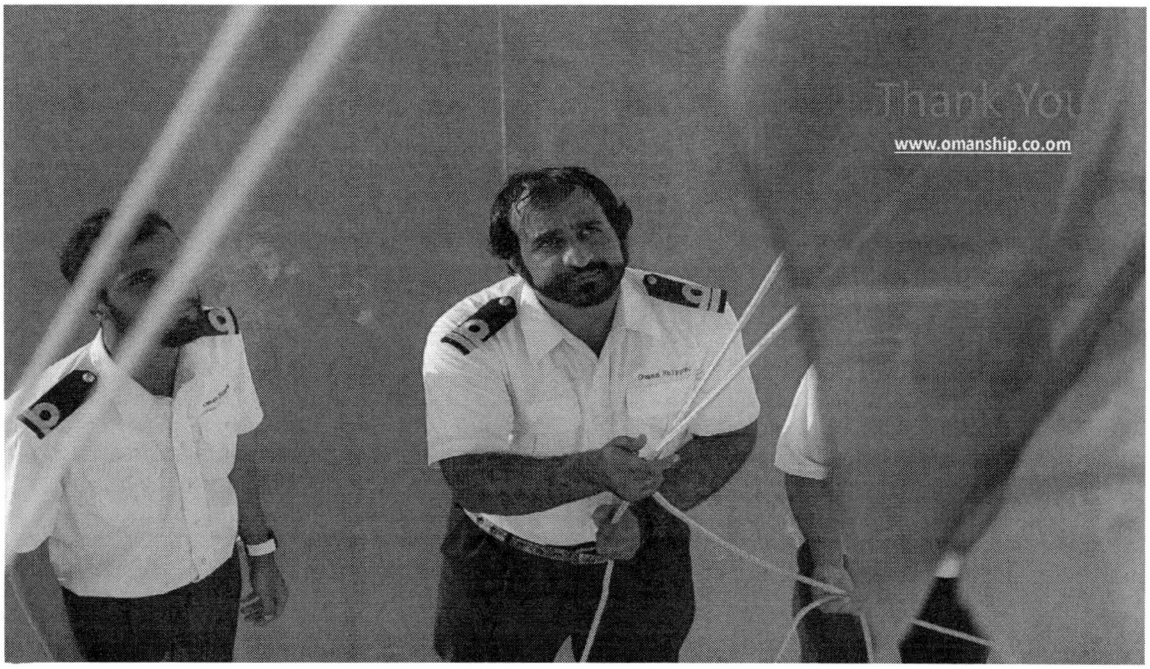
Thank You
www.omanship.co.om

IGRC2020

The development of cryogenic milling system for outsourcing business in LNG terminal

26th of February, 2020
Osaka Gas Co., Ltd.
Tetsuya Mori

大阪ガス Daigas Group

エネルギー技術研究所
ミ・ラ・イを、つくろう。

Cryogenic milling in LNG terminal

エネルギー技術研究所
ミ・ラ・イを、つくろう。

Cold energy → **Liquefied N2**

Raw material → **Freeze** → **Mill** → **Powder**

大阪ガス Daigas Group

1

Outline of Presentation

エネルギー技術研究所 ミ・ラ・イを、つくろう。

- Introduction
  - ·Outline of Daigasu (Osaka gas) group
  - ·LNG-related Technologies
- Overview of cryogenic milling
- Development of new cryogenic mill
  - ·Development overview
  - ·Computer simulation
  - ·Result
- Conclusions

大阪ガス Daigas Group

2

Outline of Presentation

エネルギー技術研究所 ミ・ラ・イを、つくろう。

- Introduction
  - ·Outline of Daigasu (Osaka gas) group
  - ·LNG-related Technologies
- Overview of cryogenic milling
- Development of new cryogenic mill
  - ·Development overview
  - ·Computer simulation
  - ·Result
- Conclusions

大阪ガス Daigas Group

3

Outline of Osaka gas

- Establishment : 1897
- Capital : 132,166 million yen
- Employees : 20,224 (Consolidated)
- Affiliated companies : 160
- Customer : Approx. 8.27 million

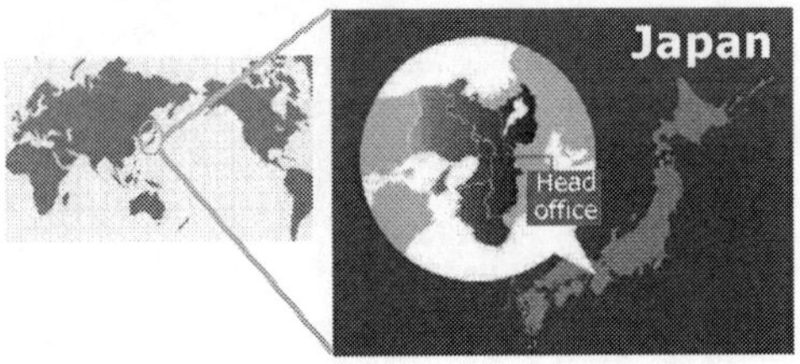

World's 3rd-largest gas utility company

Company	Country
Gas Natural SDG	Spain
Tokyo Gas	Japan
Osaka Gas	Japan
Spam S.p.A.	Italy
Hong Kong & China Gas	China
Korea Gas Corp	Korea
GAIL Ltd.	India

Source : Platts Top 250, Global Energy Company Ranking 2016 (Gas Utilities)

Outline of Osaka gas

LNG terminal
2 locations
28 LNG tanks

Pipeline
Approx. 62,700 km(*)

Power Generation
Approx. 2 GW in Japan

*) Daigas group (13th Nov. 2019)
https://www.osakagas.co.jp/company/about/business/energy16.html

https://www.osakagas.co.jp/en/business/valuechain/

Value Chain of Osaka Gas's Energy Business

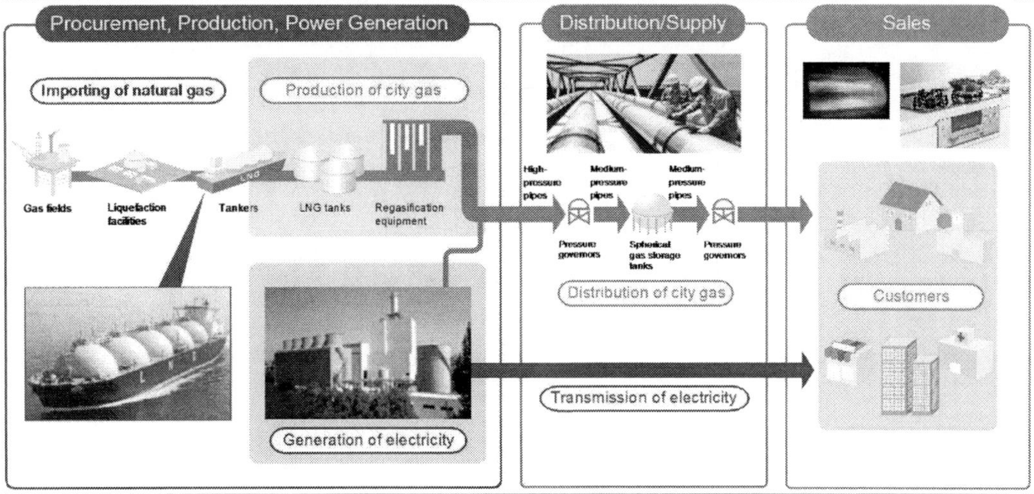

https://www.osakagas.co.jp/en/business/valuechain/

Gas business - LNG

Amount of LNG purchased
Approx. 9.5 million tons per year

From
8 countries

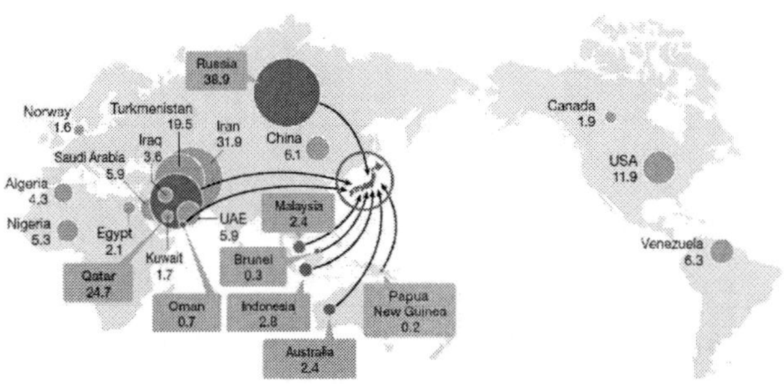

Own group carriers
7 ships

● Major country with natural gas reserves*

● Country with which Osaka Gas has concluded a long-term procurement agreement

Figures indicate the amount of natural gas reserves (Unit: trillion m³)

* Source: "BP Statistical Review of World Energy 2019" provided by BP plc.

https://www.osakagas.co.jp/en/ir/library/ar/pdf/2019/e_ar2019.pdf

エネルギー技術研究所
ミ・ラ・イを、つくろう。

LNG terminals

Owner and operator of world-class LNG terminals

- Construction and operation of LNG receiving terminals
- Technological innovations
- Overseas consulting experiences in LNG terminal and LNG cold utilization

Terminal	Senboku I & II	Himeji
Vaporizers (Capacity)	20 units (1,775 ton/h)	6 units (600 ton/h)
LNG tanks (Capacity)	22 units (1.8 mil m^3)	8 units (0.74 mil m^3)

エネルギー技術研究所
ミ・ラ・イを、つくろう。

LNG-related Technologies

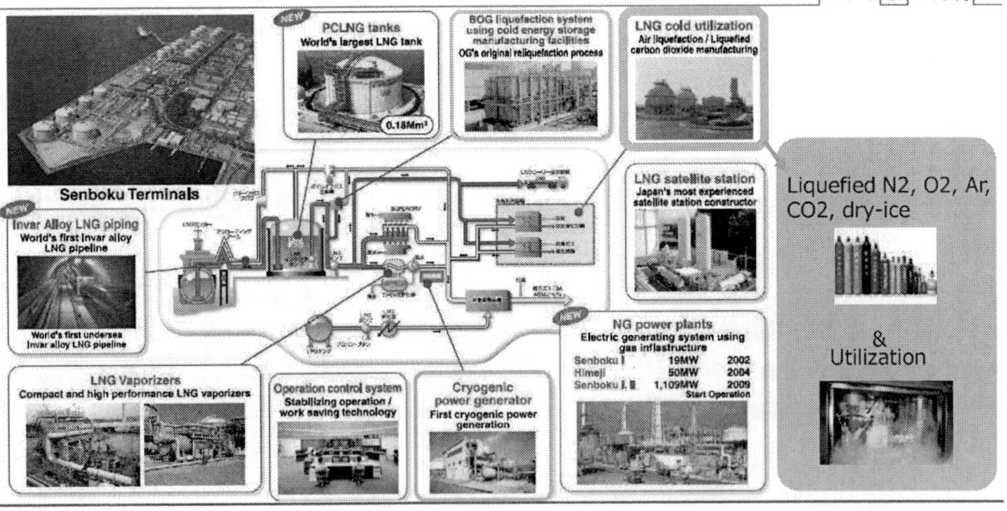

Outsourcing cryogenic milling service

エネルギー技術研究所 ミ・ラ・イを、つくろう。

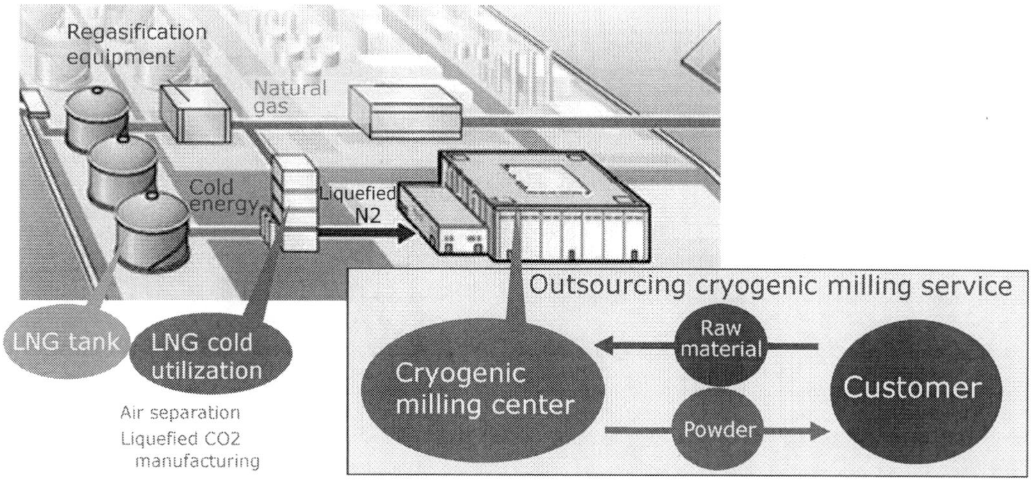

10

Outline of Presentation

エネルギー技術研究所 ミ・ラ・イを、つくろう。

- Introduction
  - ･Outline of Daigasu (Osaka gas) group
  - ･LNG-related Technologies
- Overview of cryogenic milling
- Development of new cryogenic mill
  - ･Development overview
  - ･Computer simulation
  - ･Result
- Conclusions

11

エネルギー技術研究所 ミ・ラ・イを、つくろう。

Overview of cryogenic milling

Cryogenic Mill

- Low temperature brittleness
- Liquefied N2 produced using cold energy of LNG
- Two factories for food and resin

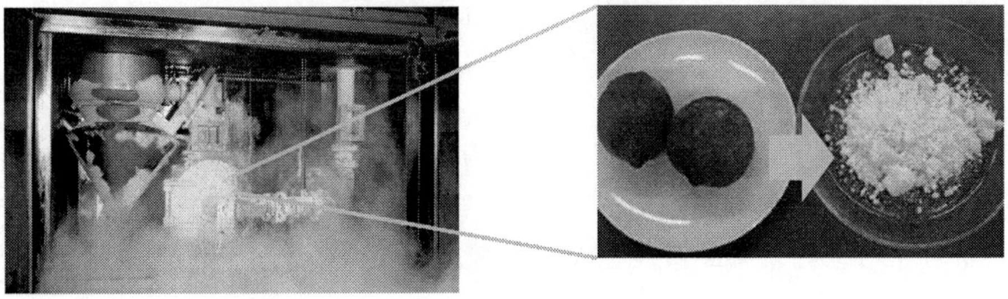

エネルギー技術研究所 ミ・ラ・イを、つくろう。

System of Cryogenic Mill

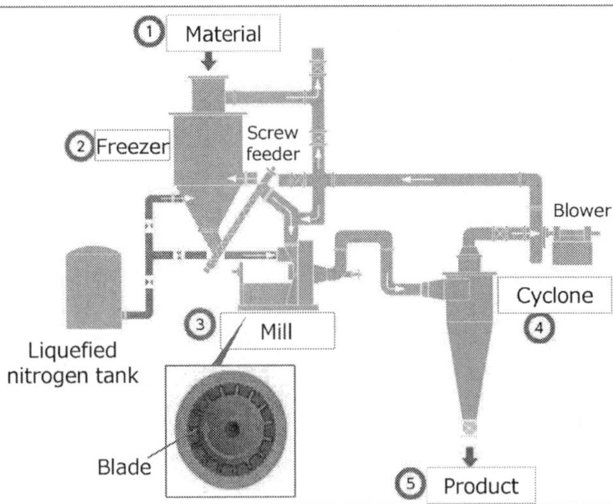

Features of cryogenic milling

- Soft materials can be easily milled.
  plastics, rubber, foods containing oil or moisture
- Cold → Suppression of deterioration due to heat
- N2 → Suppression of oxidation

Various pulverized products

-196°C Strong Zero
(Picture provided by Suntory Sprits Ltd.)

Example of cryogenic milling

- More than 800 kind of materials for over 30 year
- Two factories for food and resin

Foods

	Particle size distribution
Coffee	5~45μm(For beverages)
Roasted Sesame	180~300μm(Sesame tofu, Soup)
Gelatin	50~200μm(Food additive)
Raisins	15~60μm (confectionery)
Roasted almond	5~40μm(confectionery)

Axis: 10μm, 100μm, 1000μm

Example of cryogenic milling

Resin

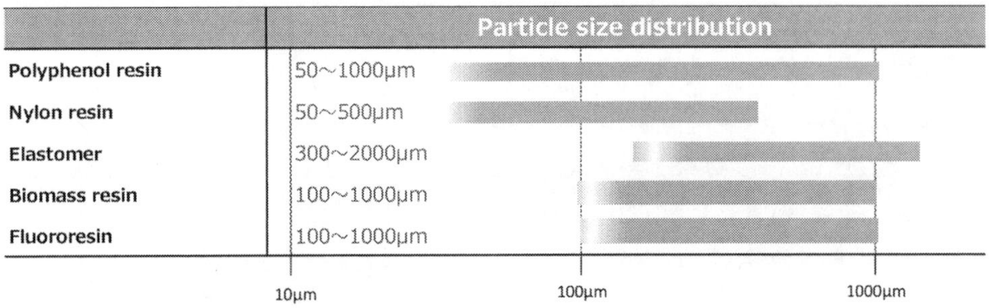

Demand for fine powder

Outline of Presentation

- Introduction
  - ･Outline of Daigasu (Osaka gas) group
  - ･LNG-related Technologies
- Overview of cryogenic milling
- Development of new cryogenic mill
  - ･Development overview
  - ･Computer simulation
  - ･Result
- Conclusions

Development of New System

The demand for fine powder has been growing in various fields.

・Advanced materials

・Resinification of vehicles and aircrafts (for weight reduction)

etc.

Development of New System

Development of Mill

Key point

- The high-speed rotation of the mill blades
- The control of the nitrogen gas flow inside the mill

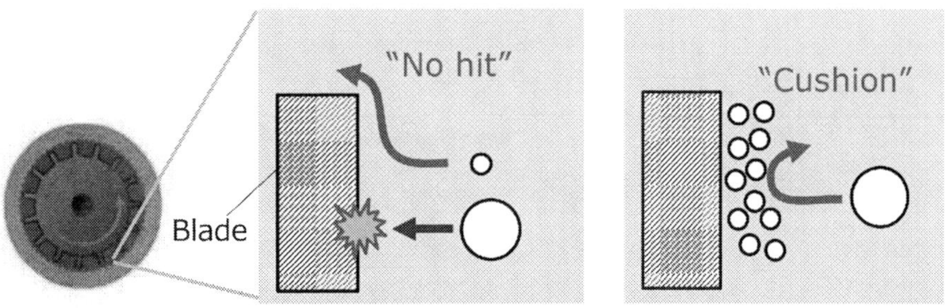

Development of Mill

エネルギー技術研究所 ミ・ラ・イを、つくろう。

- Understanding of the gas flow inside the mill
- Improvement of internal structure
- Optimization of operating conditions

↓

Computer Fluid Dynamics (CFD)

大阪ガス Daigas Group

20

Simulation in Osaka Gas

エネルギー技術研究所 ミ・ラ・イを、つくろう。

History of the simulation in Osaka Gas

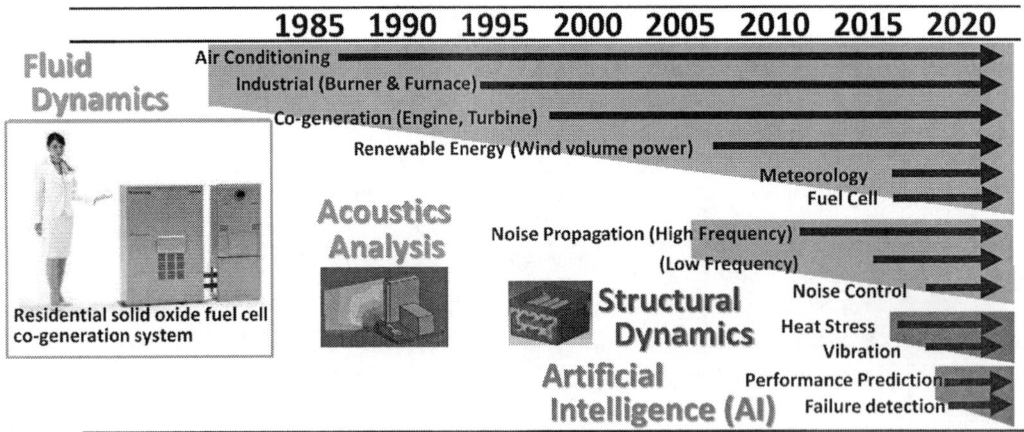

大阪ガス Daigas Group

21

Target of Simulation in Osaka Gas

- From upstream to downstream
- From small machines to meso-scale weather

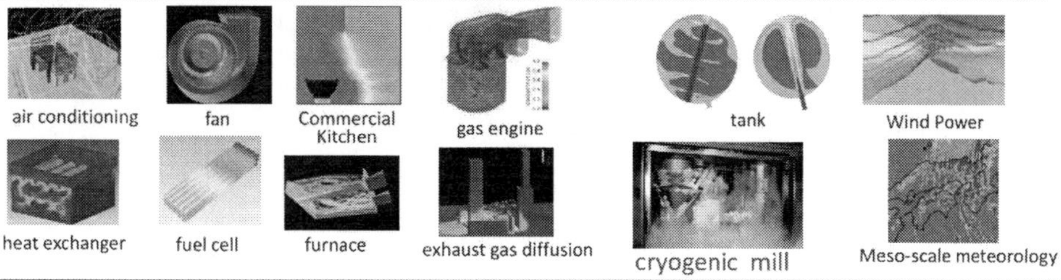

Computer Simulation

Computational Fluid Dynamics (CFD)

·Three dimensional CFD simulation
·Governing equations
Momentum, Pressure, Temperature

CFD code	ANSYS Fluent 17
Turbulence model	Realizable k-ε model
Algorithm	SIMPLE
Discretization scheme	Second order upwind
Rotation area	Multiple Reference Frame
Particle track	DPM

Example of CFD

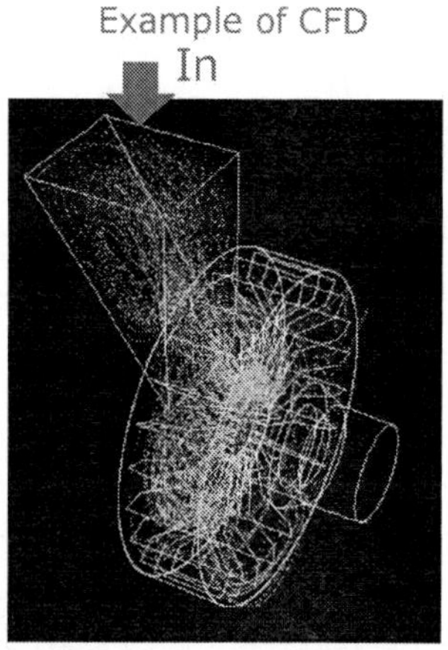

Computer Simulation

エネルギー技術研究所 ミ・ラ・イを、つくろう。

- The Simulation cannot directly obtain the particle size and throughput.
- We use the simulation to estimate the gas flow state in mill when the internal structure and conditions are changed.

Input

- Internal structure
  (Blade, Feeder, Mill body etc) Size, Shape, etc
- Operating conditions
  (Blade) Rotation speed,
  (Inlet) Gas flow rate, Temperature
  (Outlet) Pressure

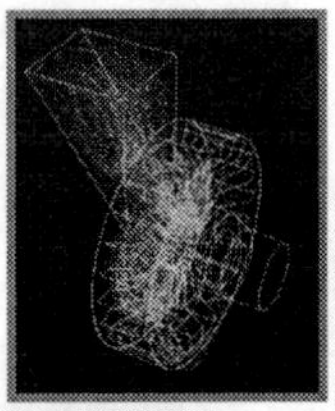

✕ Particle size, throughput

Output

Distribution of gas velocity, temperature, etc

Flow of Development

エネルギー技術研究所 ミ・ラ・イを、つくろう。

Step 1 (Prototype)
- Experiments: Particle size, throughput
- CFD: Gas flow state in the mill

Step 2
- Improvement
  - Internal structure
  - Operating conditions
- CFD: Gas flow state in the mill

Step 3 (Prototype) Change structure conditions
- Experiments: Particle size, throughput

Result

- The new system can produce much finer powder with more uniform particle size than conventional systems.

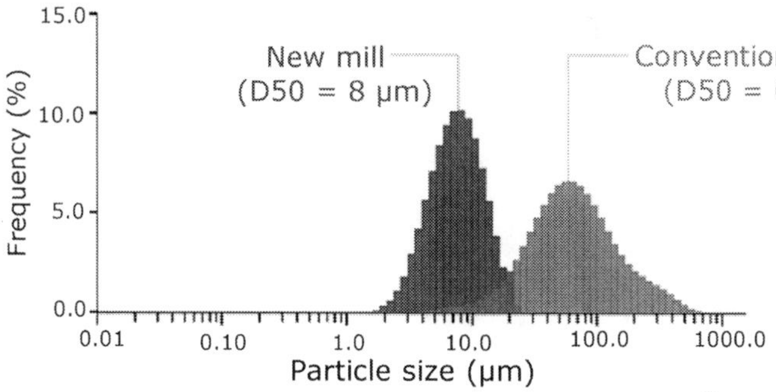

Raw material: polyethersulfone (*)

(*) Super Engineering Plastic: Highly functional plastics that overcome strength and heat resistance, which were the drawbacks of conventional general-purpose plastics.

Result

- The new system is in commercial operation and has been requested by many customers for milling.

Further improvement of various powder properties (liquidity, impregnation, etc)

High reputation
"So fine"
"I'd like a test by all means"

エネルギー技術研究所
ミ・ラ・イを、つくろう。

Conclusions

- We developed a new cryogenic milling system.
- For the development, we carried out experiments using a prototype along with CFD simulation.
- The new system can produce much finer powder with more uniform particle size than conventional systems.

大阪ガス Daigas Group 28

エネルギー技術研究所
ミ・ラ・イを、つくろう。

Conclusions

- Currently, we are also developing artificial intelligence technology to predict the size and quantity of particles that would be obtained after milling.
- We would like to support customer product development (material development) through machine learning and simulation technology.

Pioneering new markets by new cryogenic mill!

大阪ガス Daigas Group 29

エネルギー技術研究所

ミ・ラ・イを、つくろう。

Thank you very much for your attention!

大阪ガス Daigas Group

エネルギー技術研究所

ミ・ラ・イを、つくろう。

大阪ガス Daigas Group

Reference

Prediction of the size and quantity of particles

エネルギー技術研究所
ミ・ラ・イを、つくろう。

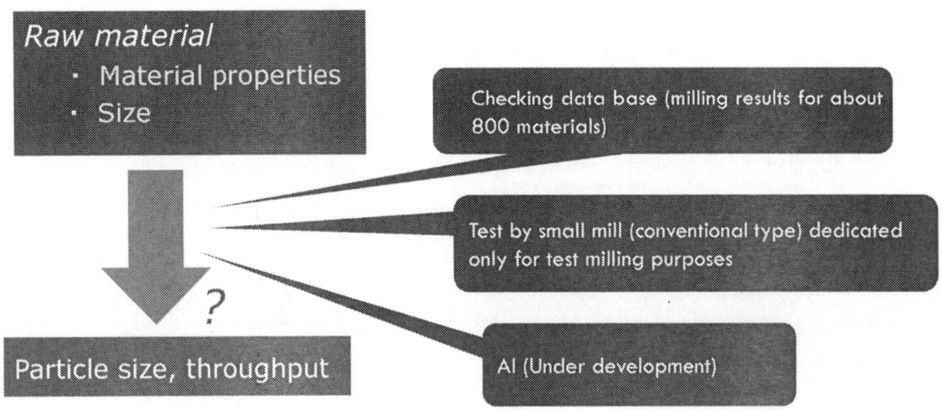

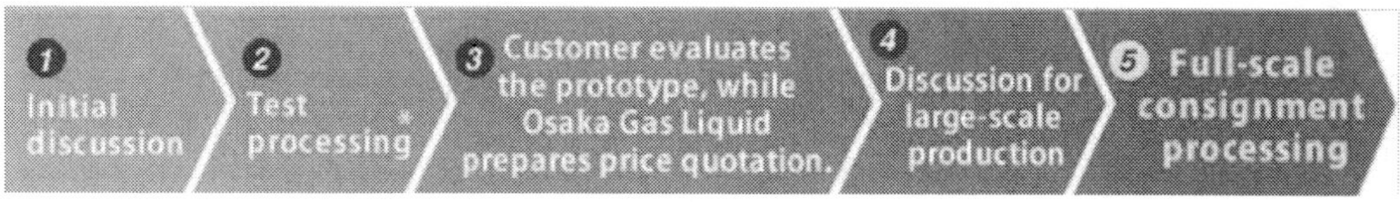

① Initial discussion to check grinding conditions and target grain size.

② Test processing of a few samples to study feasibility.

③ Acceptable raw material size: 5 mm or smaller (pellets, powder, etc.); quantity: 1-100 kg.

* Acceptable raw material criteria for foods: general viable bacteria count 100,000/g or less and negative coli group at the time of 10-time dilution.

④ Customer evaluates the prototype, while we prepare a price quotation.

⑤ Discussion of details for large-scale production, including standards requirements.

⑥ Start of full-scale consignment processing.

Classification systems

エネルギー技術研究所
ミ・ラ・イを、つくろう。

- By using various types of classifiers, we increase the added value and improve the quality and performance of products.
- By employing a wide range of machines from testing purposes to large-scale production, we can produce powder by the target grain size and particle distribution of the customer's requirements.

Types of classifiers and classification points

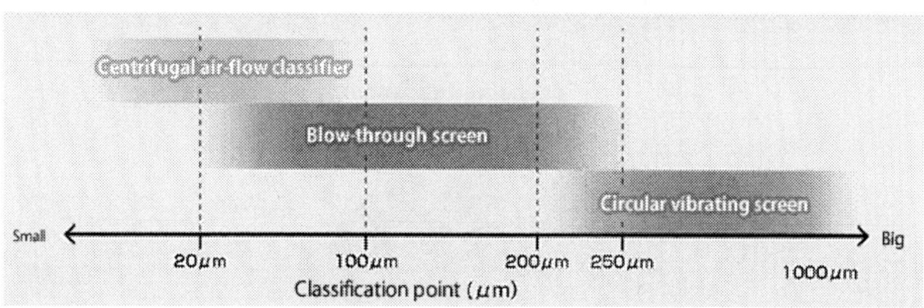

High-quality dry food powders with less thermal alteration

エネルギー技術研究所
ミ・ラ・イを、つくろう。

- A combination of cryogenic/freezer grinding and freeze-drying technologies is expected to retain more properties of the original foods such as nutrition and flavor compared with conventional methods.

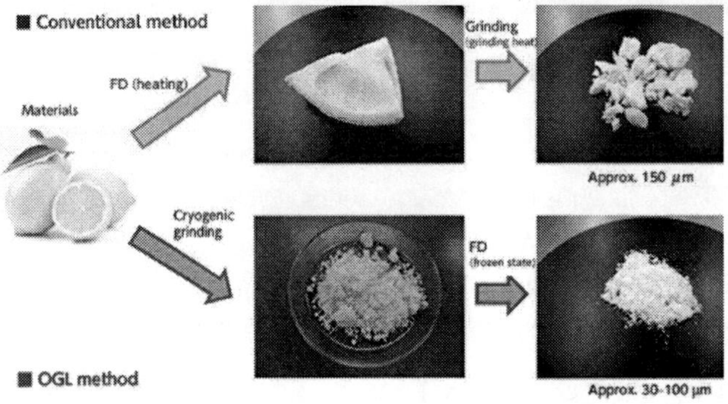

More uniform particle size of finely milled products

エネルギー技術研究所
ミ・ラ・イを、つくろう。

- Powder product with more uniform particle size (with reduced amount of fine powder waste).
- More uniform size and narrower distribution improve yield, handling, and performance.

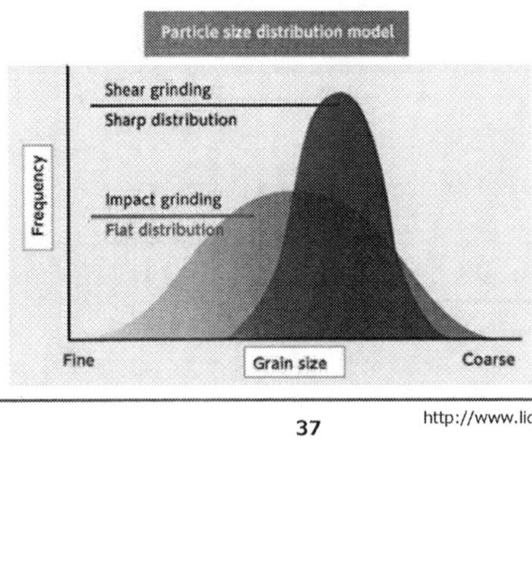

エネルギー技術研究所
ミ・ラ・イを、つくろう。

Can you grind anything if you can freeze it?

- It is effective for organic materials that become brittle at low temperature and substances commonly affected by temperature.
- Grinding is not effective for substances that are already brittle at room temperature, or ones that will not become brittle at low temperature.
- Ineffective : Thermoset resins (epoxy, etc.), inorganic compounds (ceramics, glass, minerals, etc.), fibrous materials (carbon, fiberglass, etc.)
- Effective: Thermoplastics (engineering plastics, elastomers, bioplastics, etc.), water- or oil-containing foods

大阪ガス Daigas Group

38

http://www.liquidgas.co.jp/english/product/faq.html

エネルギー技術研究所
ミ・ラ・イを、つくろう。

How finely can you grind?

- We can grind down to an average grain size of approx. 10 and several microns (25 microns pass).
- However, this is most common with brittle materials and economic efficiency requires separate consideration.

大阪ガス Daigas Group

39

http://www.liquidgas.co.jp/english/product/faq.html

エネルギー技術研究所
ミ・ラ・イを、つくろう。

What can you specifically grind?

- We have grinding results for about 800 materials and can inform you after checking the data.
- (Example) resins: PE (LD HD), PP, PMMA, Nylon 6, PLA, PTFE
  Foods: coffee, sesame seeds, gelatin, almonds, fruits

大阪ガス Daigas Group 40 http://www.liquidgas.co.jp/english/product/faq.html

エネルギー技術研究所
ミ・ラ・イを、つくろう。

Can you grind a material containing water or oil?

- Yes, we can. We can freeze and grind water and oil.
- However, the pulverized powder becomes slurry or pasty when it is returned to room temperature.

大阪ガス Daigas Group 41 http://www.liquidgas.co.jp/english/product/faq.html

What is your grinding capacity per day?

- We can handle maximum XX tons of material per day.
- However, the capacity varies depending on the properties of materials and target grain size (fineness).

What is your expected grinding yield?

- Usually, it is 95% to 98% or higher.
- However, the finer the target grain size and the narrower the distribution range, the lower the yield. On the other hand, large order volumes usually have relatively large yields.

Are there any constraints regarding the size of raw material?

- Ideally for our grinding machines, the material is in 5 mm cubes.
- Larger sizes may require another consideration of doing a coarse grinding process in advance.

Do you have a minimum order volume?

- No, we don't. We can accept orders of various sizes, from an order of a few kilograms for processing samples to several hundred kilograms or to several tons for commercial production.
- We have many types of milling machines that are adaptable to different volumes in consignment grinding

Do you accept test grinding alone?

- We have a dedicated machine and operator for test grinding and can accept orders even for just a few kilograms.
- The test machine is used for a day. Therefore, there is no restriction on volume or number of items. However, depending on the volume to be processed, you may need the machine for several days.
- Please feel free to contact us for consultation.

What are the advantages of outsourcing grinding for customers?

- You can eliminate the risk of huge capital investment and cumbersome tasks such as management of work and control of high-pressure gas and pollution (dust and noise), while easily obtaining the desired powder.

How long do you take to deliver the ground samples to us after we send you raw materials?

- Usually, we need about two weeks to deliver the samples after we receive the raw material, but it depends on how busy we are then.

Does cryogenic grinding has a bactericidal effect?

- No, it doesn't.
- The material is rapidly frozen and so there is little bactericidal effect.

- During cryogenic grinding, there is no bacterial growth.
- However, if a germ exists in the original material, it will survive in a hibernation state.

Does cryogenic grinding eliminate moisture (dry) in the material?

- No, it doesn't.
- Moisture contained in a material is frozen, so it will remain after grinding.
- However, just for your information, in grinding at room temperature, moisture may be decreased by heat generated from grinding.

How do you eliminate the risk of mixing in foreign objects (contamination)?

- We make a full effort to eliminate contamination risks by storing the equipment itself in cold storage to prevent contamination, and by disassembling and cleaning the equipment and piping after every operation.
- Our food factory is equipped with facilities with sanitary specifications.
- We pay particular attention to alcohol sterilization to prevent bacterial contamination as much as possible after cleaning the facility.

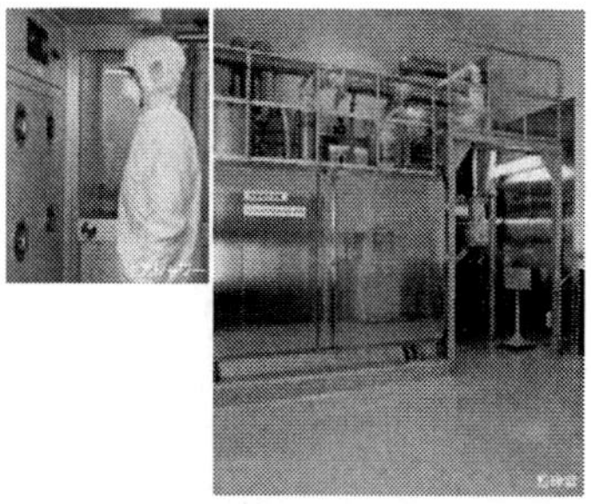

Food factory

Cryogenic power generator

エネルギー技術研究所
ミ・ラ・イを、つくろう。

- Osaka Gas has operated cryogenic power generation systems safely and stably for more than 30 years, the longest such operating period in the world.

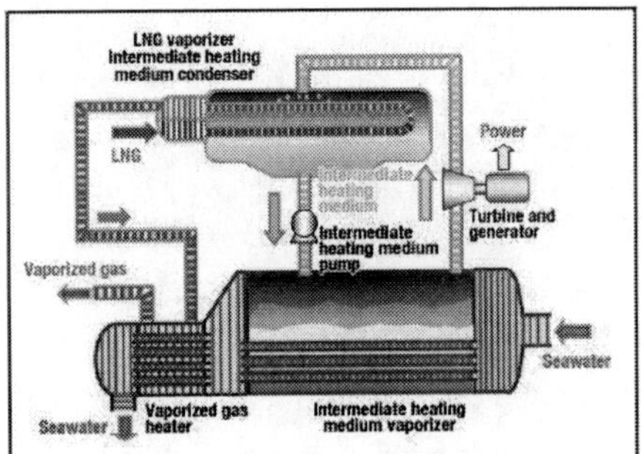

Fig. 1: Cryogenic power generation system from Osaka Gas

https://www.osakagas.co.jp/en/rd/technical/1198907_6995.html

Working principles of cryogenic power generation system

エネルギー技術研究所
ミ・ラ・イを、つくろう。

- Three types of LNG cryogenic power generation facilities are presently in operation: Rankine cycle, direct NG expansion, and Rankine cycle + direct NG expansion.
- The operating principle is explained below, taking as an example a Rankine cycle using propane as intermediate heating medium..

- The LNG pressure is increased to 3.5 to 4.5 MPaG by LNG pumps.
- The LNG vaporizer transfers heat from propane gas to LNG.
- Vaporized LNG then enters the NG trim heater and is heated to 0°C or higher by sea water.
- The liquid propane condensed in the LNG vaporizer via heat exchange with LNG is pressurized by the circulation pump and sent to the propane vaporizer where it vaporizes at a pressure in accord with the temperature of sea water.
- Vaporized propane gas drives the turbine during the expansion process and generates electricity. Subsequently, the propane gas is condensed again in the LNG vaporizer.

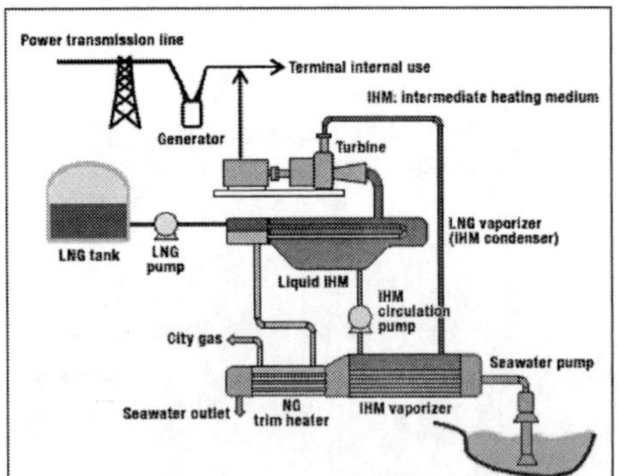

LNG cryogenic power generation system using intermediate heating medium

https://www.osakagas.co.jp/en/rd/technical/1198907_6995.html

ミ・ラ・イを、つくろう。

ミたことのない、
ライフ（暮らしとビジネス）を、
イノベーションしよう。

大阪ガス

Daigas Group

大阪ガスグループは、Daigasグループへ。

DNV·GL

'Hydrogen as a fuel for heating processes'

Special session: H2 as next step of gas, IGRC 2020

Sander Gersen

Tuesday, February 25, 2020

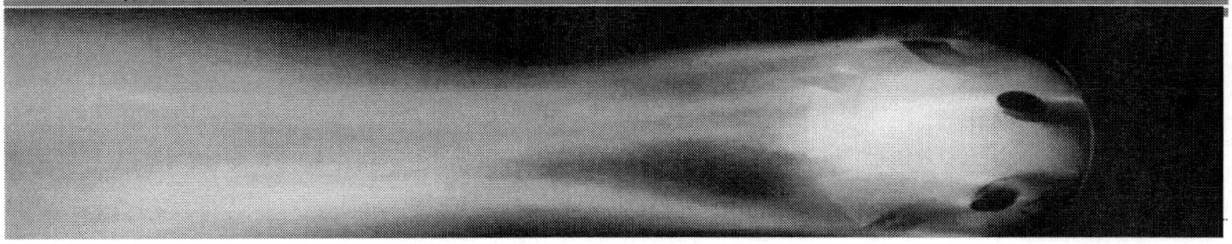

SAFER, SMARTER, GREENER

The need to reduce CO_2 emissions and enhance sustainability of energy supply are major drivers towards the introduction of renewables

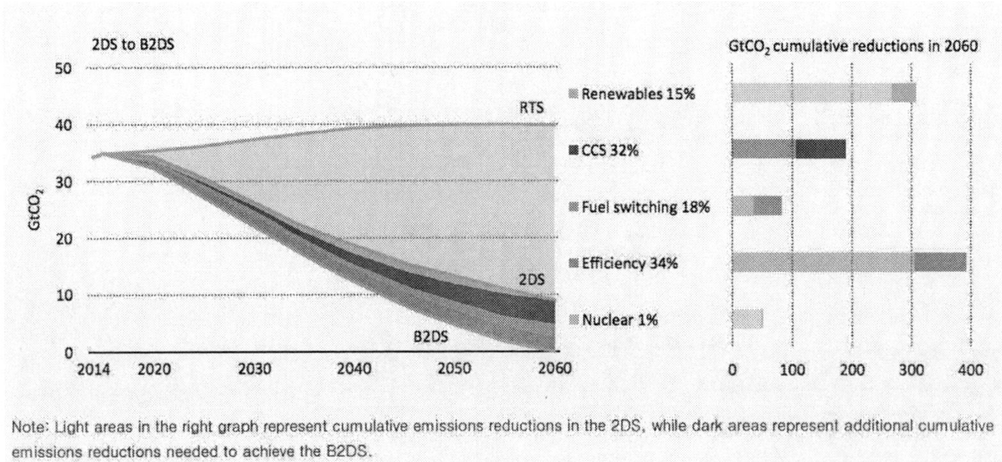

Note: Light areas in the right graph represent cumulative emissions reductions in the 2DS, while dark areas represent additional cumulative emissions reductions needed to achieve the B2DS.

Bron: https://www.carbonbrief.org/iea-world-can-reach-net-zero-emissions-by-2060-meet-paris-climate-goals

 source:https://www.iea.org/etp/explore/

Challenges to accommodate electricity produced from solar PV and wind

- Electricity produces from solar PV and wind increases substantially
- Intermittent nature of solar and wind energy puts pressure on capacity of power grids
- Excess of renewable electricity can be converted into hydrogen (Power-To-Hydrogen)
- Injecting renewable gas in existing gas grid is an effective means to avoid large investments in new infrastructures

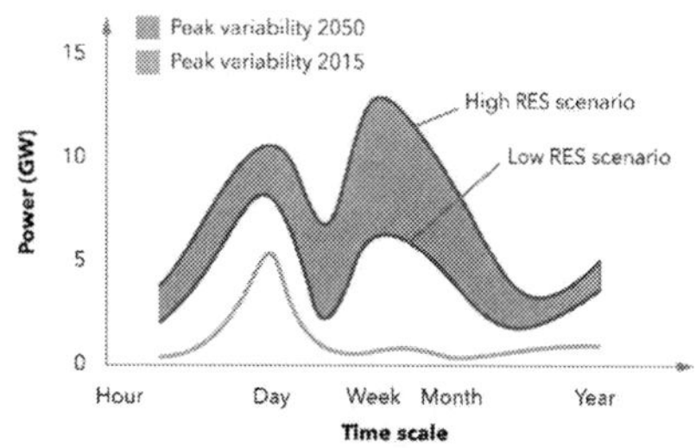

Hydrogen production technologies

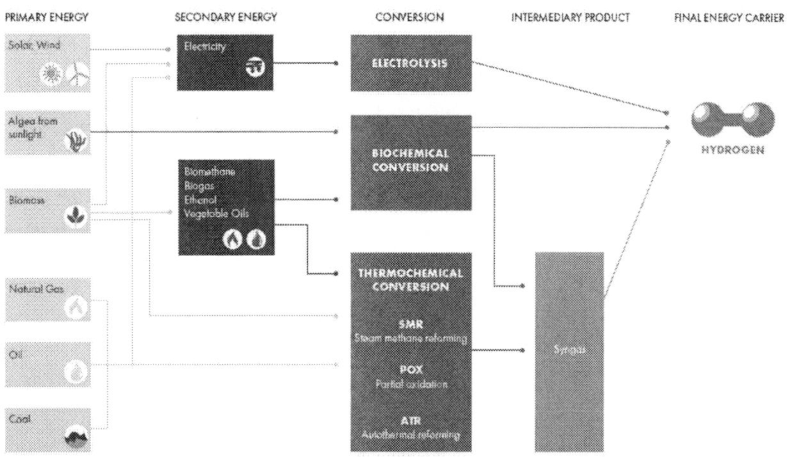

Source: Shell Hydrogen Study (2017)

Costs of hydrogen versus natural gas and electricity

- **Costs 1 MWhr heat:**
  - Natural gas (about 100nm³) € 20
  - Electricity € 50
  - H_2 from NG (SMR, € 1,65/kg) € 50
  - Green hydrogen € 70
    (elektrolysis, € 2,25/kg)
  - Hydrogen produced from electricity surplus € 35

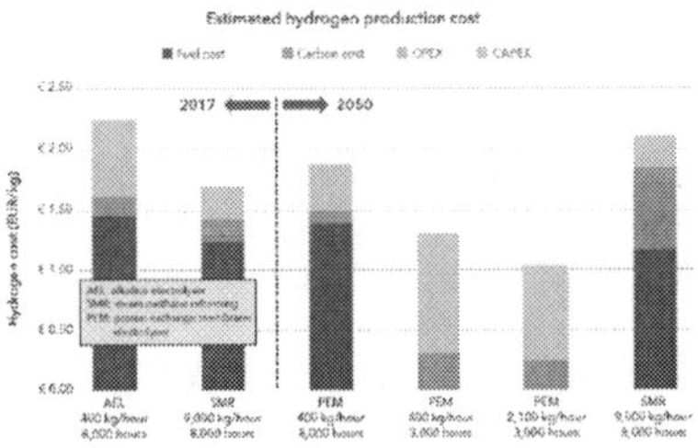

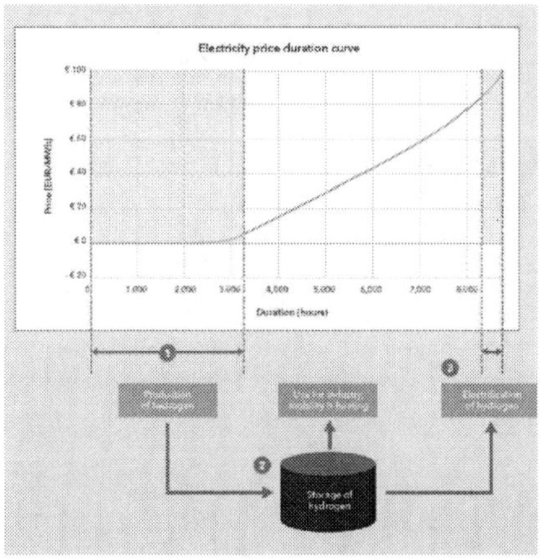

Renewable gas opportunities

- Hydrogen can be used in the mobility sector, the built environment and the industry
- Natural gas grids are already connected to the built environment and the industry
- Given the large penetration of natural gas in the total global energy mix, the increase of renewable gas in the total gas supply will result in a substantial carbon emission reduction

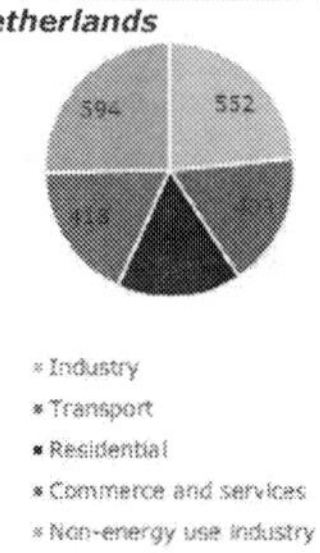

Role renewable gases in end-use equipment

- In each sector different end-use equipment is installed with a diversity of robustness towards renewable gases

Millions of end-use equipment installed:

- Wide diversity of robustness to gas quality
- Life-time equipment 15 yrs
- Competition with electricity (e.g. heat pumps)

Limited number of end-use equipment installed:

- Wide diversity of robustness to gas quality
- Often clustered with dedicated gas grids
- Gas is the preferred option in the energy-intensive industry

Hydrogen in the industry

Using hydrogen as a fuel is the fastest route to making the industry more sustainable

Advantages of using hydrogen in the industry:

1. Existing processes do not have to be adjusted drastically
2. The natural gas grid is already connected to the industry and can be used to transport hydrogen
3. Individual industrial cluster(s) can be readily converted to hydrogen
4. Shortest route towards decarbonisation for the lowest investment
5. Transport of gas is much **cheaper** than transport of electricity
6. Direct use of electricity is not always suitable for direct heating processes

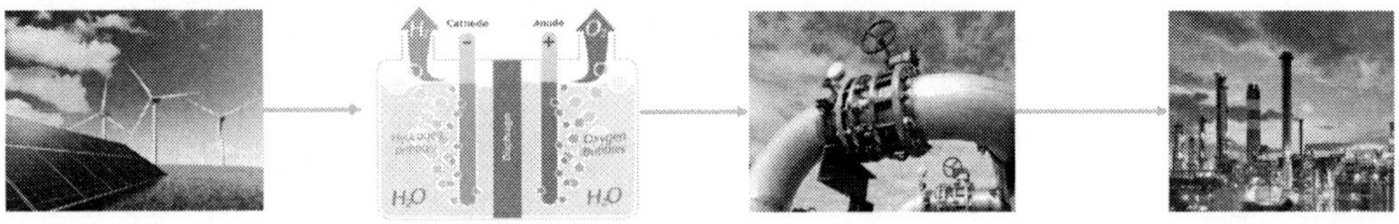

Hydrogen in the industry: example glass factories in the Netherlands

- 8 Production locations
- Total energy consumption in 2017: 9.972 TJ
  - 6.300 TJ Natural gas = 353.430 t CO_2
  - 2.700 TJ Electricity
- The annual natural gas consumption **± 200 Mln. m³** can be replaced potentially by hydrogen: equivalent to **± 600 Mln. m³** hydrogen (Economy of scale)
- As an illustration: this equals to about ca. 7500 line busses on hydrogen (the total number of line buses in the Netherlands is about 6500)

Challeges when using hydrogen in the industry (1)
Availability

- In the initial phase of the energy transition it is unlikely that there always will be enough hydrogen available to satisfy the entire industrial energy demand, whose processes usually run continuously (24/7) throughout the year
- An **attractive solution** is to apply a burner systems that can flexibly utilize the full mix of fuel compositions: 100% hydrogen, 100% natural gas and all mixtures of hydrogen and natural gas in between using a gas adaptive fuel-air ratio control system
- The major economic advantage of such a system is that it offers robust fuel flexibility with only a **limited investment**:
- **The same burner system** can be used throughout the transition, supplied initially with **varying** natural/gas hydrogen mixtures, and in the end with pure hydrogen when the supply has risen to the challenge.

Challenges when using hydrogen in the industry (2)
Combustion properties

Combustion properties of hydrogen differ substantially from natural gas:

1. Higher flame temperatures
2. Burning velocity of hydrogen is about 6 times higher than natural gas
3. Calorific value of hydrogen is three times lower than natural gas
4. Factor of four difference in the air requirement between hydrogen and natural gas
5. Potential increasing NO_x with increasing hydrogen fraction in natural gas
6. Changes in the oven atmosphere (flue gas composition)
7. Changes in the heat transfer (radiation/convection)

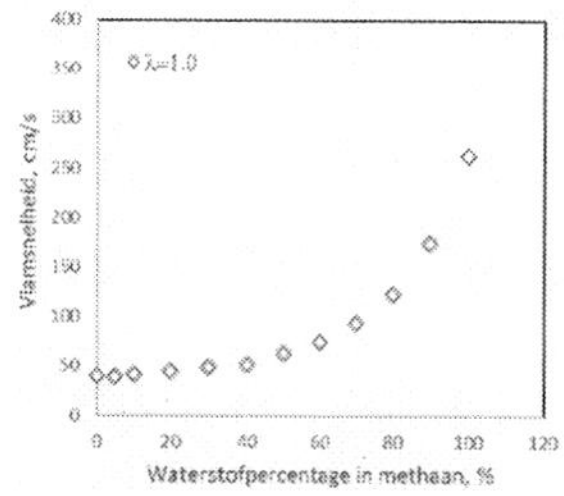

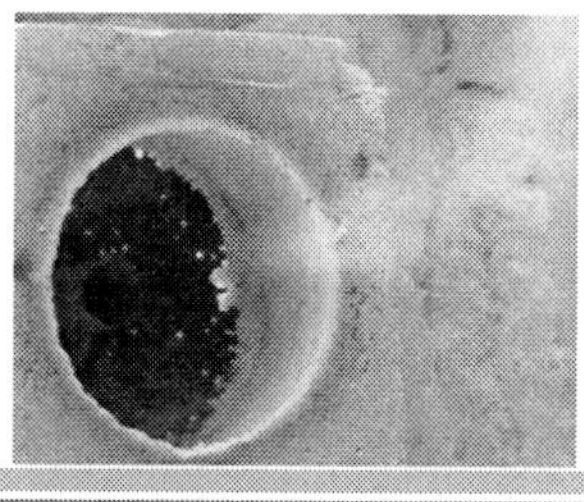

The Development of a Natural Gas/Hydrogen Boiler System

Joint project: Zantingh and DNV GL

Project: The Development of a Natural Gas/Hydrogen Boiler System

Goal: Development of a fuel flexible burner system for hydrogen/natural gas mixtures (0-100% H_2)

Experimental set-up contains:

- 475 kW Novum boiler system
- Zantingh/Unigas LowNOx-forced draft burner
- Lamtec Etamatic burner management system
- UV sensor as flame guarding system
- Hy-Optima 2710 hydrogen sensor
- Natural gas sensor (GC and Hobre Wobbe sensor)
- Fuel adaptive control system

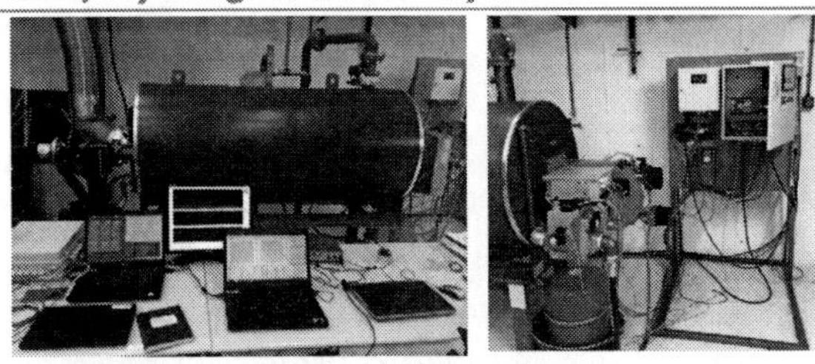

Left, 475 kw boiler system and right, forced draft burner

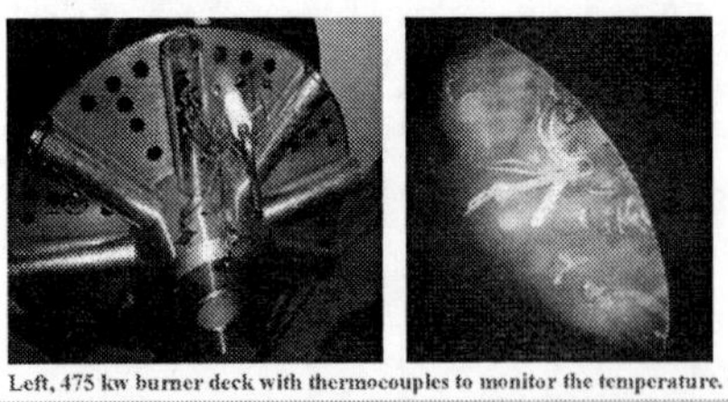

Left, 475 kw burner deck with thermocouples to monitor the temperature. Right: flame image burner

Development of smart combustion control systems to improve efficiency and the use of a wide range of sustainable fuels

- Sensors/Analyzers:
  - Real-time gas composition measurement
  - Fit for purpose, response time, accuracy
- Novel Algorithms:
  - Combustion control algorithms based on fuel gas composition and operational conditions of installation
  - Taken into account external factors (e.g. humidity, temperature, etc)
- Optimal equipment performance:
  - Fuel adaptive control for a wide range of fuel compositions
  - Safe and reliable operation within a wide range of gas compositions
  - Fuel savings
  - Acceptable emission levels

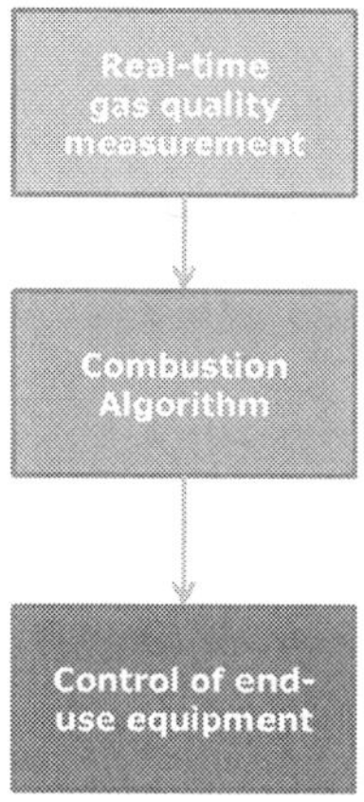

Results

- When adding hydrogen to natural gas the fuel adaptive control system keeps:
  - Fuel-to-air ratio constant
  - Thermal input (load) is constant
- Flame color changes from the typical blue color of natural gas flames to orange (hydrogen)
- No over heating of the burner is observed
- Hydrogen addition to NG results in an increase in the NO_x emission

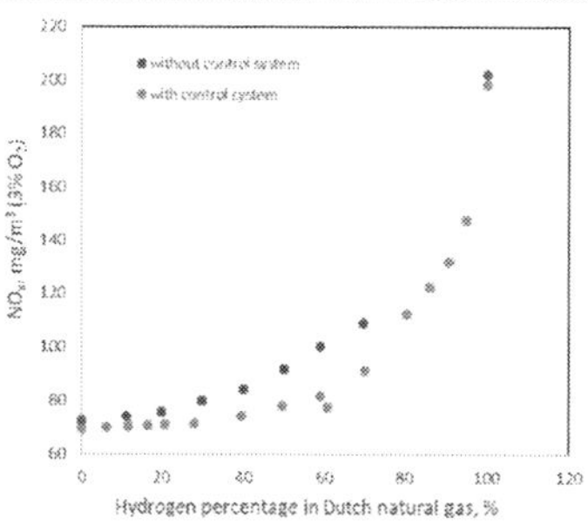

NO_x emission should be decreased by applying NO_x mitigating strategies

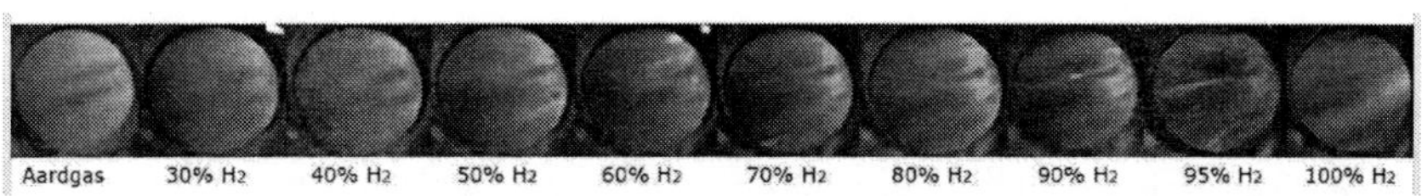

Applying Flue Gas Recirculation (FGR) as NO_X mitigating strategy
Preliminary results FGR

- A part of the flue gases is returned to the combustion air inlet.
- The dilution of combustion air with (inert) flue gases reduces the adiabatic flame temperature and consequently reduces the (thermal) NO_x formation.
- The preliminary results show that applying flue gas recirculation results in a reduction of the NO_x emission with more than a factor of 10.
- From this we conclude that flue gas recirculation is a very effective strategy to reduce the NO_x emission for hydrogen flames.

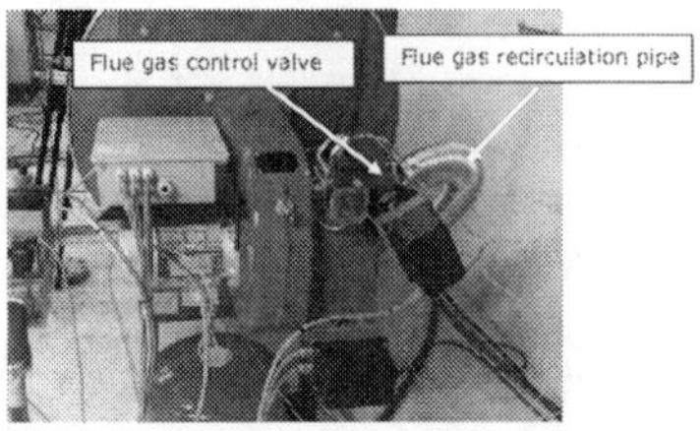

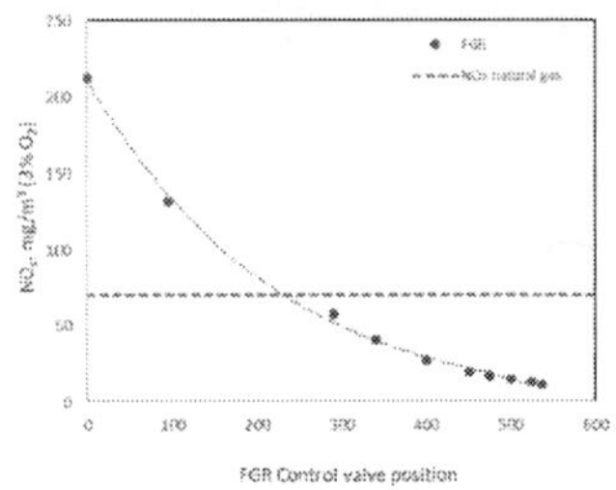

Conclusion:
the burner system developed enables the flexible introduction of hydrogen as a fuel in the natural gas grid in the energy transition period and afterwards, when natural gas is fully replaced by hydrogen

Next steps to facilitate the successful introduction of hydrogen as fuel in industrial heating processes

- Recently we started a project with 30 industrial partners to study the following aspects natural gas/hydrogen blends (0-100% H_2) in a high temperature kiln (three burner types) and a boiler (two burner types):
  - optimize flue gas recirculation for hydrogen/natural gas mixtures and pure hydrogen flames.
  - studying the changes in heat transfer to the load
  - studying oxyfuel combustion (hydrogen/oxygen)
  - CFD modelling of the heat transfer

Thank you for your attention

Questions?

Sander Gersen
Sander.gersen@dnvgl.com
+31615012156

www.dnvgl.com

SAFER, SMARTER, GREENER

ENGIE Lab CRIGEN
Research and Innovation Center
to accelerate the zero carbon transition
Hydrogen

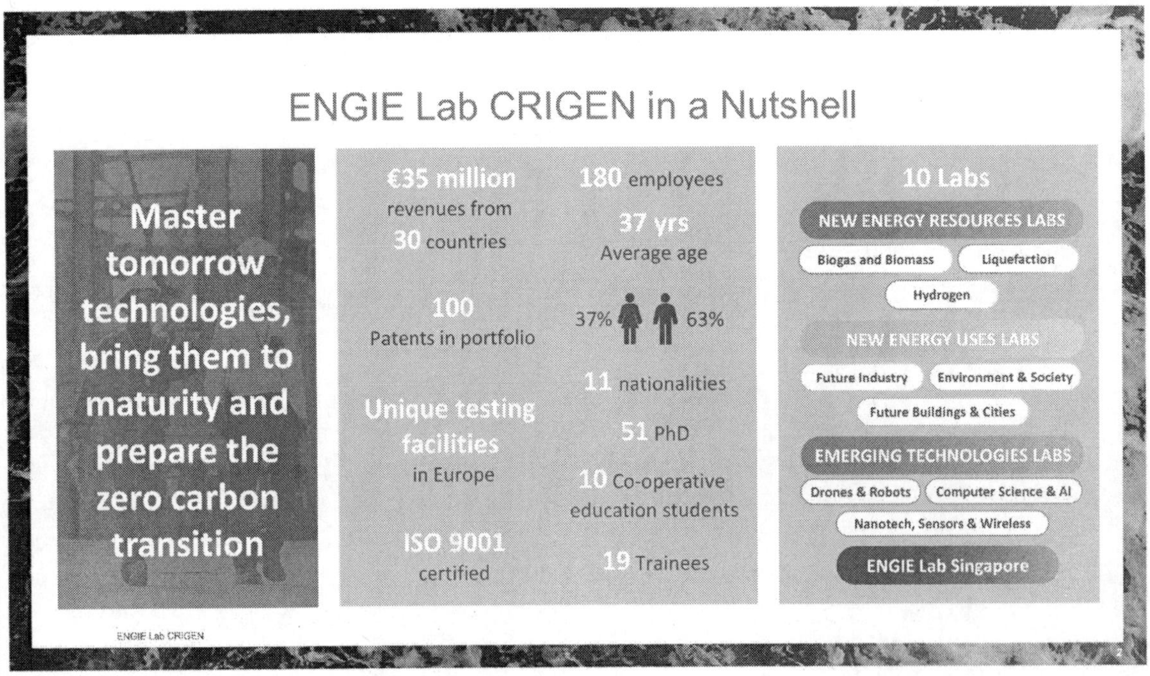
ENGIE Lab CRIGEN in a Nutshell
Master tomorrow technologies, bring them to maturity and prepare the zero carbon transition
€35 million revenues from 30 countries
100 Patents in portfolio
Unique testing facilities in Europe
ISO 9001 certified
180 employees
37 yrs Average age
37% 63%
11 nationalities
51 PhD
10 Co-operative education students
19 Trainees
10 Labs
NEW ENERGY RESOURCES LABS
Biogas and Biomass
Liquefaction
Hydrogen
NEW ENERGY USES LABS
Future Industry
Environment & Society
Future Buildings & Cities
EMERGING TECHNOLOGIES LABS
Drones & Robots
Computer Science & AI
Nanotech, Sensors & Wireless
ENGIE Lab Singapore
ENGIE Lab CRIGEN
2

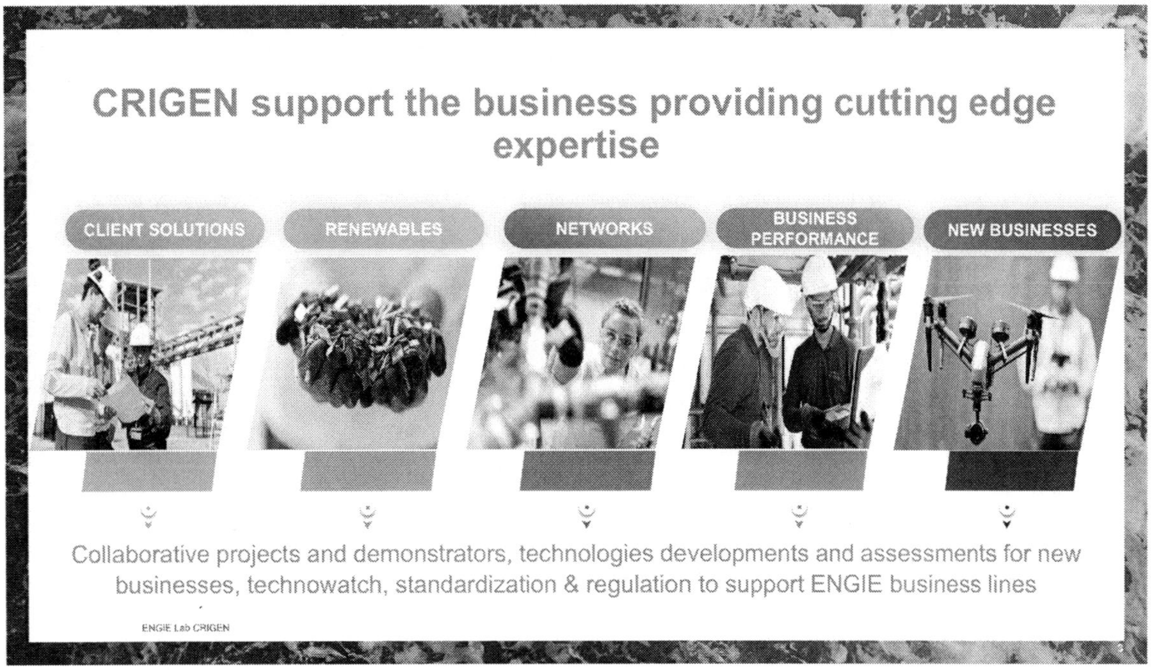
CRIGEN support the business providing cutting edge expertise
CLIENT SOLUTIONS
RENEWABLES
NETWORKS
BUSINESS PERFORMANCE
NEW BUSINESSES
Collaborative projects and demonstrators, technologies developments and assessments for new businesses, technowatch, standardization & regulation to support ENGIE business lines
ENGIE Lab CRIGEN

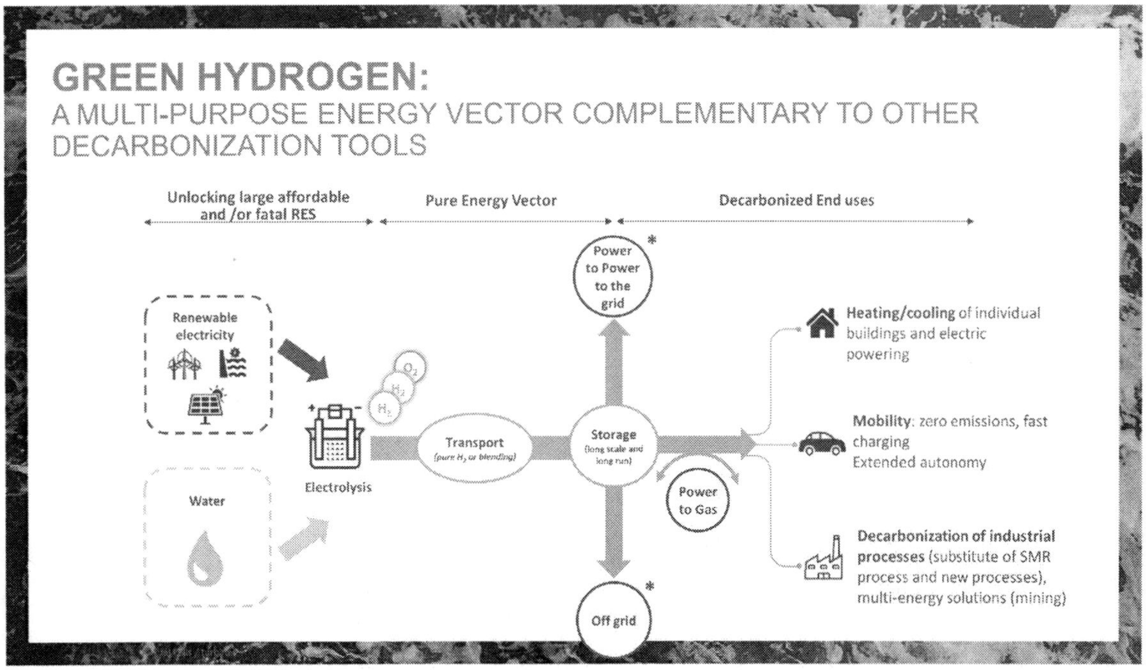
GREEN HYDROGEN:
A MULTI-PURPOSE ENERGY VECTOR COMPLEMENTARY TO OTHER DECARBONIZATION TOOLS
Unlocking large affordable and /or fatal RES
Pure Energy Vector
Decarbonized End uses
Renewable electricity
Water
Electrolysis
Transport
Storage
Power to Power to the grid *
Power to Gas
Off grid *
Heating/cooling of individual buildings and electric powering
Mobility: zero emissions, fast charging
Extended autonomy
Decarbonization of industrial processes (substitute of SMR process and new processes), multi-energy solutions (mining)

OUR STRATEGY:
DEVELOP LARGE SCALE, INTEGRATED HYDROGEN SOLUTIONS

- We **design, develop, build, operate Hydrogen systems** across the value chain using renewable sources to **deliver a comprehensive roadmap to decarbonization.**
- This means understanding customer needs, then **offering turn-key, customized services to match.**

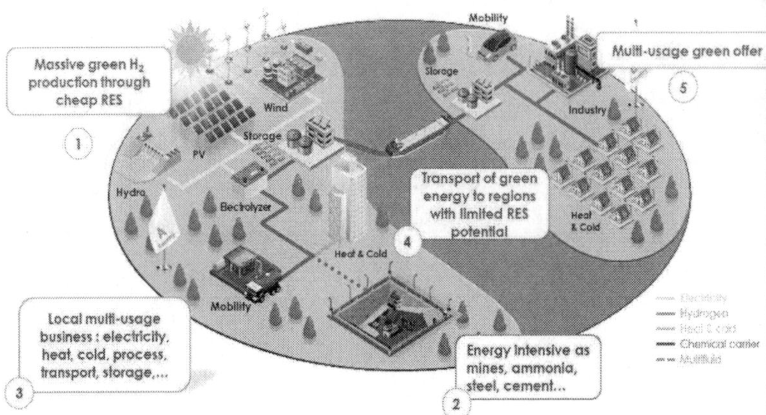

5

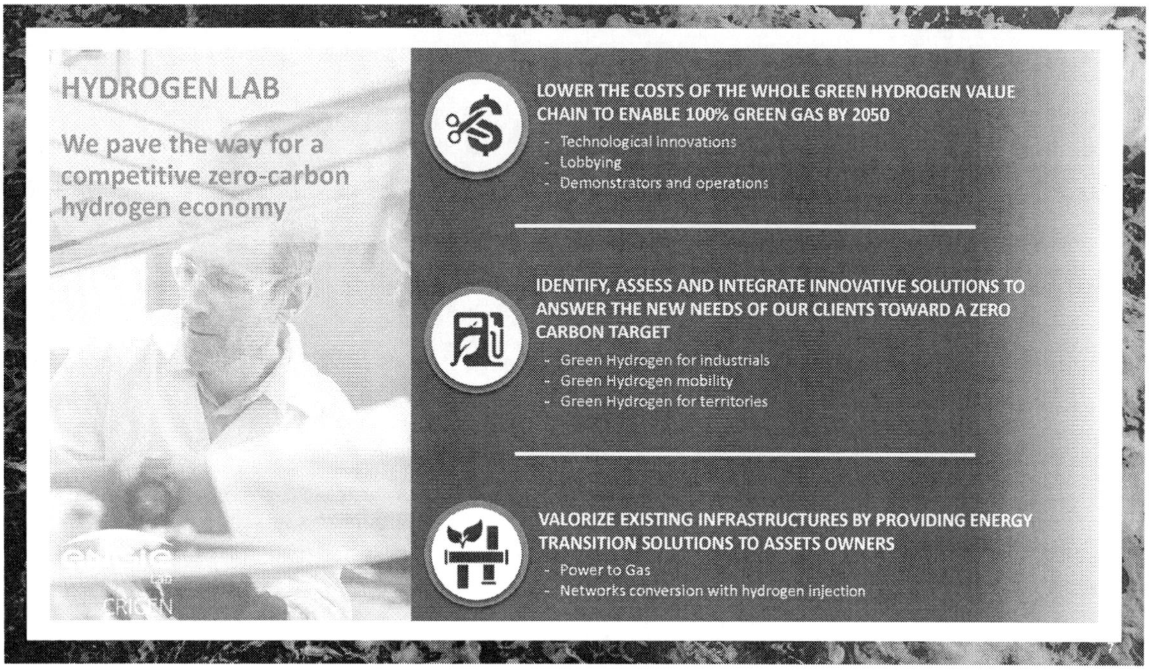
HYDROGEN LAB
We pave the way for a competitive zero-carbon hydrogen economy
LOWER THE COSTS OF THE WHOLE GREEN HYDROGEN VALUE CHAIN TO ENABLE 100% GREEN GAS BY 2050
- Technological innovations
- Lobbying
- Demonstrators and operations
IDENTIFY, ASSESS AND INTEGRATE INNOVATIVE SOLUTIONS TO ANSWER THE NEW NEEDS OF OUR CLIENTS TOWARD A ZERO CARBON TARGET
- Green Hydrogen for industrials
- Green Hydrogen mobility
- Green Hydrogen for territories
VALORIZE EXISTING INFRASTRUCTURES BY PROVIDING ENERGY TRANSITION SOLUTIONS TO ASSETS OWNERS
- Power to Gas
- Networks conversion with hydrogen injection
CRIGEN

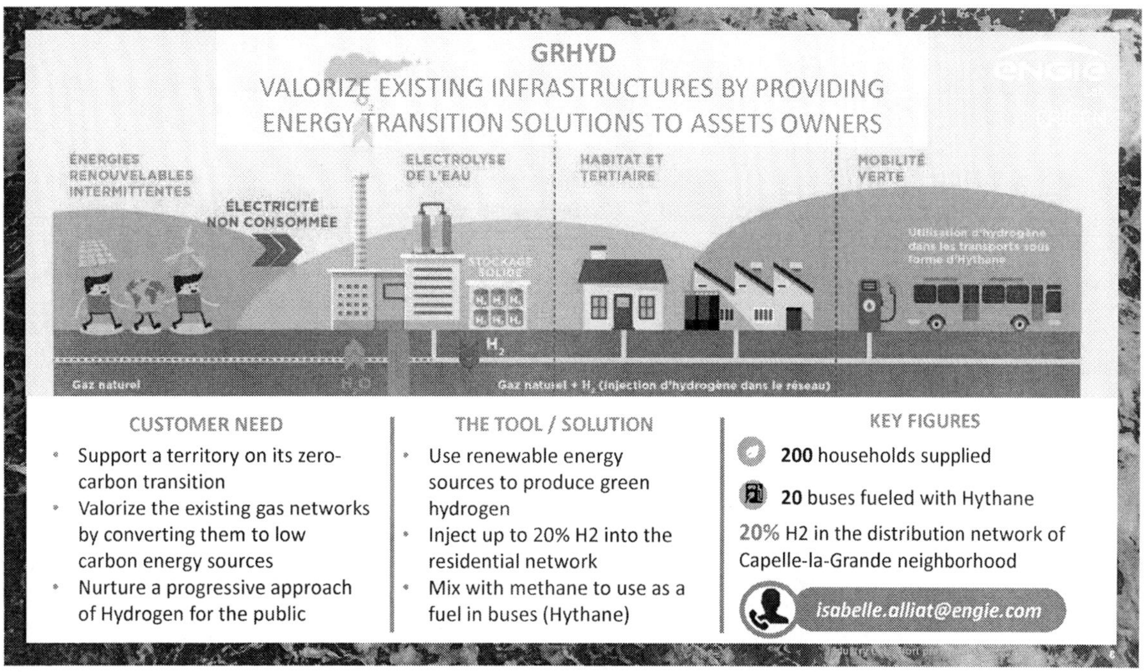
GRHYD
VALORIZE EXISTING INFRASTRUCTURES BY PROVIDING ENERGY TRANSITION SOLUTIONS TO ASSETS OWNERS
ÉNERGIES RENOUVELABLES INTERMITTENTES
ÉLECTRICITÉ NON CONSOMMÉE
ELECTROLYSE DE L'EAU
HABITAT ET TERTIAIRE
MOBILITÉ VERTE
H2
Gaz naturel
Gaz naturel + H2 (injection d'hydrogène dans le réseau)
CUSTOMER NEED
Support a territory on its zero-carbon transition
Valorize the existing gas networks by converting them to low carbon energy sources
Nurture a progressive approach of Hydrogen for the public
THE TOOL / SOLUTION
Use renewable energy sources to produce green hydrogen
Inject up to 20% H2 into the residential network
Mix with methane to use as a fuel in buses (Hythane)
KEY FIGURES
200 households supplied
20 buses fueled with Hythane
20% H2 in the distribution network of Capelle-la-Grande neighborhood
isabelle.alliat@engie.com

C2FUEL
PRODUCTION OF HYDROGEN BASED FUEL FOR SMARTER INDUSTRIAL ECOSYSTEMS

ENGIE Lab CRIGEN

CUSTOMER NEED

- Structure a circular economy at a territory scale
- Locally valorize the excess renewable energy and reduce carbon dioxide emissions
- Develop low carbon

THE TOOL / SOLUTION

- Combine disruptive technologies to produce hydrogen carriers from renewable electricity and industrial CO_2 emissions
- Test hydrogen-based fuels in real end-user systems for maritime mobility within harbor-industrial environment

KEY FIGURES

2,4 million ton of CO_2 emission avoided

2,4 million ton of formic acid/1,2 million ton of dimethyl ether produced

Up to 11 TWh of renewable energy converted and stored per year

Energy production efficiency optimized by 20%

camel.makhloufi@engie.com
secil.torun@engie.com

ENGIE LAB CRIGEN – HYDROGEN ONGOING FUNDED COLLABORATIVE PROJECT SCALE

Production · Storage & Transport · Usage

ENGIE LH2 AMBITION AND R&D ROADMAP

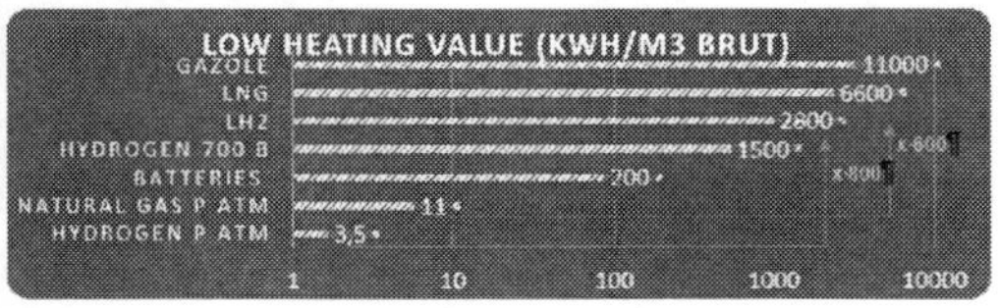

ENGIE LH2 market vision

H2 liquefaction increases the energy volumetric density of hydrogen. Due to this specificity, liquid hydrogen will likely play a major role :

- Solution for hydrogen distribution within a country (≥ 300 km)
- Solution to increase autonomy of heavy vehicles *(buses, trucks, trains, ships, planes…)*
- Solution for massive and long distance transportation

R&D program objectives

1. Develop process for hydrogen liquefaction (up to 50 TPD). *Objective is to reach the cost of 1$/kgLH2 compared to the current cost of 1.5 – 2.5 $/kgLH2*
2. Push the development of storage solutions (GTT) to reduce cost of transport *by ship at 0.6$/kgLH2and at 0.2 $/kgLH2 by truck*
3. Develop digital solutions and expertise for the management of liquid hydrogen during its storage and transport

Planning

2020 - 2022 : Process design, tests (test bench), patents

2023 - 2024 : Real condition tests, process validation

11

TechnipFMC

Technology in the LNG chain – an unfinished trail of innovation

24th February 2020
IGRC 2020, Muscat

Philip Hagyard
Group VP Gas Monetization, TechnipFMC
LNG2022 Programme Committee Chairman

LNG2022

Disclaimer

Forward-looking statements

We would like to caution you with respect to any "Forward-looking statements" made in this presentation as defined in Section 27A of the United States Securities Act of 1933, as amended, and Section 21E of the United States Securities Exchange Act of 1934, as amended. Words such as "expect," "plan," "intend," "would," "will," and similar expressions are intended to identify forward-looking statements, which are generally not historical in nature, and include any statements with respect to the potential separation of the Company into RemainCo and SpinCo, the expected financial and operational results of RemainCo and SpinCo after the potential separation and expectations regarding RemainCo's and SpinCo's respective businesses or organizations after the potential separation.

Such forward-looking statements involve significant risks, uncertainties and assumptions that could cause actual results to differ materially from our historical experience and our present expectations or projections, including the following known material factors: risks associated with the impact or terms of the potential separation; risks associated with the benefits and costs of the potential separation, including the risk that the expected benefits of the potential separation will not be realized within the expected time frame, in full or at all; risks that the conditions to the potential separation, including regulatory approvals and consultation of employee representatives, will not be satisfied and/or that the potential separation will not be completed within the expected time frame, on the expected terms or at all; the expected tax treatment of the potential separation, including as to shareholders in the United States or other countries; changes in the shareholder bases of the Company, RemainCo and SpinCo, and volatility in the market prices of their respective shares; risks associated with any financing transactions undertaken in connection with the potential separation; the impact of the potential separation on our businesses and the risk that the potential separation may be more difficult, time-consuming or costly than expected, including the impact on our resources, systems, procedures and controls, diversion of management's attention and the impact on relationships with customers, governmental authorities, suppliers, employees and other business counterparties; unanticipated changes relating to competitive factors in our industry; our ability to timely deliver our backlog and its effect on our future sales, profitability, and our relationships with our customers; our ability to hire and retain key personnel; U.S. and international laws and regulations, including existing or future environmental or trade/tariff regulations, that may increase our costs, limit the demand for our products and services or restrict our operations; disruptions in the political, regulatory, economic and social conditions of the countries in which we conduct business; downgrade in the ratings of our debt could restrict our ability to access the debt capital markets; and such other risk factors as set forth in our filings with the U.S. Securities and Exchange Commission and in our filings with the Autorité des marchés financiers or the U.K. Financial Conduct Authority.

We caution you not to place undue reliance on any forward-looking statements, which speak only as of the date hereof. We undertake no obligation to publicly update or revise any of our forward-looking statements after the date they are made, whether as a result of new information, future events or otherwise, except to the extent required by law.

TechnipFMC LNG Market Leadership

Market leadership

105 Mtpa	>20%	7.8 Mtpa
Global production delivered	*Of operating LNG capacity*	*World's largest LNG trains delivered*

50 year track record in LNG
- World's first LNG *Algeria (1964)*
- World's largest LNG trains *Qatar*
- Largest Arctic project *Yamal*

Pioneer in floating LNG (FLNG)
- World's first FLNG delivered *Petronas Satu in Malaysia*
- World's largest floating vessel *Shell Prelude in Australia*
- New frontier *Eni Coral in Mozambique*

Diversity in projects and technologies

Pioneer in modularization
- Onshore LNG trains on an unprecedented scale
- Greater cost and schedule certainty in extreme locations

Next generation mid-scale LNG
- Economic solutions for smaller reserves (1-3 Mtpa)
- Standardized, modularized design enables repeatability

Next generation FLNG
- Liquefaction engineered for minimal footprint
- Split construction to minimize module integration

TechnipFMC

LNG – a high value component of the future energy mix

TechnipFMC

Gas consumption will continue to grow

according to all energy agencies and companies

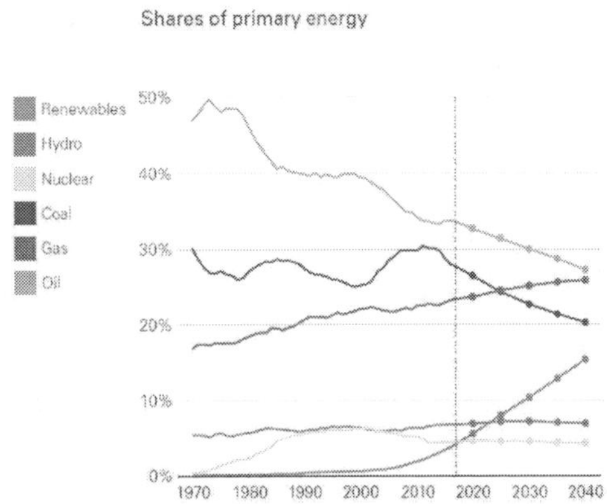

LNG contribution to global energy in 2019	2.5%
Natural gas share of primary energy	23%
Proportion of gas provided through international trade	33%
LNG share of international trade	33%

Advantages of this popular hydrocarbon, LNG

- Good air quality for fast growing urban centres
- Quick fix decarbonisation by fuel substitution
- Synergies with hydrogen
- .. and LNG is safe

Renewables and natural gas account for 85% of energy growth

BP Energy Outlook: 2019 edition
https://www.bp.com/content/dam/bp/business-sites/en/global/corporate/pdfs/energy-economics/energy-outlook/bp-energy-outlook-2019.pdf

TechnipFMC

LNG technology

What is needed and where does it come from?

TechnipFMC

LNG : the largest and fast-growing application of cryogenics

Liquefied natural gas
Up to 24 000 tpd LNG per unit

Air separation
Up to 7 000 tpd LOx per unit

Rocket launchers
Ariane 5: 25t LH2, 130t LOx

Superconducting magnets.
250kg helium per MRI

Unique features

- The largest cryogenic industry
- Capital intensive
  - Liquefaction mainly
  - High reliability > risk aversion
- Energy intensive liquefaction
- Remote locations
- Uniquely large projects

Technology drivers

- Cost reduction
  - Economies of scale
  - Construction linked to location is the biggest single cost element
- Process efficiency
- Decarbonisation

Progressive improvements in scale, efficiency, and reliability

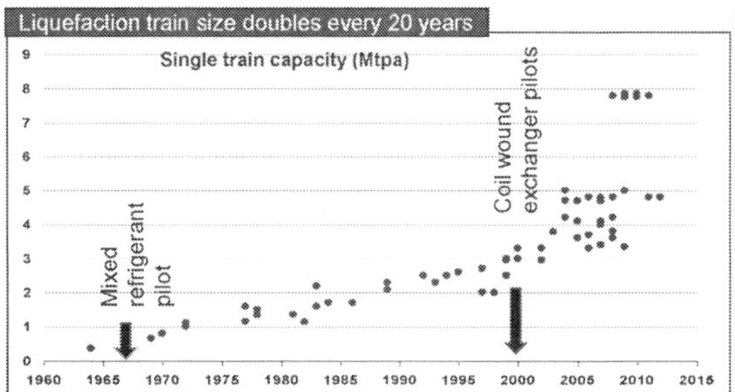

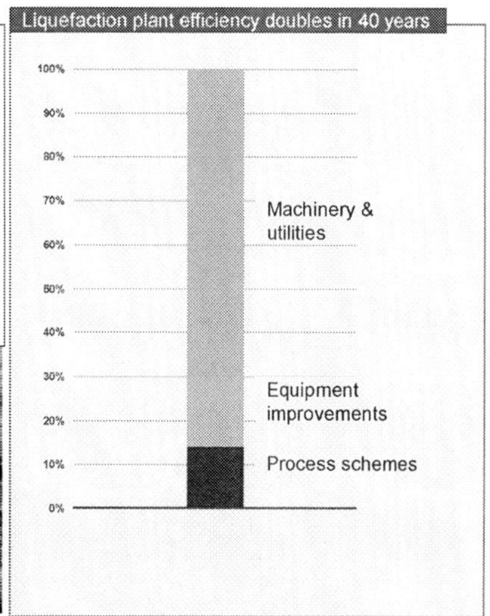

The LNG industry borrows and adapts technology

Mixed refrigerants

Refrigerant compressor drivers

CO2 removal

Compact equipment

TechnipFMC

Key areas of innovation

TechnipFMC

Innovation opportunity - small scale LNG

Recent projects, rich in innovation
- Removal of trace heavy components
- Variable speed EM driven compressors
- Nitrogen rejection

TechnipFMC

Innovation opportunity - FLNG

Prelude: First shipment: June 11th, 2019

TechnipFMC

Innovation opportunity - Arctic

TechnipFMC

What are TechnipFMC technologies in an LNG facility?

Cryomax NGL recovery

Propane exchangers with dual enhanced surface tubes:
- **GEWA-PB Propane Evaporators** wieland
- **Diesta Propane Condensors and other air coolers** Kelvion

High efficiency N2 rejection

Gas Turbine Inlet Air Chilling (IAC)

LNG Mechanical loading arms

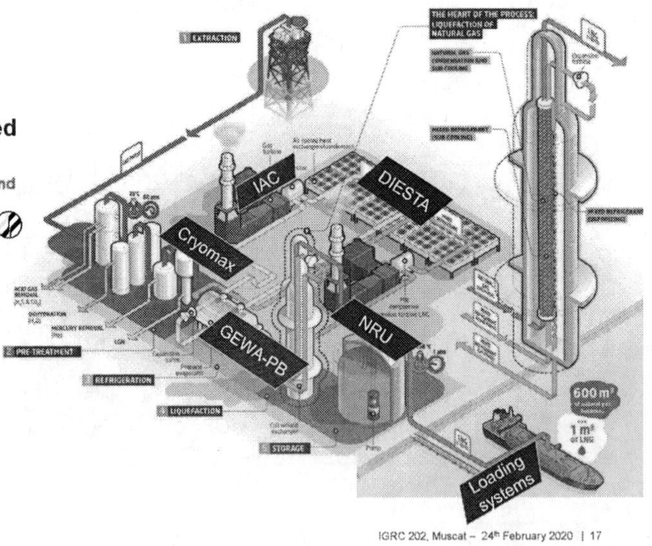

TechnipFMC

Innovation for the climate - decarbonisation across the LNG Chain

Gas field	Liquefaction	Shipping/ Loading	Regas terminals	Distribution
				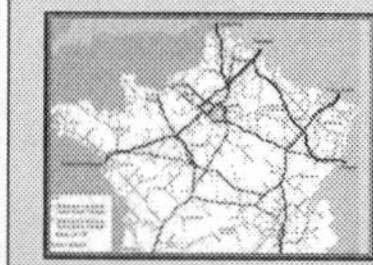

Emission monitoring; zero flaring including start-up...

Liquefaction
- Electrification
- Process efficiency
- Refrigerant inventory recovery
- CCS – process CO2 ; capture from fired equipment

Shipping/ Loading
- Zero spillage couplings
- Cargo swaps to reduce distance

Regas terminals
- Reform to H2+CCS
  -SMR, ATR or POX
  -H2 blended with NG send out
- Integrate CCGT + CCS

Distribution
- CHP
- Micro CHP

Areas requiring changes in markets and associated regulations
- Carbon capture & sequestration (a universally applied carbon tax)
- Environmental labelling of LNG
- Pure methane vs natural gas (LNG trade specification)

TechnipFMC

TechnipFMC

20th International Conference & Exhibition on Liquefied Natural Gas

Conference News @ IGRC 2020

- Programme committee
  - Moving to the next level
  - 40 industry leaders and experts
  - 6 Workgroups
    - LNG markets and advocacy
    - Operations and best practice
    - Commercial stimuli
    - Gas production, processing, liquefaction
    - Shipping, marine operations and terminals
    - Downstream infrastructure and applications
- April 2022 a new type of conference
  - Pre-conference publication
  - Sessions shaped by the delegates
  - Meet the speaker
- Launch 24 months from D-Day
  - Building on 50 years of papers
  - "Writing the next chapter in the book of LNG"

LNG 2022

Advanced BoG Management in LNG Transfer Operations

26 February 2020

ENGIE Lab

Confidentiel Restreint Libre Interne

CONTENTS

1 Introduction

2 Calculation Model

3 Example of Calculation

4 Results

5 Conclusion

01

Introduction

LNG Transfer operations in LNG Industry

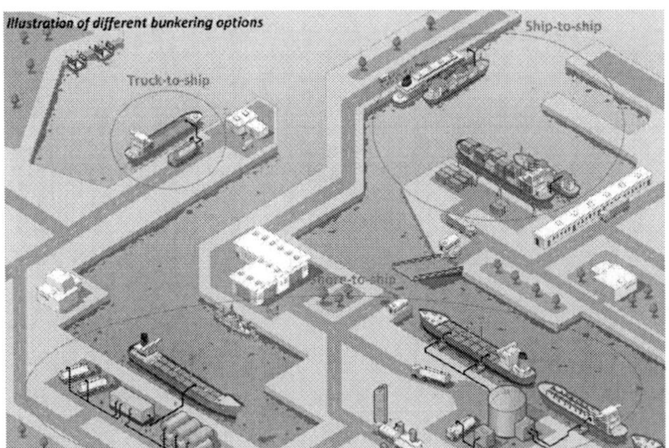

LNG Transfer operations in numbers

- A total of 5119 LNG trade voyages were completed during 2018 (8% increase with respect to the previous year)
  [source: IGU Report 2019]

- Small-scale LNG: *[source: IGU Report 2019]*
  - Global small-scale production is estimated around 25 MTPA
  - 31 Small scale LNG vessels in operation

- LNG as fuel: *[source: IGU Report 2019]*
  - A total number of 140 ships use LNG a fuel are in service
  - 160 Ships are in order
  - Approximately 3 MTPA is to be used as bunker fuel
  - By 2030, 25 MTPA of LNG will be dedicated to marine fuel
  - 7 Active bunkering ships

LNG Transfer Model
Helps Optimize design and operations

- **Operations:**
  - Day to day optimization of tank pressure and BOG management: pressure evolution, return gas, etc.
  - Manage LNG tank pressure to avoid venting and flaring
  - Evaluate the quantity of BOG generated during LNG transfer operation
  - Predict optimal tank conditions for LNG transfer
  - Minimize LNG bunkering operation time
  - Assess the evolution of bunkered LNG quality

- **Design:**
  - Choice of LNG tank maximum pressure and BOR according to several operational schemes
  - Verify LNG source compatibility with client's needs (composition, Methane Number)
  - Predict optimal LNG transfer conditions for a large portfolio of clients
  - Determine dedicated BOG management strategy

Excess Boil-off Gas in LNG transfer operations

Excess BOG =
Gas Evaporated in the Feeder + Gas Evaporated in the Receiver – Gas Consumption in the Feeder – Gas consumption in the receiver

❑ **How to Limit excess BOG:**

- Make the transfer at a low flowrate
- Limit the number of pumps in Feeder
- Do not fill all tanks simultaneously in Receiver
- Have a high pressure in Receiver to limit flash

❑ **How excess BOG can be dealt with:**

- Contribute to pressure increase in Feeder or Receiver
- Increase Gas consumption/Gas out in Feeder Receiver

02 Calculation Model

LNG Transfer Operation

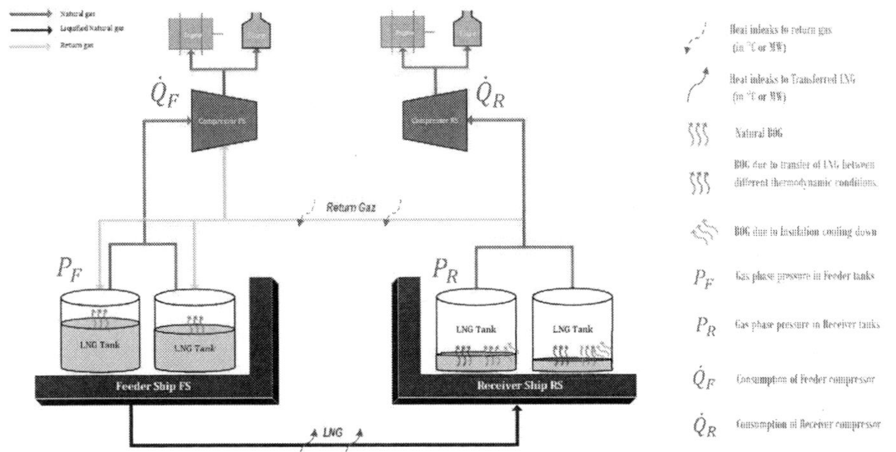

Evolution of LNG Thermophysical properties

- **Evolution of Physical properties of LNG and BOG**
  (density, Specific heat, thermal diffusivity and conductivity, Wobbe, GHV, kinematic viscosity, Molar volume, etc).
- **Liquid/Vapor thermodynamic equilibrium model**
  (Saturation temperature, latent heat of vaporization, etc)
- **Heat exchange with surroundings and between different layers inside the tank.**
- **Mass exchange between different layers (liquid/liquid or liquid/vapor)**

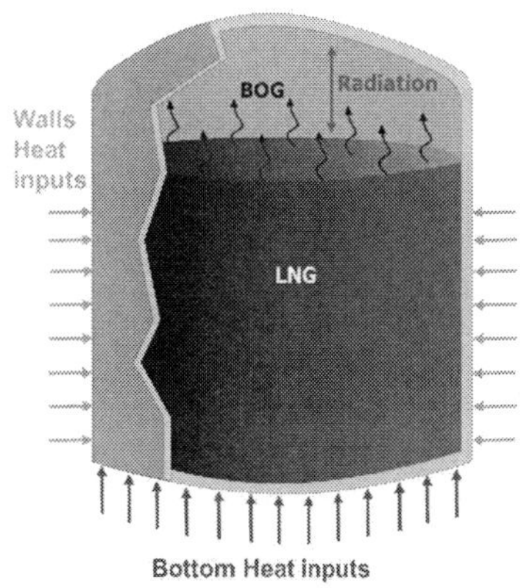

Simulation of all possible phenomena

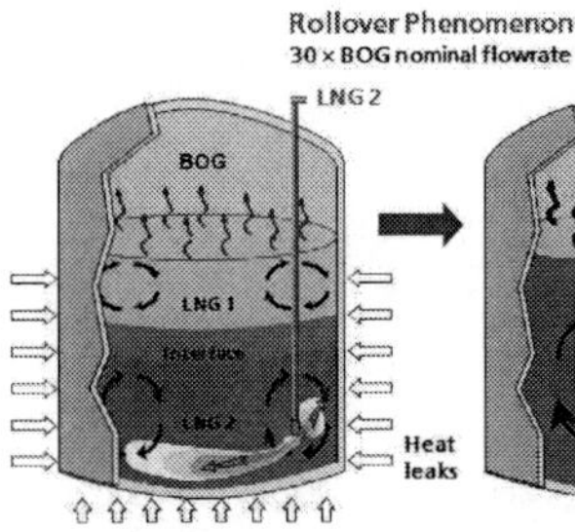

Identification of stratification and simulation of its evolution until rollover and returning to thermodynamic equilibrium

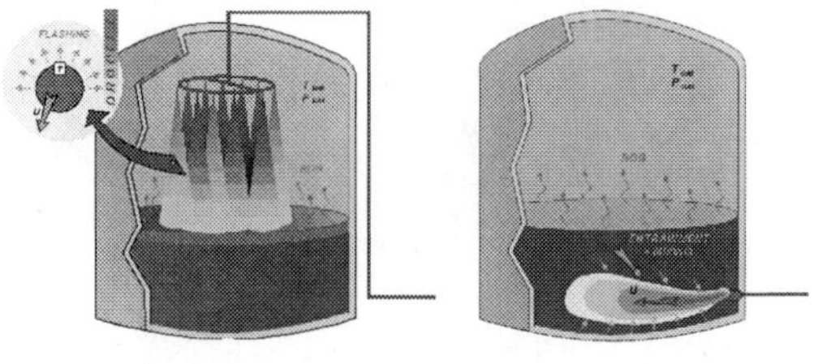

Simulation of filling scenarios, mixing, creation of different LNG layers during filling, etc

Calculation scenarios

- On each side (Feeder and Receiver):
  - Either the Pressure P is fixed and the consumption of the compressor $\dot{Q}$ is computed.
    (Max compressor capacity $\dot{Q}_{max}$ is fixed)
  - Or the consumption $\dot{Q}$ is fixed and the pressure P is calculated.
    (Max Pressure in the tanks is fixed)
- This gives the following possible scenarios:

1. Scenario A: Input $(\dot{Q}_F, \dot{Q}_R) \rightarrow \text{Output}(P_F, P_R)$
2. Scenario B: Input $(\dot{Q}_F, P_R) \rightarrow \text{Output}(P_F, \dot{Q}_R)$
3. Scenario C: Input $(P_F, \dot{Q}_R) \rightarrow \text{Output}(\dot{Q}_F, P_R)$
4. Scenario D: Input$(P_F, P_R) \rightarrow \text{Output}(\dot{Q}_F, \dot{Q}_R)$

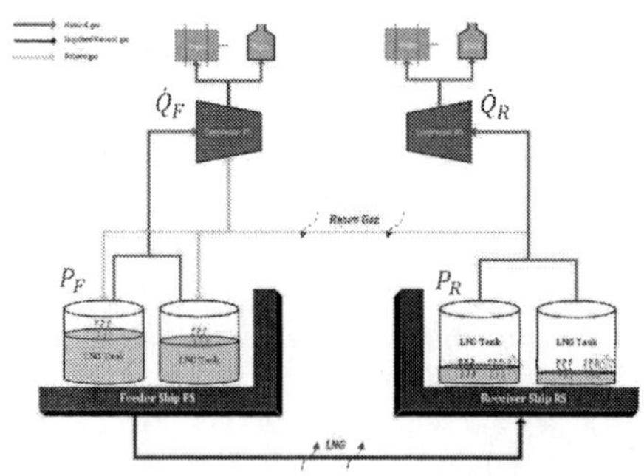

Return Gaz

- **Two possibilities for calculating return gas from the feeder to the receiver**
  - Fixed return gas module: the user fixes the return gas $\dot{Q}_{RG}$
  - Variable return gas module: the user fixes only the maximum return gas $Max(\dot{Q}_{RG})$, and the software computes the suitable return gas quantity in function of excess BOG management scenarios such as $\dot{Q}_{RG} \leq Max(\dot{Q}_{RG})$
- **The management of excess BOG is determined via the calculation of the return gas at each iteration in case of variable return gas module**

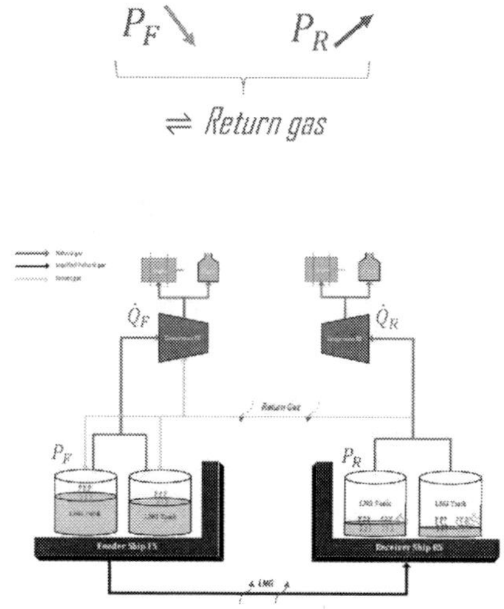

Management of Excess BOG

- **The excess BOG is managed according to the criteria below cited by priority:**
  1. BOG could be sent only from the Receiver to the Feeder
  2. $\dot{Q}_{RG} \leq Max(\dot{Q}_{RG})$ ($Max(\dot{Q}_{RG})$ is fixed by the user)
  3. $\Delta P = P_R - P_F > \Delta P_{min}$ (ΔP_{min} is fixed by the user)
  4. $\Delta P = P_R - P_F < \Delta P_{max}$ (ΔP_{max} is fixed by the user)
  5. Excess BOG management scenarios (3 different possibilities):
     - A. The pressure in the Feeder increases until a maximum $Max(P_F)$ fixed by the user
     - B. The pressure in the Receiver increases until a maximum $Max(P_R)$ fixed by the user
     - C. $\Delta P = P_R - P_F =$ constant, ΔP being fixed by the user

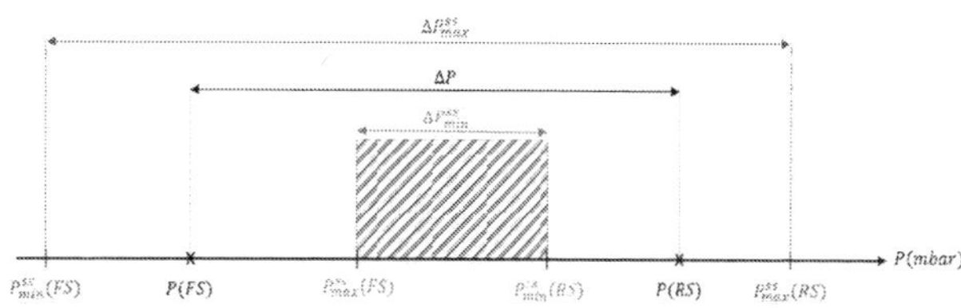

03

Example of Calculation

LNG Bunkering case study

Scenario

❑ **Hypothesis:**

1. Calculation Scenario D:
   $\text{Input}(P_F, = 1.1 \text{ bar};\ P_R = 1.16 \text{ bar}) \rightarrow \text{Output}\ (\dot{Q}_F, \dot{Q}_R)$
2. Compressors capacity range:
   $0 < \dot{Q}_F < 100\ nm^3/h$
   $0 < \dot{Q}_R < 100\ nm^3/h$
3. Return Gas flowrate range:
   $0 < Return\ Gas < 10000\ nm^3/h$
4. LNG Flowrate = $1950\ m^3/h$
5. Pressure Difference constraints:
   $0.04 \text{ bar} < P_R - P_F < 0.4 \text{ bar}$
6. Excess BOG Management scenario:
   The pressure in the Receiver increases until $P_R(max) = 1.55$ bar

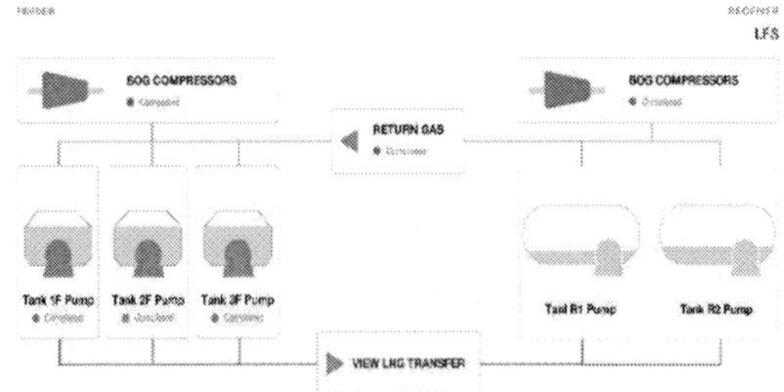

LNG Bunkering Case Study

Results-Pressure and return gas evolution

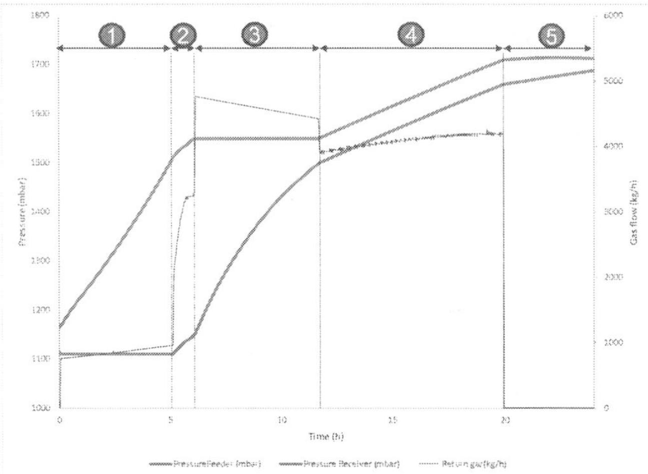

1. $P_R \uparrow$ (Excess BOG Management scenario)
2. $P_R - P_F = 0.4\ bar$
3. $P_R = 1.55\ bar$
4. $P_R - P_F = 0.04\ bar$
5. Disconnection

04

Sensibility of operational parameters on excess BoG

Sites Characteristics

- LNG transfer from LNG carrier (LNGC) to an FSRU
- Each ship has 4 Membrane tanks
- BoG consumption:
  - — No Gas consumption onboard of the LNGC: $\dot{Q}_{max}(LNGC) = 0$
  - — Maximum FSRU's compressor capacity : $\dot{Q}_{max}(FSRU) = 3000\ kg/h$

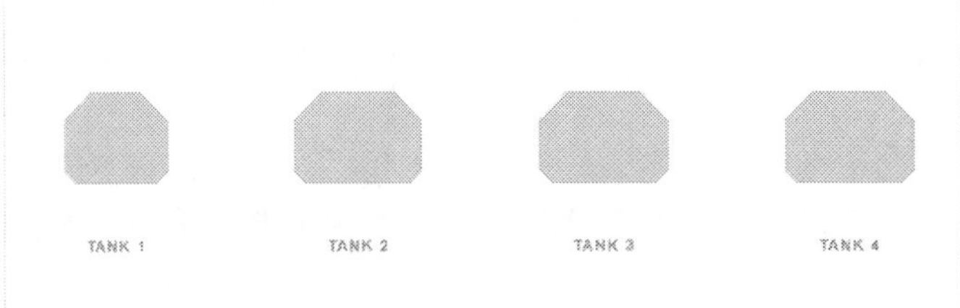

LNG's characteristics

LNG Composition (%mol)

	Lean	Rich
CH_4	99.6	85
C_2H_6	0.2	9.5
C_3H_8	0.1	3.0
iC_4H_{10}	0.1	1.0
nC_4H_{10}	0	1.0
iC_5H_{12}	0	0.13
nC_5H_{12}	0	0.12
C_6+	0	0
N_2	0	0.25

- Same LNG composition initially in the LNGC and FSRU
- Initial LNG conditions:
  - — FSRU: LNG at equilibrium at FSRU tank pressure
  - — LNGC: LNG saturation pressure = tank pressure + 10 marg
- LNG temperature increases 0.73°C during the transfer
- Return gas temperature increases 10°C in return gas pipe

Results

Transfer rate (m^3/h)	FSRU Pressure ($mbar$)	LNGC Pressure ($mbar$)	Transfer time (h)	Lean LNG			Rich LNG		
				Total return gas ($tons$)	Total Excess BOG ($tons$)	Peak Excess BOG (kg/h)	Total return gas ($tons$)	Total Excess BOG ($tons$)	Peak Excess BOG (kg/h)
3000	1190	1100	43	65.74	56.82	1506	78.06	37.15	1174
6000	1190	1100	21.5	120.88	29.41	1619	130.66	28.39	1579
9000	1190	1100	14.33	140.26	27.58	2303	149.78	27.58	2103
3000	1190	1130	43	73.09	128.04	3384	86.22	105.06	2654
6000	1190	1130	21.5	130.31	73.96	3871	139.75	64.18	3204
9000	1190	1130	14.33	148.99	56.17	4304	157.93	50.44	3722
3000	1190	1160	43	80.25	203.33	5440	93.93	178.90	4612
6000	1190	1160	21.5	139.63	128.21	6794	149.43	116.85	6003
9000	1190	1160	14.33	158.48	98.56	7676	167.67	91.44	6953

Return Gas: Effect of transfer rate

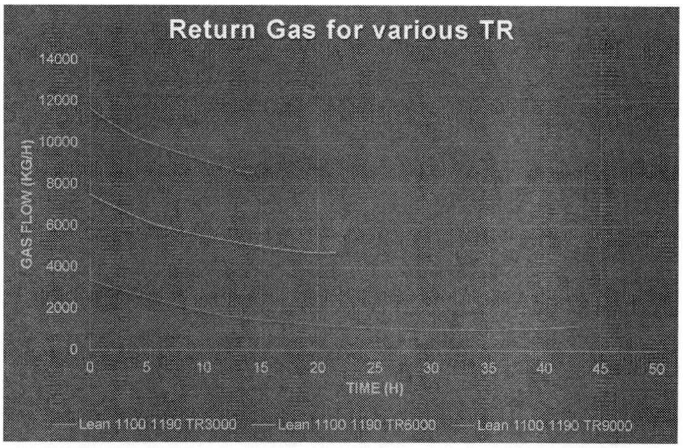

Return Gas: Effect of Pressure difference between LNGC and FSRU

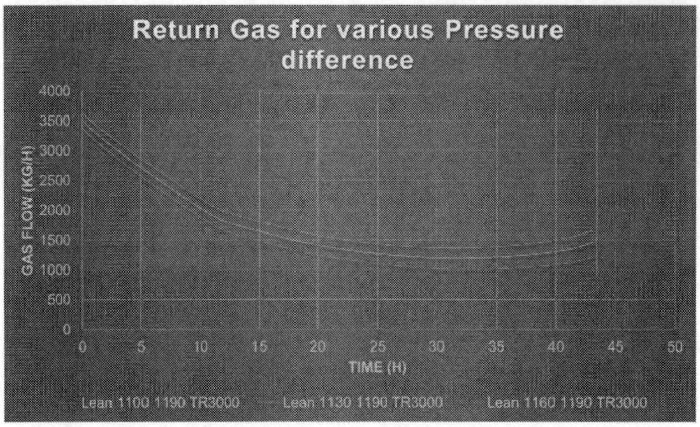

Return Gas: effect of LNG quality

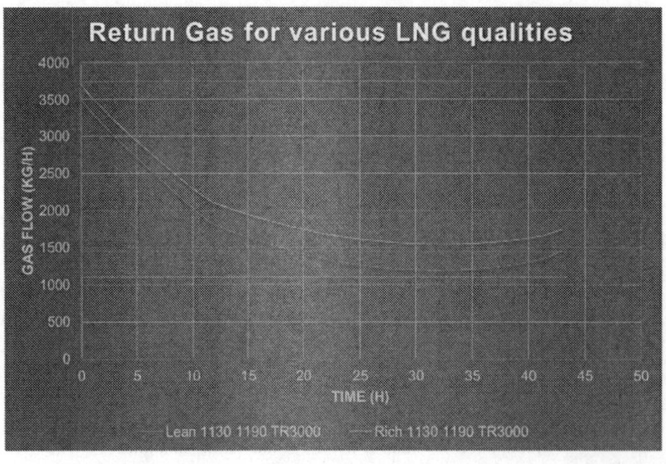

Pressure increase due to Excess BOG

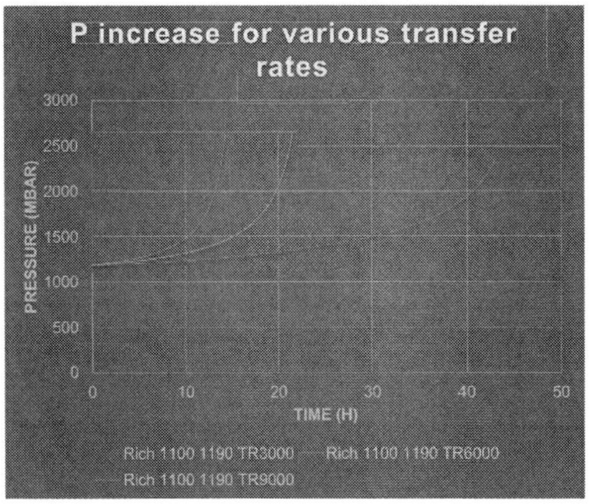

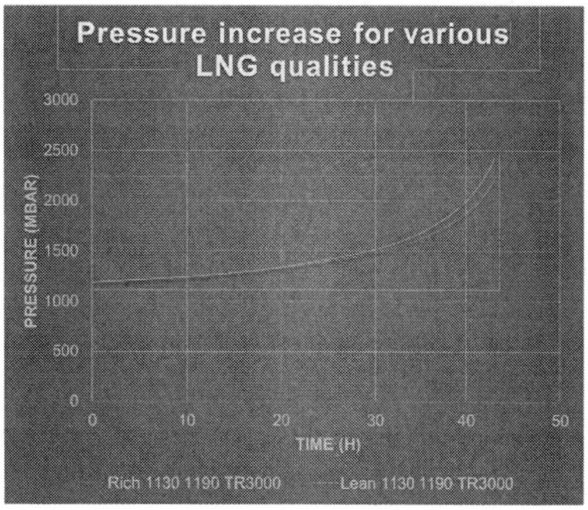

05

Conclusion

Conclusion

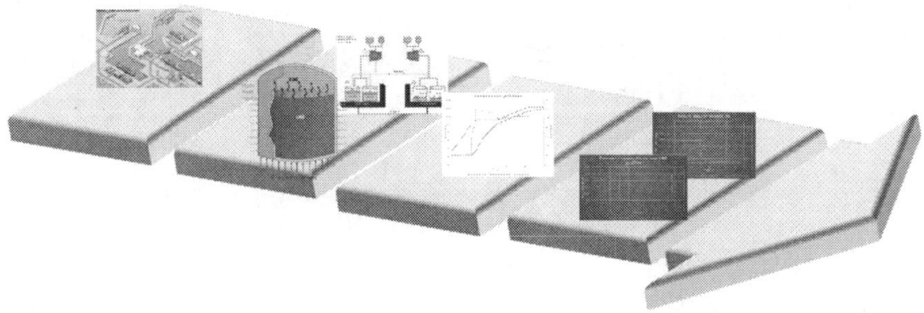

CRIGEN
361, avenue du Président Wilson
93210 Saint-Denis La Plaine
FRANCE
Tél : +33 (0)1 44 22 00 00

engie.com

AGH

AGH UNIVERSITY OF SCIENCE AND TECHNOLOGY

PGNiG

Polish Oil and Gas Company

Combining artificial intelligence and reservoir simulation for optimal control of underground gas storage

Jerzy Stopa

Polish Oil and Gas Company, Warsaw, Poland
AGH University of Science and Technology, Krakow, Poland

IGRC 2020 MUSCAT

24-26 FEB | WWW.IGRC2020.COM

AGH

PGNiG

What is optimization?

Common sense: the action of making the best or most effective use of a situation or resource

What does „the best or most effective" mean ?

In engineering and mathematics: in the simplest case - maximizing or minimizing a real function that is a measure of effectiveness

Given: **a function** $J: A \rightarrow R$ **from some set** A **to the real numbers**

Sought: **an element** X_{opt} **such that** $J(X_{opt}) \leq J(X)$ **for all** X ***from*** A **(minimization)**
or such that that $J(X_{opt}) \geq J(X)$ ***for all*** X ***from*** A **(maximization)**

J – objective function, X – „decision variable" or „control"

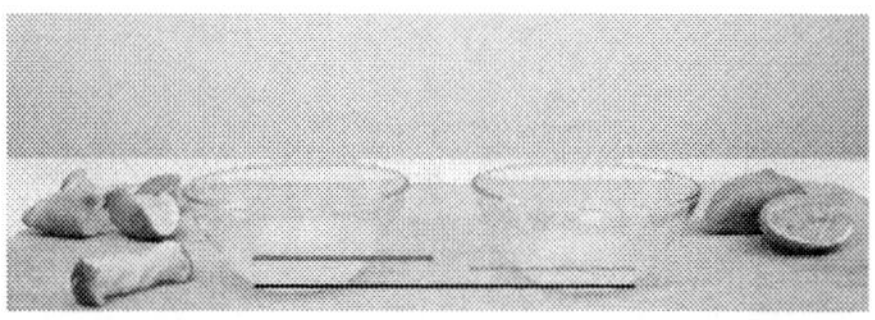

Examples of J:
- ***Max NPV***
- ***Min Water Production***
- ***Max Energy Production***
- ***other***

AGH

Reservoir simulation vs reservoir optimization

PGNiG

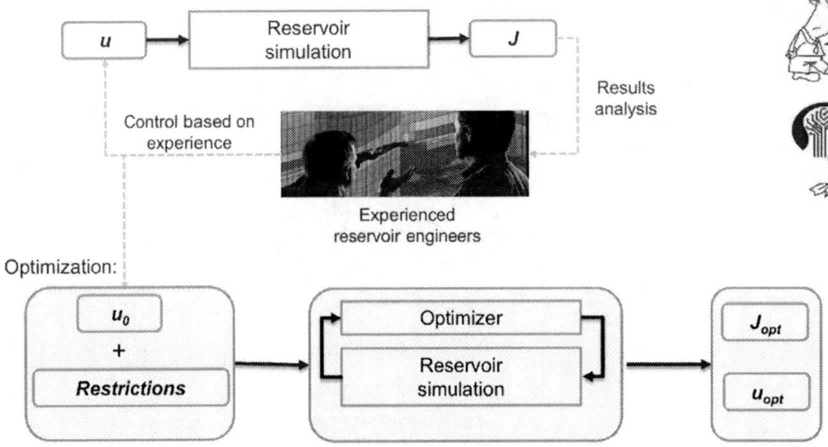

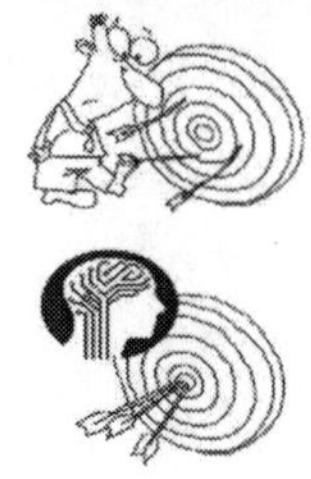

AGH

Complexity of optimization in petroleum engineering

PGNiG

- No explicit relation between objective function and decision variables – necessity of using reservoir simulators
- Complex mathematical model – set of nonlinear PDEs
- Complex geological model – millions of cells for heterogeneous reservoir
- Simulation is computationally expensive, unreasonable number of runs necessary for optimization using classical methods
- Problem of local extrema for multimodal functions
- Results difficult to interpret and sometimes to understand

Novel approach:
combination of AI based optimal control and reservoir simulation

Intelligence vs. artificial intelligence

AGH PGNiG

Intelligence:

The ability to **understand**, **learn** from past experience and **use gained knowledge** in new situations

Artificial Intelligence

The ability of a machine to perform tasks commonly associated with intelligent beings

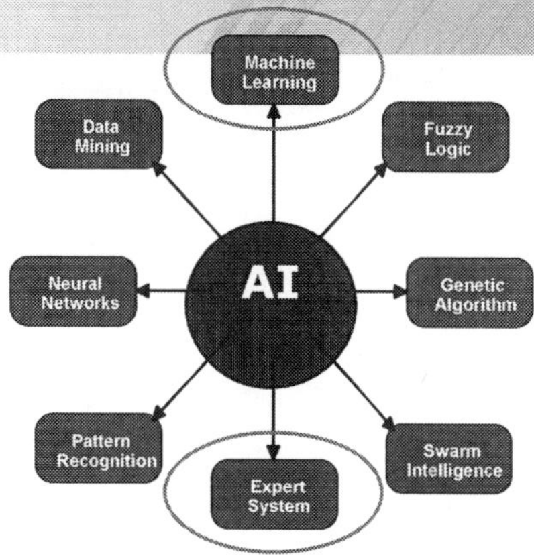

The capability of a machine to imitate intelligent human behavior can be used for automatic determination of the best scenario for optimal control of the underground gas storage

Case study: Wierzchowice UGS using lower quality gas as a cushion gas

AGH PGNiG

- **Working volume: 1.2 bln scm**
- **Maximal injection rate: 6.0 mm c.m./day**
- **Maximal withdrawal rate: 9.6 mm c.m./day**
- **Developed in a depleted reservoir of natural gas containing 29% of nitrogen**
- **Gas remaining in reservoir: 4.1 bcm, part of that gas is employed as a cushion gas**
- **Energy remaining in reservoir: 113,5 billions of MJ. The economic value of that energy is about 679 millions euros.**
- **The same volume of a high quality gas would contain 160,3 bln MJ of energy that is equivalent to 959 millions euros.**
- **Difference: 280 millions euros**

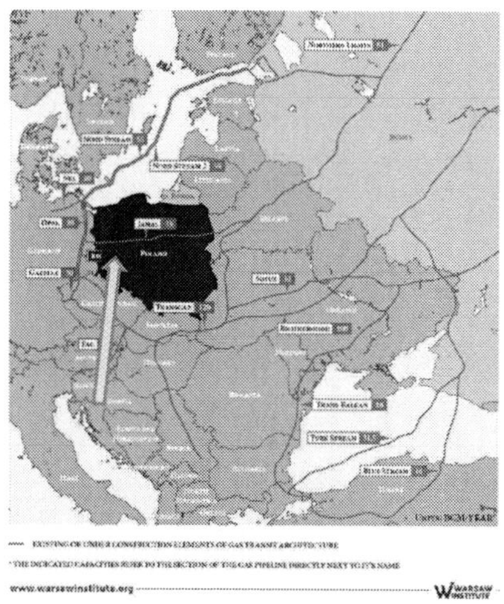

Gas composition problem

Component	**Native gas** mole fraction	**Injected gas** mole fraction
Nitrogen	0.29	0.01-0.03
Methane	0.70	0.96-0.985
...	...	...

- Mixing of the injected gas and native gas and therefore variable composition of gas extracted from the storage.
- Controlling the injection and withdrawal operations by use of reservoir simulation technique

7

History of the injection/withdrawal cycles for Wierzchowice UGS.

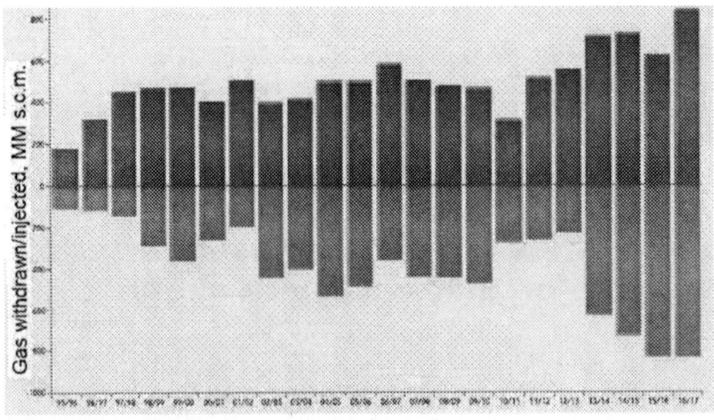

withdrawal   injection

8

AGH

PGNiG

Visualisation and reservoir simulation

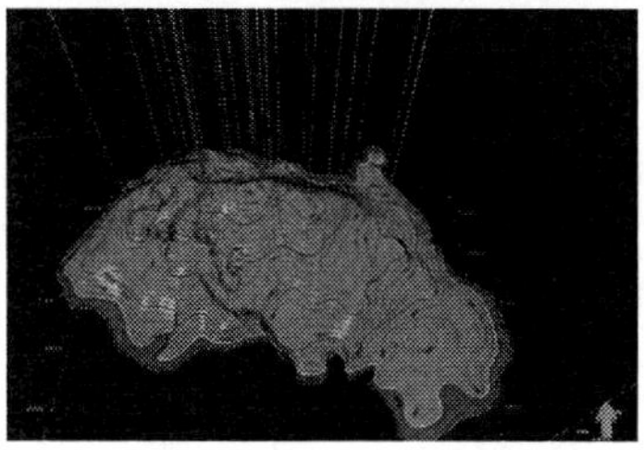

Structure

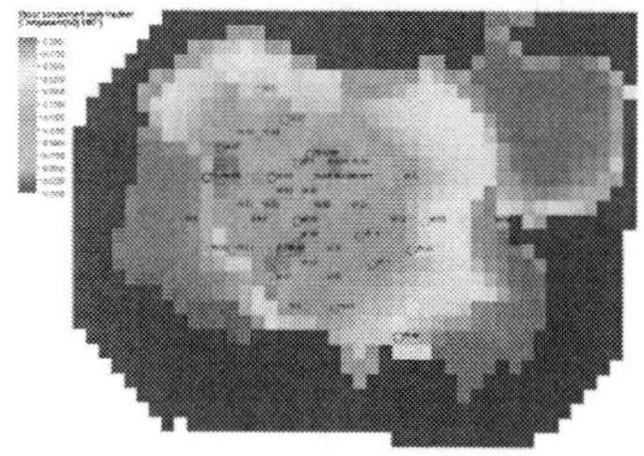

Simulated N2 concentration

No. Of wells:

- **Operational wells: 12 horizontal, 1 vertical**
- **Observation wells: 15**

9

9

AGH

PGNiG

Typical N2 concentration profiles for selected wells, Wierzchowice UGS

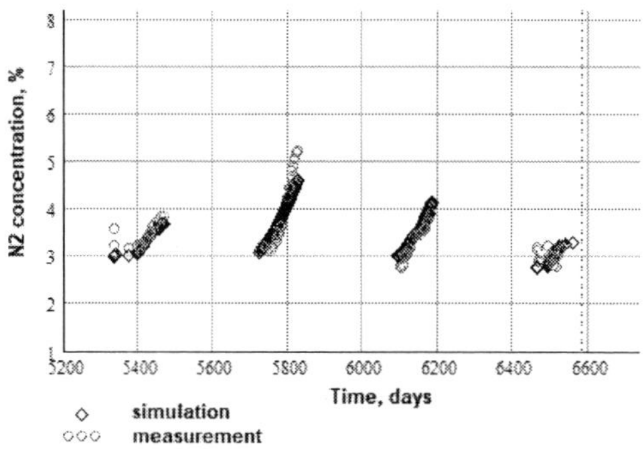

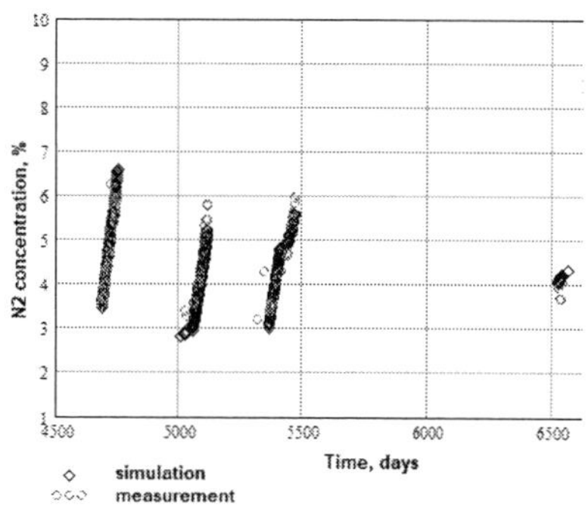

10

AGH

Control of the Wierzchowice UGS

PGNiG

- Reservoir simulation model in ECLIPSE 300 simulator
- Early approach: decisions based on the experience – effective but not guarantee an optimal decision
- Present approach: **to find time depended well controls that maximize energy produced from the storage**
- **Method - combination of:**
  - **parameterized decision tree to respect experience of past and present generations of engineers**
  - **machine learning tool to determine the optimal controls automatically**
  - **simulation model to respect physics of the reservoir**

11

AGH

Optimal control of UGS

PGNiG

To determine such well controls that maximize the total energy produced from the storage, respecting the physical and technical constrains, including quality of the produced gas.

Energy produced from a single well

$$E_k(\boldsymbol{u}) = \sum_{n=1}^{T}\left(\sum_{i=1}^{N} C_i\, y_{i,n,k}\, u_{n,k}\right) \quad (1)$$

Objective function – total energy

$$J(\boldsymbol{u}) = \sum_{k=1}^{W} E_k(\boldsymbol{u}) \quad (2)$$

Optimal control problem

$$J(\hat{\boldsymbol{u}}) = \max_{\boldsymbol{u}} J(\boldsymbol{u}) \quad (3)$$

Respecting physics and geology

$$\frac{dx}{dt} = f\left(x, \nabla x, \nabla^2 x, u\right) \quad (4)$$

AGH

Handling of time depended flow rates

PGNiG

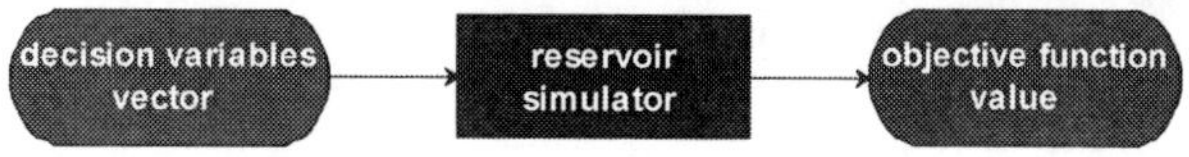

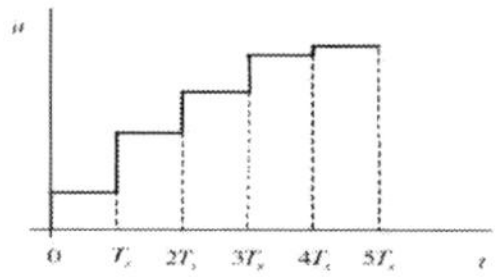

$$\hat{u}_{PMG} = \begin{pmatrix} \hat{u}_{11} & \hat{u}_{12} & \dots & \hat{u}_{1T} \\ \hat{u}_{21} & \hat{u}_{22} & \dots & \hat{u}_{2T} \\ \vdots & \vdots & \ddots & \vdots \\ \hat{u}_{W1} & \hat{u}_{W2} & \dots & \hat{u}_{WT} \end{pmatrix}$$

- $u_{n,k}$ - stepwise representation of the flow rate of the k-th well at the n-th day of the production cycle
- thousands of decision variables
- a black-box problem with a high evaluation time

AGH

Parameterized decision tree for automatic generation of the controls for production wells

PGNiG

Combination of reservoir simulation and optimization tool - Sequential Model-based Algorithm Configuration (SMAC) to optimize the parameters p1,p2,p3

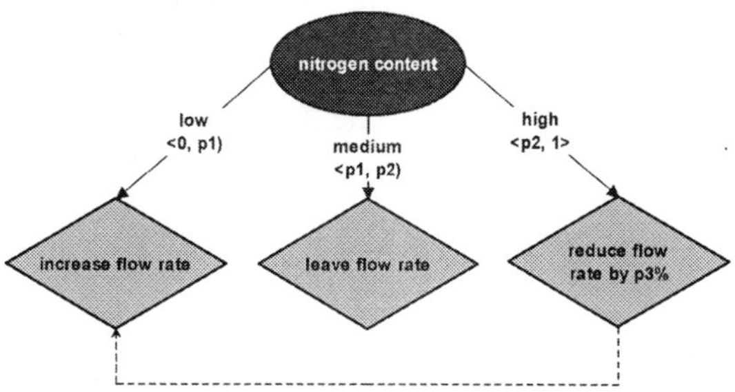

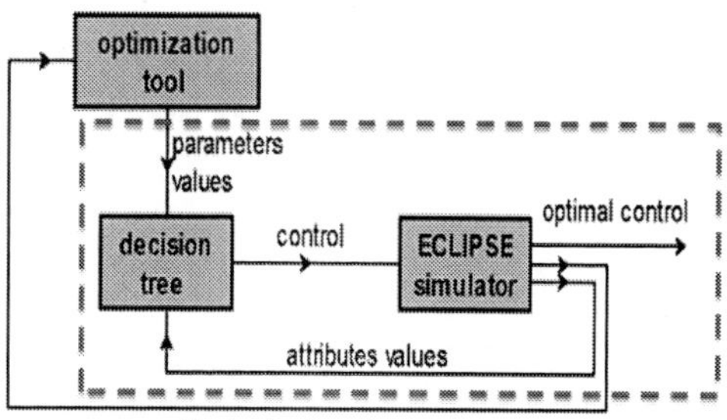

Arbitrary p1, p2, p3 replaced by optimized values 0.0217, 0.0374 and 0.2762 respectively (not intuitive)

AGH

Testing on the examples, one historical cycle – comparison of historical and optimized well controls

PGNiG

Well 1

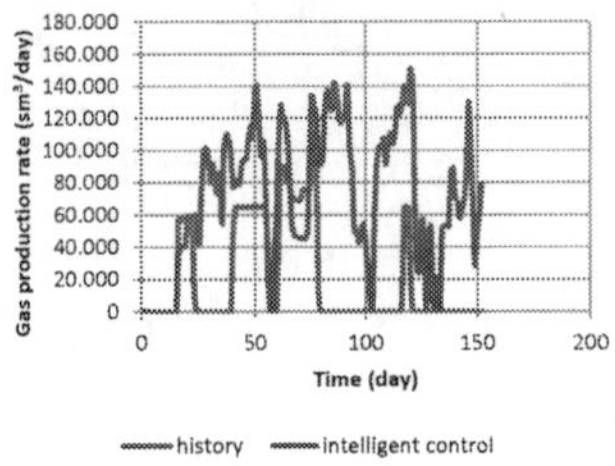

Well 2

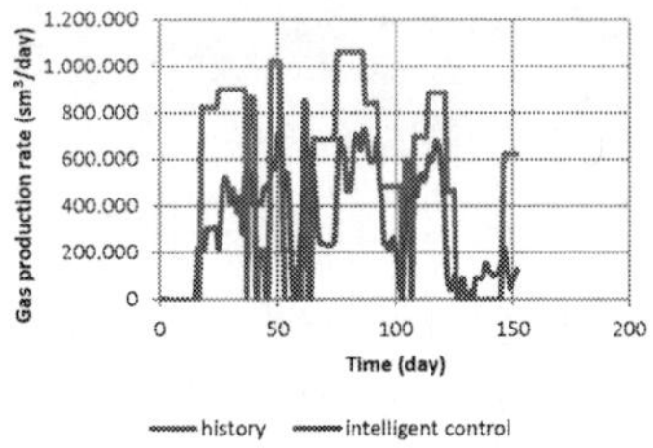

15

AGH

Comparison of the simulated nitrogen content in the produced gas obtained by the use of historical and optimized well controls

PGNiG

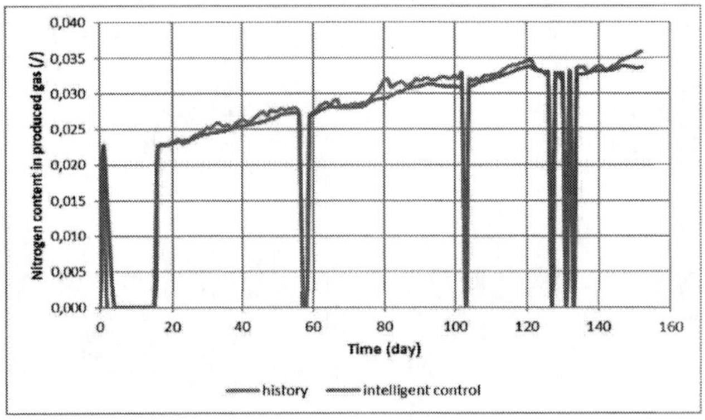

The cumulative nitrogen content in the produced gas was reduced by 2% 16

Energy production in one withdrawal cycle. Comparison of historical and optimized controls

PGNiG

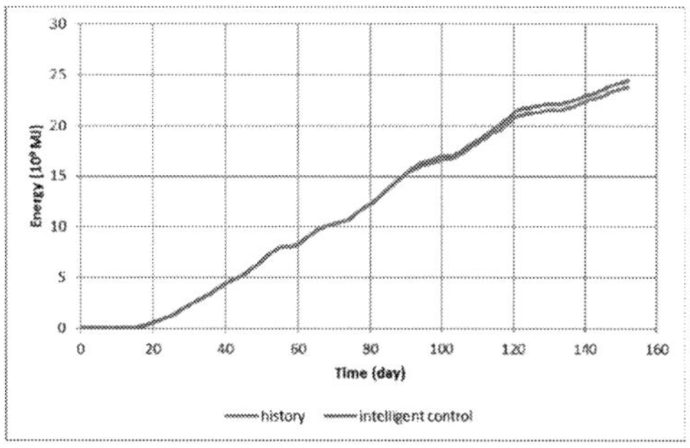

The energy production was increased by 2.4% .
Economic value of the additional energy is 5.8 million €
(If the energy price is 0.01 €/MJ)

17

The next generation decision scheme

PGNiG

- The operation wells are clustered into two groups according to location on the structure: GC- central wells and GE-external wells
- The algorithm was intended to mimic the decision schematic used by a human operator for both injection and withdrawal cycles
- New decision scheme was developed
- 11 decision variables x1-x11 were employed
- The decision tree has been parametrized and then optimized using the same numerical procedure as before

www.agh.edu.pl

The next generation decision scheme

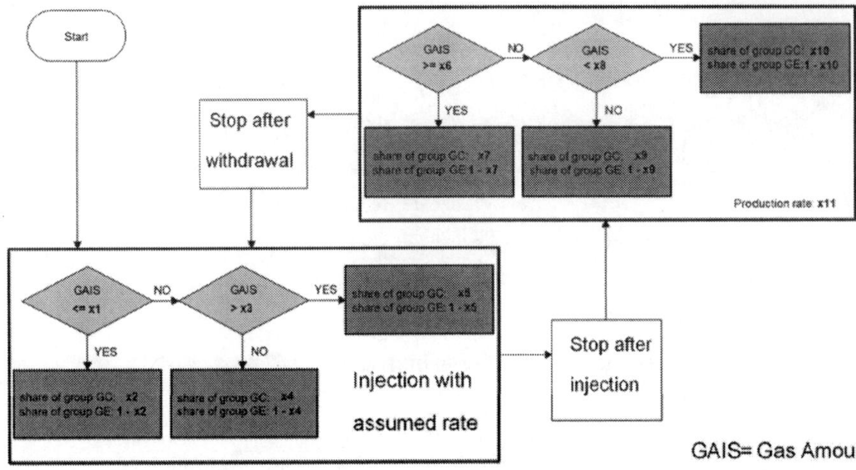

GAIS= Gas Amount In Storage

AI generated controls vs. operator controls

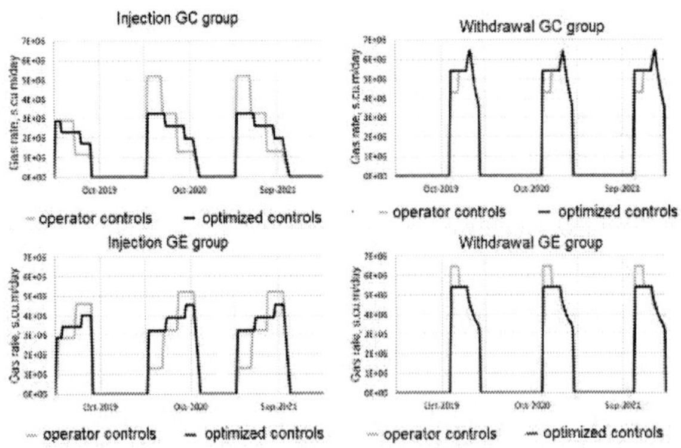

- **ΔJ=0.2· 10^9 MJ of energy**
- **0.16% more Energy in total**
- **2.55% more Energy in last 40 days of the cycle**

- **AI can learn**
- **AI is effective in new situations**

The additional energy resulting from optimization for the three simulated cycles would be about 0.2×10^9 MJ for the same volume of gas withdrawn from the storage. Moreover, the amount of energy withdrawn in the last 40 days of the third simulated cycle was increased by 2.55% as compared to the reference scenario

Conclusions

- Performance of the UGS can be improved (almost for free) by use of optimal control as combined with reservoir simulation and AI
- The algorithm integrates professional experience/best practices and machine learning. This results in synergy of the human intelligence and artificial intelligence
- The optimization software linking reservoir simulator and AI has been created in cooperation of Polish Oil and Gas Company and AGH University of Science and Technology. The software automatically generates optimal controls for each well
- The similar approach can be easily implemented for other problems related to optimal control of reservoirs.

Acknowledgements

The author would like to express thanks to Polish Oil and Gas Company PGNIG S.A. for permission to prepare and present this paper.

21

OQ

Accuracy of Fiscal metering system Gas Network

Faisal Al Raisi

Care Share Dare

OQ

OQ Gas Network

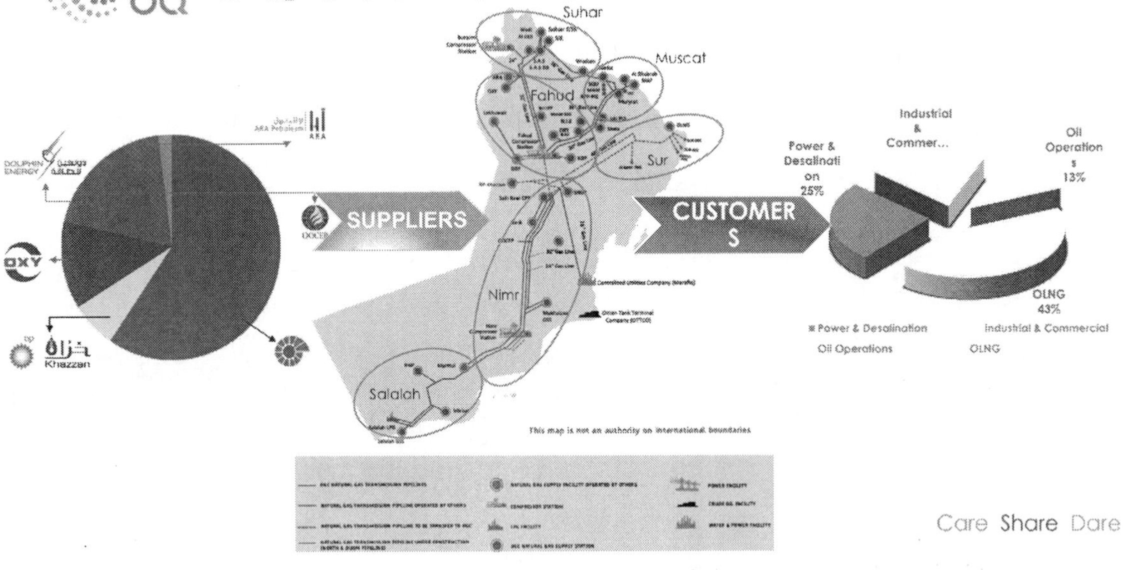

Care Share Dare

Introduction

- OQ, gas network division, operates a pipeline network of over 4000 km.
- Yearly sales amount approximately 37 billion standard cubic meters which equivalent to approximately1.33 thousands trillion British thermal unit .
- OQ operates 51 Flow Metering Stations (FMS), and the number is growing.
- All OQ measurements are calibrated and traceable to international physical standards.
- Each FMS consist of dual stream connected with Z-proving line, and equipped with redundant :

1- gas quality analyzer ,

2- Ultrasonic flow meter

3- pressure and temorature measurement devices

4- flow computer

Care Share Dare

Importance of accurate measurement

- Importance illustration by calculation example:
  - Fiscal Metering System measures 7 MMSCMD.
  - Calculating with 3 USD/MMBTU, it is worth about 780,000 USD/day
  - An error of 0,1% means 780 USD/day, or 285,000 USD/year!
  - Error can be plus and minus, so someone suffers from error!
  - Therefore, OQ build a strong measurement strategy , policies and Standards to:
    - ➢ Determine the delivered amounts with zero systematic error (bias)
    - ➢ Will actively search for systematic errors and provide measures to eliminate them.

Care Share Dare

OQ Measurement Policy

Measurement policy

- In the process of the determination of delivered quantities, OQ strives to achieve this task with zero systematic error (bias) and with the lowest possible, though affordable, uncertainty. OQ will actively search for systematic errors and provide measures to eliminate them

Care Share Dare

OQ Measurement Strategy

Eliminate errors

1- provides innovative digitalization solutions (Agile metering system)
2- 24/7 auto validations and verification (AGR)
3- Metrological confirmation Quarterly comprehensive validation

Reduce uncertainty

1- Upgrading the FMS to latest highest accurate equipment's
2- Using high accurate test equipment's
3- introducing live uncertainty calculation (Agile metering system)

Traceability

1- performing UFM wet calibration at internationally approved accredited lab Accurate; Measurement system according to OQ standard
2- Re-certifying test equipment's at accredited labs (Iso-17025)

Accurate Measurement

1- Introducing 4eys check mechanism using AGR and Agile metering system
2- Perform auto- validation every 15 minutes by AGR

Care Share Dare

Innovative solutions in gas measurement

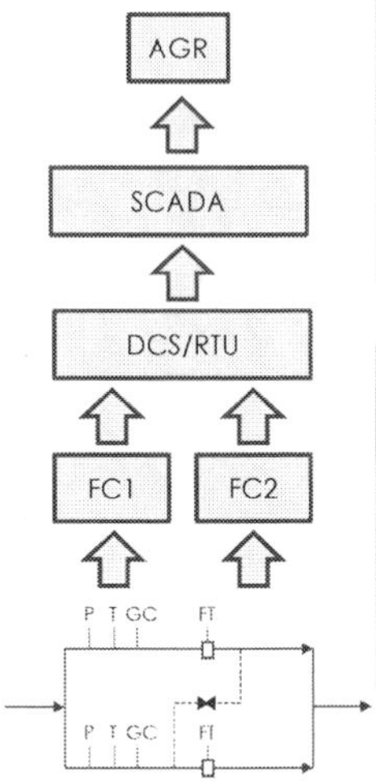

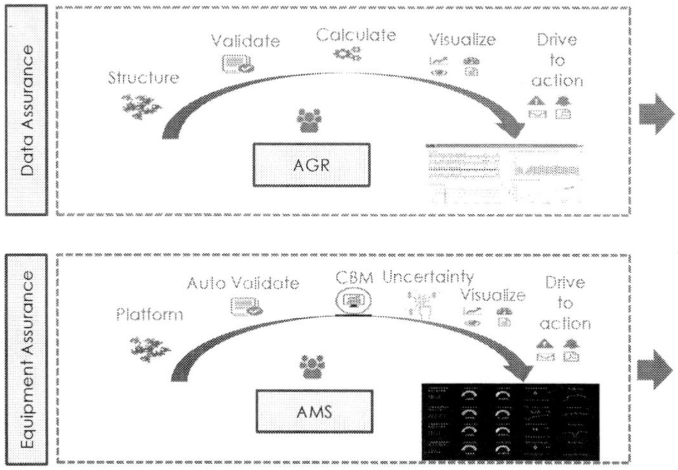

- Digitalize gas reporting process.
- Early notifications
- Over utilization detection
- Data Reconciliation
- Auto-approval levels
- Data Security.
- Automated Gas Delivery reports (invoicing).
- Digital data archiving.

- Eliminate Human Errors
- Extending maintenance program (CBM)
- Optimizing working hours
- Digitalization of custody transfer application
- Compliance with international standard (ISO 10012).
- Digitalizing all validation processes including reporting.

Care Share Dare

WATHIQ واثق

2 Functionalities in 1 single application

Nomination:

enables Company customers and suppliers as well Company personal to collaboratively establish nomination figures

Inquiry:

supports and helps to deal with customer inquiries, managing the incidents from the moment they're captured through to their resolution

- Internet access + OQ network access
- Hosted on OQ virtual server
- Disconnected from AGR / Scada

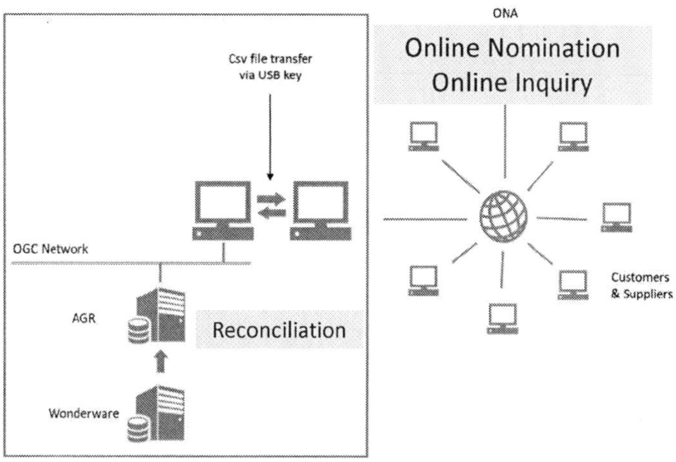

Care Share Dare

OQ Summary

- OQ derive innovative solutions/technologies in gas measurement such as Agile metering system, AGR and Wathiq to achieve:
- Zero disputed notes on measurements
- 91 % Customer satisfaction on measurement which conducted by 3rd party
- Delivered 37 billion standards cubic meters accurately and within an uncertainty of > 1%.
- Being the reference in gas measurement in the region.
- Being the certifier body for producers report
- Reconciliation for supply Vs demands reached to zero difference

Care Share Dare

OQ

Care Share Dare

IGRC 2020 MUSCAT

24-28 FEB | WWW.IGRC2020.COM

TERĒGA storengy UNIVERSITÉ DE PAU ET DES PAYS DE L'ADOUR

RINGS

Research on the Injection of New Gases into Storages

IGRC 2020, OMAN

1 INTRODUCTION : Energy transition in Europe

The gas mix is changing in Europe, with renewable gas slowly replacing Natural Gas.

European objective: Increase **renewable gas** in gas mix to **10% by 2030**.

→ **Biomethane** from methanization of wastes
→ H_2 from power to gas

- **Biomethane** is already injected in the natural gas grid, with more than 80 injection plants in France.
- H_2 is locally injected on pilot projects (such as Jupiter 1000, south of France).

Underground Gas Storages must evaluate their compatibility to biomethane and H_2 in order to adapt gas quality guidelines.

TERĒGA IGRC 2020, OMAN 2

1 INTRODUCTION : Presentation of an aquifer UGS

In Underground Gas Storages (UGS), the injected gas drains the porous rock, creating a water-gas contact.

→ Trace elements from natural gas can migrate into the water phase.
→ They are systematically degraded by microorganisms naturally present in deep anoxic environments [1].

What would be the impact of H_2 or biomethane injection in this deep, anoxic environment?

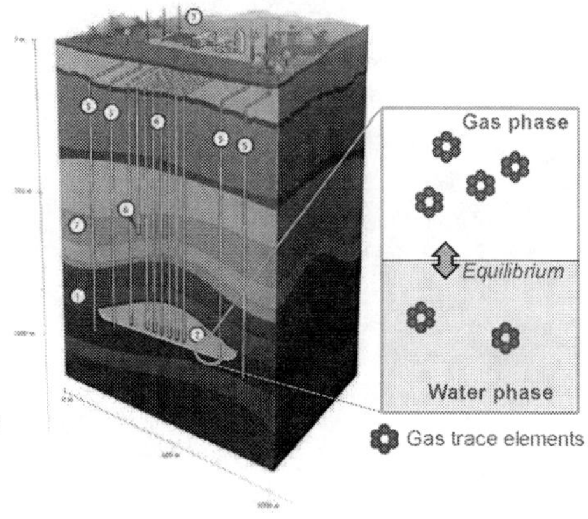

→ **H_2** can be an energy source for bacteria and is highly reductive

→ **Biomethane** is very similar to natural gas (> 97% CH_4) but may contain up to **1% O_2** .

[1] Berlendis S, et al. *Environ Sci Technol*, **44**(9) (2010) ; Aüllo T, et al. *Front Microbiol*, **7**:122 (2016) ; Ranchou-Peyruse M, et al., *MicrobialBiotechnology*, (2016)

1 INTRODUCTION : Presentation of of the RINGS project

Re-creating the UGS *in-situ* conditions in a labratory reactor

→ RINGS reactor can work up to 150°C and 150 bars
→ Water and rock phases are sampled in the real UGS
→ The initial gas phase is composed of CH_4 (99%), CO_2 (1%) and traces of BT[1].

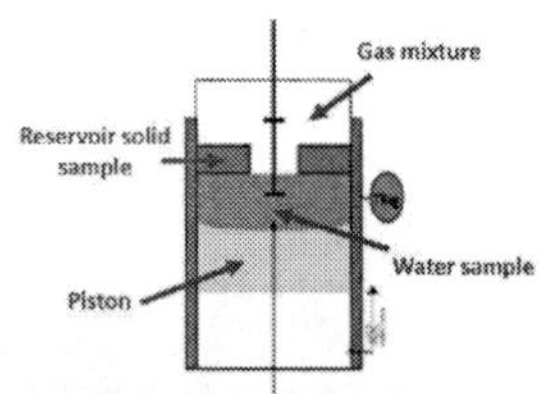

→ Renewable gas are injected in the reactor to monitor the biological, geological and chemical impact of renewable gas storage in UGS.

[1] BT = Benzene, Toluene

2 RINGS project : MATERIEL and METHODS

Rock and Water phase sampling

- Rock samples are composed of reservoir cuttings collected during drilling operations.
- Water samples containing in-situ bacteria were collected using a specific downhole sampler.

Chemical analysis in the RINGS reactor

→ On-line analysis of the **gas phase**: P, T, H_2, O_2, CH_4, N_2, H_2S
→ On-line analysis of **water phase**: P, T, **anions** (mainly fluoride, acetate, chloride, nitrate, carbonate, sulfate and sulfide) **cations** (mainly sodium, potassium, magnesium and calcium).
→ Off-line analysis of **microbial diversity** and **BT** concentration in water samples.

X-Ray analysis of the rock phase

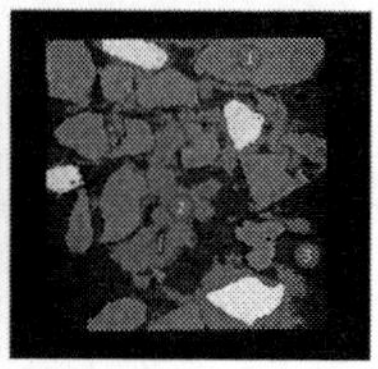

→ **X-Ray tomography** analysis allows the identification of biofilms in the rock porosity. (1: rock 2: biofilm 3: water)

→ **X-Ray diffraction** analysis gives information on the mineralogy composition of the rock phase

3 RINGS project : first results on the biomethane storage experiment

3.1. Re-creating the *in-situ* conditions of the UGS

The first 130 days of this 6-month experiment were used to re-create the *in-situ* conditions in the reactor.

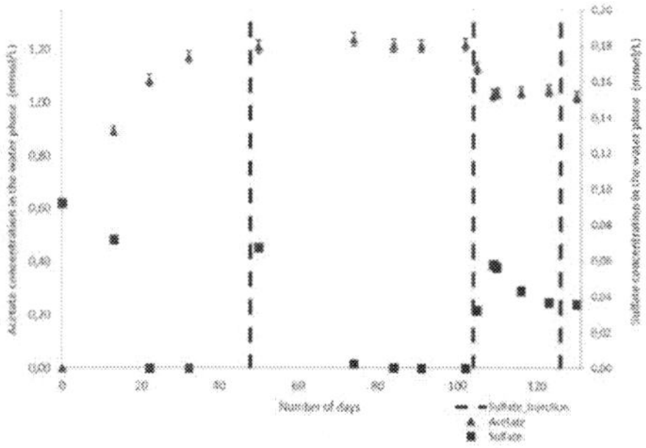

Key microorganisms in UGS are sulfate-reducing bacteria which degrade organic matter (i.e. BT) and respire sulfate.

Sulfates in the reactor were systematically consumed .

Microbial *live and dead* analysis showed that the organisms in the reactor were alive and active.

3 RINGS project : first results on the biomethane storage experiment

3.2. Biomethane injection in the reactor

Once the chemical and biological conditions in the reactor were stable, BIOMETHANE was injected in the reactor.

→ **Biomethane** = CH_4 (98%) + 1% CO_2 + **1% O_2**

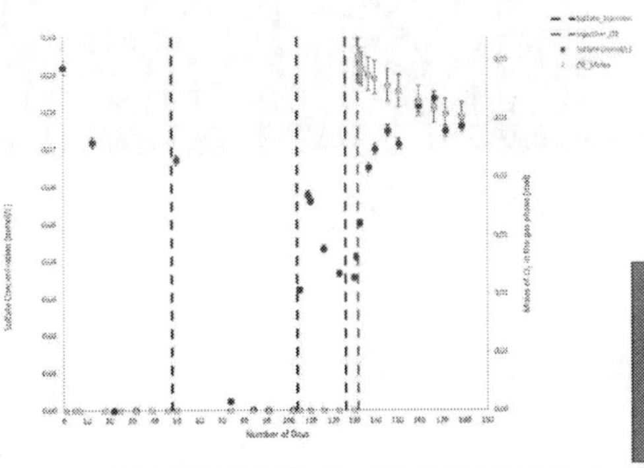

- Once O_2 was injected, the sulfates were not consumed anymore.
- The H_2 concentration in the reactor increased.
- The live and dead microbial analysis showed that some bacteria were still alive in the reactor.

→ These data show that sulfate-reducer bacteria that degrade the BT compounds **are not active anymore**. But other bacteria (fermentative?) are active.

→ *To confirm this hypothesis we are waiting for the microbial diversity analysis, the BT analysis and the X-Ray analysis of the rock phase.*

4 Conclusion and next steps

The RINGS reactor can successfully re-create the *in-situ* conditions of aquifer Underground Gas Storages.

This first experiment on simulating a BIOMETHANE storage in UGS tells us that:
- Biomethane with 1% O_2 has a real impact on the UGS microbial diversity
- Interpretation of microbiology and X-Ray data will help understand the exact impact

→ The next experiment planed in the RINGS reactor for 2020 will be the storage of H2NG blend in an aquifer UGS from northern France.

TERĒGA

guilhem.caumette@terega.fr

IGRC 2020 MUSCAT

STOCHASTIC APPROACH TO MINIMIZING MAINTENANCE COST INCORPORATING ATTITUDE TOWARDS RISK

Mahmood Al-Kindi and Nasser Al-Azri

Department Of Mechanical And Industrial Engineering,
Sultan Qaboos University ,
Muscat PC 123 P.O.Box 33 Al Khoud

جامعة السلطان قابوس
Sultan Qaboos University

Outline

- Introduction
- Decision Analysis
- Decision Model
- Managerial Insight
- Recommendations

جامعة السلطان قابوس
Sultan Qaboos University

IGRC 2020 MUSCAT

Introduction

Sultan Qaboos University

- Testing energy efficiency is very essential to quality and reliability of utility services.
- Reliability can be kept intact with continuous maintenance; however unnecessary maintenance can be costly.
- Maintenance frequency can be optimized by optimizing tradeoff between cost of energy faults and cost of maintenance.
- Here, energy faults is assumed as an increasing function of time as in *reliability bathtub model*.

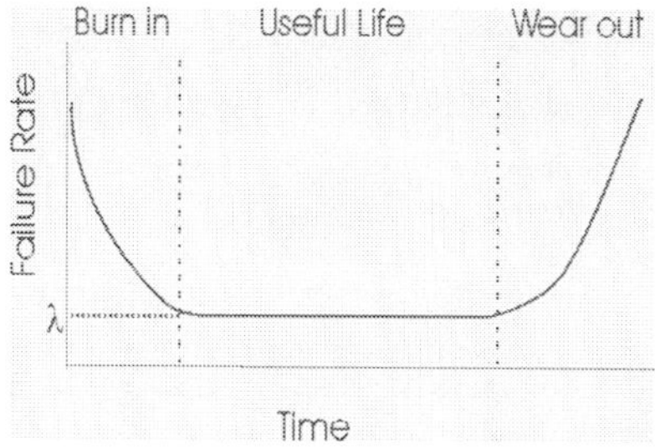

Reliability/failure bathtub model

IGRC 2020 MUSCAT

Decision Analysis

Sultan Qaboos University

- Decision analysis is a logical procedure for the balancing of the factors that influence a decision.
- The analysis procedure integrates uncertainties, monetary and non-monetary values, and preferences in a basic structure that models the decision.
- The core bases for any decision are:
  1. The alternative or *what can we do?*
  2. The information or *what do we know?*
  3. The preferences or *what do we like?*

IGRC 2020 MUSCAT

Decision Analysis

- Any economic planning is influenced by three main players:
  - The number of variables
  - The static/dynamic nature of theses variables
  - The number of uncertain variables
- The complexity level (search domain) indicated by the cube volume increases as these three dimensions increase.

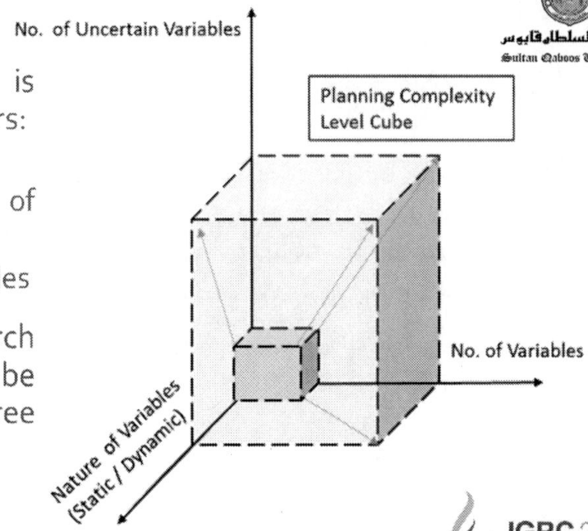

What is the problem

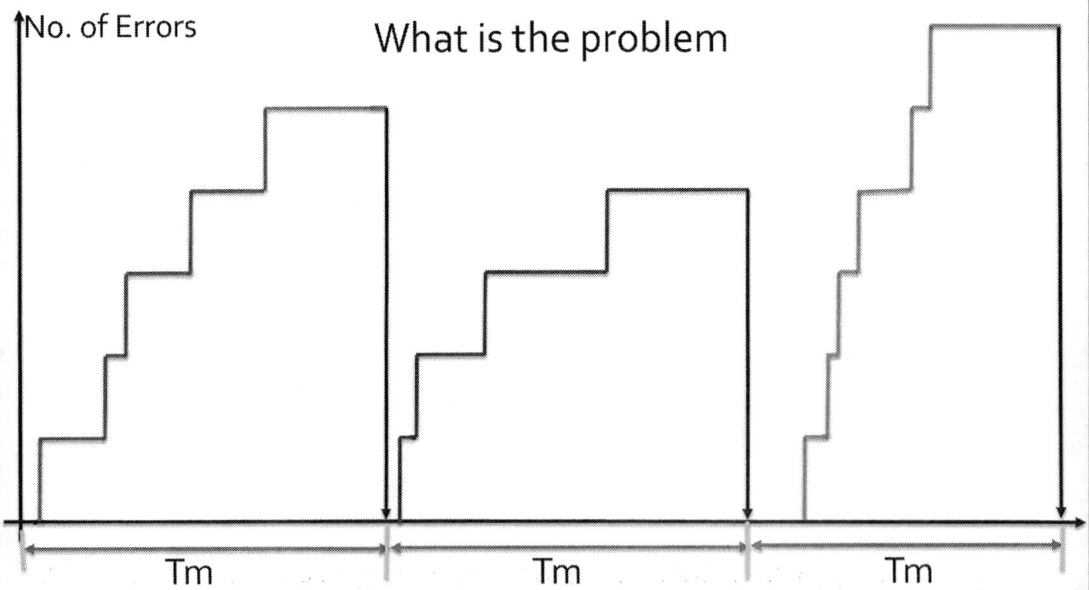

Decision Model

- Energy/system errors and problems occur during running phase over *T* period of time where fault cost takes place during running times.
- The magnitude and frequency of errors follow the bathtub norm.
- Cumulative errors are recorded for the period [0,T] at each time T.
- Errors are generated according to Poisson process with mean, λ and *X is defined as* the number of errors at time *T* and m is a constant setup cost paid for each maintenance visit. The cost paid for errors is defined as c(T). Hence, the total testing cost is:

$$C_{testing} = m + Xc(T)$$

- The expected total test cost is:

$$E\left(C_{testing}\right) = m + \lambda T\, c(T)$$

Decision Model

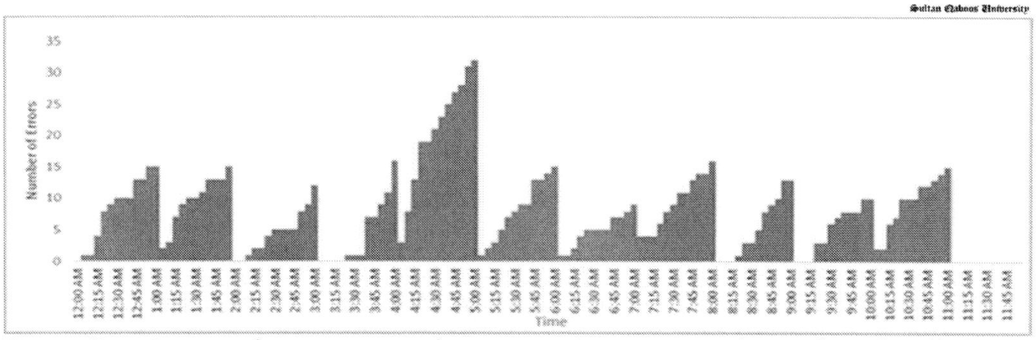

- Every 60 minutes the team conducts a maintenance where they fix all the errors that were observed after the last performed maintenance.
- Obviously each maintenance cycle has a different number of errors. So running a maintenance every 60 minutes might not be optimal.

Decision Model

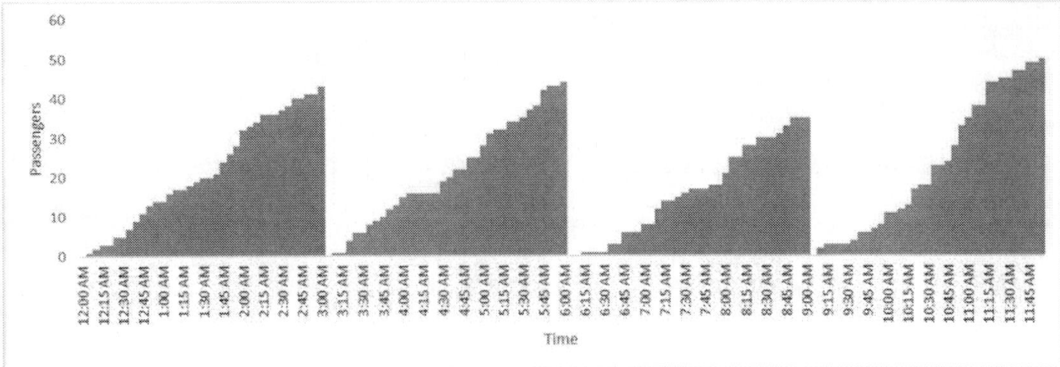

- Only four tests in 12 hours period at 180 minutes. Thus, there exists an optimal strategy for maintenance testing.

Decision Model

- If the team decides to perform maintenance every t_m time –unit. Then the total cost C is summation of setup cost and fixing costs will be:

$$C = m + X_1 c(t_m) + m + X_2 c(2t_m)\mathrm{K} + m + X_n c(nt_m)$$

- When the fixing cost is at time t is similar to it at time t_{i+1}

$$c(t_i) = c(t_{i+1}) = \mathrm{L} = c(t_n)$$

- Therfore we can express the total cost as

$$C = nm + \sum_{i=1}^{n} X_i c(t_m) = nm + c(t_m) \sum_{i=1}^{n} X_i$$

Case1: Maintenance Cost is increasing function of running time

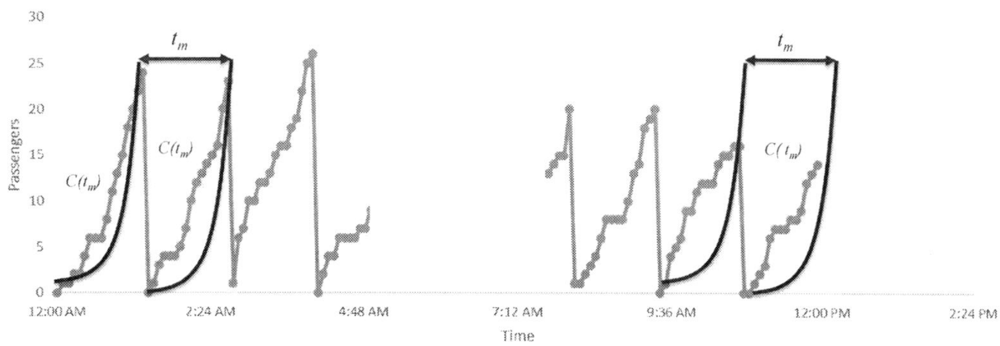

Decision Model

جامعة السلطان قابوس
Sultan Qaboos University

- Since the total number of errors X follow a Poisson distribution then to total number of error Y in the expression in the previous equation further reduced to

$$Y = \lambda T$$

- The total number of tests is obviously T/t_m. Therefor the expected saving Π, which Revenue or Allocated Budget (B) minus the maintenance cost, and its variation can be expressed as

$$E(\Pi) = B - m\frac{T}{t_m} - \lambda T\, c(t_m)$$

$$Var(\Pi) = \lambda T\, c(t_m)$$

IGRC 2020 MUSCAT

Decision Model

- For a decision maker with exponential utility function facing a Gaussian deal, the certain equivalent *CE* is given as

$$CE = E(\Pi) - \frac{\gamma}{2} Var(\Pi)$$

- Where γ is the risk tolerance which takes the value from -1 (risk seeking) to +1 (risk averse). The value of γ= 0 means that the decision maker is risk neutral. For the cost function $c(t_m) = ct_m^a$ then the optimal time interval t^*_m and number of text n^* are obtained as respectively

$$t_m^* = \left(\frac{m}{ac\lambda\left(1+\frac{1}{2}\gamma\right)} \right)^{\frac{1}{1+a}} \qquad n^* = T\left(\frac{m}{ac\lambda\left(1+\frac{1}{2}\gamma\right)} \right)^{\frac{-1}{1+a}}$$

IGRC 2020 MUSCAT

Decision Model

- The maintenance cost also vary with time as the shape parameter a captures this. Suppose that *m*=$30 per test, *T*=1, *c*= $40 and *λ*=100 per period *T*. Then Figure 5 below shows the optimal number of tests at difference cost nature.

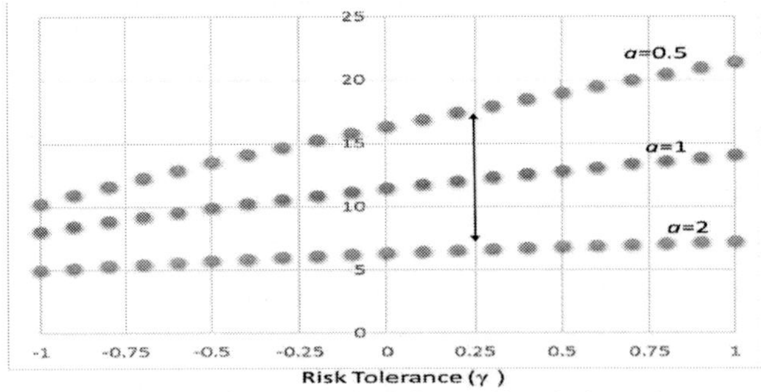

Optimal number of tests (n^*) for different risk attitudes and maintenance cost functions.

IGRC 2020 MUSCAT

Managerial Insights

The optimal strategy reveals two interesting insights:

- Risk seeking conducts less tests compared to other risk behavior (neutral and averse). Which may suggest that every decision maker or team leader has different optimal schedule.
- Risk averse is most affected by the nature or rework cost (i.e. concave, linear, or convex). There is a tendency to run more test when the fixing cost is concave the reason is to avoid the increase of maintenance cost as the team wait longer .

Recommendations

Maintenance should be run economically by knowing :

1. Errors or issues rate or intensity
2. Cost of setup cost of maintenance
3. Nature of fixing costs
4. Team Attitude toward Risk behavior

Future Work

- Dashboard to obtain optimal schedule Automatically.
- Different Cost and Issues-Types Analysis
- Incorporating Learning Effect
- Introduce Machine Learning and IoT in the Model

* A Person Risk Aversion Coefficient

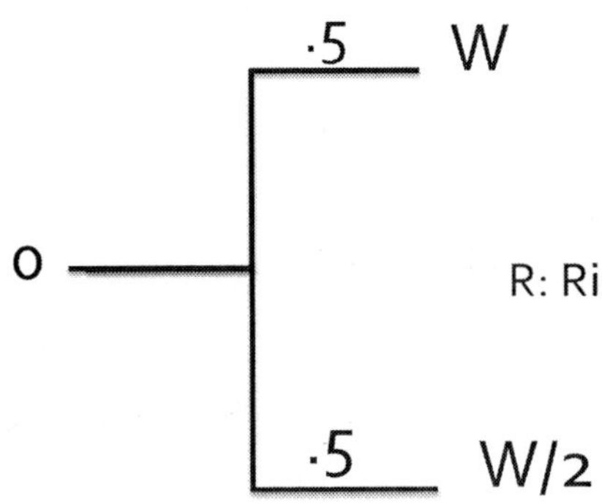

$$U(x) = a + b\,e^{-\gamma x}$$
$$r = e^{-\gamma x}$$

R: Risk Odds (winning/ losing)

26.02.2019

EJAAD

"Contributing to a Knowledge Based Economy"

Dr. Sausan Al Riyami
e-mail: sausan.riyami@ejaad.om

Why EJAAD ?

OEF 2015

Addressed 5 Key streams :

ENERGY DEMAND | ENERGY SUPPLY | R&D | LABOR | W.F.E NEXUS

Outcome : Oman Energy Master Plan 2040

OEIA R&D Summit 2016

Recommendation : Produce **protocol** with guiding principles form national and international energy stakeholders operating in Oman

The 2017 Oman Energy IA R&D **Protocol** —— EJAAD Lunching ——→

2

Ejaad Governance

Founders

Vision

To become a sustainable R&D and Innovation collaboration hub linking business requirements to local research capabilities towards boosting Oman's knowledge base economy

Mission

To leverage efficient and sustainable partnerships through:

(1) Seamless e-systems and e-platforms

(2) Establishment of focused Center of Excellences

(3) Become the national knowledge power-house in the areas for focus

3

Ejaad Journey

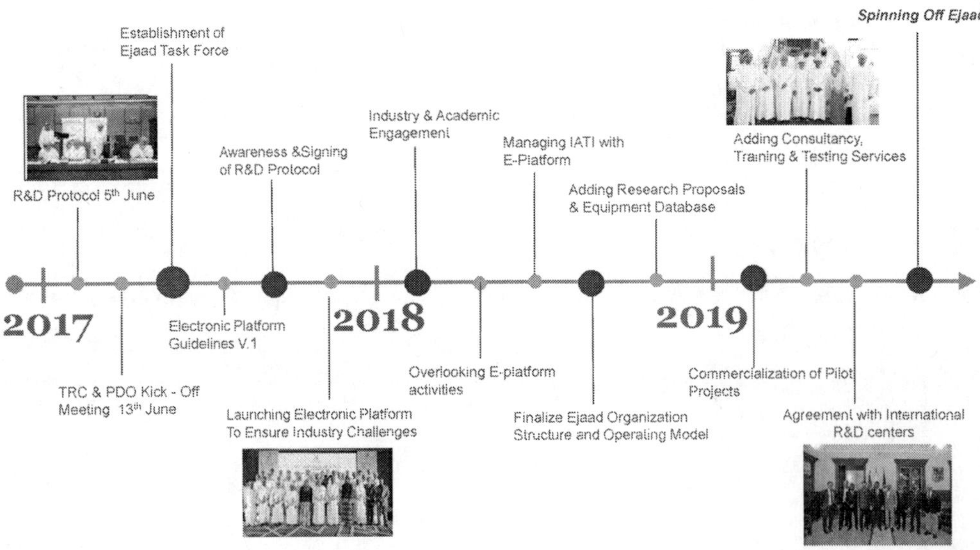

4

Ejaad focus on 8 areas:

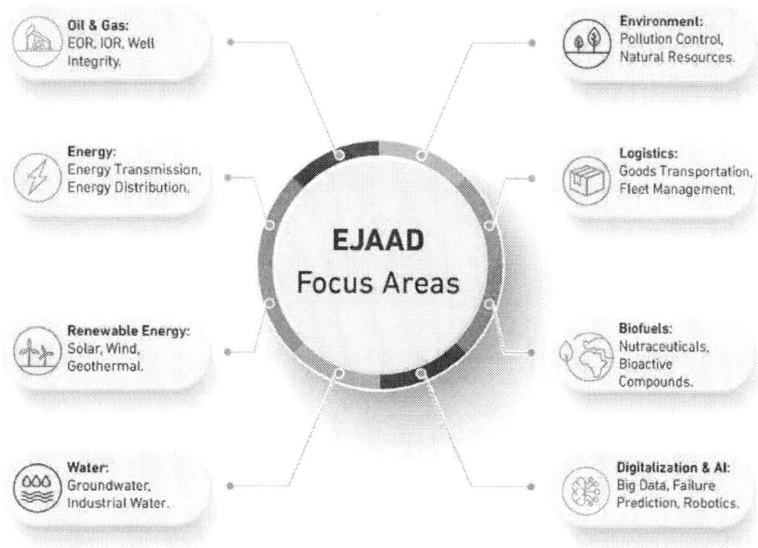

5

EJAAD Improved Framework: Innovation through Integration

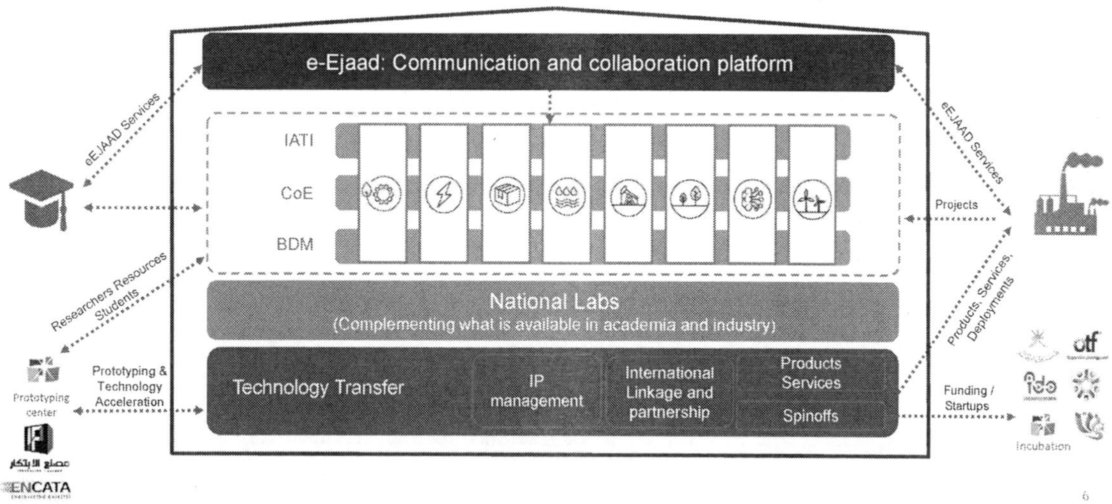

6

EJAAD Services

ايجاد EJAAD

Industrial Challenges
Marketplace for industry to present challenges and for academia to propose solutions

Researcher Network
Access to +1000 researchers based in Oman

Research Ideas
Leverage fund / support for groundbreaking research ideas & solutions

Equipment & Facilities Database
List of R&D equipment in academia & industry

Staff Mobility
Secondment of expertise from academia to industry and vice versa

Studies
A hub of expertise for seeking an advice within a particular field

Tech. capability Program
One-stop-shop for training, short-courses & workshops

Int'l Collaboration
Access to international academia and industry

7

R&D Landscape & EJAAD

ايجاد EJAAD

Research Classification	Basic Research			Applied Research			Application & Commercialization		
Activities	Discovery & Feasibility			Development, Testing & Validation			Integration & Deployment		
Technology Readiness Levels	TRL1	TRL2	TRL3	TRL4	TRL5	TRL6	TRL7	TRL8	TRL9
	Basic Principles Observed & Reported	Technology Concept Formulated	Experimental Proof of Concept	Technology Validated in lab	Technology Validated in Relevant Environment	Technology Demonstrated in Relevant Environment	System Prototype Demonstration in Operational Environment	System Complete and Qualified	Actual System Proven in Operational Environment

← **Ejaad** → (TRL4 – TRL9)

- **Developing new marketable products, made in Oman**
- **Adding national value by creating spin-offs, entrepreneurs and high-skilled jobs**
- **Realization of technology, from research labs to field deployment**
- **Advancing a sustainable knowledge-based economy**

8

Signees of R&D Protocol

9

Thank you

TÉCNICO LISBOA

MIT Portugal

IN+ CENTER for INNOVATION, TECHNOLOGY and POLICY RESEARCH

Impact of CO_2 Content on the Performance of Biogas/H_2 Inverse Diffusion Flames

Filipe M. Quintino[a], Edgar C. Fernandes[a]

[a] Center IN+, Instituto Superior Técnico, University of Lisbon
1049-001 Lisbon, Portugal

International Gas Union
Research Conference 2020

IGRC 2020 MUSCAT

Muscat, Oman || February 2020

Contents

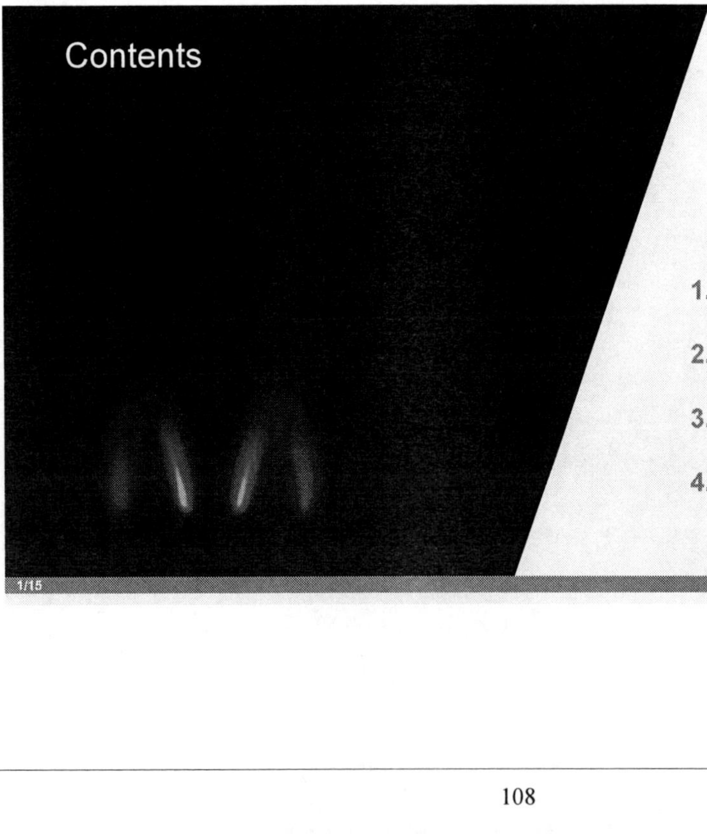

1. Introduction
2. Experiment
3. Results
4. Conclusion

1/15

Introduction

Currently the world is moving from fossil fuel towards renewable energy sources.

Global climate target:
- limit global warming prior to pre-industrial levels to 2 °C[1];
- current fossil-fuel share indicate that this limit might be surpassed in 20 years[2].

Renewable gases such as Biogas and Hydrogen will be key to decarbonise gas-powered systems.

[1]Meinshausen, Malte, et al. "Greenhouse-gas emission targets for limiting global warming to 2 C." Nature 458.7242 (2009). 1158.
[2]A.S. Denning. Combustion to concentration to warming: What do climate targets mean for emissions? Climate change and the global carbon cycle. in Dominick A. Dellasala and Michael I. Goldstein, editors, Encyclopedia of the Anthropocene, Elsevier, Oxford, pages 443-452., 2018.

Introduction

Complete overhaul of delivery infrastructure and appliances is expensive and unrealistic.

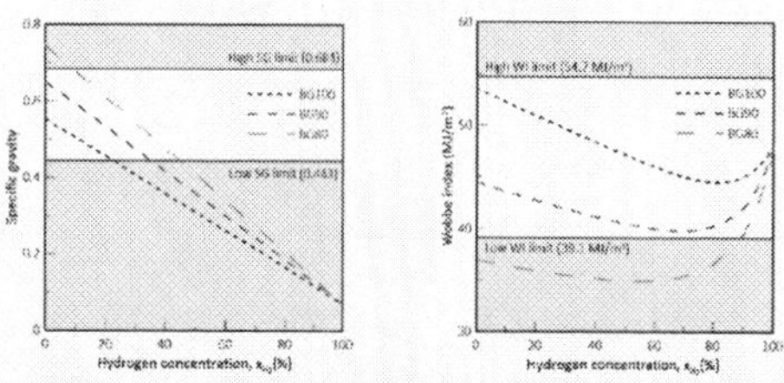

Fig: Variation of specific gravity and Wobbe index with H_2 concentration for BG100, BG90 and BG80[3]. Limits for the 2nd gas family under EN 437:2003+A1[4] norm.

[3]Quintino, F. M. and Fernandes, E. C., "Interchangeability Analysis of Biogas and Hydrogen Blends ", 27th European Biomass Conference & Exhibition (EUBCE) 2019 pp. 736-741
[4]EN 437:2003+A1, Test gases - Test pressures - Appliance categories. European Committee for Standardization (2009).

Introduction

A gradual transition is necessary, and appliances must be able to deal with higher variability in blend composition as market structure changes.

Problem:

Premixed configurations are **sensitive to changes in fuel feed** and unable to deal with highly variable blend composition.

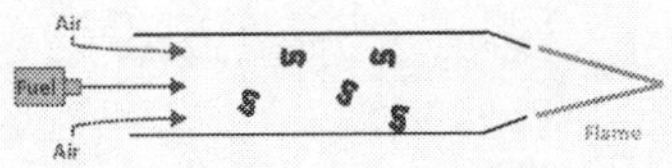

Prone to **flashback** and **blowout** phenomena.

Our approach:

Inverse Diffusion Flame (IDF)

- Not prone to flashback – **Safe;**
- **Extended stability range,** particularly in lean regime;
- **Reduced soot** production;
- **Low pollutant** emissions.

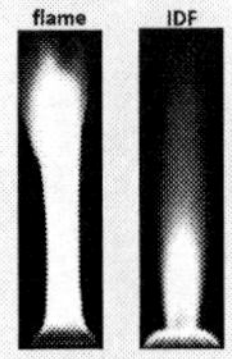

Fig: Comparison between normal diffusion flame and IDF[5].

[5]L. Dong, C. Cheung, and C. Leung. , International Journal of Heat and Mass Transfer, 50(25-26):5124–5153, 2007

Introduction

Inverse Diffusion Flames (IDF)

Principle: Central air jet with adjacent outer fuel co-flow.

- Coaxial Burners[6]

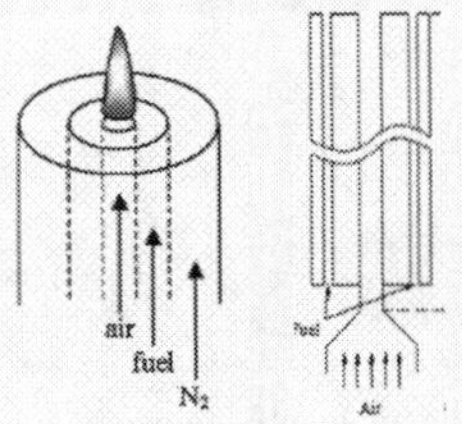

- Centrally Arranged Ported (CAP) – Top view[6]

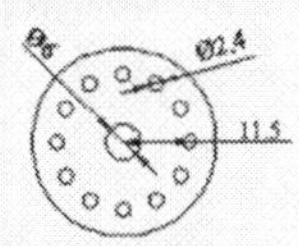

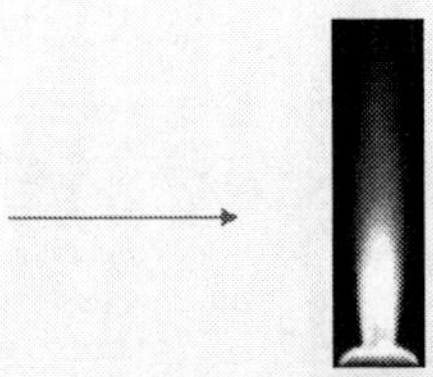

- Lamella or multi-slit configuration[7]

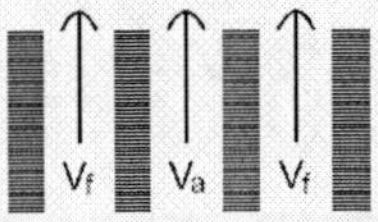

[6]J. Miao, C. Leung, C. Cheung, Z. Huang, and H. Zhen. , Energy, 104:284 – 294, 2016
[7]Vaz, F. Characterization of inverse diffusion fames with methane and hydrogen. (University of Lisbon, 2019)

Introduction

Previous research on CH_4/H_2 IDF in multi-slit burner

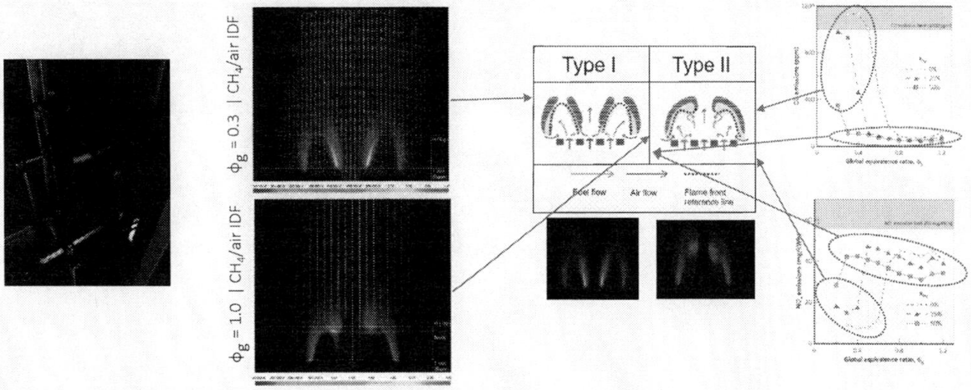

[7]Vaz, F. Characterization of inverse diffusion fames with methane and hydrogen. (University of Lisbon, 2019).
[8]Ribeiro, M. Characterization of methane inverse diffusion flames in a multi-slot burner. (University of Lisbon, 2019).

Introduction

Objectives

The unique features of the **IDF configuration** make it a **prime candidate** to be employed in a **changing market structure** scenario with **hydrogen** and **biogas**.

This work intends to **analyse the impact of CO_2** in the performance of **$CH_4/CO_2/H_2$ inverse diffusion flames in a multi-slit burner.**

Since **performance is directly related with flame structure**, it was studied with two experimental techniques:

- flame inspection through **photography**;
- flame **chemiluminescence** to evaluate the variations of OH*, CH* and C_2* in the flame.

Experiment

Setup

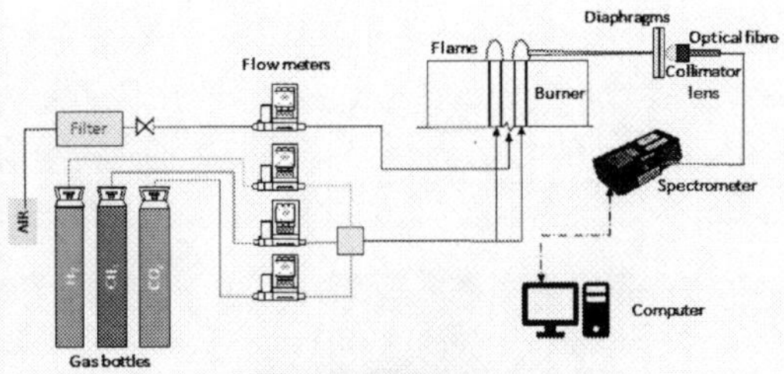

[redacted]

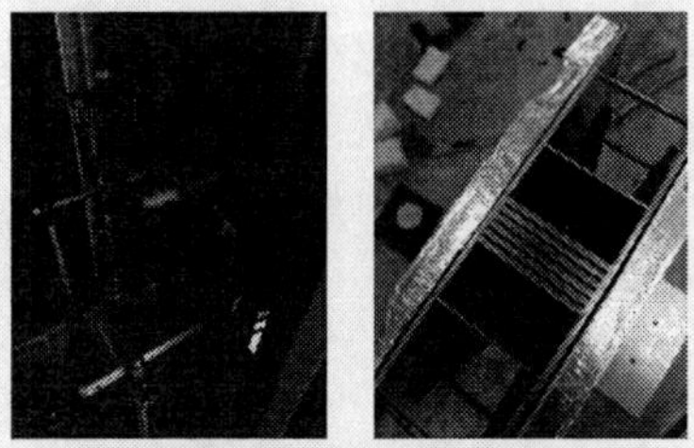

[redacted]

-**Spectroscopy** with QE65000 (Ocean Optics).
-**Photography** with Canon EOS 7D & EF-S 18-135 mm lens.

Experiment

Procedure

Four blends tested:

BG100 – 0% H_2 **BG80** – 0% H_2

BG100 – 25% H_2 **BG80** – 25% H_2

BG## → ## = %CH_4

Flame power: **P = 238 W**

Global equivalence ratio:

$$0,2 \le \varphi \le 1,0$$

$$\varphi = \frac{\left(Q_{air}/Q_{fuel}\right)_{st}}{\left(Q_{air}/Q_{fuel}\right)}$$

Chemiluminescence:

$\varphi = 1,0$

10 acquisition points

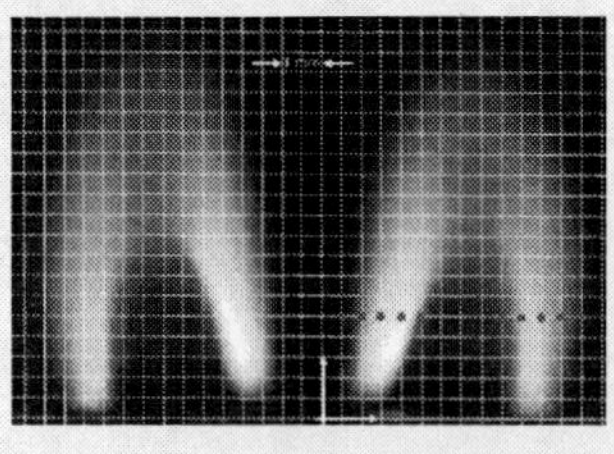

Results

Flame structure

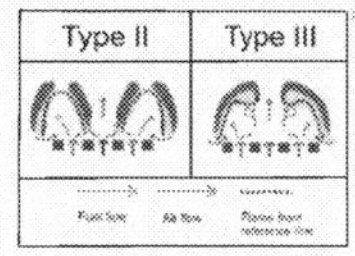

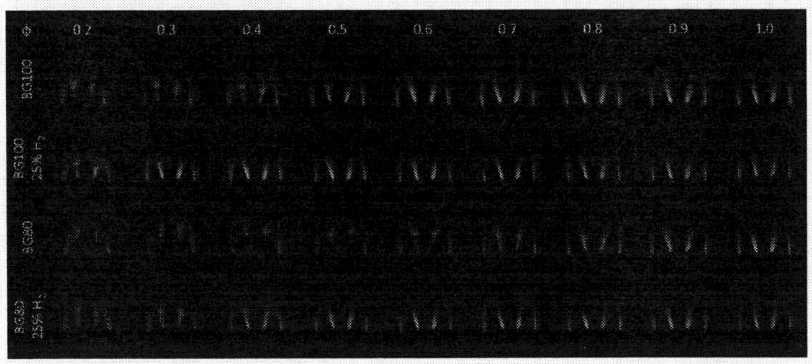

The **introduction of carbon dioxide (CO_2) does not produce flame structures other than those already known**: Type I and Type II.

Leaner flames are more prone to stabilize in a **Type II** structure.

Results

Flame structure & impact on pollutants

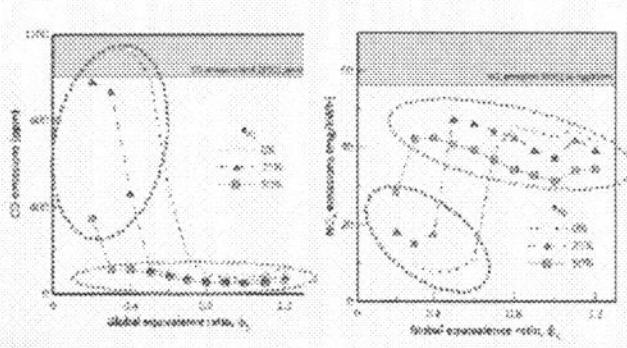

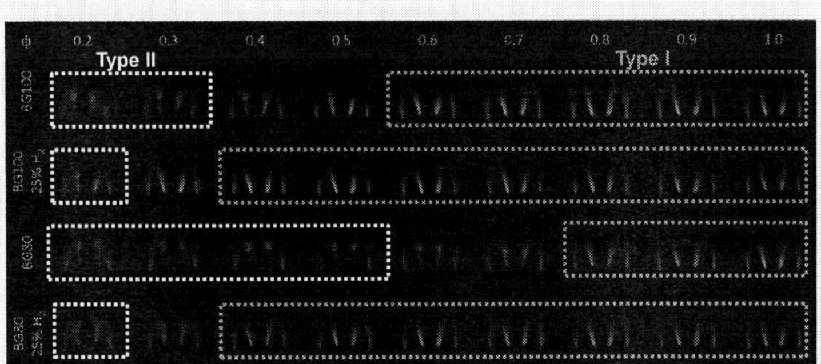

The positive **impact of hydrogen is dominant** over the influence of carbon dioxide.

The **impact of CO_2** in flame structure is **expected to increase CO** emissions. However, **hydrogen can be employed to suppress this effect**.

Results

OH*, CH* and C_2* chemiluminescence

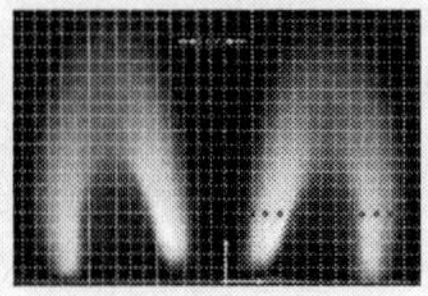

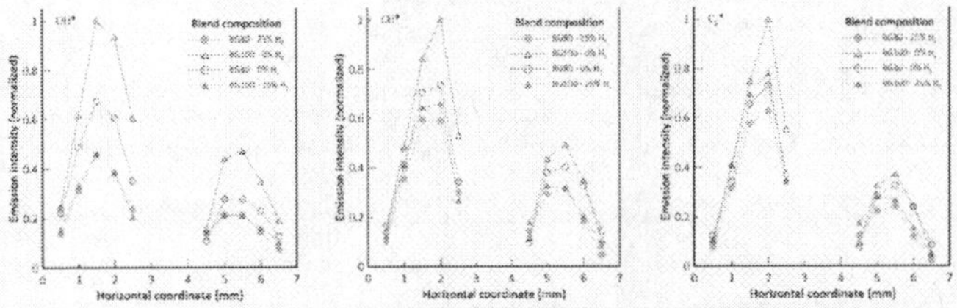

Fig: OH*, CH* and C_2* normalized emission intensity for the four blends under analysis. Each symbol is a spectroscopy measurement for a horizontal flame coordinate.

- Both CO_2 **and** H_2 **lower light emissions** associated with all radicals.
- The **inner flame region** (left-hand side) **has a higher emission intensity**, indicating **enhanced combustion** in this region.

Results

OH*, CH* and C_2* chemiluminescence

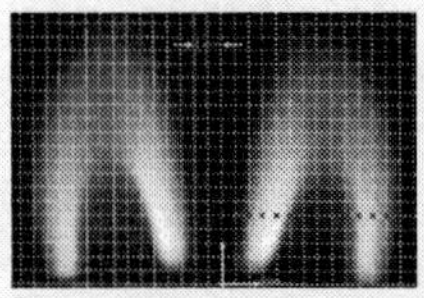

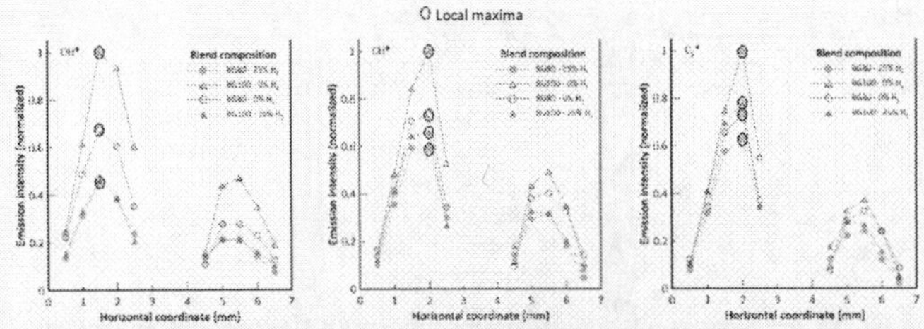

Fig: OH*, CH* and C_2* normalized emission intensity for the four blends under analysis. Each symbol is a spectroscopy measurement for a horizontal flame coordinate.

- **OH*, CH* and C_2* light emission is associated with local air/fuel (ϕ) variations**[9,10]. Higher CH* and C_2* prevalence are associated with rich combustion.
- OH* local maxima located at a collection point left of that of CH* and C_2* indicate that **combustion is richer as we move to the right-hand side of the inner flame region.**

[9]Quintino, F. M., Trindade, T. P. & Fernandes, E. C. Biogas combustion: Chemiluminescence fingerprint. Fuel 231, (2018).
[10]Panoutsos, C. S., Hardalupas, Y. & Taylor, A. M. K. P. Numerical evaluation of equivalence ratio measurement using OH* and CH* chemiluminescence in premixed and non-premixed methane-air flames. Combust. Flame 156, 273–291 (2009).

Results

Local equivalence ratio

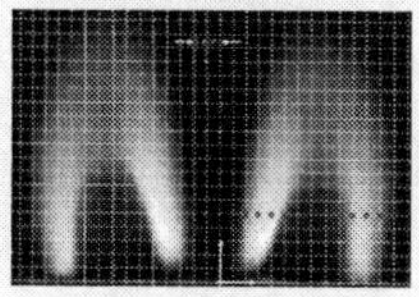

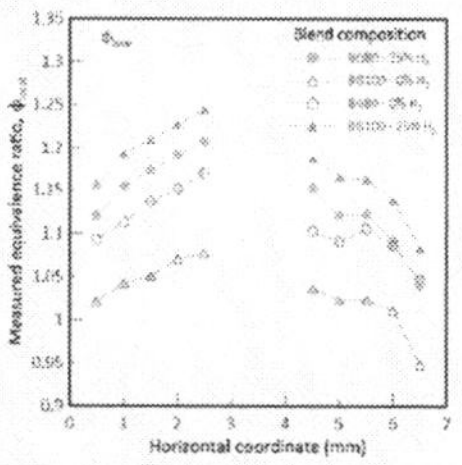

Fig: Local equivalent ratio variation on IDF for the blends under analysis. Global equivalence ratio was set at ϕ = 1.0 and power at P = 238 W.

- **CH*/C$_2$* ratio** was employed to infer local ϕ. Curve trends support the hyphotesis withdrawn from local OH*, CH* and C$_2$* emissions.
- To obtain more precise absolute values **new calibration curves will be performed accounting for temperature variations.**

Conclusions

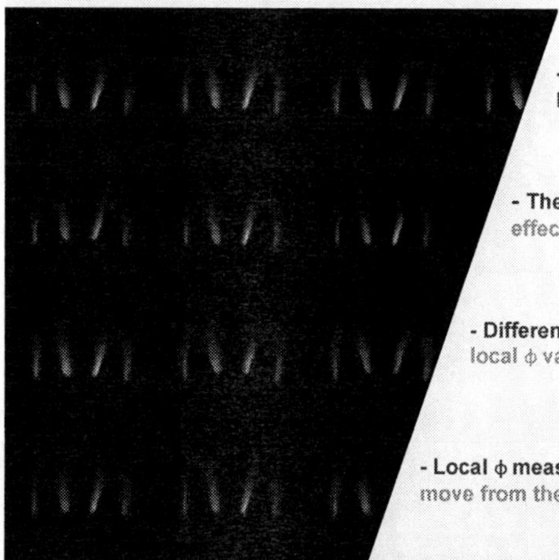

- Carbon dioxide significantly impairs IDF performance **when H_2 is not present.**
- **The presence of 25% (v./v.)** H_2 enrichment cancels the negative effect of CO_2.
- **Different monotony in OH*, CH* and C$_2$* curves indicate** significant local ϕ variations.
- **Local ϕ measurements demonstrate that** flames are richer as we move from the centre towards the fuel jet.

TÉCNICO LISBOA

MIT Portugal

IN+ CENTER for INNOVATION, TECHNOLOGY and POLICY RESEARCH

THANK YOU FOR YOUR ATTENTION!

FILIPE QUINTINO
filipe.marques.quintino@tecnico.ulisboa.pt
(+351) 91 812 74 25

FEBRUARY 2020

IGRC 2020 MUSCAT

Achieving Wellbore Integrity in High H_2S and CO_2 Environment in Oman

Hamada Youssef
Cementing SME
Petroleum Development Oman

February, 2020

شركة تنمية نفط عمان
Petroleum Development Oman

Agenda

- Field Overview
- Introduction
- Lab Testing & Results
- Field Trial
- Overview of Cement Job: On Job Trials
- Cement Bond Assessment

شركة تنمية نفط عمان
Petroleum Development Oman

Field overview

- Drilling wells in the southern part of Oman with challenging production targets in CO_2 rich conditions
- Reserves captured in a dolomite stringer A2C floating and encased in the Ara Salt.
- High sour and high-pressure floaters with ± 3% H_2S and 15% CO_2.
- Multi-disciplinary effort has been performed to optimize the well construction without compromising HSE.
- Improving well delivery process and maximizing gas recovery.

شركة تنمية نفط عمان
Petroleum Development Oman

Introduction

- Corrosive gases are known to chemically attack the Portland cement (Carbonation).
- Cementing in harsh CO_2 and H_2S environments is common when exploring for productive reservoirs.
- The challenges:
  - ✓ Low Portland slurry was designed with elastomeric self-healing materials.
  - ✓ Zonal isolation not compromised during CO_2 acid attack.
  - ✓ Long-term cement sheath integrity maintained during the life of the well.

شركة تنمية نفط عمان
Petroleum Development Oman

Laboratory Testing Procedure

1. CO2 corrosion testing: was performed as following
   - 90 days at 250°F and 1200 psi. (exposed) 2 samples
   - 90 days at 250°F and 3000 psi. (non-exposed) 2 samples
2. Testing performed
   1. Phenolphthalein Dye Test
   2. XRD Analysis
   3. Crush strength Test

شركة تنمية نفط عمان
Petroleum Development Oman

Lab Testing & Results

1. Phenolphthalein Dye Test:

   To understand the extent of carbonation in the sample (pink on Alkali and becomes colorless on acidic).

Neat Cement Slurry exposed

CO_2-Resistant slurry unexposed

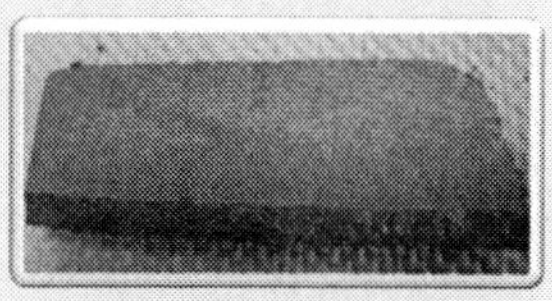

CO_2-Resistant slurry exposed

شركة تنمية نفط عمان
Petroleum Development Oman

Lab Testing & Results

4. XRD Analysis:
   - Slurries exposed to CO_2 have shown formation of Calcite and Aragonite which are forms of calcium carbonates.
   - Formation of carbonates in the neat slurry is almost three times higher than that in CO_2-resistant slurry.
   - CO_2-resistant slurry undergoes less corrosion as compared to the neat slurry.

Phases	Neat Slurry Unexposed %	Neat Slurry Exposed %	CO_2 resistant Unexposed %	CO_2 resistant Exposed %
Calcium Hydroxide	15	-	-	-
Tobermorite	11	4	5	11
C-S-H gel	74	69	83	68
Calcite	-	17	-	4
Aragonite	-	10	-	5
Unknown @ 2θ=27.0	-	-	12	12

شركة تنمية نفط عمان
Petroleum Development Oman

Lab Testing & Results

1. Crush strength Test:
   - Compressive strength increased for both exposed samples with the non-modified recipe increasing more.
   - As a part of the corrosion process, calcium carbonate is formed and deposited in the cement matrix thus increasing the density and the compressive strength (due to presence of solids).

شركة تنمية نفط عمان
Petroleum Development Oman

Lab Testing & Results

5. Young's Modulus,

    - Poisson's Ratio and Brazilian Tensile Strength

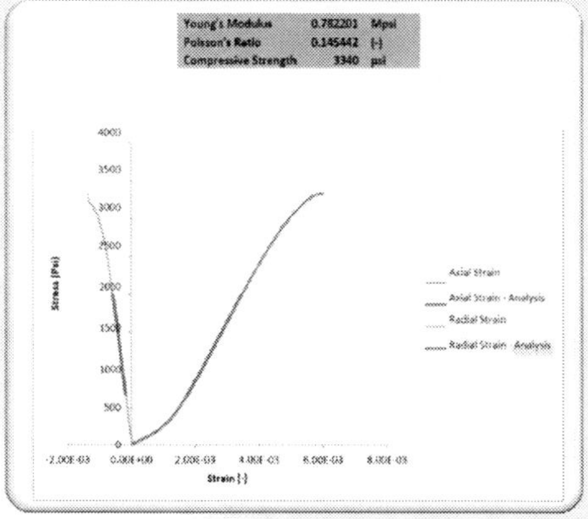

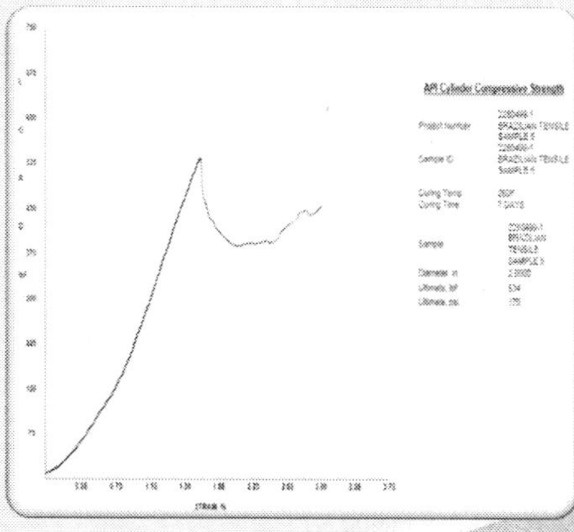

Petroleum Development Oman

Field Trials

- New cement design completed and validated through an extensive lab testing.
- To determine the mixability and pumpability a field trial need to be conducted.
- Blends, chemicals and equipment prepared and transported to nearby location.
- A 10 bbls of cement slurry smoothly mixed and safely pumped to the waste pit.
- Cement Homogeneity, fluidity and density was confirmed.
- All planned parameters were found in line with the lab testing.

Petroleum Development Oman

Overview of Cement Job – On Job Trials

- After completion of lab testing and a successful field trial was conducted.
- 1st & 2nd trials: 4-1/2-inch production liner
  - All Best practices considered.
  - All required software simulations.
  - Wellbore cleanout best practices applied.
  - All fluids (mud wash, cement and displacement) carried out by cement unit.
  - Plug was bumped and liner was pressure tested.
  - Approximately 6 bbls cement returns were received on surface.
  - Cement evaluated later using a cement mapping tool, and good quality cement was seen.

شركة تنمية نفط عمان
Petroleum Development Oman

Cement bond assessment

- New cement slurry has a low acoustic impedance.
- A cement mapping tool that is commonly used for light and foam cements has been used instead of conventional CBL-VDL.

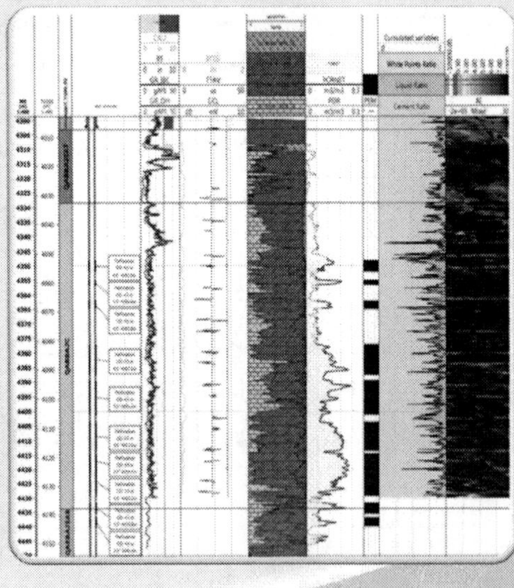

شركة تنمية نفط عمان
Petroleum Development Oman

Cement bond assessment

Thanks

Questions ??

Petroleum Development Oman

Qatargas Flare Reduction Journey

Hilal Saad H A Al-Mohannadi - Head of Environmental Operations, Qatargas

25 February 2020

The World's Premier LNG Company

QATARGAS

Outline

- Qatargas Facility Overview
- Flare Performance
- Elements of Qatargas Flare Management Program
- Key Flare Reduction Projects and Initiatives
- Conclusion

Qatargas Facility Overview

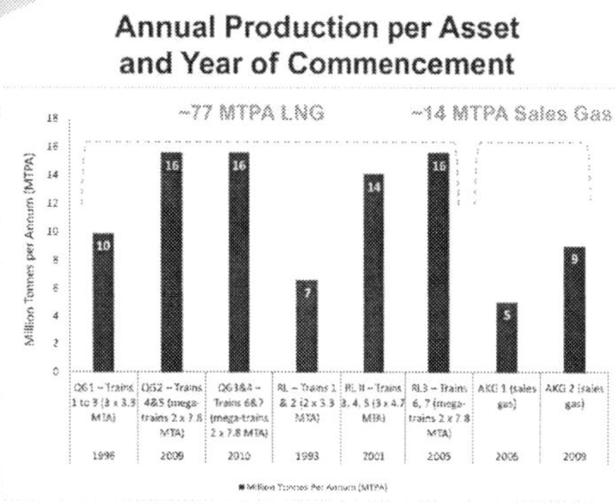

QATARGAS

Qatargas Flare Minimization Journey (2011-2019)

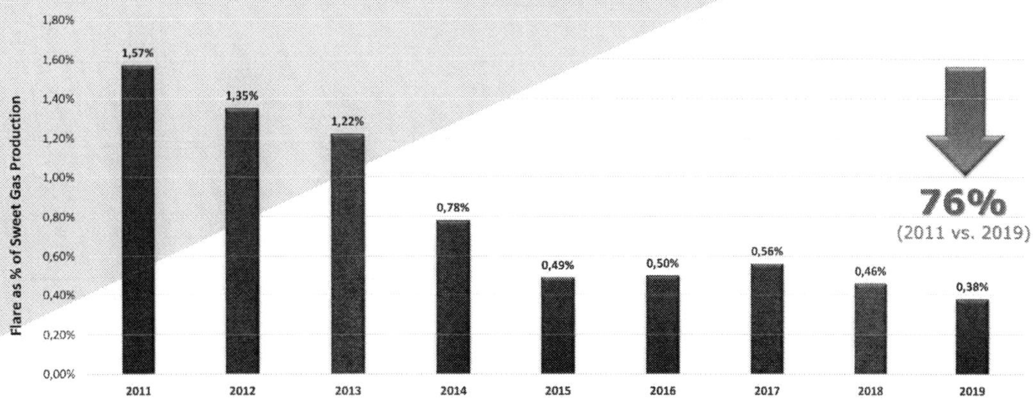

Key Flare Reduction Drivers:

- Robust Flare Management Program
- Operational Excellence Initiatives
- Shutdown / Turnaround Management
- Flare Reduction Engineering Projects

QATARGAS

Qatargas Flare Management Program

Flare Management is a key component of:

- Qatargas Management System (QGMS) 'Ensure Protection' Process
- Qatargas Environmental Management System

Main Elements of the Qatargas Flare Management Program

- Regulatory Drivers
- Management Commitment
- Dedicated Resources
- Monitoring & Reporting Program
- Verification and Certification

Regulatory Drivers and Management Commitment

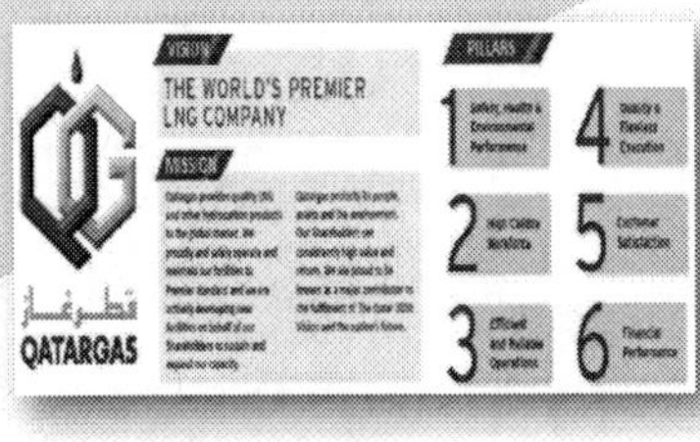

Industry Standards

QNV 2030

- **NDS (2011-2016):** Reduce flaring (per million tonnes of energy) by 50%
- **NDS (2018-2022):** Focus on climate change

Qatargas Direction Statement

- Drives Management Commitment for flare reduction
- Flare reduction captured in Company strategic objectives

Regulatory Requirements (Permits)

Flaring Target: 0.3% of Production

Qatar Petroleum

Manage hydrocarbon losses

GGFR

Dedicated Resources – Flare Minimization Teams

Main Focus: Enhanced Awareness, Monitoring and Reporting

- Flare data discussed in-light of planned KPIs and targets.
- Flare performance tracking and validation conducted.
- Surveillance and mitigation of 'bad actors' and action items raised, tracked and mitigated.
- Operational and reporting procedures and flare calculations reviewed and optimized.
- Operational, reliability and maintenance initiatives progressed.
- Flare meter performance assessed.

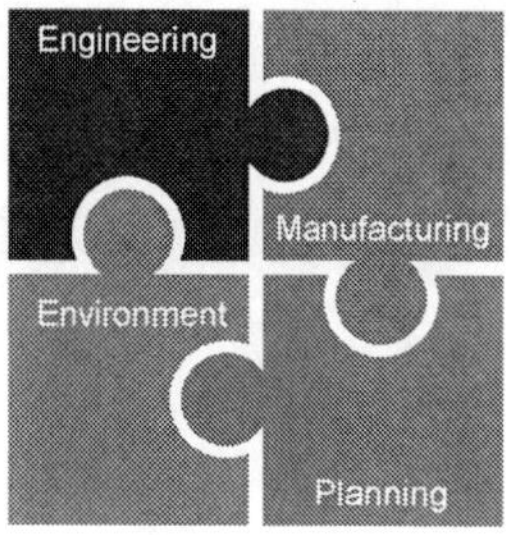

Page 7

QATARGAS

Monitoring and Reporting Programs

- Dry – Wet – Sour flare headers have Ultrasonic Meters (USMs)
- **Flaring Breakdown:**
  - Purge, Pilot, Valves (control, relief, passing)
  - Planned (shutdowns, start-up)
  - Unplanned (trips, upsets)
  - Tankage and Jetty flaring
- USMs validated annually with uncertainty assessment
- USMs basis for flaring data – all data available through plant data historian
- **Reporting:**
  - Daily trend monitoring
  - Weekly performance reporting
  - Monthly QA/QC and flare performance report
  - Annual performance review and KPI setting

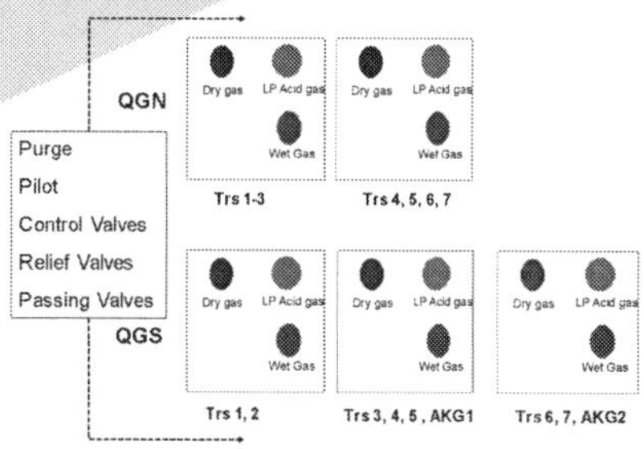

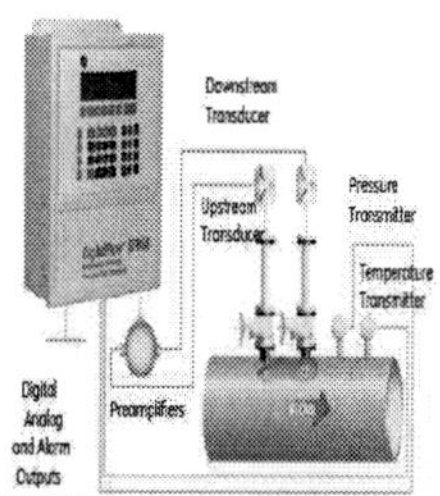

Page 8

QATARGAS

Qatargas Flare Profile (2011 vs 2019)

Key Flare Reduction Categories

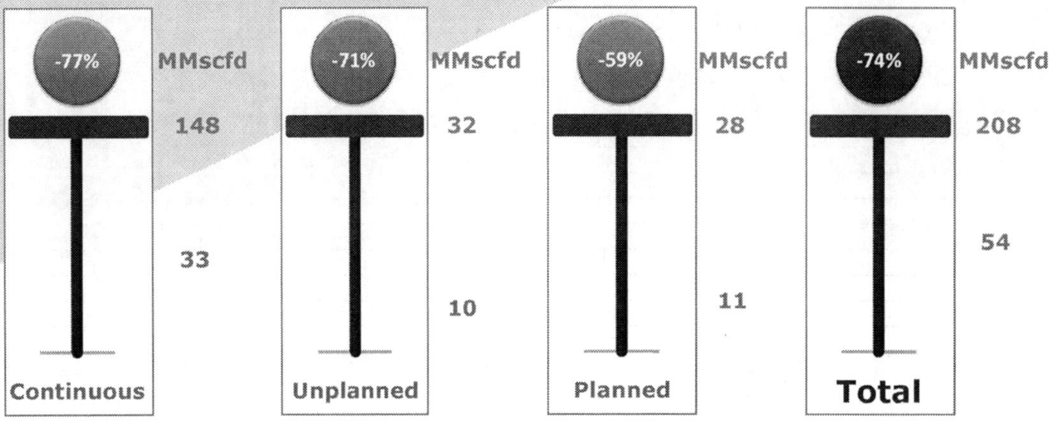

QATARGAS

Key Flare Reduction Projects / Initiatives

- ❑ Purge gas reduction project
- ❑ Passing valve monitoring program
- ❑ Excess fuel gas management
- ❑ Jetty flare recovery project

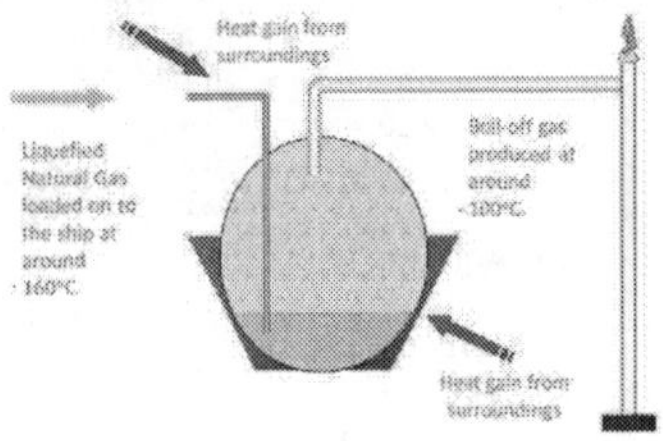

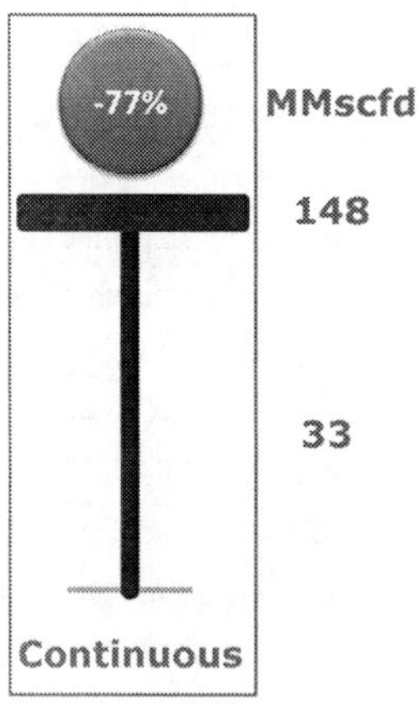

QATARGAS

Key Flare Reduction Projects / Initiatives

- ❑ Implementation of robust root cause failure analysis (RCFA)
- ❑ FOPAM – Field Operator Pro-Active Monitoring
- ❑ Equipment Strategy, Bad Actor and Operator Care Programs.

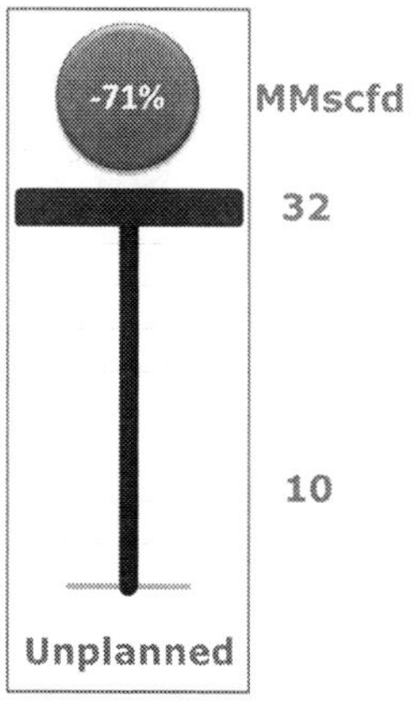

Enhanced Plant Reliability

Key Flare Reduction Projects / Initiatives

- ❑ Use of N2 for purging
- ❑ Utilization of system pressure to drain the system.
- ❑ Optimized usage of defrost gas in all units and recycling dry gas to defrosting system.
- ❑ Recover and process all liquids from liquefaction, fractionation and nitrogen rejection units.
- ❑ Produce fuel gas instead of flaring during dehydration unit startup.
- ❑ Reduce startup time/flare by injecting cold liquid Propane (C3) to the overhead reflux drum as external reflux.

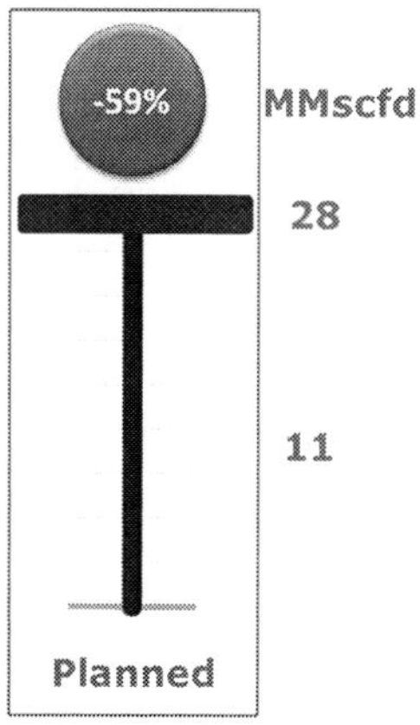

Verification and Certification

- Shareholder Monthly / Quarterly Reports
- Annual Benchmarking
- GHG A&R Verification
- Annual Sustainability Reporting
- Qatargas Management System Audits
- ISO14001 Audits
- Regulatory Reports / Inspections

Conclusions

- **Flaring optimization and reduction is an ongoing and continuous improvement process and a critical Oil & Gas Sector program.**
- **Regulatory drivers and management commitment are key enablers.**
- **Plant Resources control the success of the program.**
- **Accurate monitoring and reporting affect the integrity of the program.**
- **Future Qatargas Plans:**
  - **Multiple options and engineering solutions are currently under study.**
  - **New Flare Reduction Project (FRP) in progress at 4 LNG mega-trains – installation of gas interconnections to help divert gas to other operating trains instead of flaring.**

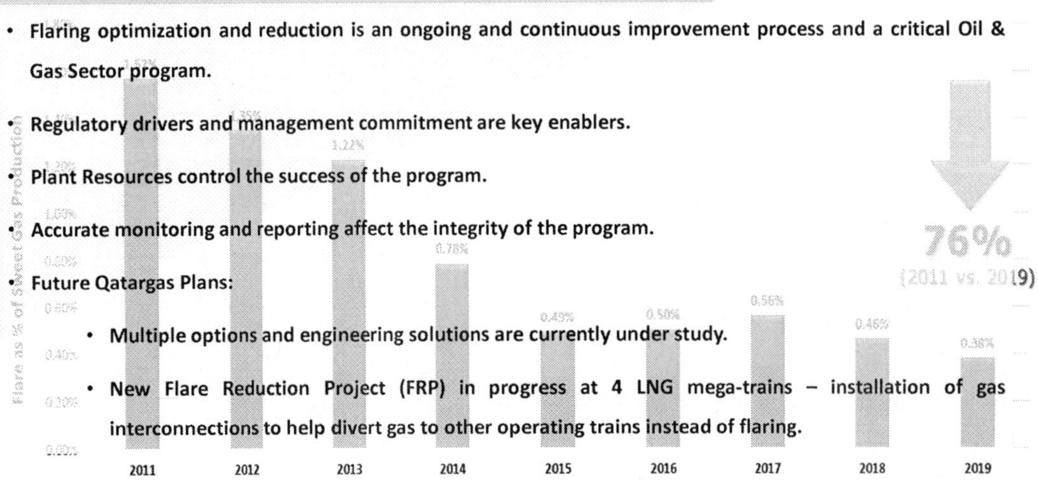

IGRC 2020
"Solutions for the future, technology for future"
ENGIE RESEARCH CENTER CRIGEN
Hydrogen Lab
26th February 2020 – Oman
mures.zarea@engie.com

engie
OUR WORLDWIDE PRESENCE*
167,000
employees
€60.6 billion
in revenue
€2.5 billion
recurring profit after tax
€11 billion
of planned investment in growth in 2019-2021
EUROPE
€47.2 billion
in revenue
OF WHICH, FRANCE
€24.9 billion
in revenue
AND BENELUX
€6 billion
in revenue
NORTH AMERICA
€3.9 billion
in revenue
LATIN AMERICA
€4.2 billion
in revenue
AFRICA
€0.4 billion
in revenue
NEAR AND MIDDLE EAST
€1.9 billion
in revenue
ASIA AND OCEANIA
€3 billion
in revenue
*on 31/12/2018
ENGIE LAB HYDROGEN
RESTRICTED – NO REPRODUCTION ALLOWED - PROPERTY OF ENGIE
IGRC 20202 OMAN
2

OUTLINE

- ENGIE's hydrogen vision
- From H2 R&D projects…
- … to disruptive businesses for ENGIE

ENGIE

ENGIE RESEARCH CENTRE CRIGEN IN A NUTSHELL

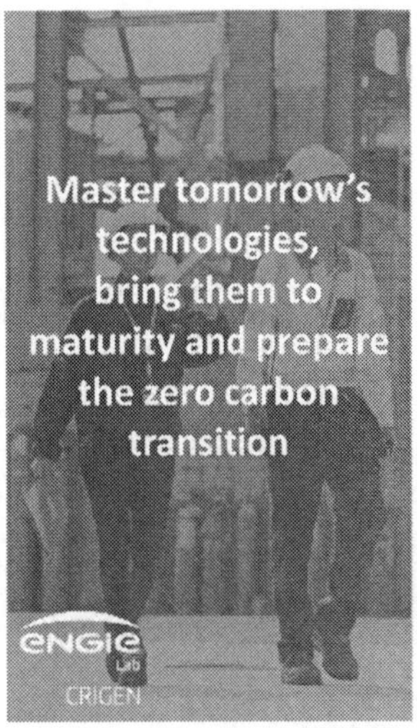

€35 million revenues from

30 countries

120 Patents in portfolio

Unique testing facilities in Europe

ISO 9001 certified

180 employees

37 years Average age

37 % 63 %

11 nationalities

51 PhD

10 Co-operative education students

19 Trainees

10 Labs

NEW ENERGY RESOURCES LABS
- Biogas and Biomass
- Small Scale Liquefaction
- Hydrogen

NEW ENERGY USES LABS
- Future Industry
- Environment & Society
- Future Buildings & Cities

EMERGING TECHNOLOGIES LABS
- Drones & Robots
- Computer Science & AI
- Nanotech, Sensors & Wireless

ENGIE Lab Singapore

ENGIE

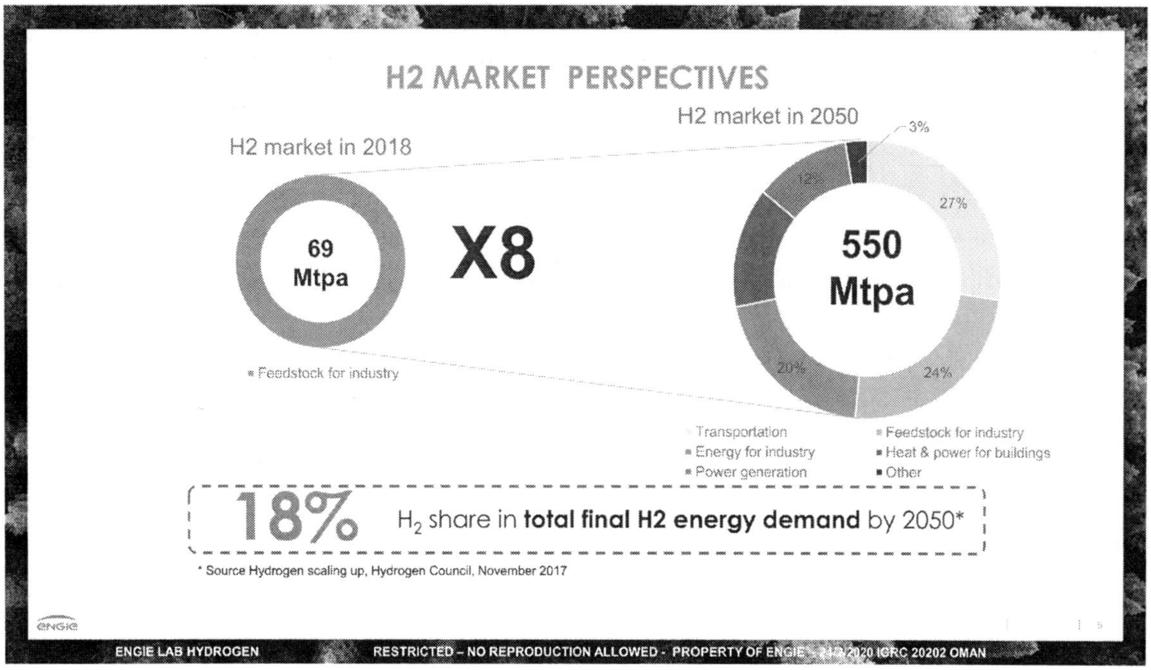
H2 MARKET PERSPECTIVES
H2 market in 2018
H2 market in 2050
69 Mtpa
X8
550 Mtpa
3%
12%
27%
20%
24%
Feedstock for industry
Transportation
Energy for industry
Power generation
Feedstock for industry
Heat & power for buildings
Other
18% H2 share in total final H2 energy demand by 2050*
* Source Hydrogen scaling up, Hydrogen Council, November 2017
ENGIE LAB HYDROGEN
RESTRICTED – NO REPRODUCTION ALLOWED - PROPERTY OF ENGIE

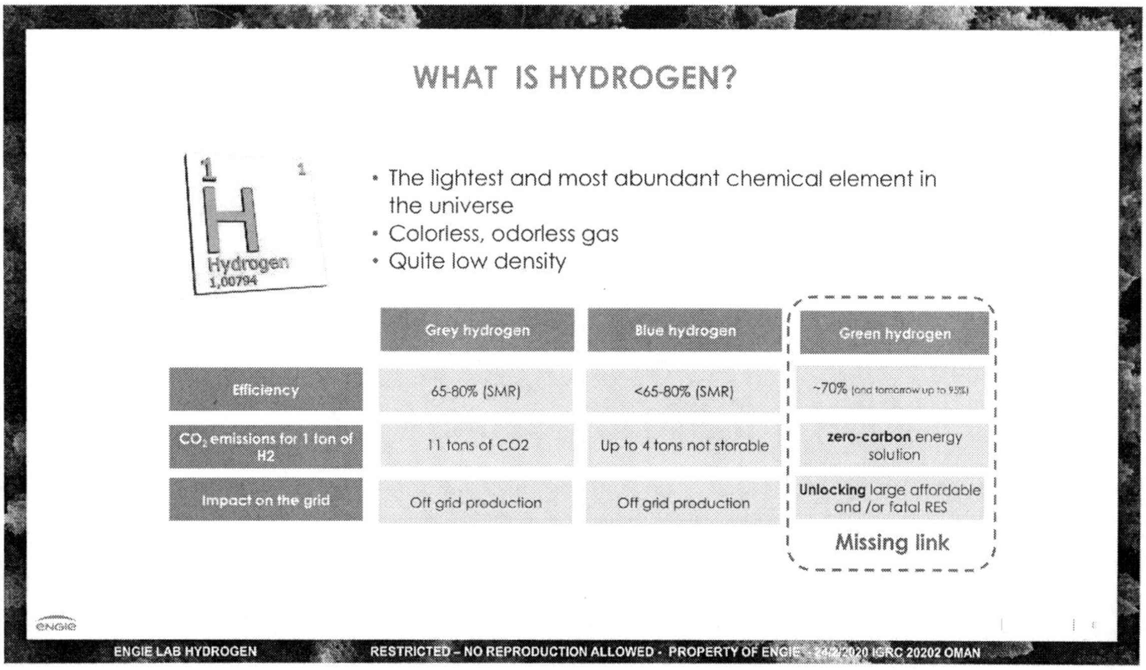
WHAT IS HYDROGEN?
1
H
Hydrogen
1,00794
The lightest and most abundant chemical element in the universe
Colorless, odorless gas
Quite low density
Grey hydrogen
Blue hydrogen
Green hydrogen
Efficiency
65-80% (SMR)
<65-80% (SMR)
~70% (and tomorrow up to 95%)
CO2 emissions for 1 ton of H2
11 tons of CO2
Up to 4 tons not storable
zero-carbon energy solution
Impact on the grid
Off grid production
Off grid production
Unlocking large affordable and /or fatal RES
Missing link
ENGIE LAB HYDROGEN
RESTRICTED – NO REPRODUCTION ALLOWED - PROPERTY OF ENGIE

RENEWABLE H_2 CAN CONTRIBUTE TO EFFICIENTLY REACHING ENVIRONMENTAL GHG EMISSIONS REDUCTION GOALS

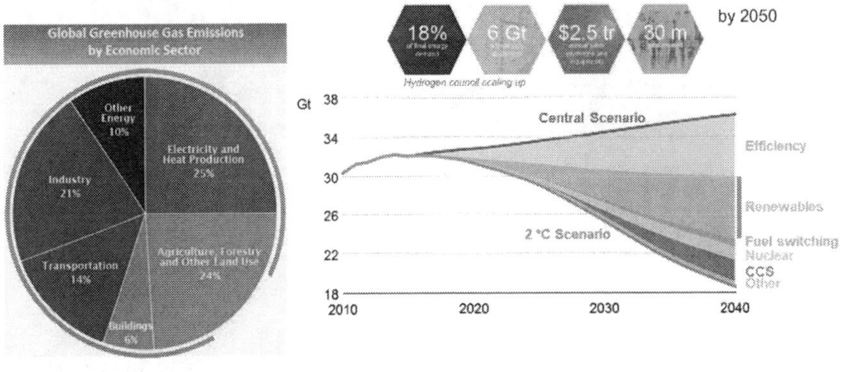

H_2 could directly or indirectly contribute to decarbonizing almost all business areas

GIECC climate change 2014 synthesis report

ENGIE'S STRATEGY ON HYDROGEN FROM LOCAL TO GLOBAL MARKET : ACCELERATE, AGGREGATE AND INTEGRATE

ENGIE will architect, build and operate large scale H2 solutions
ENGIE will provide turnkey solutions across the value chain

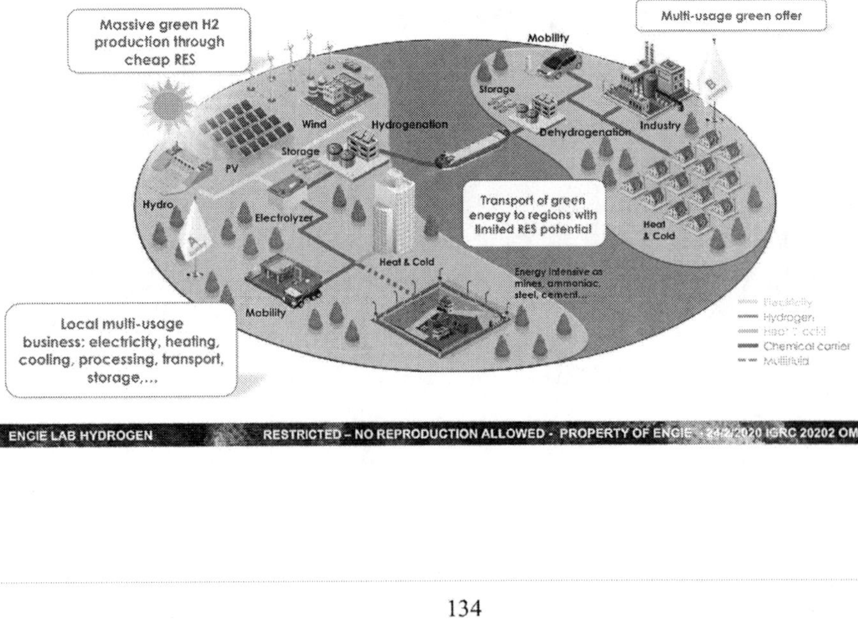

ENGIE Lab CRIGEN – Hydrogen
Ongoing Funded Collaborative projects at Global Scale
NEPTUNE (FCHJU)
REFLEX (FCHJU)
MULTIPLHY (FCHJU)
MACBETH (SPIRE)
HYCARE (FCHJU)
HARWELL (UK BEIS)
C2FUEL (H2020)
GRHYD (ADEME)
HYCAUNAIS (ADEME)
LIVING H_2
Storage & Transport
Production
Usage
ENGIE LAB HYDROGEN
12/12/2019

FROM H2 R&D PROJECTS...

TWO COMPLEMENTARY VISIONS FOR HYDROGEN

ENGIE Lab Hydrogen supports two complementary and necessary visions for hydrogen

Decentralized hydrogen production close to the end user

- # Green hydrogen generation plants generally up to 20MW for renewable electricity surplus reuse, industry and mobility
- # Green liquid commodity produced locally for mobility or industrial usage
- # Substituted natural gas for grid injection

Hydrogen mass production in countries with high renewable energy potential and transportation to user

- # Large scale production (GW scale) for market creation (industry and mobility) and cost competitive hydrogen production
  - Supported by major drop in PV and wind auction bids
- # Hydrogen corridors with key challenges

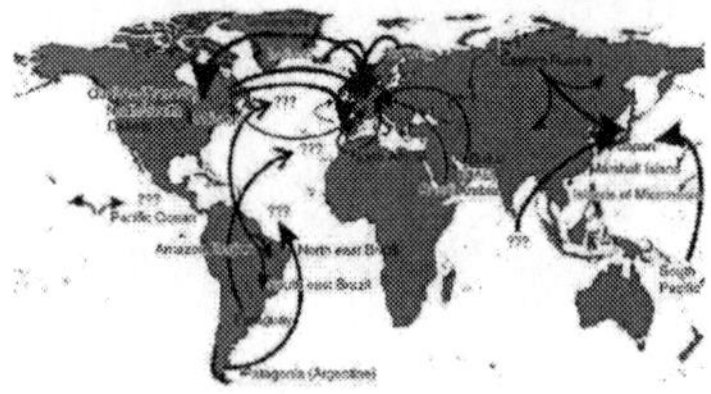

Ball and al, 2009

ENGIE LAB HYDROGEN R&D ACTIVITIES SUPPORT THESE TWO VISIONS

On production to **mature or improve H2 production technologies** with

- # Higher electrical efficiency especially for large scale production
- # Decrease investment cost especially for decentralized production

On **hydrogen transportation and storage** to promote key technologies providing profitable business cases both for inland and maritime transport

On **hydrogen usages** to favor hydrogen penetration for stationary and mobility applications (up to HRS)

In **power-to-gas** through development and collaboration in methanation for grid injection

In **power-to-liquid** for industrials and carbon captured fuel

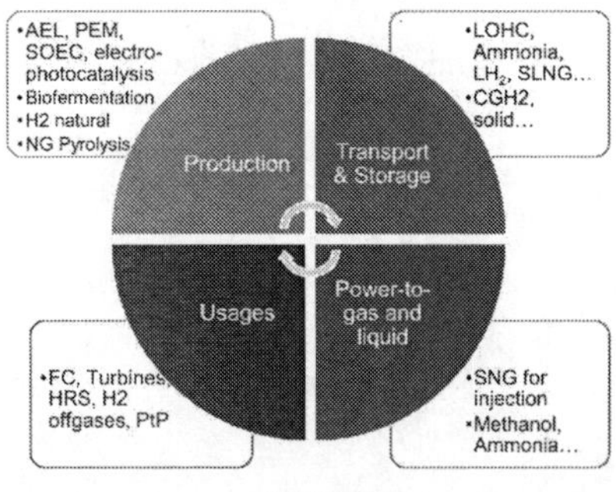

Ball and al, 2009

HYDROGEN Thematic

We pave the way for a competitive zero-carbon hydrogen economy

LOWER THE COSTS OF THE WHOLE GREEN HYDROGEN VALUE CHAIN TO ENABLE 100% GREEN GAS BY 2050
- Technological innovations
- Tools
- Demonstrators and operations

IDENTIFY, ASSESS AND INTEGRATE INNOVATIVE SOLUTIONS TO ANSWER THE NEW NEEDS OF OUR CLIENTS TOWARD A ZERO CARBON TARGET
- Green Hydrogen for industrials
- Green Hydrogen mobility
- Green Hydrogen for territories

VALORIZE EXISTING INFRASTRUCTURES BY PROVIDING ENERGY TRANSITION SOLUTIONS TO ASSETS OWNERS
- Power to Gas
- Networks conversion with hydrogen injection

engie

DECISION-MAKING TOOLS FOR EFFICIENT R&D

COSTHY is an **in-house dynamic decision-making tool** which allows :

to self-design the architecture of the process (Electrolysers / Storage / Compressors…)

to study multi end-use applications (H_2, Gas, Liquid, Power)

to consider variable electricity sourcing

HYTAC is an in-house add-on linked with COSTHY which allows to determine the levelized cost related **to multi-segment transportation of hydrogen**

Consider up to 6 conditioning methods

Up to three terrestrial or maritime segments

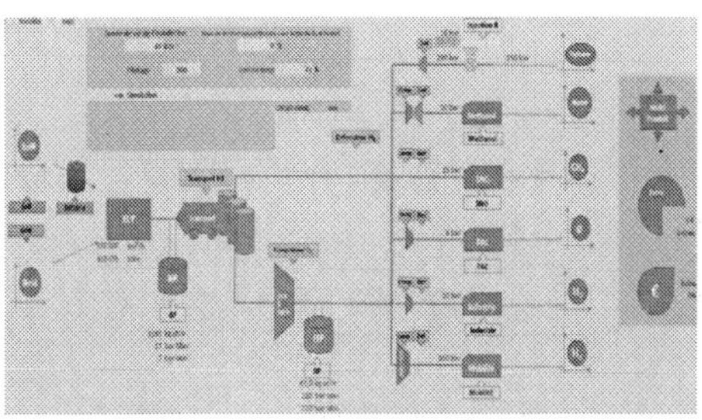

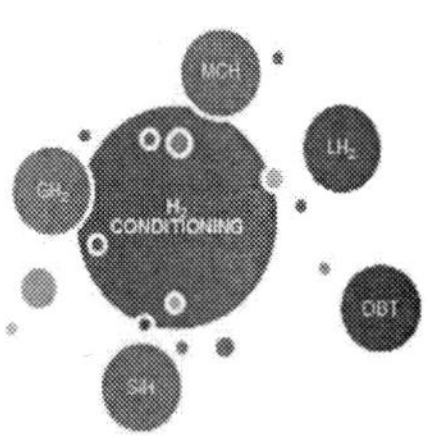

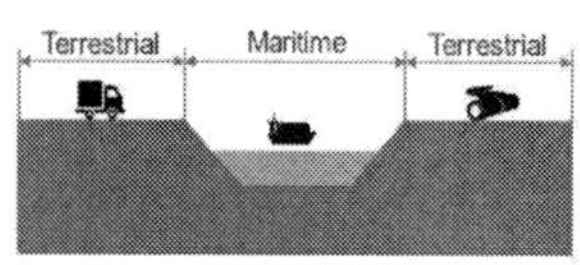

FROM H2 R&D PROJECTS...
ENGIE

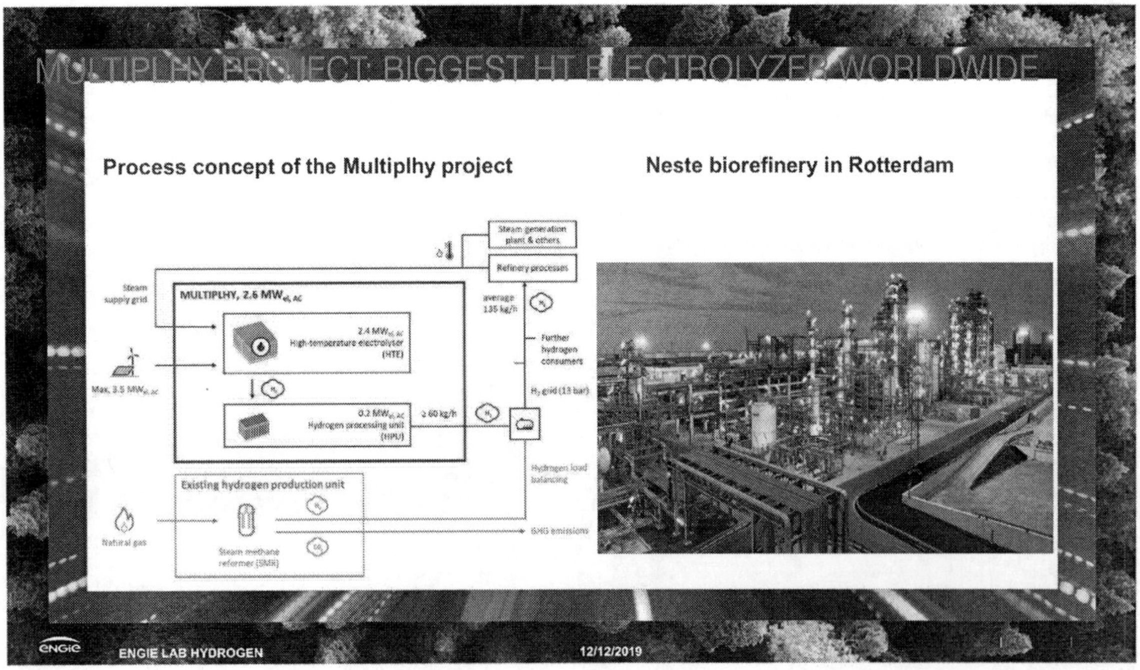
MULTIPLHY PROJECT: BIGGEST HT ELECTROLYZER WORLDWIDE
Process concept of the Multiplhy project
Neste biorefinery in Rotterdam
Steam generation plant & others
Refinery processes
Steam supply grid
MULTIPLHY, 2.6 MW el, AC
2.4 MW el, AC High-temperature electrolyser (HTE)
Max. 3.5 MW el, AC
0.2 MW el, AC Hydrogen processing unit (HPU)
≥ 60 kg/h
average 135 kg/h
Further hydrogen consumers
H2 grid (13 bar)
Hydrogen load balancing
Existing hydrogen production unit
Natural gas
Steam methane reformer (SMR)
GHG emissions
ENGIE LAB HYDROGEN
12/12/2019

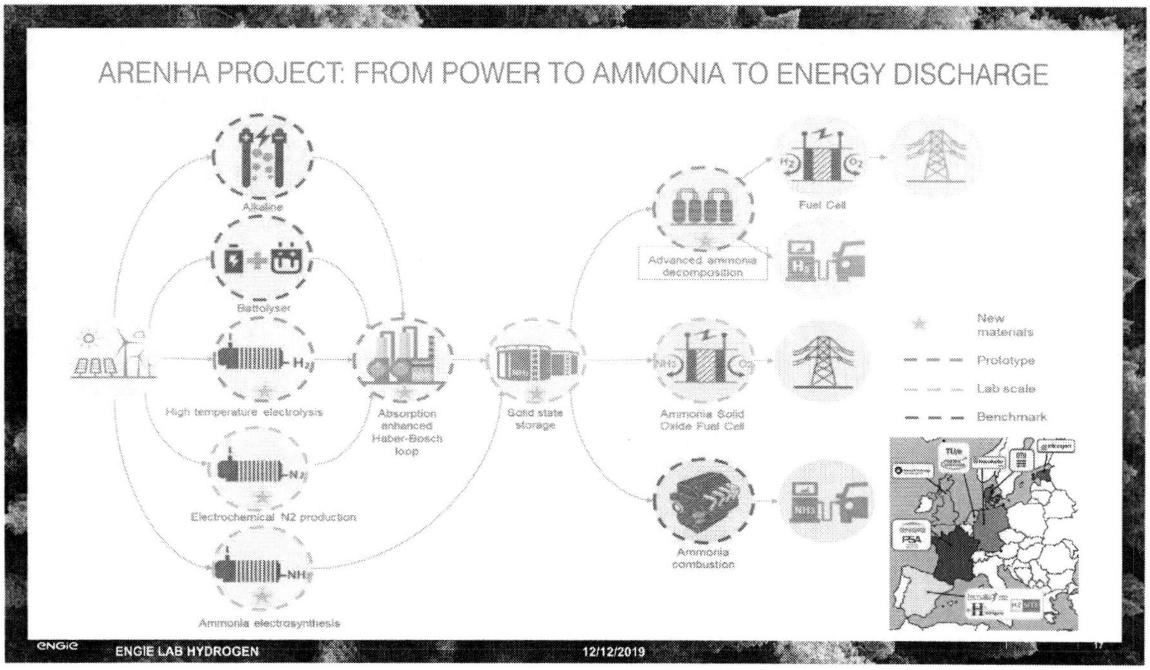

ARENHA PROJECT: FROM POWER TO AMMONIA TO ENERGY DISCHARGE
Alkaline
Battolyser
High temperature electrolysis
Electrochemical N2 production
Ammonia electrosynthesis
Absorption enhanced Haber-Bosch loop
Solid state storage
Advanced ammonia decomposition
Fuel Cell
Ammonia Solid Oxide Fuel Cell
Ammonia combustion
New materials
Prototype
Lab scale
Benchmark
ENGIE LAB HYDROGEN
12/12/2019

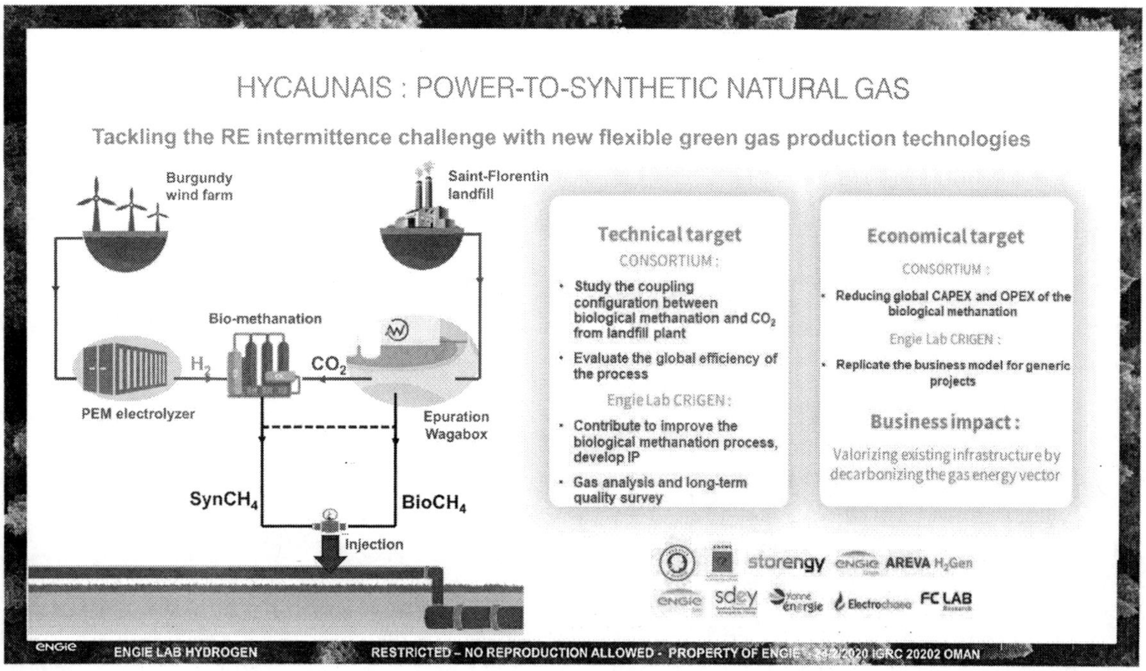

HYCAUNAIS : POWER-TO-SYNTHETIC NATURAL GAS
Tackling the RE intermittence challenge with new flexible green gas production technologies
Burgundy wind farm
Saint-Florentin landfill
Bio-methanation
PEM electrolyzer
Epuration Wagabox
SynCH4
BioCH4
Injection
Technical target
CONSORTIUM :
Study the coupling configuration between biological methanation and CO2 from landfill plant
Evaluate the global efficiency of the process
Engie Lab CRIGEN :
Contribute to improve the biological methanation process, develop IP
Gas analysis and long-term quality survey
Economical target
CONSORTIUM :
Reducing global CAPEX and OPEX of the biological methanation
Engie Lab CRIGEN :
Replicate the business model for generic projects
Business impact :
Valorizing existing infrastructure by decarbonizing the gas energy vector
storengy
AREVA H2Gen
sdey
Electrochaea
FC LAB
ENGIE LAB HYDROGEN
RESTRICTED – NO REPRODUCTION ALLOWED - PROPERTY OF ENGIE
IGRC 20202 OMAN

LIVING H2 : PURE H2 FC-BASED MCHP AT CRIGEN TO ENABLE GREEN H2 WITHIN BUILDINGS FOR HEAT AND POWER

Consortium :

- Engie Crigen (F)
- InHouse Eng. (D)
- DBI (D)
- Regensburg Univ. (D)
- CEA Liten (F)

Start date
Oct. 1st 2019
Duration : 3 years

Co-financed by ANR (F) and PTJ (D)
Sustainable energy French – German call of projects 2019

Objectives :

Optimization and development of a 5 kWe H_2 fuel cell mCHP
- New catalytic H_2 burner for both off-gas and peak load
- Improved PEMFC MEAs for improved stability and lifespan

Installation and test in real conditions within Crigen's mechanical room
- Connection to a "green" hydrogen supply
- Implementation of H_2 odorization

Identify and qualify markets
- Technical, economical and social acceptances

Main activities :

System development and integration (inHouse)

PEMFC MEAs for improved stability and lifespan (CEA Liten)

H_2 burner and odorization systems development (DBI)

Technical, economical and social acceptances (Regensburg + Engie)

Installation and in-field test (Engie + InHouse + DBI)

PRHYDE – PROTOCOL TO REFUEL HYDROGEN HEAVY DUTY VEHICLE

Unlock HD H2 vehicle market through safe and efficient refueling

Specify need and constrains for large vehicle such as truck, buses and train.

Identify technical bottleneck for high flow filling of large tank

Risk analysis to share common and standard view with vehicle manufacturers

Protocol development and validation through simulation and experimental campaign

Develop refueling protocol through ENGIE lab H2 simulation tool

ENGIE Lab H2 will be in charge of simulation through 2 models :

- **HYFILL simplified model for H2 tank refueling simulation**
- **3D Computational Fluid Dynamic model for more complex cases :**

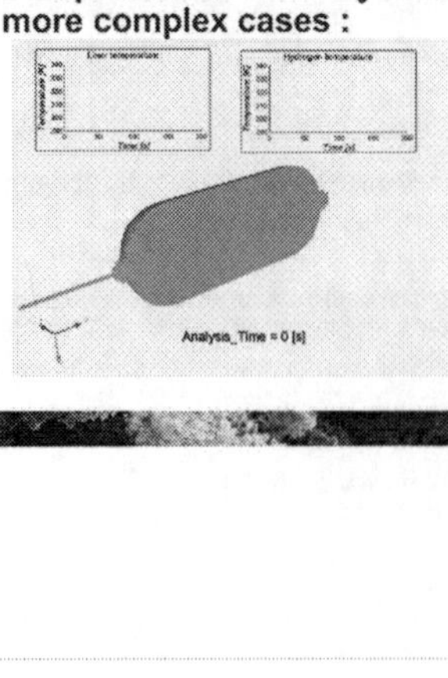

METROHYVE 2 – METROLOGY FOR HYDROGEN VEHICLE

Address the metrological challenges associated with H2 mobility

Ensure accurate flow metering and reduce control cost

Develop gas reference material and methodology for quality analysis of hydrogen

Develop and test methodology for representative sampling of hydrogen at vehicle refueling station

Evaluate the real impact of impurities on the fuel cell of the vehicles

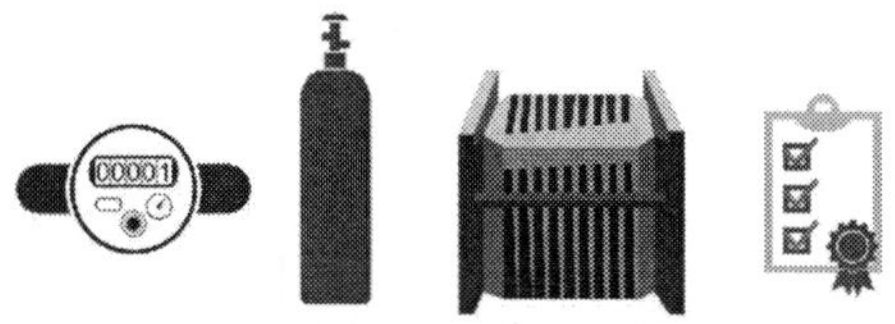

Collect → Sample → analyze

Develop an efficient methodology of H2 quality control at refueling station

Development of the methodology and sampling device

Validation in laboratory (St Fons)

Validation at refueling station of GNVert

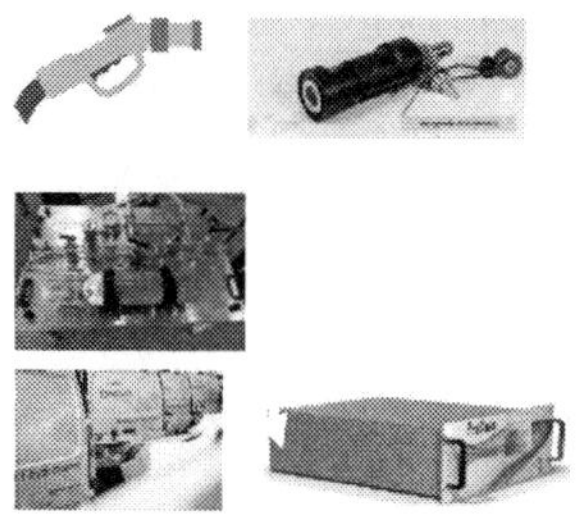

GRHYD
VALORIZE EXISTING INFRASTRUCTURE BY PROVIDING ENERGY TRANSITION SOLUTIONS TO ASSETS OWNERS

ENGIE Lab CRIGEN

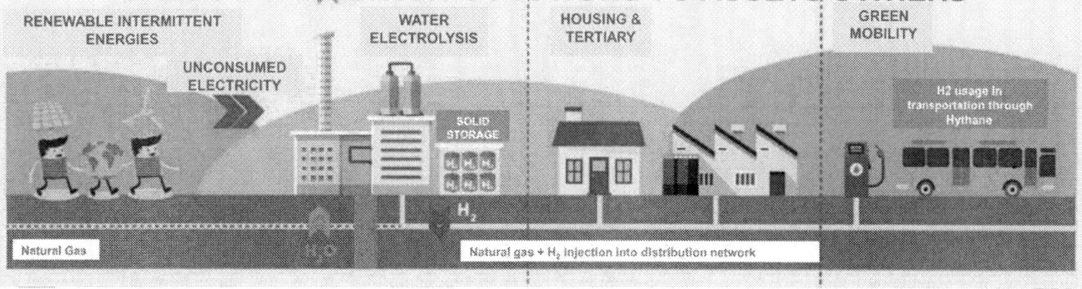

CUSTOMER NEED

- Support a territory in its zero-carbon transition
- Valorize the existing distribution gas network by converting it to low carbon energy sources
- Nurture a progressive approach of Hydrogen for the public

THE TOOL / SOLUTION

- Use renewable energy sources to produce green hydrogen
- Inject up to 20% H2 into the residential network
- Mix with methane to use as a fuel in buses (Hythane)

KEY FIGURES

200 households supplied

20 buses fueled with Hythane

20% H2 in the distribution network of Capelle-la-Grande district

HYDROGEN AS DISPATCHABLE FORM OF ENERGY STORAGE FOR ENERGY SUPPLY DIVERSIFICATION

Hydrogen is key for new energy corridors and energy supply diversification through international trading

Main fossil fuel exporters to Europe

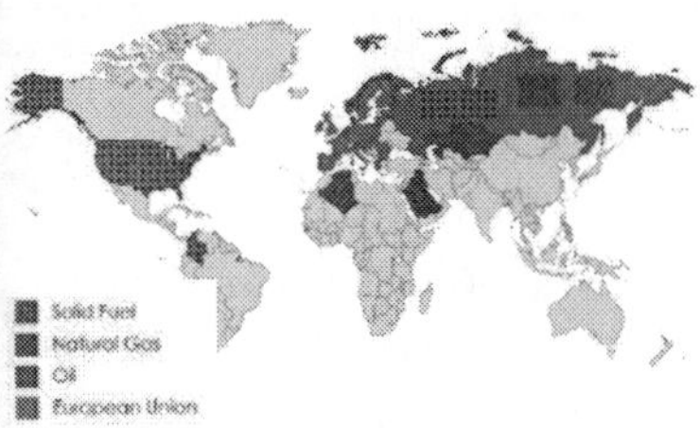

Lowest PV auction bids worldwide in 2018 and 2019

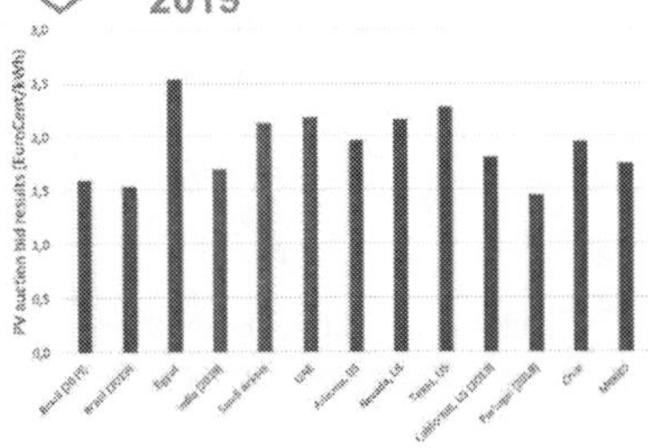

Hybrid Wind and PV cumulative full load hours – Source VTT

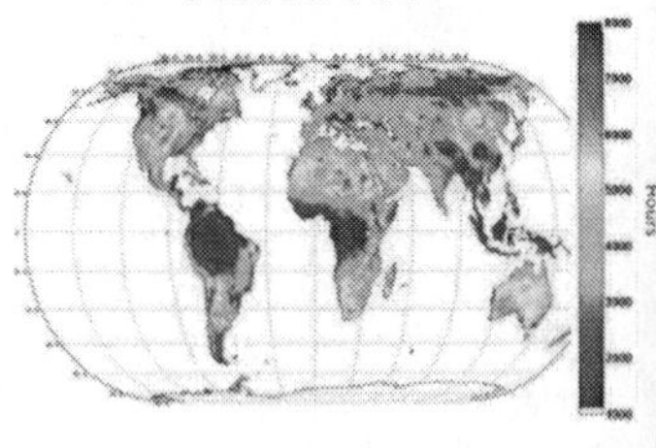

New energy corridors with countries showing electricity economic potential lower than 30€/MWh with cumulative load hours higher than 5000 hours are possible!

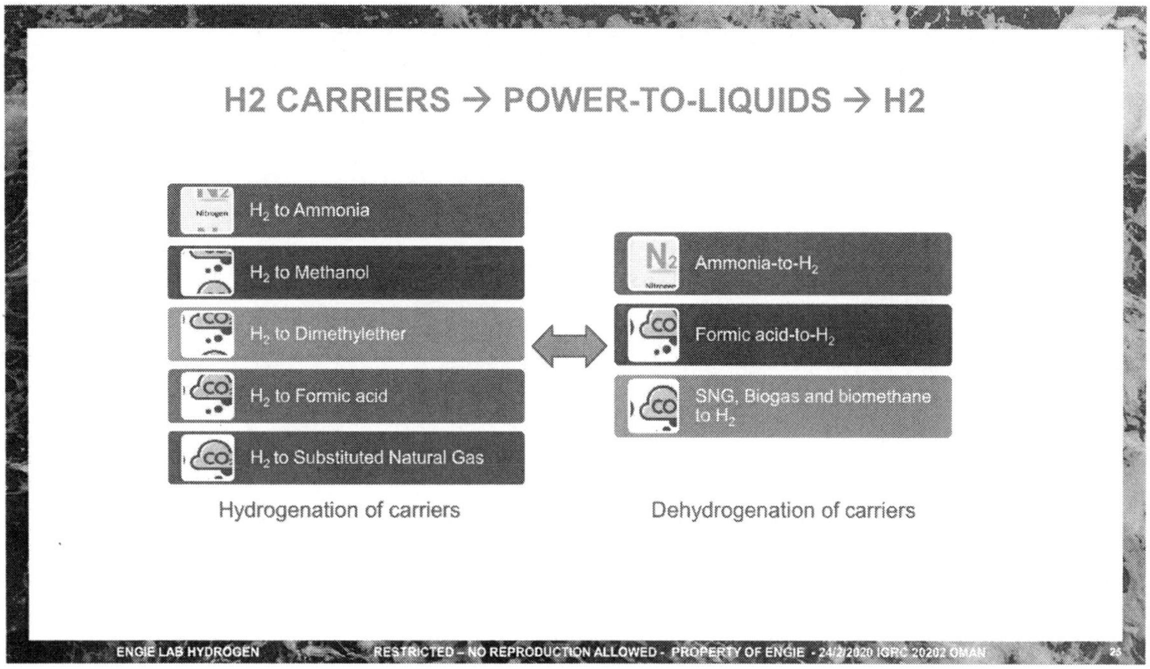
H2 CARRIERS → POWER-TO-LIQUIDS → H2
H2 to Ammonia
H2 to Methanol
H2 to Dimethylether
H2 to Formic acid
H2 to Substituted Natural Gas
Hydrogenation of carriers
Ammonia-to-H2
Formic acid-to-H2
SNG, Biogas and biomethane to H2
Dehydrogenation of carriers
ENGIE LAB HYDROGEN
RESTRICTED – NO REPRODUCTION ALLOWED - PROPERTY OF ENGIE - 24/2/2020 IGRC 20202 OMAN

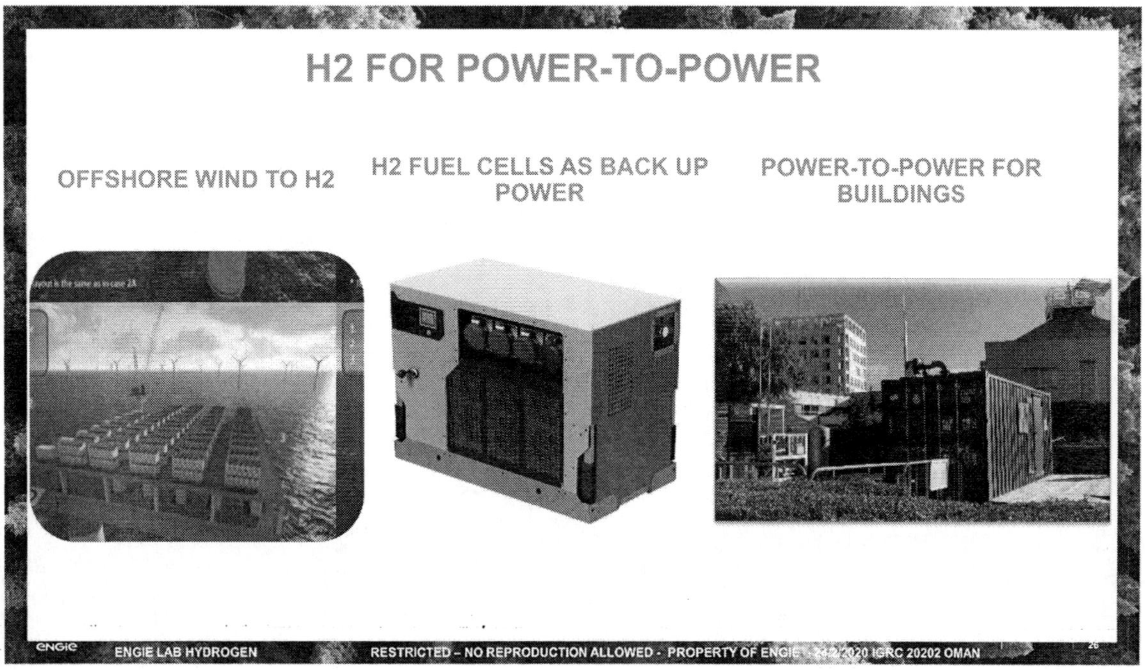
H2 FOR POWER-TO-POWER
OFFSHORE WIND TO H2
H2 FUEL CELLS AS BACK UP POWER
POWER-TO-POWER FOR BUILDINGS
ENGIE LAB HYDROGEN
RESTRICTED – NO REPRODUCTION ALLOWED - PROPERTY OF ENGIE - 24/2/2020 IGRC 20202 OMAN

FROM HYDROGEN COMMUNITY TO HYDROGEN SOCIETY
Renewable energy
Hydrogenation (electrolysis etc.)
Hydrogen from unused renewable digester gas, biomass, etc.
Production
Compressed hydrogen
Liquid hydrogen
MCH Ammonia
CH_3
NH_3
Transportation
Hydrogen powered generation
Hydrogen PEFC Ammonia FC
Dehydrogenating
The use of waste heat from FC systems for air conditioning hot water supply etc.
FC Bus
FC Boat
FCV
ENGIE LAB HYDROGEN
RESTRICTED – NO REPRODUCTION ALLOWED - PROPERTY OF ENGIE
27

CHALLENGES IN ZERO CARBON TRANSITION

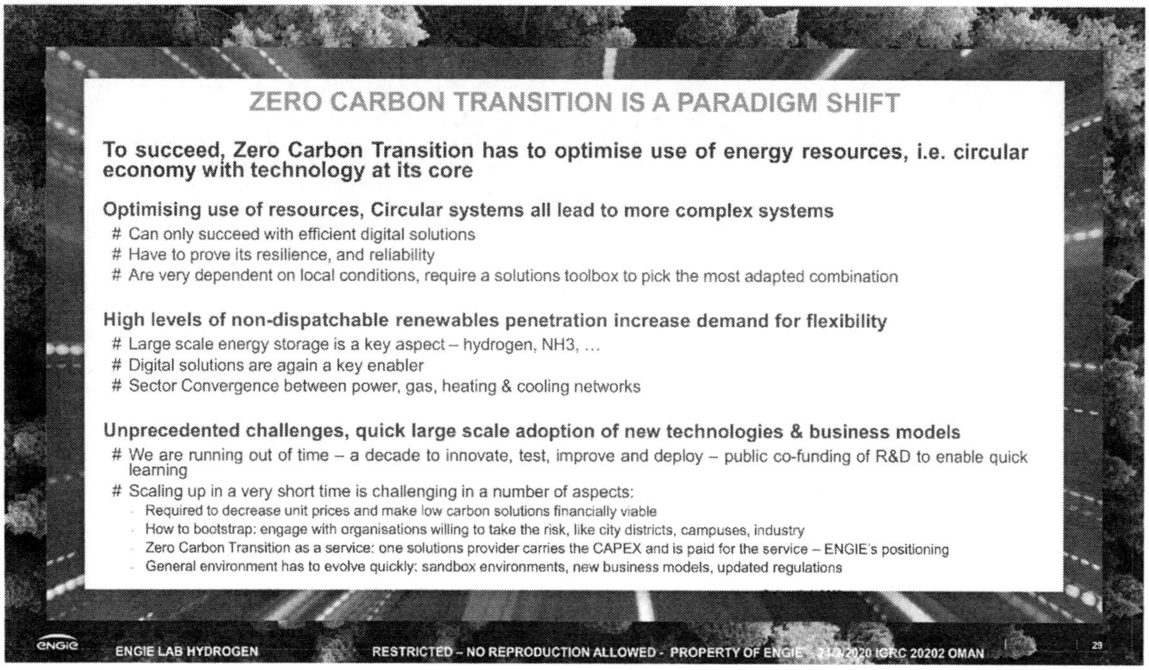
ZERO CARBON TRANSITION IS A PARADIGM SHIFT
To succeed, Zero Carbon Transition has to optimise use of energy resources, i.e. circular economy with technology at its core
Optimising use of resources, Circular systems all lead to more complex systems
Can only succeed with efficient digital solutions
Have to prove its resilience, and reliability
Are very dependent on local conditions, require a solutions toolbox to pick the most adapted combination
High levels of non-dispatchable renewables penetration increase demand for flexibility
Large scale energy storage is a key aspect – hydrogen, NH3, …
Digital solutions are again a key enabler
Sector Convergence between power, gas, heating & cooling networks
Unprecedented challenges, quick large scale adoption of new technologies & business models
We are running out of time – a decade to innovate, test, improve and deploy – public co-funding of R&D to enable quick learning
Scaling up in a very short time is challenging in a number of aspects:
Required to decrease unit prices and make low carbon solutions financially viable
How to bootstrap: engage with organisations willing to take the risk, like city districts, campuses, industry
Zero Carbon Transition as a service: one solutions provider carries the CAPEX and is paid for the service – ENGIE's positioning
General environment has to evolve quickly: sandbox environments, new business models, updated regulations
ENGIE
ENGIE LAB HYDROGEN
RESTRICTED – NO REPRODUCTION ALLOWED - PROPERTY OF ENGIE - 24/2/2020 IGRC 20202 OMAN
29

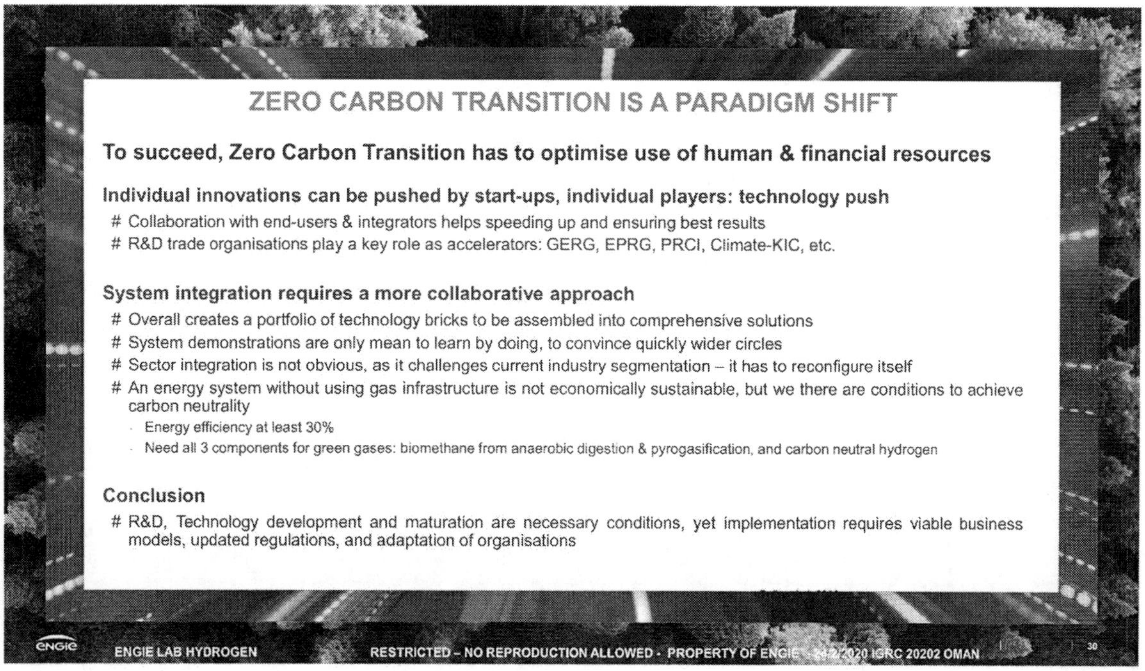
ZERO CARBON TRANSITION IS A PARADIGM SHIFT
To succeed, Zero Carbon Transition has to optimise use of human & financial resources
Individual innovations can be pushed by start-ups, individual players: technology push
Collaboration with end-users & integrators helps speeding up and ensuring best results
R&D trade organisations play a key role as accelerators: GERG, EPRG, PRCI, Climate-KIC, etc.
System integration requires a more collaborative approach
Overall creates a portfolio of technology bricks to be assembled into comprehensive solutions
System demonstrations are only mean to learn by doing, to convince quickly wider circles
Sector integration is not obvious, as it challenges current industry segmentation – it has to reconfigure itself
An energy system without using gas infrastructure is not economically sustainable, but we there are conditions to achieve carbon neutrality
Energy efficiency at least 30%
Need all 3 components for green gases: biomethane from anaerobic digestion & pyrogasification, and carbon neutral hydrogen
Conclusion
R&D, Technology development and maturation are necessary conditions, yet implementation requires viable business models, updated regulations, and adaptation of organisations
ENGIE
ENGIE LAB HYDROGEN
RESTRICTED – NO REPRODUCTION ALLOWED - PROPERTY OF ENGIE - 24/2/2020 IGRC 20202 OMAN
30

THANK YOU
FOR YOUR ATTENTION
ACKNOWLEDGMENT
secil.torun@engie.com

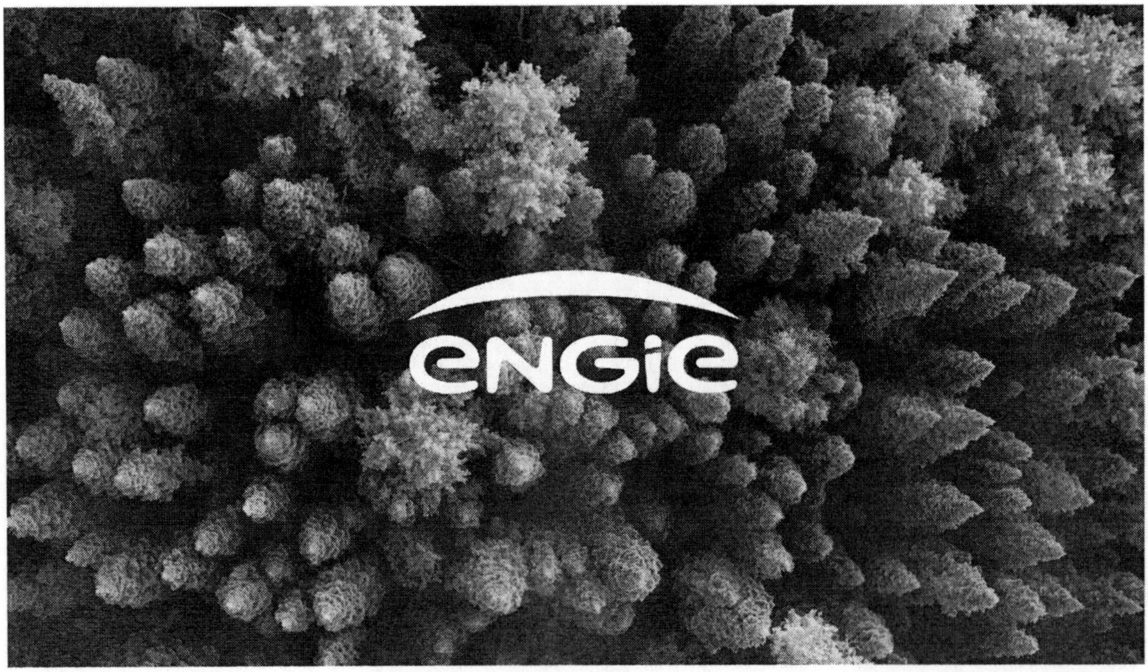

SUSTAINABLE GAS TECHNOLOGY DGC

CH4 emissions from end use

Jean Schweitzer, DGC
IGRC 2020

Version 17-02-2020

Dansk Gasteknisk Center a/s

SUSTAINABLE GAS TECHNOLOGY DGC

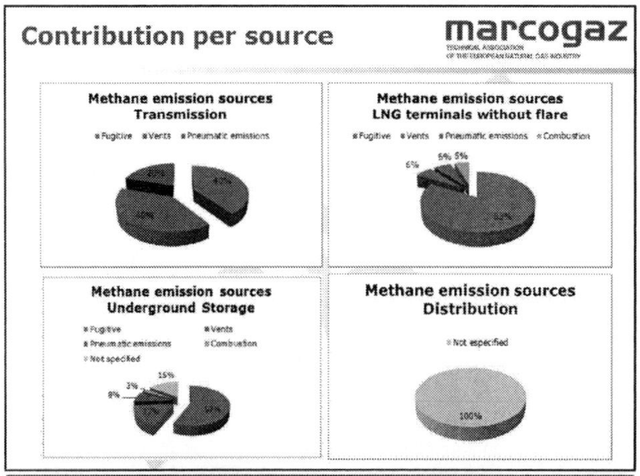

About 0.3%

CH4 emissions from end-use = **the unknown part of the picture**

Methane emissions Utilisations

?

Probably > 0.3%

.... but probably the highest contributor to emissions

Dansk Gasteknisk Center a/s

SUSTAINABLE GAS TECHNOLOGY

DGC

DK investigations in the last few years

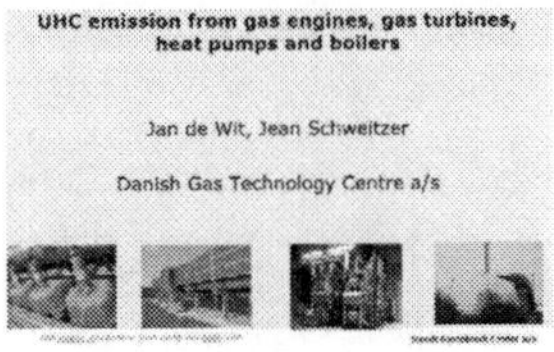
UHC emission from gas engines, gas turbines, heat pumps and boilers

Jan de Wit, Jean Schweitzer

Danish Gas Technology Centre a/s

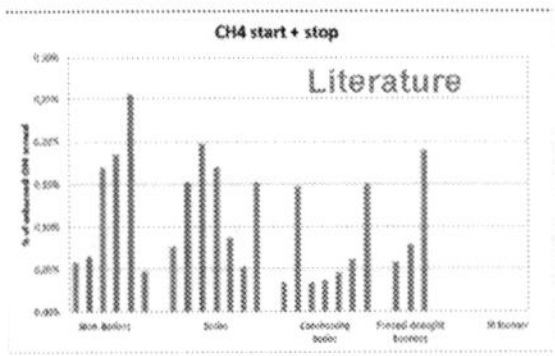

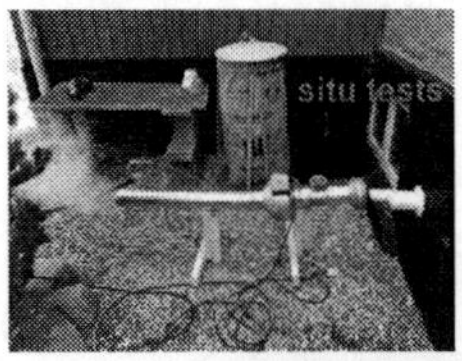

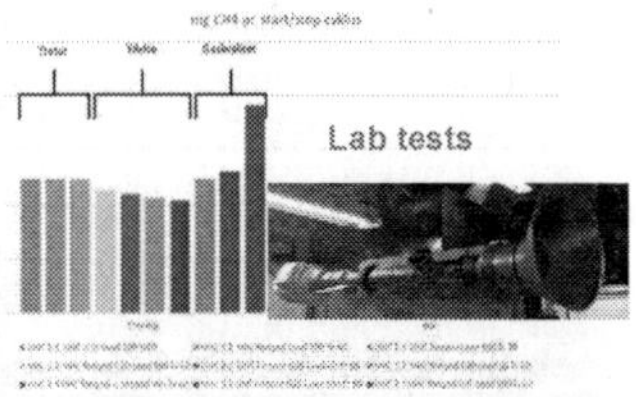

DGC site tests for CHP-plant owners etc. (50 – 100 every year since 1990)

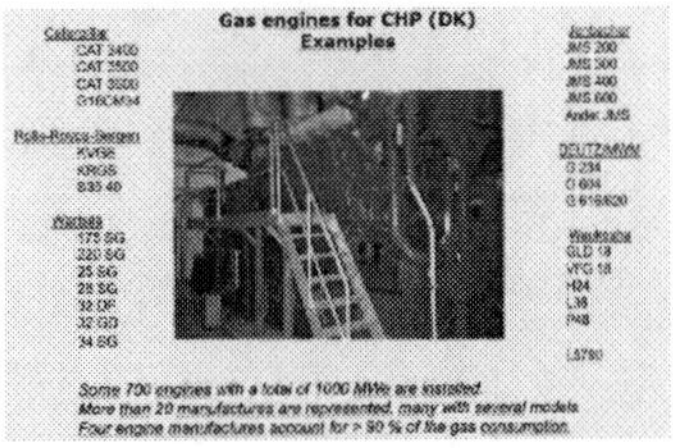

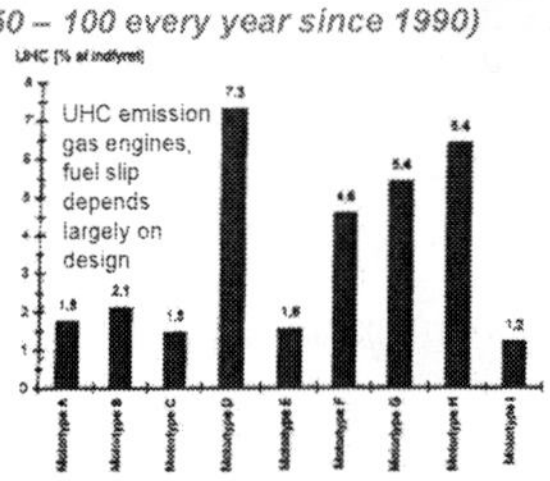

- Still many segments are not investigated
- Wide spreading of emissions in the same segment

SUSTAINABLE GAS TECHNOLOGY

DGC

Understanding CH4 emissions from end-use

How/when does CH4 emission occur?
(for many of the technologies)

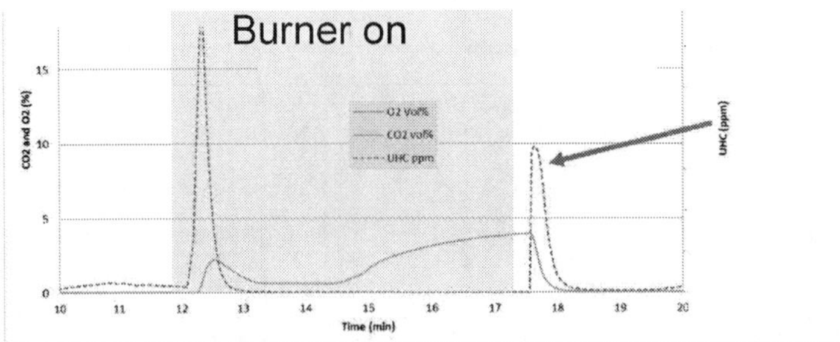

Overview of CH4 emissions from end-use

CH4 emissions vs. unburned hydrocarbons (CxHy)

(Only CH4 is a greenhouse gas)

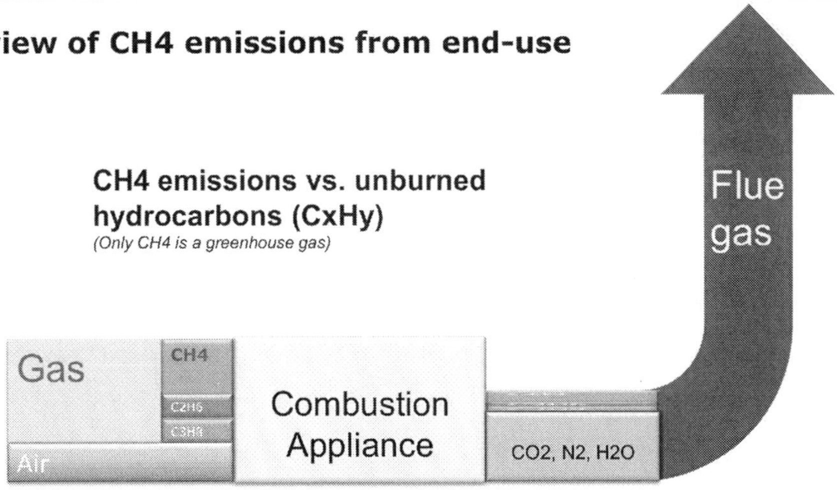

SUSTAINABLE GAS TECHNOLOGY DGC

FOCUS ON BOILERS. Emissions during burner start stop.

Literature investigation

- No big differences between segments, BUT **large differences within a given segment.**

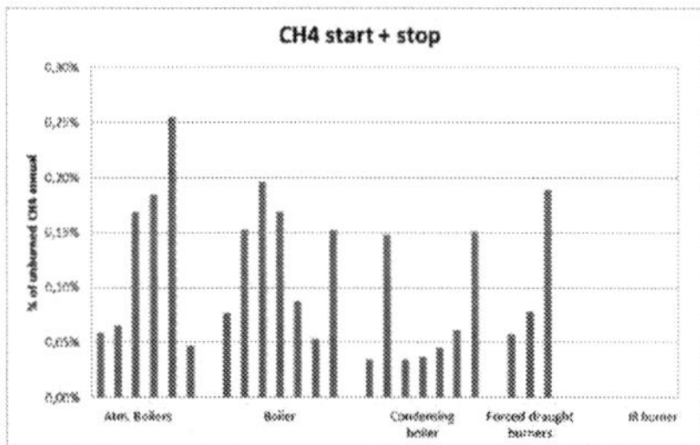

(Literature Data for existing markets < 2005 or so)

Results confirmed by more recent lab tests (DGC and more 2019-2019)

Domestic Boilers = from 0,1 to 0,3%

Data above converted in annual CH4 emission in % for boilers for
- *20,000 kWh annual need for heating*
- *14,000 start-stop annually*

Other domestic and commercial segments: no data

- Cookers - ovens
- Radiant heaters.
- New gas technologies (mCHP, fuel cells, GHP, etc.).

SUSTAINABLE GAS TECHNOLOGY

Project – Measurement campaign on boilers (2018-2019)

Test campaign on CH4 measurement on domestic boilers (> 50 measuring points)

Why gas boilers:
Very high number of appliances installed!

Objective

- Gathering information (testing in lab) on unburned hydrocarbons and CH4 from NEW CONDENSING GAS BOILERS in order to get a clear view on actual emissions

Partnership

- DGC/GRDF/ENGIE/APPLUS/CATIM

Dansk Gasteknisk Center a/s

SUSTAINABLE GAS TECHNOLOGY

Boiler tested = domestic condensing boilers (about 20 to 30 kW)

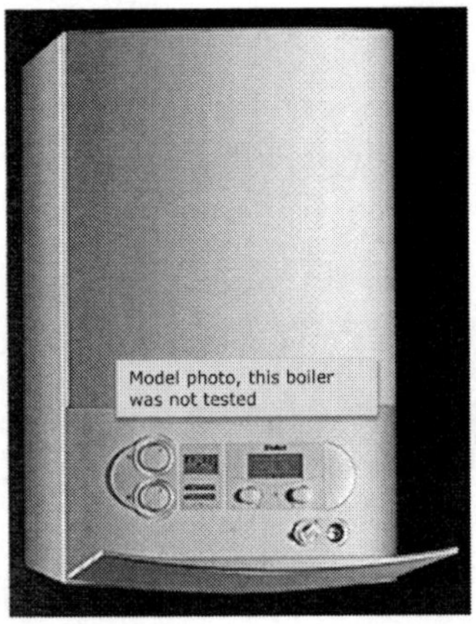

Results from the literature were confirmed:

- Similar level obtained on new condensing boilers
- Wide spreading of test results
- The start-stop (ss) frequency is one of the essential parameters for boilers and similar technologies.
- Reduction of the start-stop frequency is, however, a parameter that can be managed (electronics, modulation, correct sizing)
- The gas quality has a strong influence on the CH4 emissions

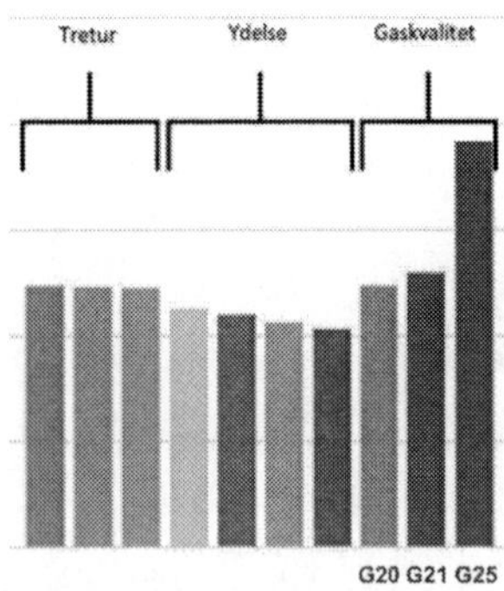

Dansk Gasteknisk Center a/s

Evaluation of EU CH4 emissions from end-use (with known data 2020)

Gas turbines have very low emissions (references suggest average value as low as 0.005% of the gas used) and are rather homogenous. Stationary engines can have rather high emissions (about 2.4% of the gas used as average), and is very inhomogeneous.

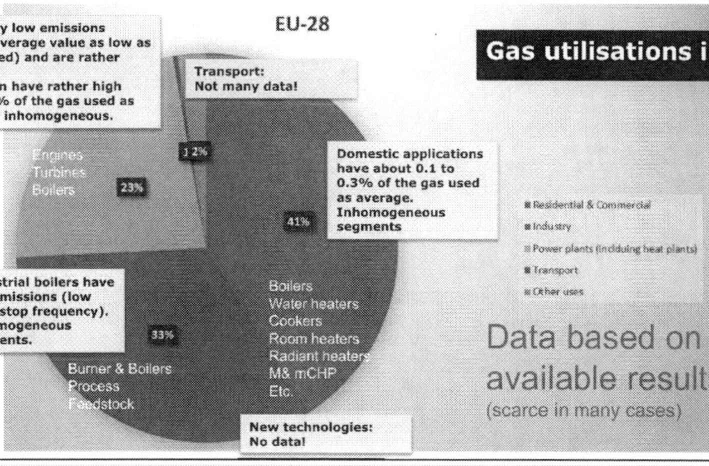

Industrial boilers have low emissions (low start stop frequency). Inhomogeneous segments.

Summary

What we know

- Measurement procedure for end-use is available (field & lab. tests)
- Range of emissions for some of the market segments
- Some existing solutions can already help to reduce emissions for boilers
- Very wide spreading of emissions for the same segment

What we don't know

- Overall impact of end-use not known or inaccurate: too many technologies/segments not investigated. Too wide spreading of emissions. Impact of new gases. Impact of maintenance operations etc.
- Mechanism for the emission only partly understood
- Mitigation solutions need to be developed for the most sensitive segments

COLLABORATIVE PROJECT. Assessment & reduction of methane emission from gas utilization

A gas industry effort
- to accurately document the loss of methane from end-use
- to understand CH4 emissions
- to propose mitigation solutions

1- Test of appliances. Lab & in-situ + literature

Understanding the mechanisms of CH4 emissions.

4-Proposing Mitigation solutions to reduce CH4 emissions

2- Emission factors

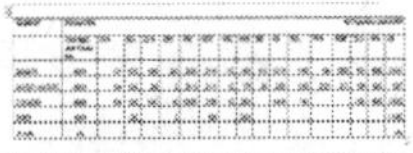

Appliances market data

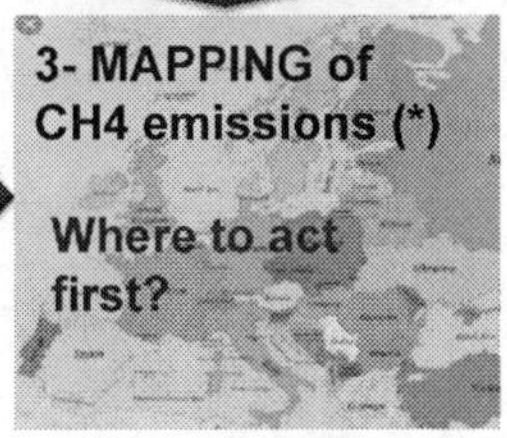

Coordination with other initiatives world wide (DVGW; GTI)

(*) Combining Number of appliance, Avg consumption and EF)

Content & results

- AIM1 = Establish a solid documentation of the emission of methane from the different segments of gas utilisation.
- AIM2 = Understanding the mechanisms behind CH4 emissions (influencing parameters on the emissions factors of existing equipment, e.g. boilers, engines, turbines and stoves that are operating today). *This includes also the impact of injecting new gases.*
- AIM3 = Investigating mitigation solutions.

- RESULT1: An accurate mapping of methane from end use (country by country).
- RESULT2: Recommendations for achieving reduction of the methane emission (replacing existing appliances or retrofit solutions).

SCOPE

APPLIANCES
- Existing installed appliances
- New appliances

GASES
- NG (L & H) (incl LNGs)
- NG - H2 mix
- NG - Biomethane mix

To make the present picture of CH4 emiss.

To make projections: will replacement of old appliances with new ones change the picture of CH4 emiss. ?

Overall organisation

WP1: Literature work and desk work

WP2. Testing protocols & Test program. Choice of appliances to test

WP 3 Testing

WP4. Mitigation solutions

WP 5 Use of results. Mapping of emissions. Recommendations for reduction

WP 6 Communication. Dissemination. Liaison with other initiatives (US, GE, etc.)

SUSTAINABLE GAS TECHNOLOGY

Organisation: Simple structure proposal

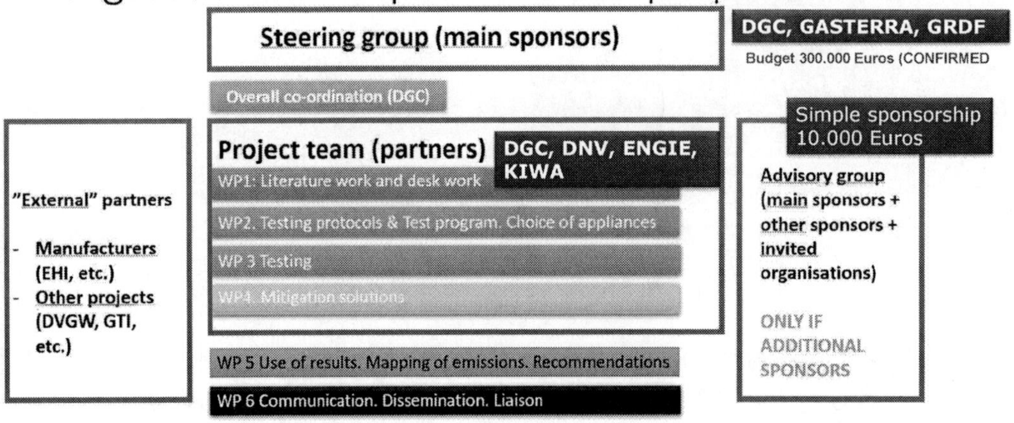

Dansk Gasteknisk Center a/s

SUSTAINABLE GAS TECHNOLOGY

Conclusion

Understanding the CH4 emission is fundamental in order to:

- Get the right picture of today's emissions from end-use
- Propose mitigation solutions to reduce the emissions

The research programme proposed should help our industry to reduce the impact of CH4 emission from gas utilisation on global warming.

We invite more interested parts (including non- EU countries) to join us!

Mail: Jean Schweitzer: Jsc@dgc.dk

Dansk Gasteknisk Center a/s

Qatargas Flare Reduction Project

Ahmed Benhamidcha
Lead Process Engineer –Qatargas Operating Company ltd.

25/02/2020

IGRC 2020 MUSCAT

The World's Premier LNG Company

QATARGAS

LEGAL DISCLAIMER

Page 2

QATARGAS

Agenda

- ❑ Project Background
- ❑ Introduction
- ❑ LNG Mega Train Overview
- ❑ Project Objective & Overview
- ❑ Flare Reduction Opportunity#1 – Feed Gas Interconnection
- ❑ Flare Reduction Opportunity#2 – Off-Specification Gas Recycle
- ❑ Flare Reduction Opportunity#3 – Fuel Gas Compressor end Flash Gas Interconnection
- ❑ Conclusion

QATARGAS

Project Background

Qatargas-North operates seven Liquefied Natural Gas (LNG) trains producing 42 Million tons per annum (Mtpa). Three of these seven trains (Trains 1, 2, and 3) use the conventional C3-MR process, each producing 3.3 Mtpa, and four (Trains 4, 5, 6, and 7) use APX process are the world's largest mega-trains, each producing 7.8 Mtpa.

The four mega-trains (Trains 4, 5, 6 and 7) were originally designed based on "no flaring concept" during normal operational conditions. However, flaring occurs during transient operations accompanying events such as shutdown or start-up of trains.

Qatargas is committed to:

- Contribute to the achievement of Qatar national flare reduction targets, part of the National Development Strategy (Qatar National Vision 2030)
- Follow applicable regulations and minimise impact on environment through finding a balance between the environmental/economic needs of the State of Qatar and sound international practice.

QATARGAS

Introduction

Reduction of flaring is one of the most ambitious objectives set by Qatargas in order to reduce the CO_2 emission from their LNG trains through a reduction of flaring events. Qatargas has set targets to reduce the overall LNG process flaring to 0.30 percent of sweet gas production by 2020.

Over 70% reduction of flaring from 2011 to 2016 from on-plot LNG facilities has been achieved. This has been done through operational initiatives on source reduction, increased plant reliability, reduced flaring during shutdown and start-up of the plant.

Despite a several initiatives which have been implemented to minimise the flaring, the flare reduction challenges are being addressed in mega-trains.

QATARGAS

LNG Mega Train Overview

The LNG mega trains consist of two inlet facilities, one for Trains 4 and 5 (QG 2 asset) and the other for Trains 6 and 7 (QG 3&4 asset). Each train consists of the following systems/units:

- Unit 02: Acid Gas Removal (AGR)
- Unit 03: Dehydration, mercury and Mercaptans removal
- Unit 04: NGL (Natural Gas Liquid) recovery
- Unit 05, 06: Gas chilling and liquefaction
- Unit 07: NGL fractionation
- Unit 08: Nitrogen rejection
- Unit 09 : Sulfur recovery unit
- Unit 10 : Helium extraction

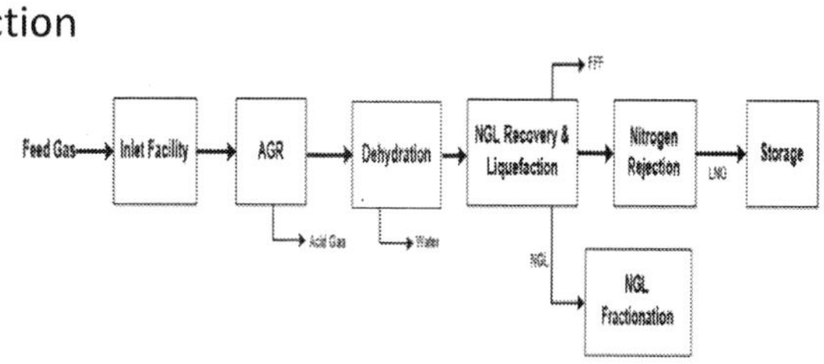

QATARGAS

Project Objective & Overview

The overall objective for the Qatargas Flare Reduction Project is to explore and study all the possible opportunities to reduce routine flaring and minimise its impact on the environment.

Explore Opportunities > Select & Study > Implement

A three significant flare reduction opportunities of eight have been studied and implemented in LNG Mega-Trains, that are:

1. Flare Reduction Opportunity #1: Feed Gas Interconnection.
2. Flare Reduction Opportunity #2: Off-Specification gas recycle.
3. Flare Reduction Opportunity #3: Fuel Gas Compressor K0801 end Flash Gas Interconnection

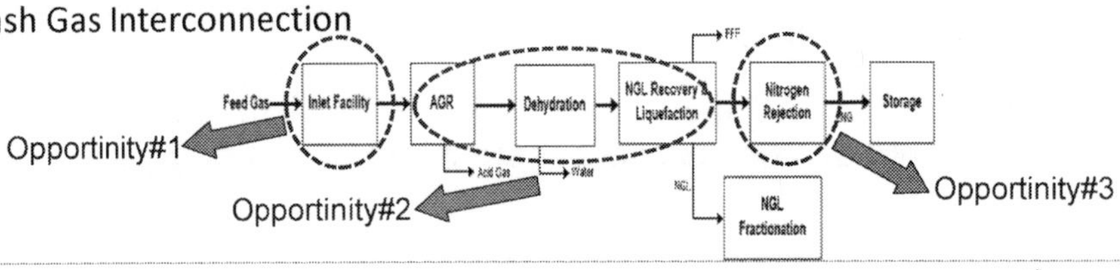

Flare Reduction Opportunities – Overview

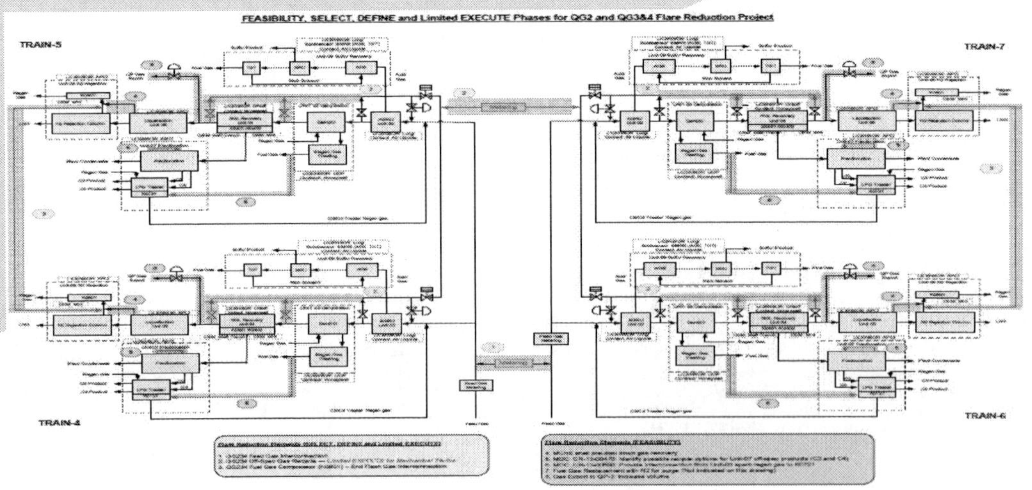

Flare Reduction Opportunity#1 : Feed Gas Interconnection

- QG2 and QG3&4 trains start-up load is less than the minimum flowing capacity of offshore platforms. Therefore, excess feed gas during start-up is being flared. The balance (400 to 600 MMSCFD) would be routed to the inlet facility of the other asset.
- Transfer QG2 excess Feed Gas to QG3&4 (and vice versa) when one of the Asset's LNG trains is shut down while the other one is starting-up.
- The existing 24"feed gas interconnection line (receiving facilities of QG2 and QG3&4) has been modified to allow bidirectional use.
- New royalty metering station provided.
- All mechanical works have been performed during normal operations.

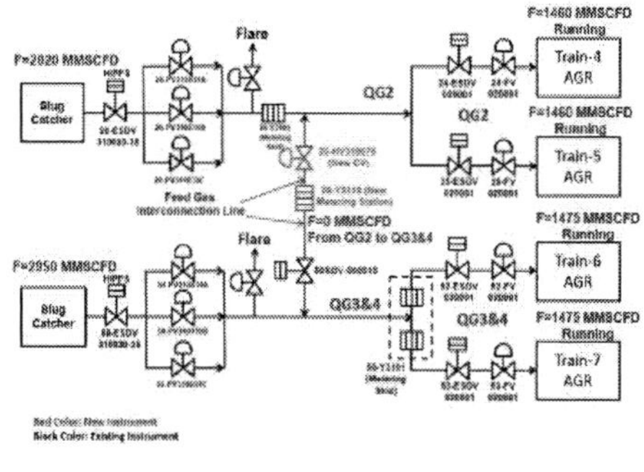

Feed Gas Interconnection Line in Inlet Receiving Facility in QG2 and QG3&4 (Normal Operation)

Flare Reduction Opportunity#1 : Operating Mode

Operating Mode	Description	Feed gas flow rate (MMSCFD)
A	QG2 Excess Feed Gas is transferring to QG3&4	600
B	QG3&4 Excess Feed Gas is transferring to QG2	400 (max 600)

To allow the Feed Gas transfer, 2bar differential pressure between the inlet facilities is required. This can be achieved by adjusting the set pressure at the Inlet Pressure Reduction Facilities.

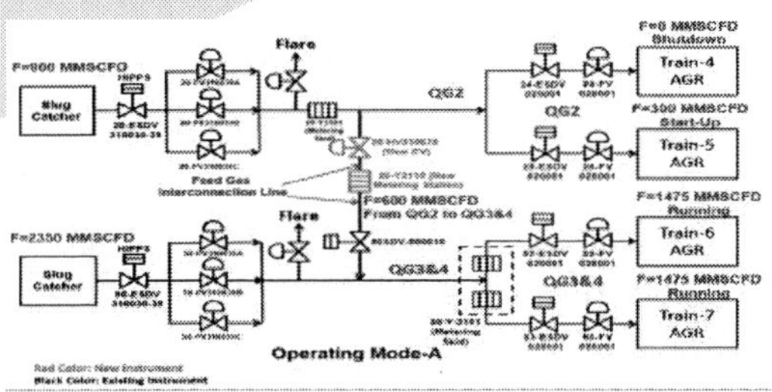

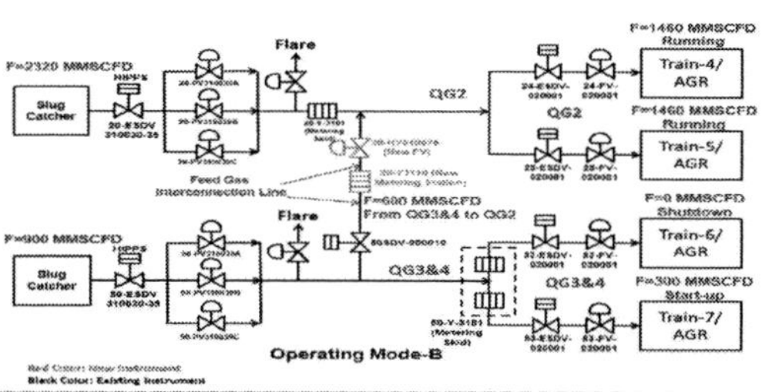

Flare Reduction Opportunity#2 – Off-Specification Gas Recycle

- Design Concept

During start-up of LNG train, the product gas does not meet the required specification for gas processing and liquefaction (off-spec product from AGR, Dehydration and NGL recovery units).

Start-up activities typically take several hours in sequence from the upstream units to downstream units. Consequently associated flaring also takes several hours until the product gas becomes on-spec, therefore, a huge amount of gas is wasted and CO_2 is emitted.

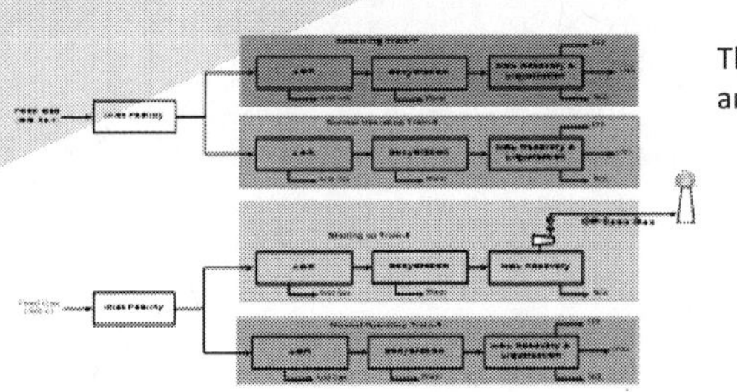

Off-Spec Gas to Flare

The minimum turndown required for startup of AGR, Dehydration and NGL recovery units are indicated in below table.

Unit Name	Produced Gas	Minimum Turndown Limit (MMSCFD) Recommended by Process Licensor
Acid Gas Removal Unit	Sweet gas	250
Dehydration, Mercury and Mercaptans Removal Unit	Dry sweet gas	320
NGL Recovery Unit	Lean sweet gas	600

Page 11

QATARGAS

Flare Reduction Opportunity#2 – Off-Specification Gas Recycle

- Modification Concept

- Divert, during the startup of one train, the off specification gas from AGR, Dehydration and NGL Recovery units outlet to the inlet of AGR unit of a running train instead of flaring it.
- Installed recycling lines within trains and an interconnection line with a fiscal metering system between QG2 & QG3&4 (red in sketch)
- The modified facilities able to perform recycling off-spec gas Intra-Asset or Inter-Asset .
- The maximum envisaged off spec gas flow rate recycled is 600 MMSCFD.
- Several studies such as steady state simulations (by Unit's Licensors) as well as dynamic simulation were carried out to assess the impact of the proposed modifications on the existing facilities.

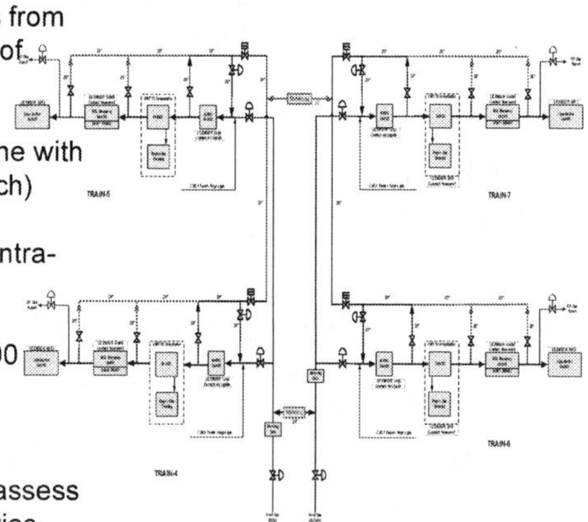

Page 12

QATARGAS

Flare Reduction Opportunity#2 – Off-Specification Gas Recycle

Impact of the Modified Concept: Loss of Recycle Gas from Start-up Train → Dynamic Simulation to evaluate :

AGR Unit

- If the system and controller performance in AGR unit is stable and the pressure and flow rate are maintained.

NGL Recovery Unit

- The disturbances in feed flow & composition
- The Performance of Deethaniser (C_3+ recovery and lean gas specification).
- Temperature excursions in Aluminum Brazed Heat Exchangers.
- Stability of Controllers (Multiple controllers & Deethaniser liquid level controller)

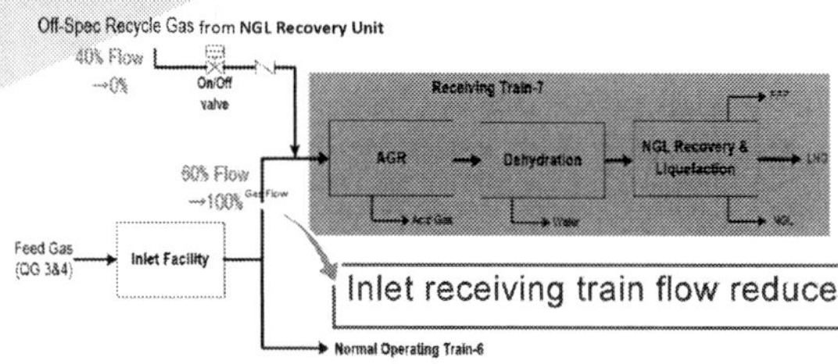

Inlet receiving train flow reduced → Recovered by increasing flow from Inlet facility

QATARGAS

Flare Reduction Opportunity#2 – Off-Specification Gas Recycle

Dynamic Simulation Results →

- Operational disturbances of the receiving train could not be avoided.
- Modifications to controller configuration and operation are recommended & implemented to stabilise operation of the receiving train.

System	Results	Recommendations (software modification)
AGR Unit	AGR Absorber pressure decreases below the allowable minimum	Control system should be updated
NGL Unit	Lean gas pressure decreases, which opens the expander IGV leading to JT operation mode.	
NGL Unit	Deethaniser bottom (NGL) flow reduces to zero for a while (below the minimum turndown rate of 40% flow)	Deethaniser reboiler operation should be modified to respond to the rapid change of feed gas composition
NGL Unit	Deethaniser bottom liquid level fluctuates	
Liquefaction Unit	Heavy component in LNG exceeds the allowable range	

QATARGAS

Flare Reduction Opportunity#2 – Off-Specification Gas Recycle

- Key Process Parameters

Original Controller

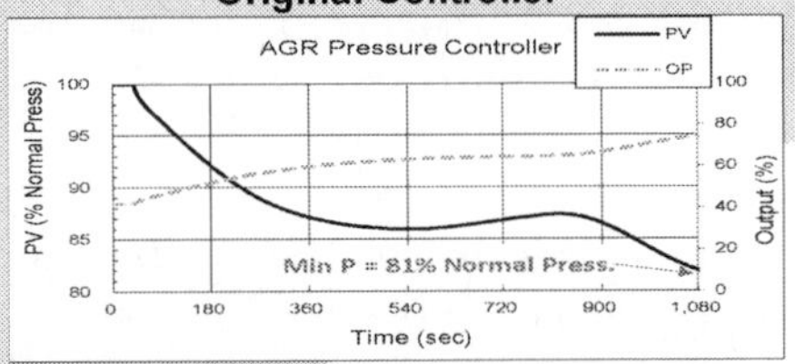

AGR pressure decreased below its minimum allowable pressure, leading to unacceptable disruption

Modified Controller

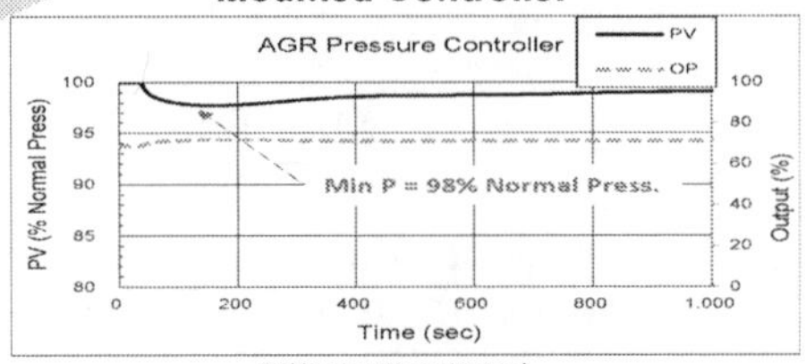

Stable operation maintained

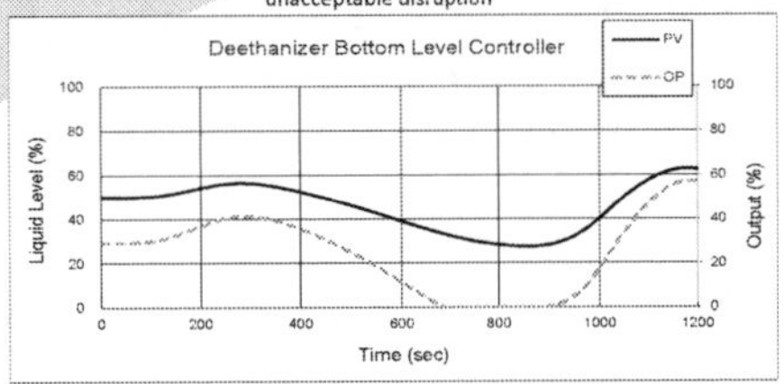

Unstable Operation

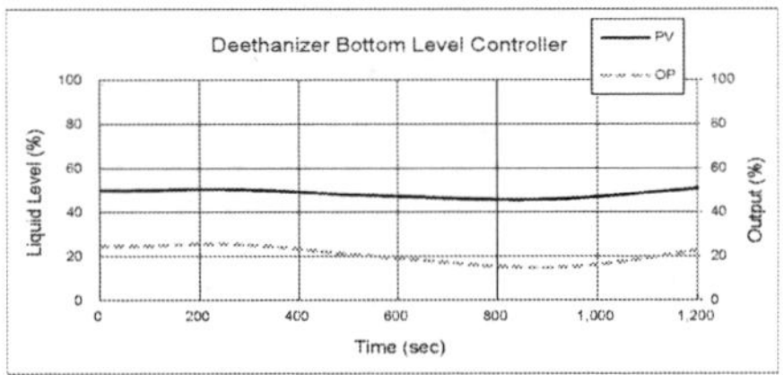

Stable operation maintained

QATARGAS

Flare Reduction Opportunity#3 – Fuel Gas Compressor end Flash Gas Interconnection

- Design Concept

Each LNG Mega train has its own Fuel Gas Compressor (K0801) which is used to supply end flash gas to the fuel gas system (Utility). The feed gas to K0801 is the combined end flash gas from the Nitrogen Rejection Column (C0801) and fuel gas from LNG Flash Drum (V0801).

When one fuel gas compressor (K0801) is tripped while LNG production is ongoing, all the end flash gas on the suction of the tripped K0801 is sent to the dry gas flare. Consequently, the end flash gas supply to the Fuel Gas System is also disturbed.

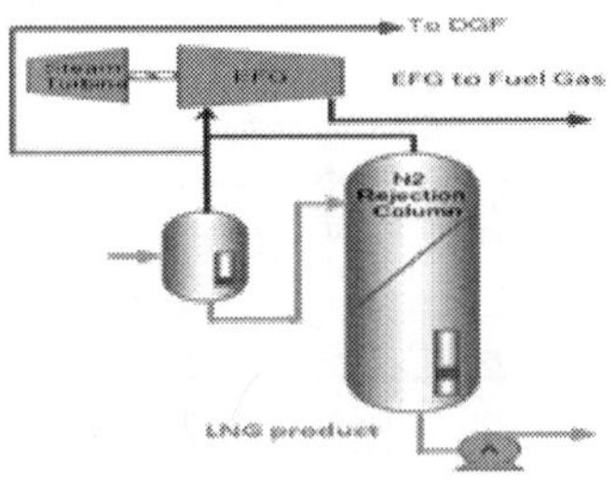

Flow Diagram Nitrogen Rejection Unit

QATARGAS

Flare Reduction Opportunity#3 – Fuel Gas Compressor end Flash Gas Interconnection

- Modification Concept

- Recover part of the end flash gas when a fuel gas compressor (K0801) trips or during cool down of Nitrogen Rejection Unit.
- Interconnection lines provided with the required valves and instrumentations between end flash gas compressors (K0801) of the same asset (two trains) at both the suction and discharge ends.
- 42-inch line, new HV and new ESDV for the suction end interconnection and a new 12-inch line, new FV and new ESDV for the discharge end interconnection
- The verification of the additional capacity of fuel gas compressor K0801 was carried out.

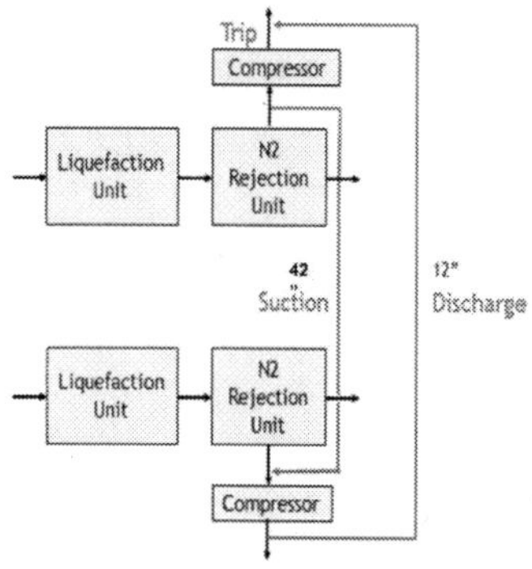

Overview of Opportunity#3

Flare Reduction Opportunity#3 – Fuel Gas Compressor end Flash Gas Interconnection

Fuel Gas Compressor K0801 Analysis (EFG Recovery)

The Fuel Gas Compressor is a four-stage centrifugal machine, with inter stage coolers to cool the fuel gas after each compression stage using the FCW. The Fuel Gas Compressor is driven by a steam turbine.

A simulation model HYSYS has been developed to study all the considered cases for the maximum compressor capacity adequacy:

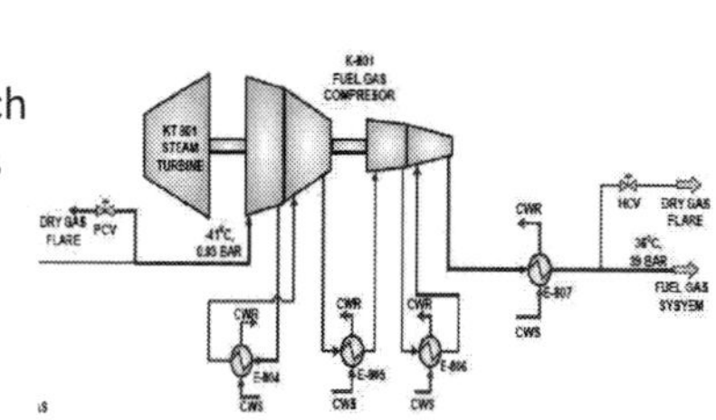

Simulation Results

- Maximum allowable EFG flow can be diverted through inter-connecting piping to the running fuel gas compressor is around 46 ton/hr (35.5%).
- The additional flow consumption required of FCW and HHP steam for Fuel Gas Compressor system with the new capacity has been assessed and identified.

Conclusion

By implanting this project, Qatargas has achieved :

- ❑ Zero gas flaring during startup of AGR, dehydration and NGL Recovery units,
- ❑ Gas flaring reduced by 35% during the cool down of Nitrogen Rejection Unit (Startup) / or when Fuel gas compressor K0801 trips (Normal Operation),

The facilities have been tested against the design criteria and the results shows that all the key specifications and performance parameters have been met and achieved.

Thank you

العمانية للغاز الطبيعي المسال
Oman LNG

Fit For the Future: Major Projects at Oman LNG

Tuesday, February 25th, 2020

The best at what we do.

The LNG Business in Brief

Video Link: (What is LNG?)

LNG Growth Outlook

Supply and Demand

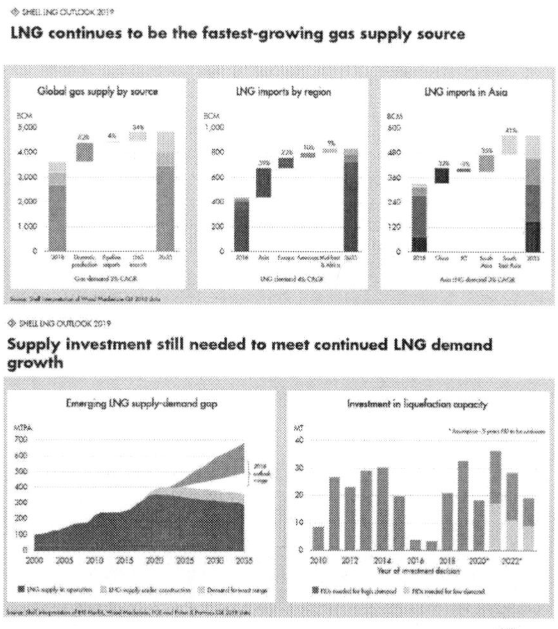

1. Footnote

About Oman LNG

- Oman LNG is a three train LNG plant in Sur, Oman. The company was established by the Royal decree of Sultan Qaboos of Oman in 1994.
- The construction was launched in November 1996, and the plant was commissioned in September 2000.
- Oman LNG operates three LNG trains with a total capacity of 10.4 million tonnes per year.

3

Capital Projects in Oman LNG

Power Project : Currently electrical power to the LNG facilities is provided through open cycle gas turbine driven power plant located within the facility. Oman LNG has selected the option of 12 Gas Engines driven power plant (200 MW) to replace the existing gas turbines power plant to increase the overall energy efficiency of the LNG plant

Top Project Risks:

1	Soil Conditions (Pressure Grouting)
2	Contractor & sub-contractor performance
3	Scope creep
4	SIMOPS

RP18 : The RP18 is comprised of approximately 350 activities. The premises are a combination of ARP scope, minor plant changes that support reliability and Asset Integrity and Process Safety Management (AIPSM), focused on re-establishing nameplate capacity and increasing reliability.

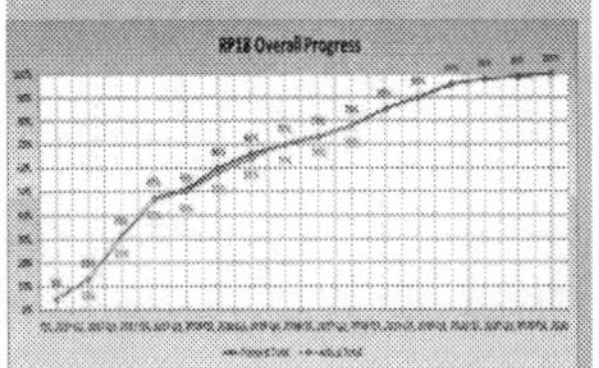

DBN : Oman LNG is developing a debottlenecking project to increase the plant capacity. The Debottlenecking main scope consists of upgrading the LNG trains helper motors together with associated brownfield integration scope. Target capacity uplift of 1 MTPA aligned with surplus gas.

Top Project Risks.

1	C&P Strategy needs to be finalized and approved in line with the execution strategy
2	Award and (proper) specification of LLI due to ongoing FEED developments
3	Key resources not all mobilized
4	Large brownfield component can lead to scope creep and cost/schedule overruns

4

Collaboration with Shell P&T

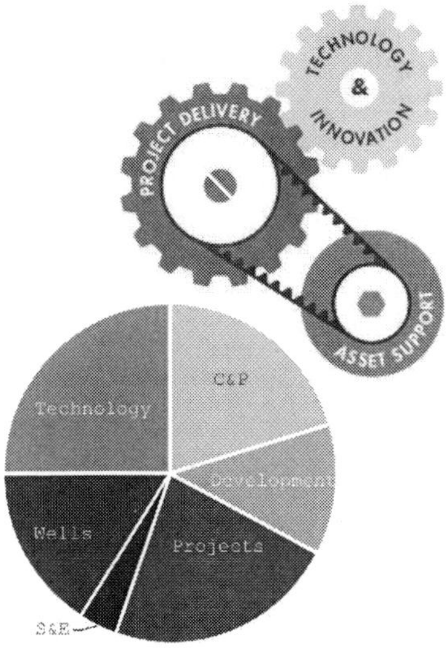

Adapting the Opportunity Realization Standard (ORS) to OLNG Major Projects

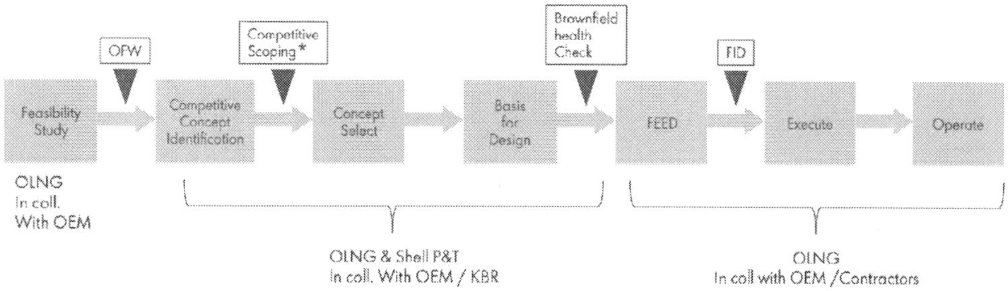

*Competitive scoping approach helped optimize project scope, identifying and re-focusing the project on key value generating scope elements.

Power Plant

Business drivers:
Increase overall energy efficiency & Reduce Fuel gas consumption

Parameter	UOM	GTG Power Plant	GEG Power Plant	Comment
Efficiency GTG Power Plant:	%	~27%	~ 44%	
Fuel Gas Consumption:	kton/a	433	276	Based on 105 MW electrical load

Reduce GHG emissions

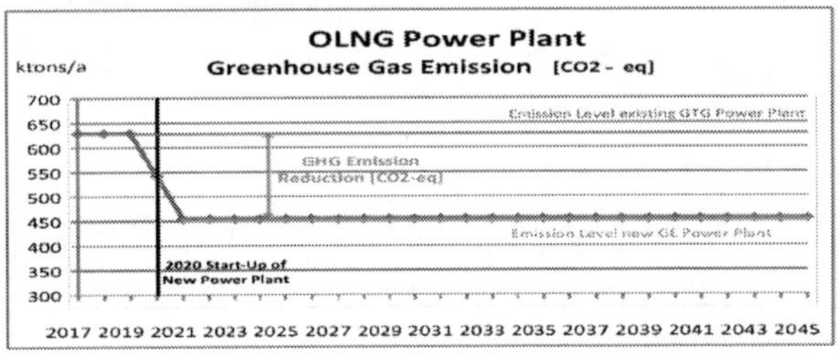

The reduction in GHG footprint measured as CO_2 equivalent is about ~27% GHG emission for the existing power plant (180Kton of CO_2-e per annum) for a project life of approximately 20-25 years.

Power Plant

Concept Select Options:

1. Combined Cycle
2. Gas Engine Generators
3. Gas Turbine Generators

- **Scope (12 Gas Engine Generators)**

Challenges:

- Operation and Maintenance of GEGs in LNG application (Once per month).
- Lube oil consumption and methane slippage
- Sensitivity to Fuel Gas composition and composition changes.
- GTG to GEG change-over on a live plant.
- Adapting with international standards.

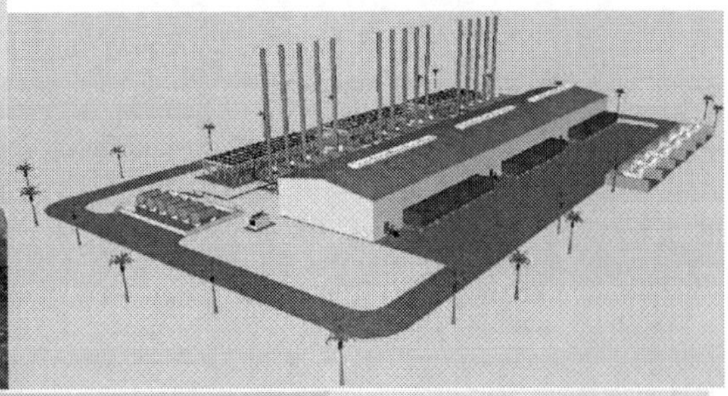

Readiness Plan 2018

Oman LNG has been in operation for more than 15 years. Many of the instrumentation and other vendor packaged technologies are obsolete. Oman LNG is expecting new gas by 2018 with potentially leaner gas. The new gas composition will impact the plant capabilities to process and meet the nameplate capacity.

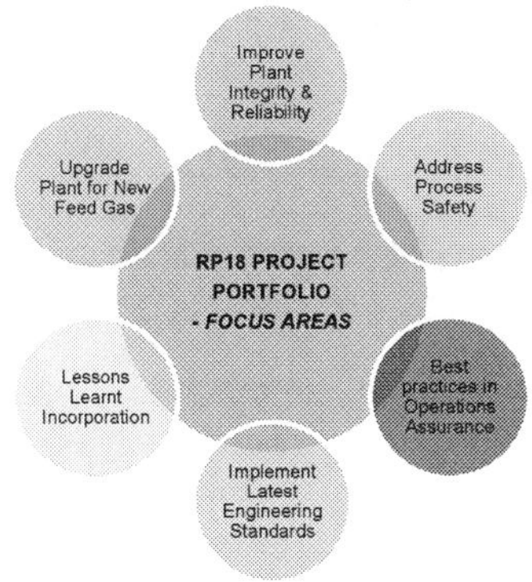

RP18 has around ~350 activities which are to be executed in order to ensure plant readiness for maximum availability and utilization.

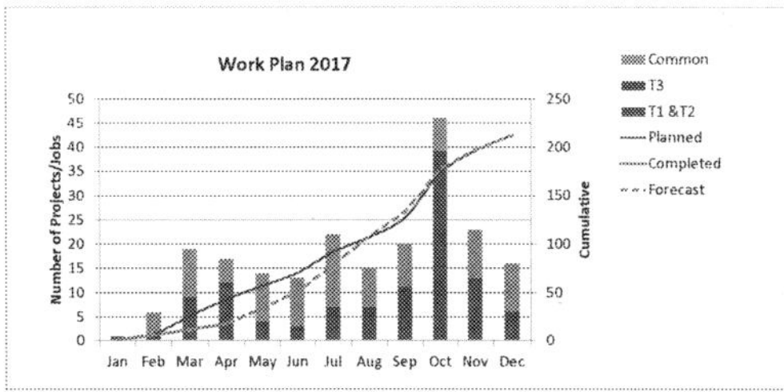

Debottlenecking Oman LNG

"OLNG has the opportunity to unlock additional plant capacity from the existing three LNG trains through developing specific de-bottlenecking opportunities at competitive unit rates, by leveraging Oman LNGs established position in Oman's gas infrastructure and by implementing a phased development approach, which targets available gas volumes in Oman."

Competitive Concept Identification:

- Inlet Air Chilling for Gas Turbines
- Upgrade of Train 1&2 Gas turbines helper motor to match Train 3
- Upgrade of gas turbines helper motor power to a higher capacity
- Mixed flow impellers for HP MR Machine
- Increase in Gas Turbine Firing Temperature
- Increase in feed gas pressure

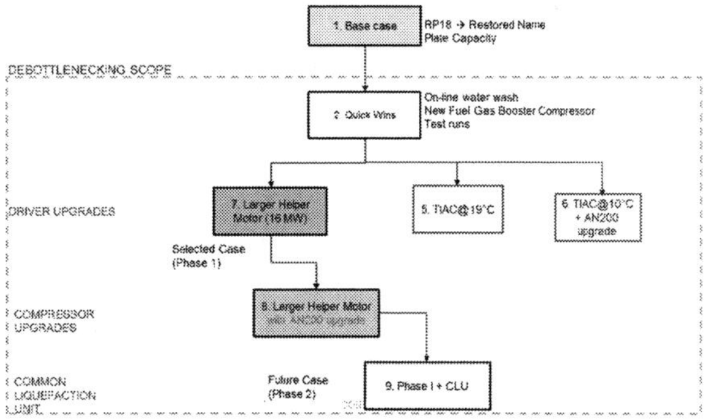

Scope and Schedule of DBN

MR Helper Motor & VSDS upgrade to 18 MW (LLI)

T-1/2 PR Helper motor rerating to 8.3 MW

STRATEGIES:

- **Target Summer:** upgrade driver's power enabling to produce more LNG in summer
- **Quick wins** stretch existing equipment leveraging design margins

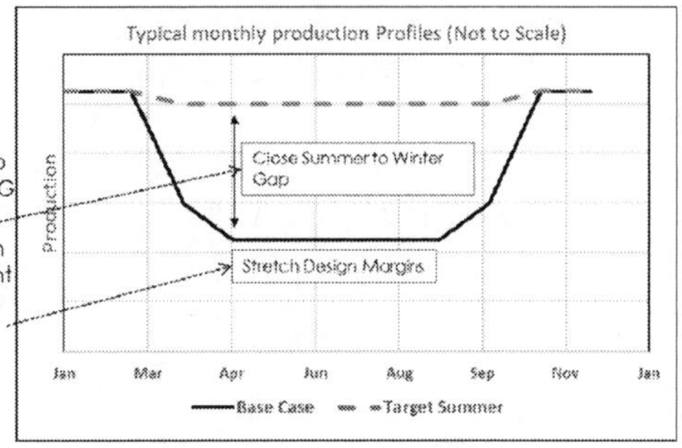

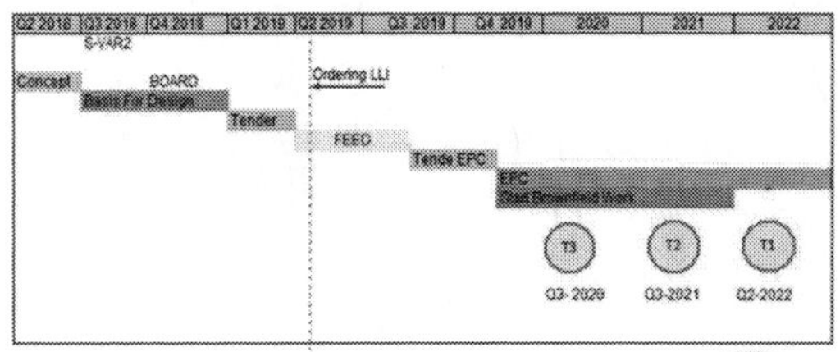

Brownfield integration Scope (Column, Vessel, Kettle, CV, RV, metering run)

3 additional Gas Engines

11

Your challenges?

Could you take advantage of the strong global demand by **increasing capacity with a new LNG train or facility**?

What is the right balance for you between **economy of scale** and the trend **towards mid-scale projects**?

Do you need to control capital expenditure while providing high plant **availability**, low **operating expenditure** and a short **schedule**?

Is your site **footprint constrained**? Do you need a liquefaction technology that takes up less space?

If you plan to use conventional technology, does your licence leave you free to **engage in competitive equipment tenders** to improve your project economics significantly?

Could you harness the latest technology to **increase efficiency** through larger-capacity trains and/or by taking advantage of low and/or seasonally low temperatures?

Are you under pressure to reduce **carbon dioxide (CO_2) emissions**?

Are you looking for gas processing technologies that can **simplify and reduce the cost** of your LNG line-up while enabling you to use more contaminated feed gas?

12

21 February 2020

Thank you

Friday, February 21, 2020

Baker Hughes

Modularization for LNG Plants

Yousef Tamimi
Regional Sales Director, Gas Processing, MENAT

INTERNATIONAL GAS UNION RESEARCH CONFERENCE 2020

24-26 February | Oman Convention & Exhibition Centre

9. Oral session - B. LNG Projects

25th of Feb, 2020

Index

Slide

3 Modular Design overview
4 LNG Modularization evolution
5 Why Modular LNG plants
6 Modular LNG plant - The technology
7 Modular LNG plants – Construction
9 Modular technology for Small & Mid scale LNG plants

Modular Design overview

- Modules are complete pre-assemblies of equipment, bulk materials and components, which are fabricated in an offsite facility into a steel structure, which can be transported and installed at the main construction site.
- *The addition of plug and play to this design approach will greatly reduce field installation times which will bring the overall cost of the entire project down(see figure 1).*

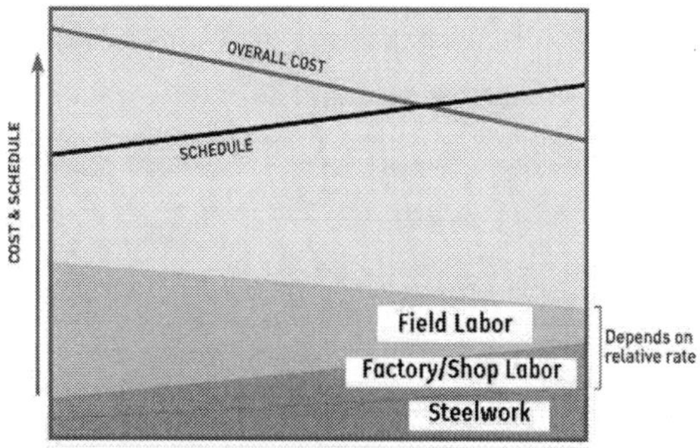

Figure 1: Typical impacts on cost and schedule of increasing the degree of modularization for development of a new liquefaction plant – source: www.lngjournal.com

3

Baker Hughes

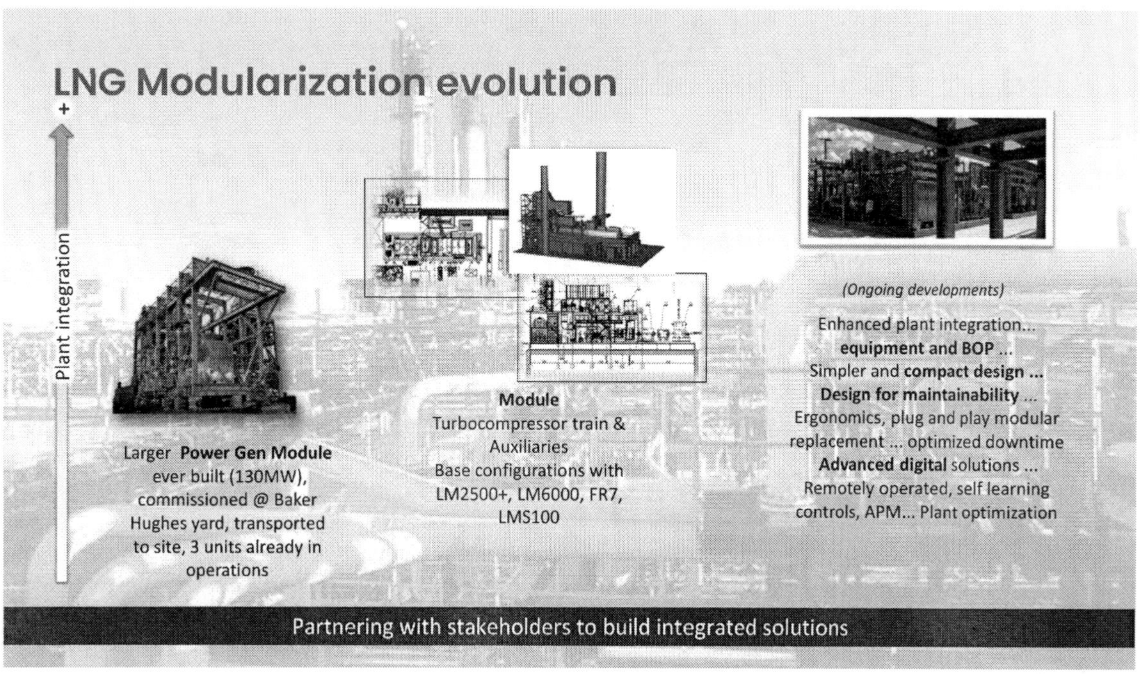

Why Modular LNG Plants

Utilizing different Gas resources

- Conventional Gas fields
- Associated petroleum gas (APG)
- Pipeline gas
- Flare gas

Modular LNG trains advantage

- Flexible design to meet different feed gas specs
- Sizing train capacity to optimize CAPEX/OPEX and plan future expansions to mitigate upfront investment risks
- Optimize Commercial operation of the facility by utilizing selective trains for production

Overcome conventional challenges

- Limit resources needed at site by using plug& play design – critical for high risk countries& fragile ecosystems
- Incremental CAPEX to accommodate market dynamics, phased approach for expansion to enhance risk-return spectrum for owners
- Fast tracking the Project cycle of 3 years compared to min 5 years of conventional stick built plants

Baker Hughes

Modular LNG – The Technology

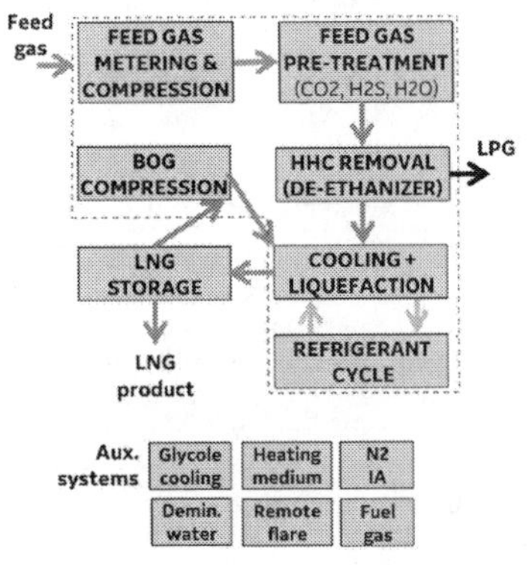

6

Four non-licensed liquefaction processes:

- **N2-TEXP** - Closed-loop N2 expander
- **CH4-TEXP** - Closed-loop CH4 expander
- **SCMR** - Single-cycle Mixed Refrigerant
- **PCMR** - Pre-cooled Mixed Refrigerant.

Baker Hughes

- **Modular LNG plants – Construction**

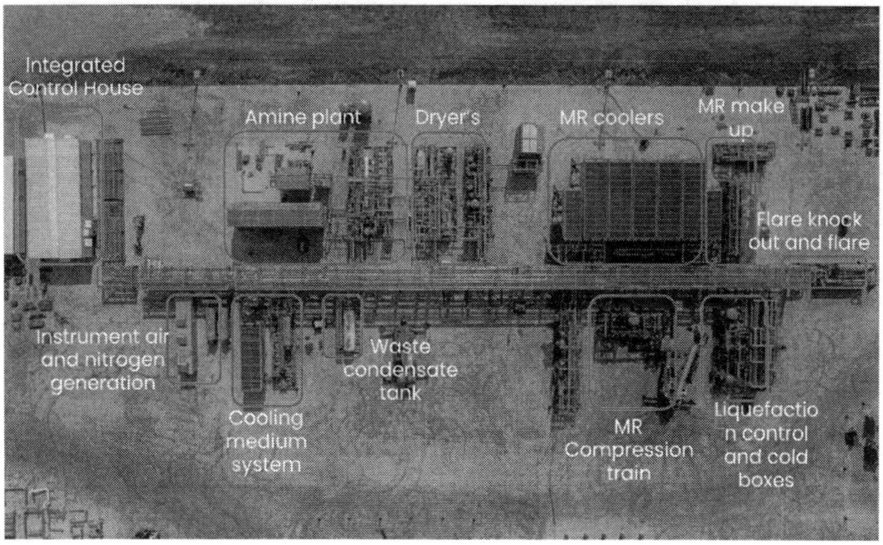

Truckable design

Baker Hughes

The entire plant from inlet battery limit to outlet liquid to storage will be fabricated at approved fabrication facility that incorporates the Baker Hughes HSE and Quality Systems.

The LNG plant will be laid out as it would be in the field and interconnecting piping will be flanged and bolted together. A pressure test of the system will be done, witnessed by the client and signed off.

All skids electrical, 24VDC 120/480 VAC will be completed and a 90% FAT will be accomplished, witnessed and signed off by the client.

Prior to removal all spool pieces will be marked to the corresponding flange for easy field installation.

Modular technology for Small & Mid scale LNG plants

Rig fueling, flare capture, stranded gas

Trucking, marine, mining, concrete

Power generation, peak shaving

Flare capture, stranded gas

Gas		Liquid		
MMscfd	m³/day gas	GPD	m³/day LNG	mTPD
2	53,400	25,000	94	40
4	107,000	50,000	189	80
8	214,000	100,000	377	159
10	267,000	122,000	471	194
20	534,000	244,000	943	390
25	668,000	300,000	1,200	480
36	974,000	450,000	1,700	716
50	1,298,000	600,000	2,250	954
110	2,862,000	1,300,000	5,000	2,067

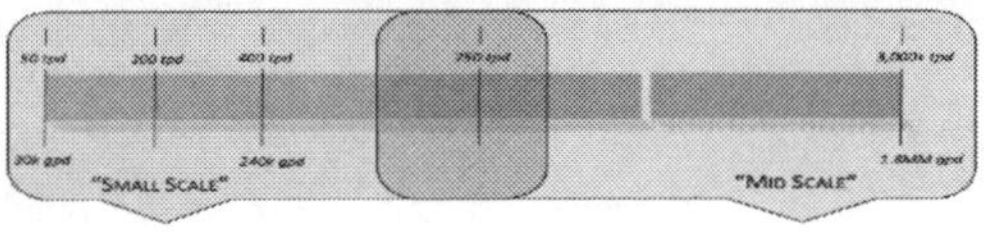

Baker Hughes

Baker Hughes

Novel Activated Carbons for CO_2/CH_4 Adsorption

Khalil AL HANASHI

IGRC2020 Muscat

Motivation

- Natural gas is replacing coal for power generation due to its low CO_2 emission, 50% less compared to coal [1].
- The demand for natural gas is expected to grow from the existing sweet natural gas fields. Hence, sour gas fields, currently unfavorable, will need to be produced.
- 40% of the world's natural gas reserve contains high CO_2 content prohibiting its monetization due to high processing cost [2]
- Amine scrubbing is the current industry standard for CO_2 removal from natural gas but can cost up to 50% of the CAPEX in sour natural gas fields [3].
- 22% of the total natural gas used in Oman is consumed by the oil and gas industry, there is a need to reduce the energy intensity of the oil and gas industry itself.
- Alternative, low cost and low energy penalty CO_2 removal technology is needed.

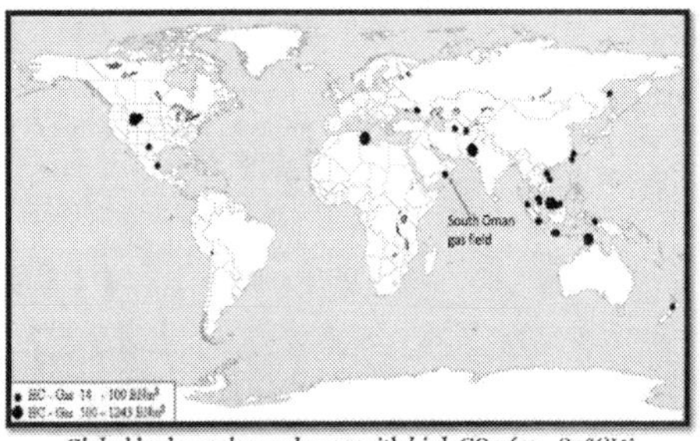

Global hydrocarbon volumes with high CO2 (15 - 80%)[4]

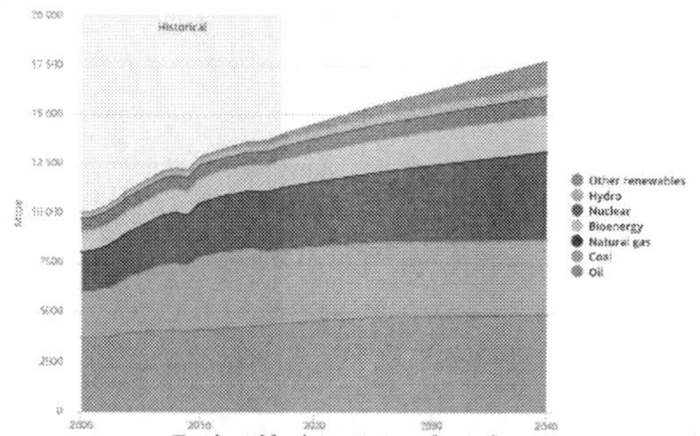

Total world primary energy demand

[1] US Energy Information Administration. https://www.eia.gov/tools/faqs/faq.php?id=73&t=11
[2] Total E&P (2007) Sour Gas, A History of Expertise, http://www.total.com/sites/default/files/atoms/file/total-sour-gas-history-expertise
[3] Shell Global Solutions (2015), *Sour gas processing*, http://s06.static-shell.com/content/dam/shell/static/globalsolutions/downloads/mediacentre/industry-commentary/sour-gas-processing.pdf.
[4] Burgers, W. F. J. et al. (2011) 'Worldwide development potential for sour gas', Energy Procedia, 4, pp. 2178–2184.

Aim

The overall aim of this research was to investigate the feasibility of using activated carbons as solid adsorbents for pressure swing adsorption process to remove CO2 from high pressure and high CO2 natural gas.

- South Oman field conditions were considered at an early stage of the adsorbent development.
- Utilization of a widely available date pits biomass waste in Oman to produce activated carbons for high pressure CO2 adsorption.
- Utilising a novel coal tar pitch carbon beads for high pressure CO2 adsorption.

How we went about it

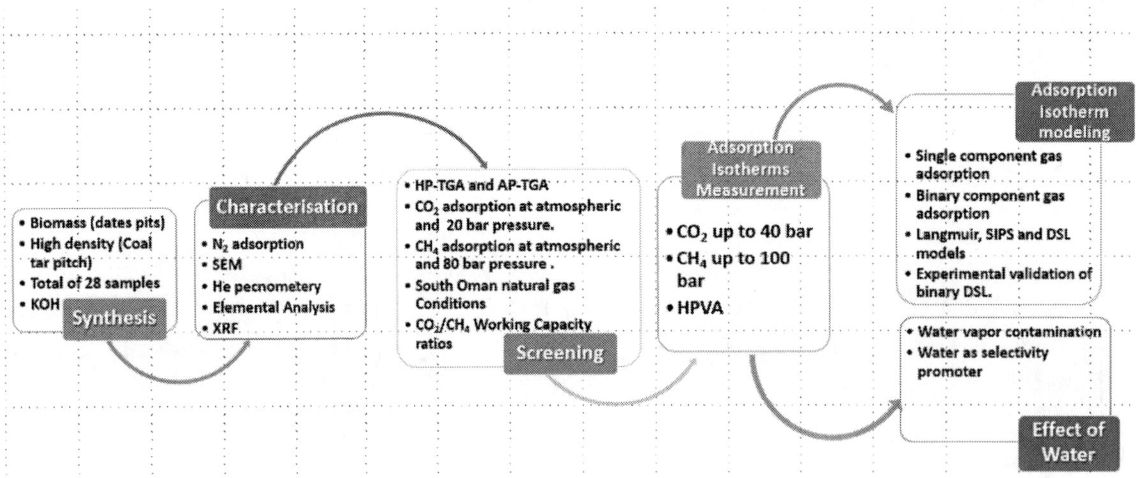

Materials Synthesis

- Two Precursors:
  Date pits biomass local to Oman.
  High density coal tar pitch spheres.
- KOH solution impregnation for activation.
- KOH amount
  1 ,2,3 and 4 KOH/Carbon mass ratio
- Activation temperature
  600, 700, 800 and 900°C

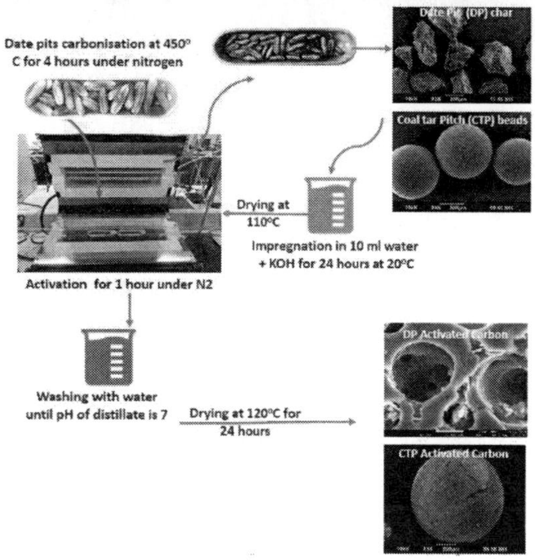

Pore network in Date Pit Activated Carbons

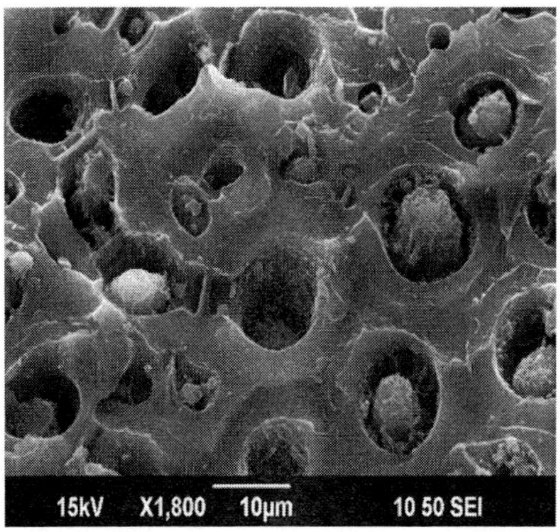

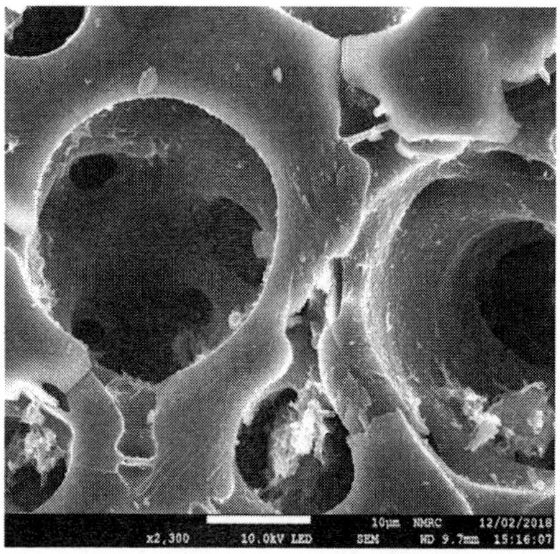

Materials Characterisation

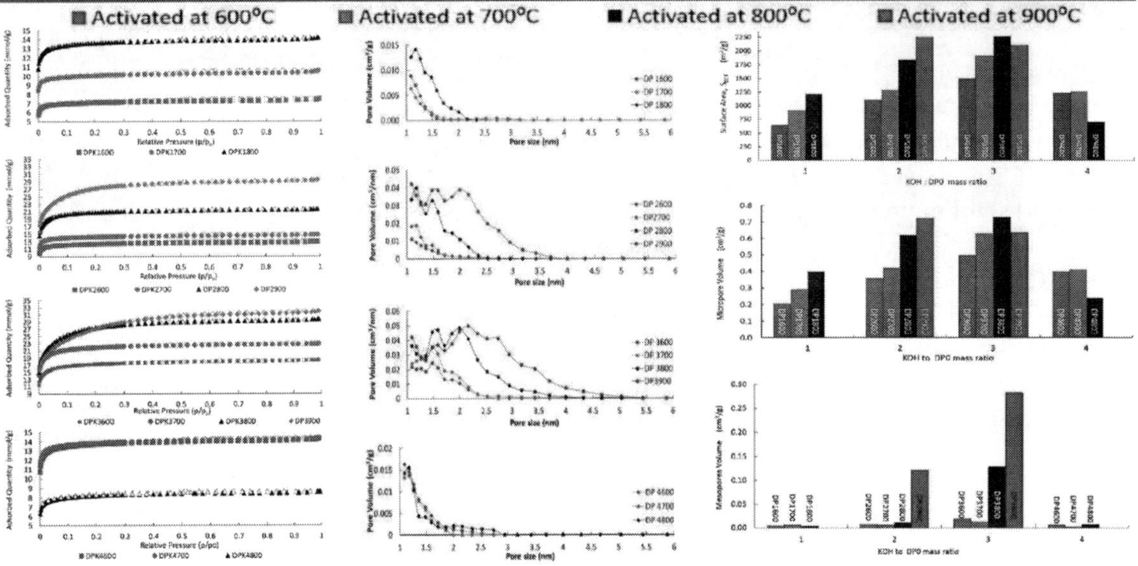

Activated Carbons Screening (1)

- HP-TGA was used at CO_2 and CH_4 partial pressure in typical South Oman gas mixture.
- 20 bar for CO2 and 80 bar for CH4 at 50°C
- Continues Dynamic adsorption measurement.
- Helium was used as purging and pressurizing gas.

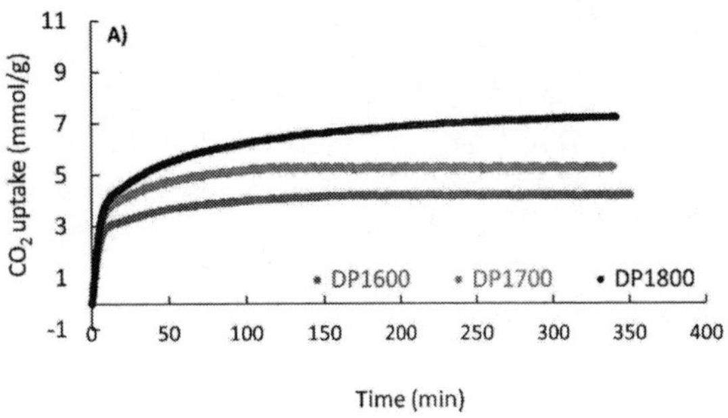

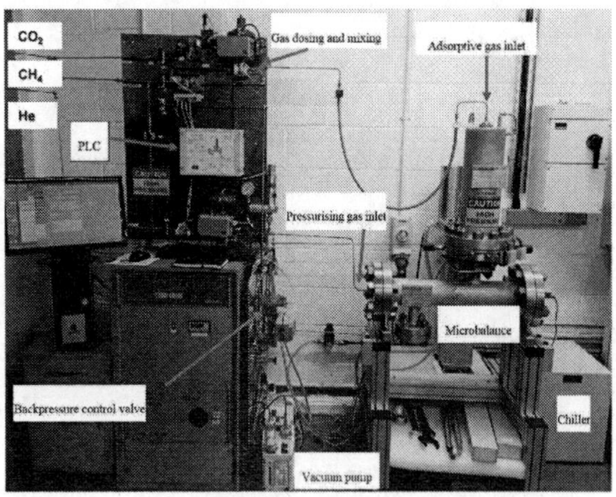

Activated Carbons Screening (2)

CO_2 Adsorption (20 bar)

CH_4 Adsorption (80 bar)

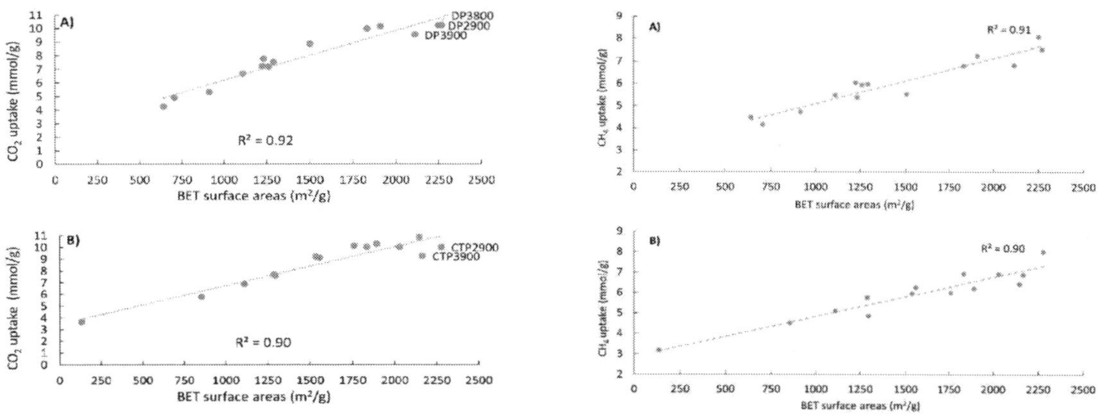

- Linearly correlates with BET surface Area.

Activated Carbons Screening (3)

CO_2/CH_4 working capacity ratios:

- Used as adsorbents screening criterial.
- Working capacity is the difference in adsorption capacity between adsorption and regeneration pressures where adsorption pressure is taken as the partial pressure of each component and regeneration pressure is atmospheric.
- Showed relation to the PSD for the high performing samples from both precursors.

Sample	CO_2 working capacity*	CH_4 working capacity**	CO_2/CH_4 working capacities ratio
DP1600	2.63	3.65	0.72
DP1700	3.43	3.73	0.92
DP1800	5.62	4.93	1.14
DP2600	4.96	4.52	1.10
DP2700	5.53	5.21	1.06
DP2800	8.11	5.77	1.41
DP2900	9.23	7.10	1.30
DP3600	7.21	4.80	1.50
DP3700	8.44	6.32	1.34
DP3800	8.94	6.64	1.35
DP3900	8.73	5.99	1.46
DP4600	6.11	4.50	1.36
DP4700	5.44	4.80	1.13
DP4800	3.46	3.10	1.12

*Measured at 20 bar (CO_2 partial pressure in South Oman natural gas).
**Measured at 80 bar (CH_4 partial pressure in South Oman natural gas).

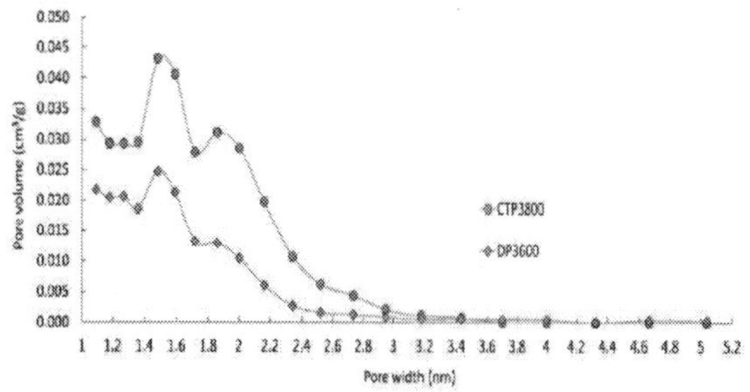

CO_2 Adsorption Isotherms

- ❑ Measured using HPVA up to 40 Bar at 25 and 50 °C
- ❑ Langmuir model poor fit due to not accommodating heterogeneous adsorption.
- ❑ SIPS and DSL showed good fit.
- ❑ Higher number of parameters for DSL (4) could have contributed to better fitting.

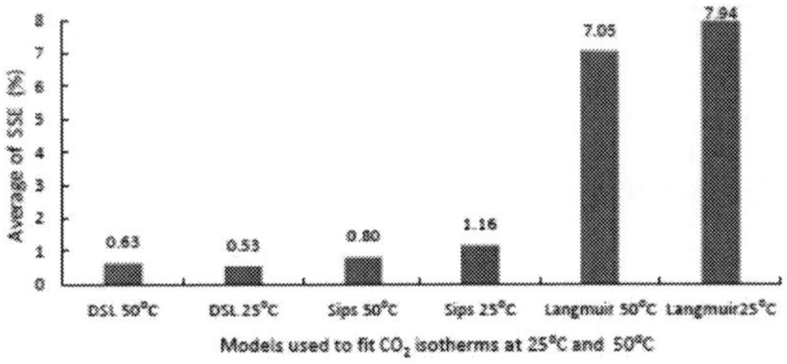

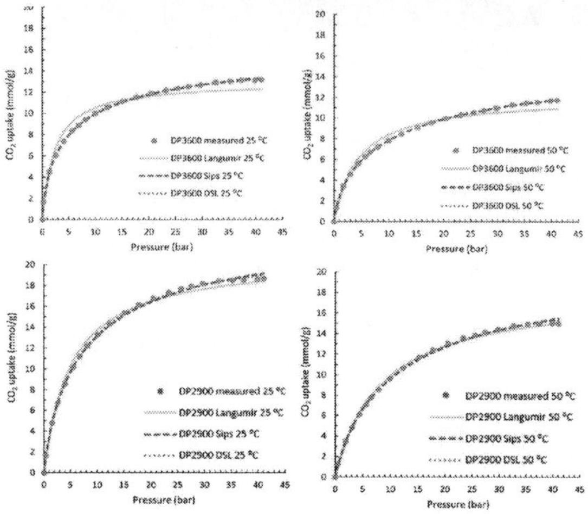

CH_4 Adsorption Isotherms

- ❑ Measured using HPVA up to 100 Bar at 25 and 50 °C
- ❑ Langmuir model poor fit due to not accommodating heterogeneous adsorption.
- ❑ SIPS and DSL showed good fit with DSL slightly better performance, higher number of model parameters for DSL (4) could have contributed to better fitting.

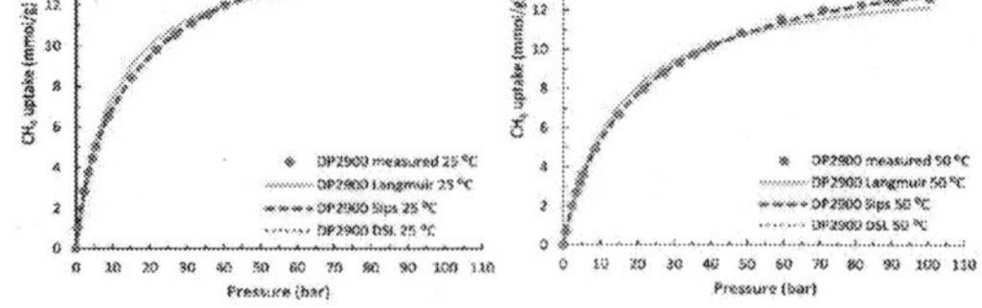

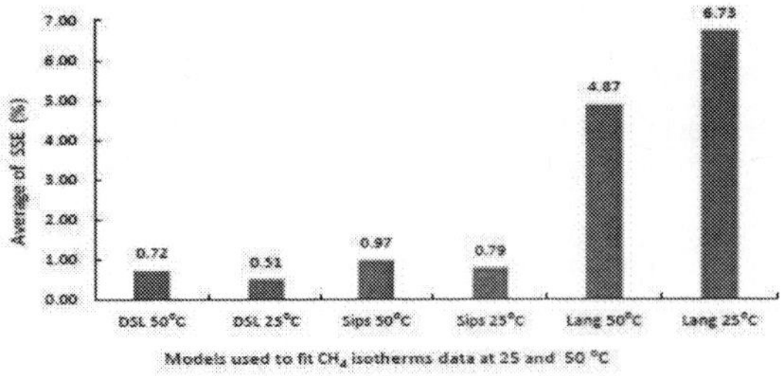

Binary Component Gas Adsorption Prediction

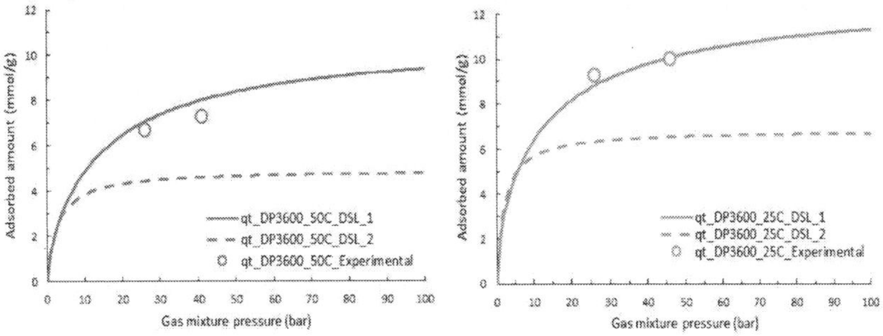

- ❑ Extended DSL model was used to predict mixed gas (CO_2 and CH_4) adsorption.
- ❑ The extended DSL model in binary gas adsorption scenario could have 2 pairs of equation (DSL1 and DSL2) to predict two values for CO_2 and two values for CH_4 based on similar or opposite gas –site interaction "site matching" first discussed by Ritter *et al.* (2011)
- ❑ Experimental HP-TGA was used to measure total mixed gas adsorption to aid the selection of the extended DSL model equation pair.

Mixed Gas prediction

- ❑ The adsorbed amount of both CO_2 and CH_4 increases with increasing each component fraction.
- ❑ High CH_4 fraction in binary (CO_2 andCH_4) gas mixture (80%) causes significant CH4 co-adsorption as per DSL model prediction

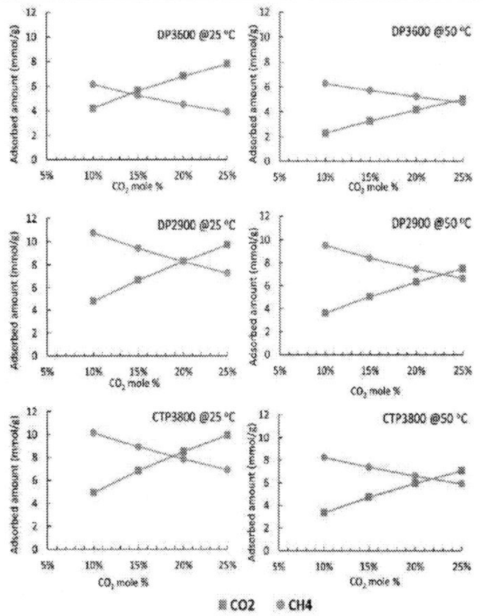

Effect of Water

- 5% of pre-adsorbed water reduces CO_2 uptake but 50% and 70% could positively impact CO_2/CH_4 selectivity on the expense of the CO2 uptake capacity.

- Water occupies some of the micropores hence reducing CO2 adsorption capacity.

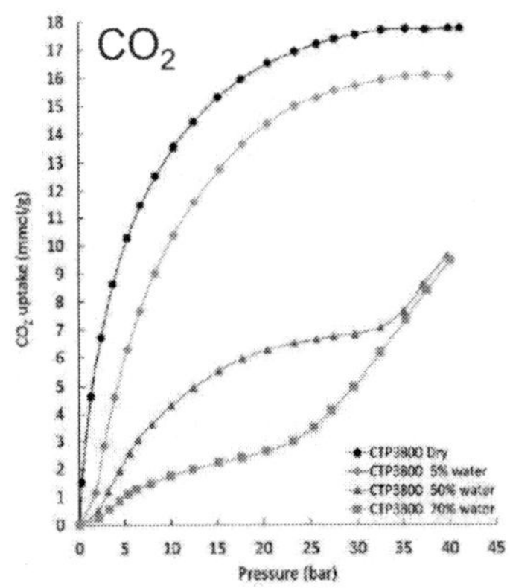

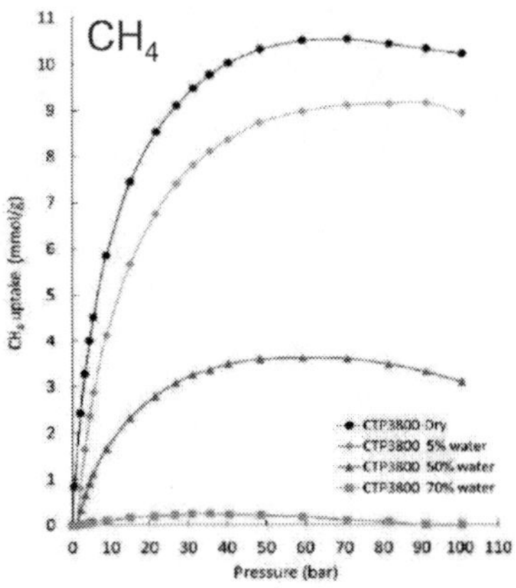

Conclusions

1. Novel Activated Carbons were prepared by carbonisation then KOH activation of date bit biomass (max S_{BET} 2269 m^2/g). DP3600 showed the best CO2/CH4 working capacity ratio
2. Preferential CO_2 adsorption over that of CH_4 demonstrated by CO_2/CH_4 working capacity ratios above 1 (measurement from HP and LP TGA) at CO_2 and CH_4 partial pressure in a typical natural gas mixture of South Oman.
3. PSD and local electrical charge on the surface of the carbon due to potassium intercalation into the carbon matrix contributed to the preferential CO2 adsorption.
4. Micropores volume and BET surface area showed good correlation with both CO2 and CH_4 adsorption at 50 °C and pressure of 20 bar for CO_2 and 80 bar for CH_4.
5. Compared to Langmuir , DSL and SIPS adsorption isotherm models showed a better fit to both CO_2 and CH_4 adsorption data due to accommodating heterogenous adsorption.
6. Mixed gas (CO_2 20% and CH_4 80%) adsorption was predicted using the extended DSL model at 100 bar and 50°C.
7. Activated Carbons have good potential for PSA application where the required purity of the rejected CO_2 is not high

Acknowledgment

شركة تنمية نفط عمان
Petroleum Development Oman

University of Nottingham
UK | CHINA | MALAYSIA

إيجاد
EJAAD

Thank You

Materials Characterisation (CTP)

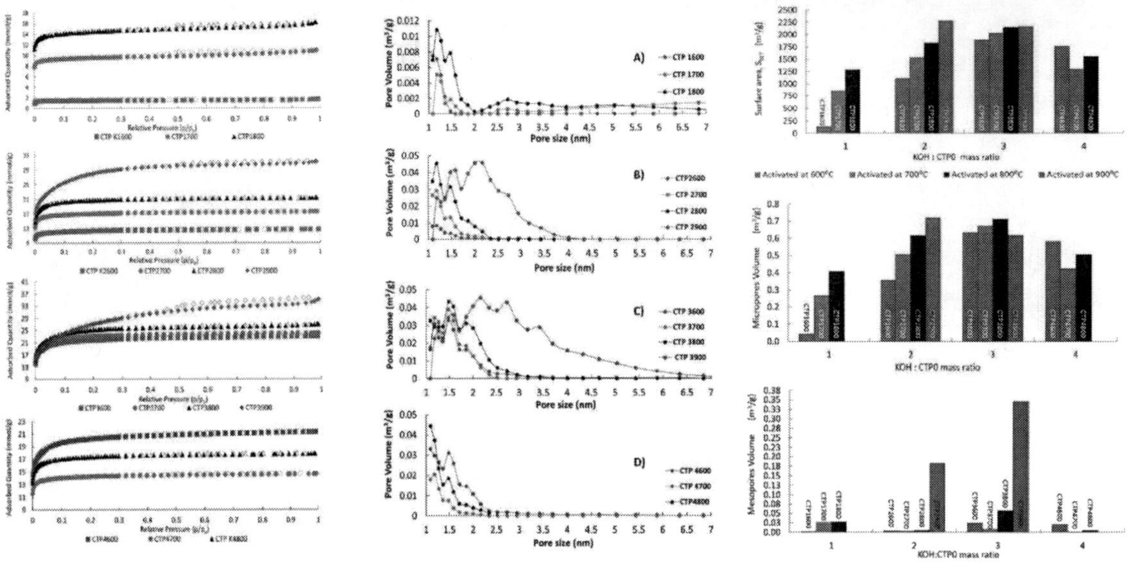

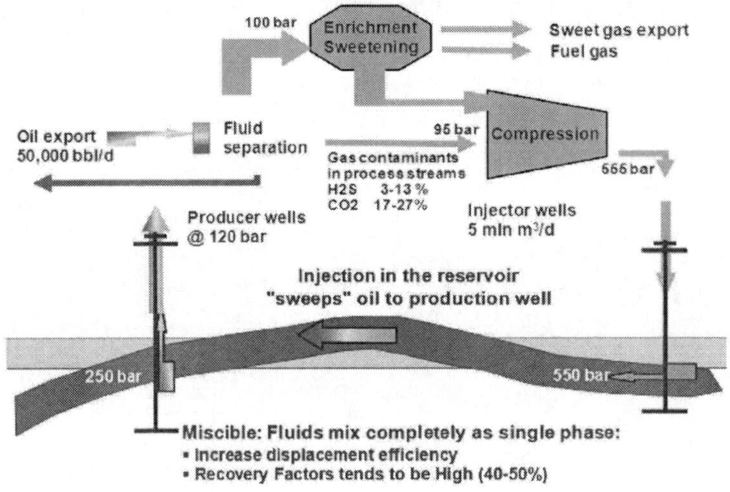

South Oman oil and gas field

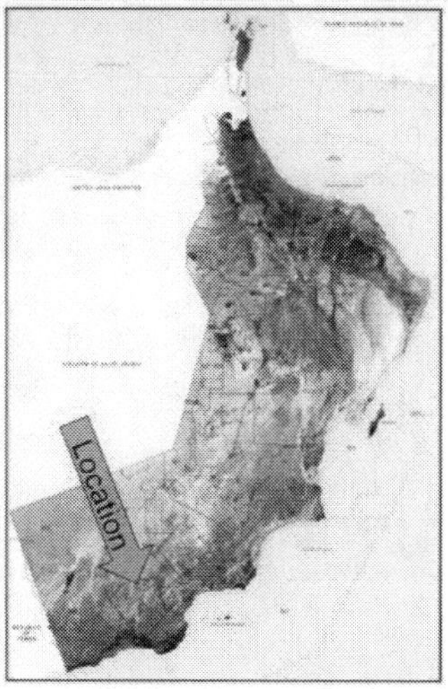

- Discovered:1997-2002
- Nearly 600 million years old, one of the oldest commercial oilfields on earth
- Key Challenges:
  - Deep: 3-5 km deep
  - High Pressures: 400 to 1000 bar
  - Sour reservoir fluids
    - H2S: 1% to 10%
    - CO_2: 1% to 25%
  - *Other non-conventional components: e.g. Mercaptans, Chlorides, Wax, COS*

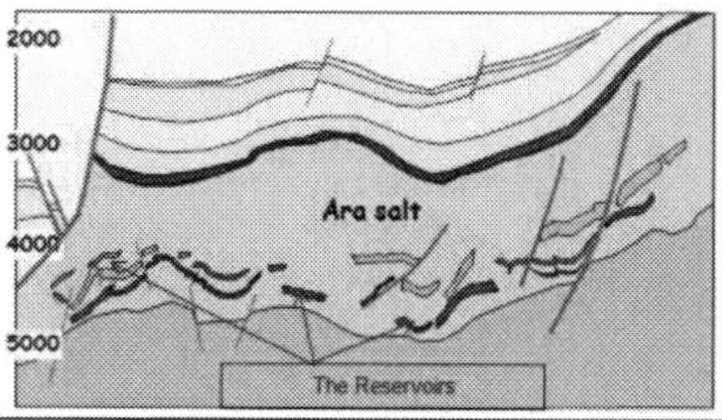

IGRC 2020 MUSCAT

24-25 FEB | WWW.IGRC2020.COM

Visualization / IoT of Industrial Equipment

Miki Yanaka

Osaka Gas Co., Ltd., Japan

m-yanaka@osakagas.co.jp

Overview of Osaka Gas Co., Ltd.

IGRC 2020 MUSCAT

Key Facts

- ✓ Headquarters in Osaka, Japan
- ✓ Started business operations in 1905
- ✓ Supplying over 8.7 million customers
- ✓ Imports about 9.5 MTPA of LNG
- ✓ Developing 11,600 GWh of power business

Gas sales Volume by Use (million m³)(2017.3)

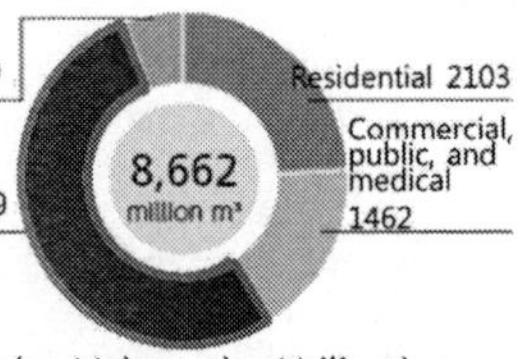

Gas Sales Volume by Utilization (Million m³)

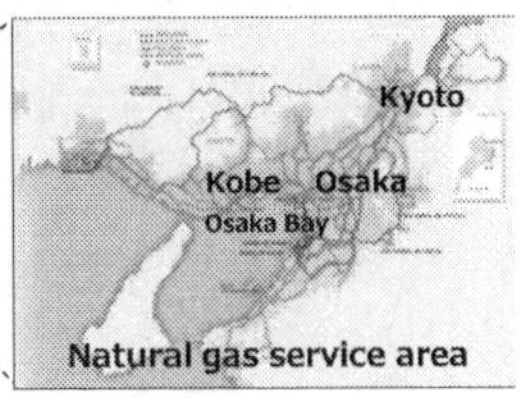

Various services for industrial customers

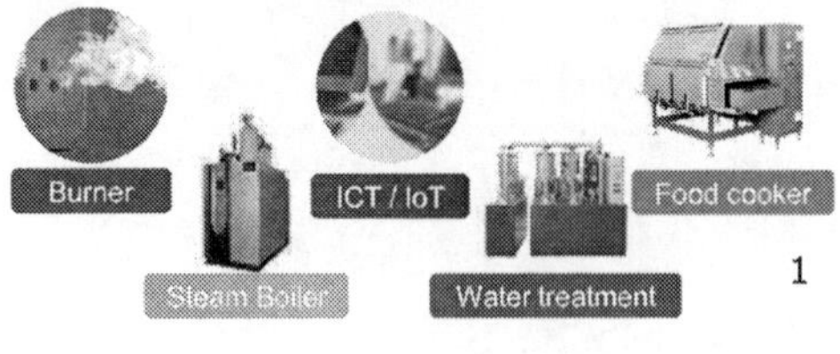

1

Summary of presentation

Development of IoT service named **D-Fire**

- ✓ For manufacturing factories such as small and medium companies
- ✓ Service characteristics
  - (1)Energy and equipment management functions
  - (2)Low price for easy introduction
- ✓ With experience in proposing energy saving and optimal operation of manufacturing factories
- ✓ Promoting introduction of IoT services to help solving issues for entire factory

2

Issues in manufacturing industry and IoT utilization

Issues in manufacturing industry

- Labor saving to cope with declining birthrate and aging population
- Preventive maintenance of equipment to stabilize operations
- Energy saving for environmental conservation and cost reduction
- Efficient operation by linking data between departments

Solved by utilizing IoT

Ideal factory operation by utilizing IoT

- Connect all equipment and systems in factory
- Optimal operation of entire factory, not each department

Issues of IoT service introduction

- Less IoT application of energy management and equipment management
- Hard to introduce without cost performance

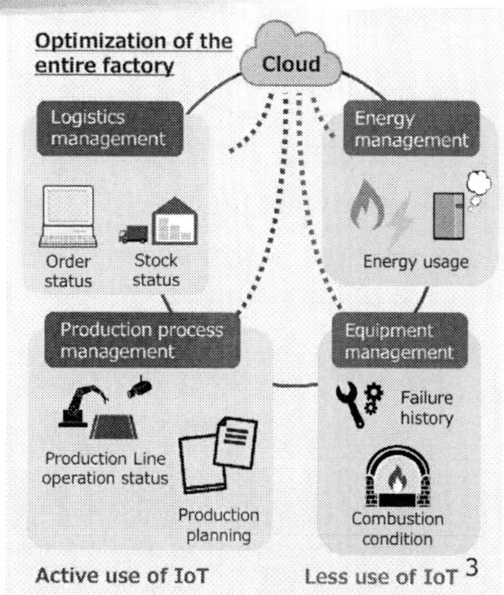

3

Why Osaka Gas is working on IoT services

Conventional

Relationship with customer

- ✓ Supplying natural gas as clean energy
- ✓ Challenge to develop various devices
- ✓ Development of high efficiency equipment
- ✓ Introducing energy saving technique

Strength of OSAKA GAS

- ✓ Professional of combustion technology
- ✓ Abundant equipment diagnosis and proposal for energy saving based on data
- ✓ Optimal operation proposing on manufacturing process

Proposal

Osaka Gas IoT services named D-Fire

- For manufacturing factories
- Including energy and equipment management functions
- Low price for easy introduction

4

Osaka Gas IoT service for manufacturing factory "D-Fire"

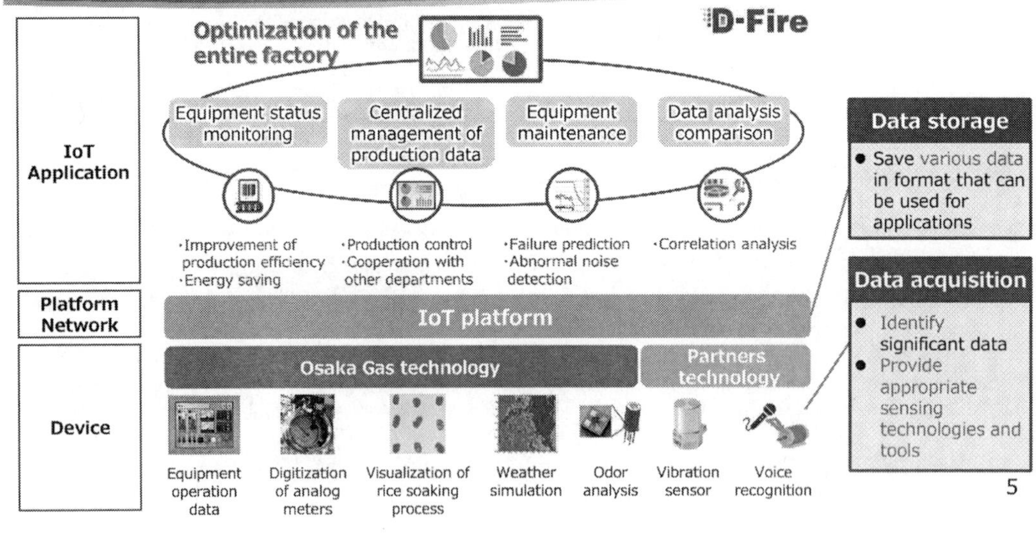

5

D-Fire original functions

IGRC 2020 MUSCAT

Issue Less IoT application of energy management and equipment management

Initiative

	Function	Effect	Industries can be deployed
Energy management	Energy intensity unit management	-Quick draft of energy saving improvement	All industries
	Defect rate trend management (Cross analysis of energy data and quality data)	-Quality improvement by reviewing processing conditions	Metal industry
	Steam load management	-Energy saving improvement -Operation improvement	Food industry Chemical industry Textile industry
Equipment management	Equipment efficiency trend management	-Early detection of aging equipment	All industries
	Operational efficiency trend management	-Energy saving improvement -Increase efficiency by reviewing production processes	All industries
	Trend management of combustion conditions	-Energy saving improvement -Prevention of sudden failure	Metal industry

6

Realization of easy introduction price

IGRC 2020 MUSCAT

Issue Hard to introduce without cost performance

Initiative

Consideration	Comparison item	Conventional service	D-Fire
Lowering cost of service itself	Service features	Original system for users	Has versatile application and can be deployed
Reducing initial investment	Data storage	User's server	Osaka Gas IoT Platform
	Contract method	Sale	Multi-year contract

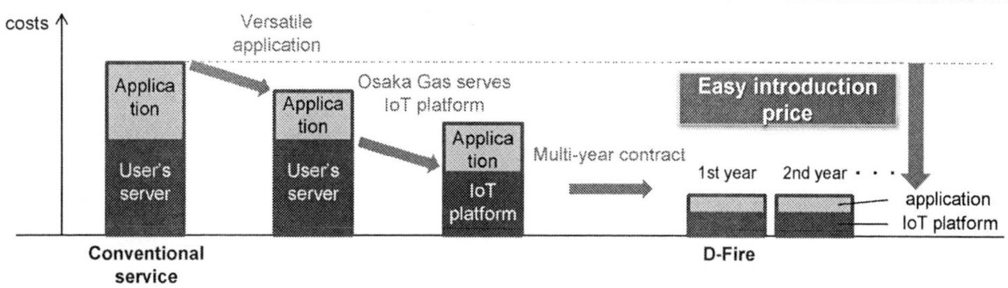

7

Introduction example 1 - Metal industry-

IGRC 2020 MUSCAT

Issue
- ✓ Data collecting and analyzing take time by hand
- ✓ Comparative analysis is not served, and energy saving PDCA cycle has not been turned
- ✓ Production suspension due to sudden failure

Initiative

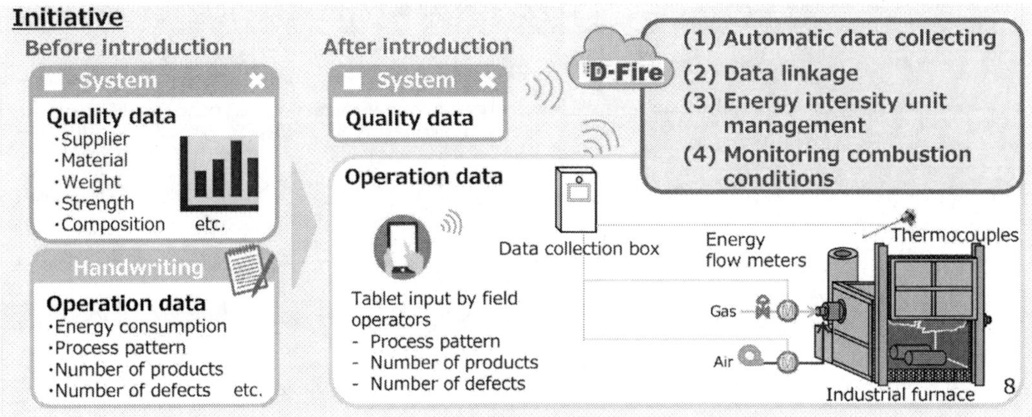

8

Introduction example 1 - Metal industry-

IGRC 2020 MUSCAT

Energy intensity unit management

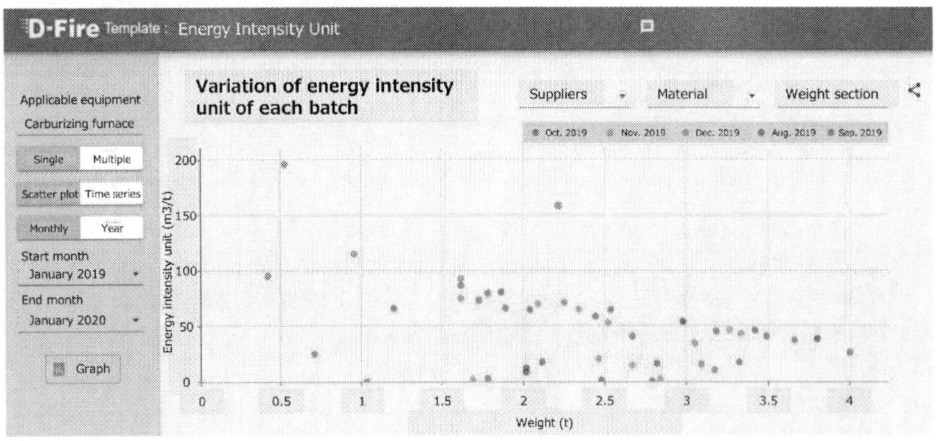

9

Introduction example 1 - Metal industry-

Energy intensity unit management

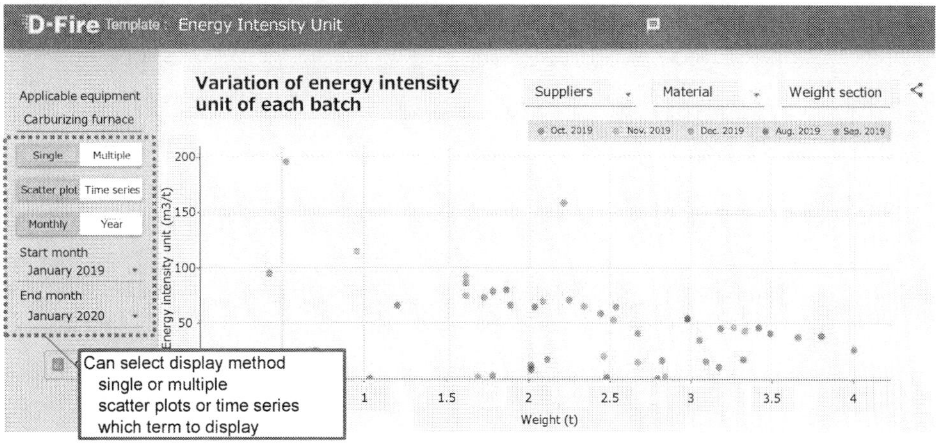

9

Introduction example 1 - Metal industry-

Energy intensity unit management

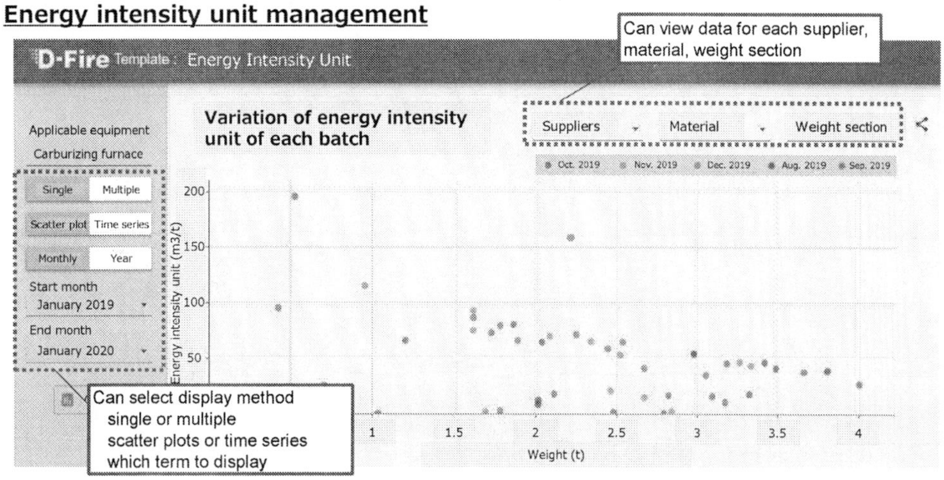

9

Introduction example 1 - Metal industry-

Energy intensity unit management

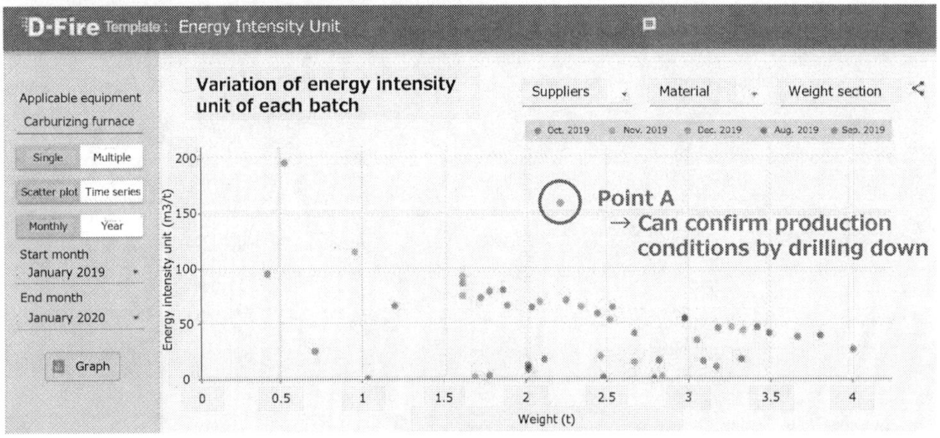

9

Introduction example 1 - Metal industry-

Operation status details

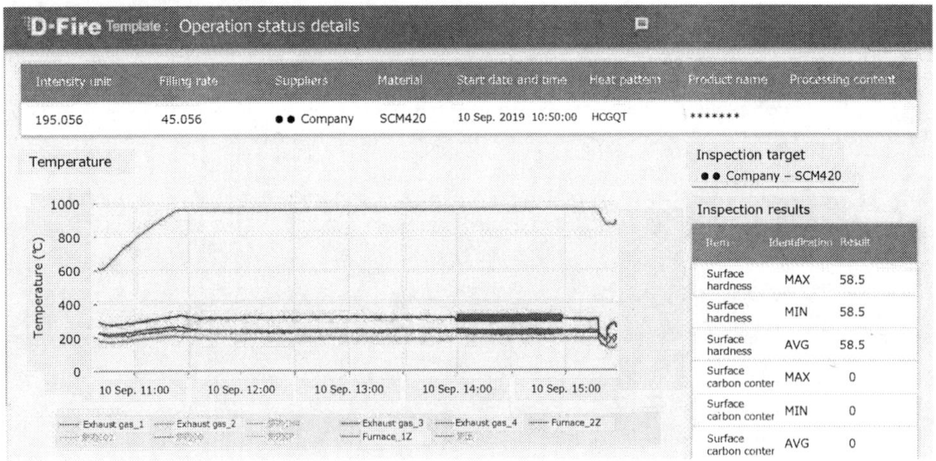

10

Introduction example 1 - Metal industry-

Operation status details

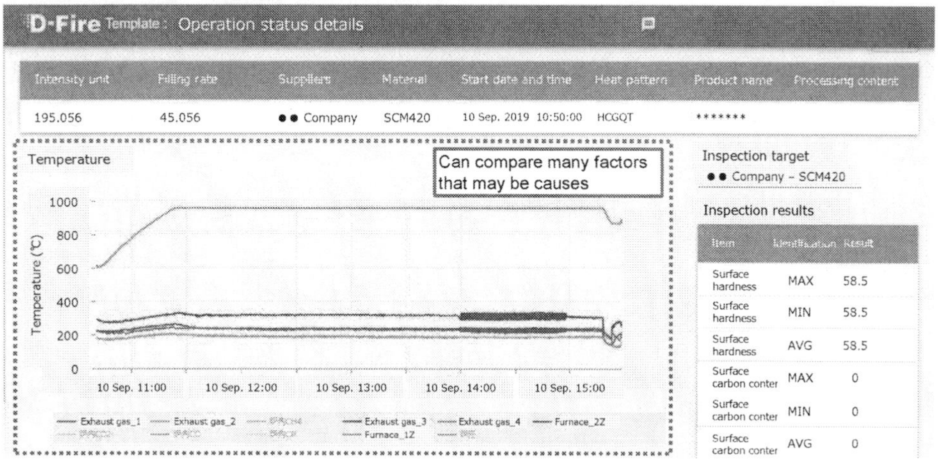

10

Introduction example 1 - Metal industry-

Operation status details

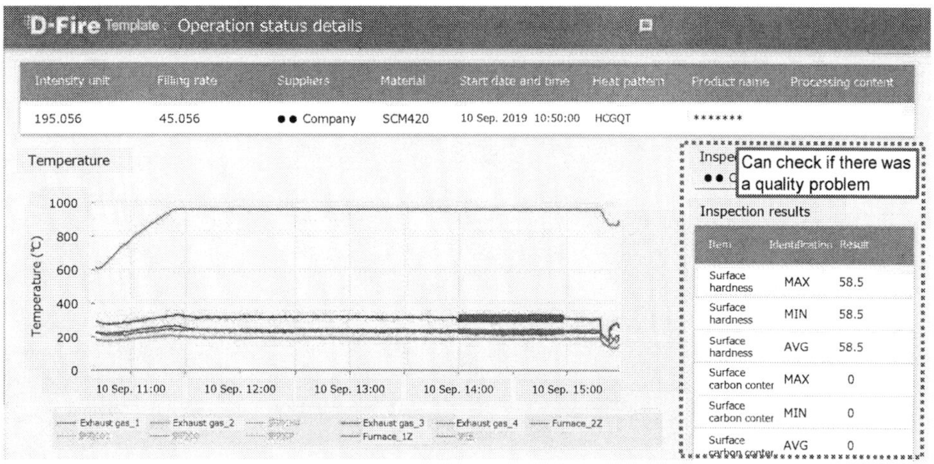

10

Introduction example 1 - Metal industry-

Operation status details

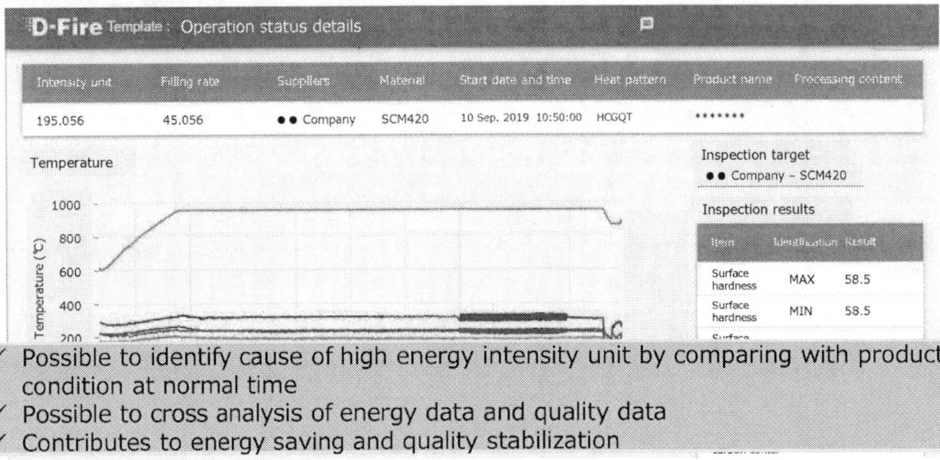

- ✓ Possible to identify cause of high energy intensity unit by comparing with production condition at normal time
- ✓ Possible to cross analysis of energy data and quality data
- ✓ Contributes to energy saving and quality stabilization

Introduction example 1 - Metal industry-

Expected effect

		Issue	Function	Effect
Common IoT service functions	Automated data collection	Takes time because data is collected by hand	-Automatic data acquisition	-Labor saving
	Data linkage	Data analysis takes time due to lack of centralized management of production data	-Linkage with existing systems	-Labor saving
D-Fire original functions	Energy management	Comparative analysis of production data for each treatment has not been performed, and energy saving PDCA cycle has not been turned	-Energy intensity unit management for each condition -Cross analysis of energy data and quality data	-Quick PDCA cycle -Energy saving -Quality stabilization
	Equipment management	Production suspension due to sudden failure	-Trend management of combustion conditions	-Reduction of lost profits -Lower maintenance costs

11

Introduction example 2 - Food industry-

IGRC 2020 MUSCAT

Issue ✓ Steam usage status is not known precisely

Initiative

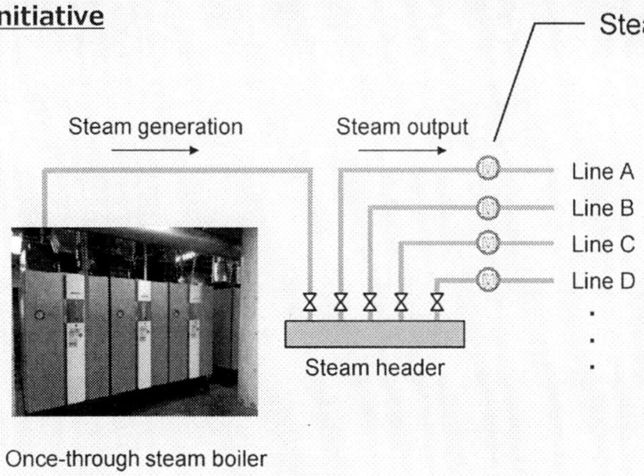

Before introduction

- Recorded by hand at once a day
- Only integrated value could be grasped

After introduction

- Monitoring in real time
- Continuous changes in steam flow rate can be grasped

12

Introduction example 2 - Food industry-

IGRC 2020 MUSCAT

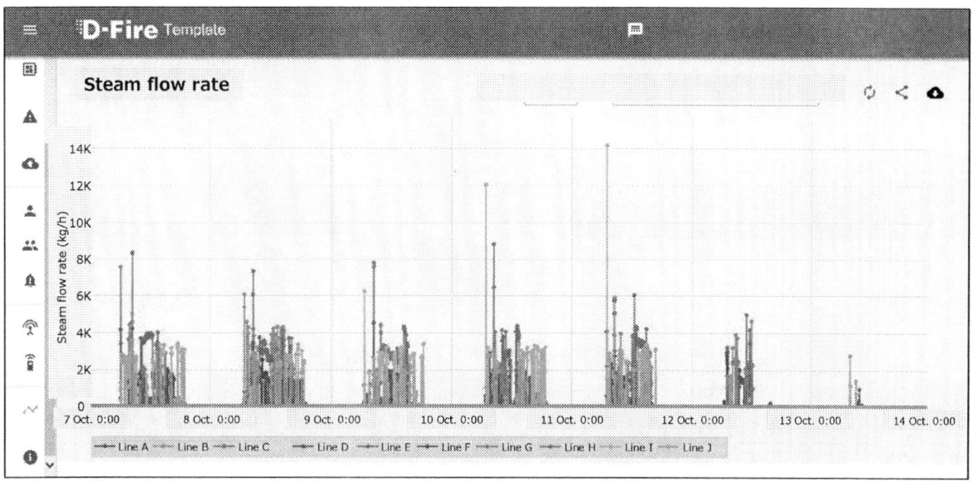

13

Introduction example 2 - Food industry-

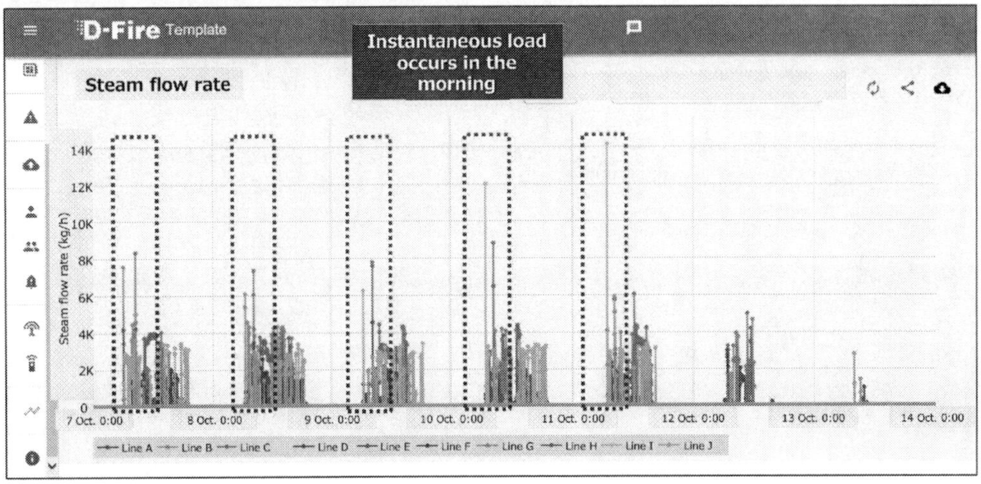

13

Introduction example 2 - Food industry-

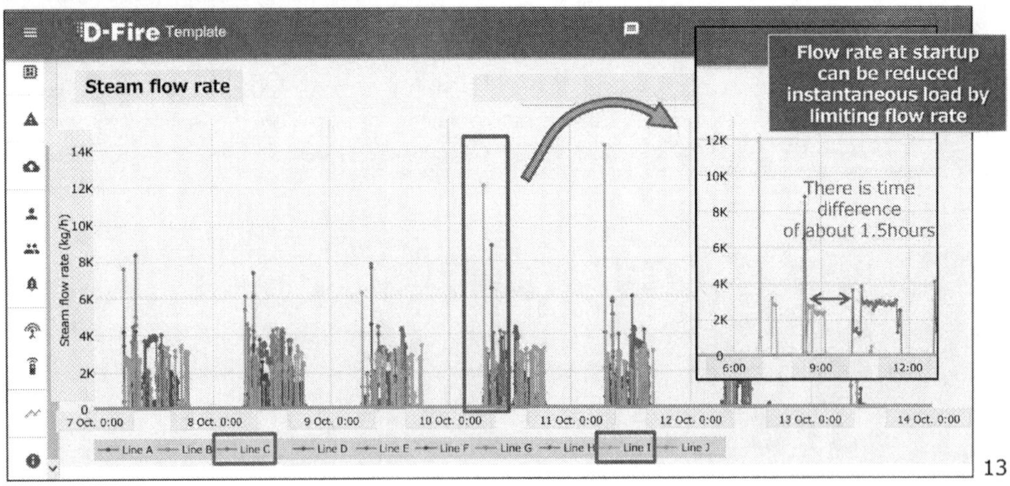

13

Introduction example 2 - Food industry-

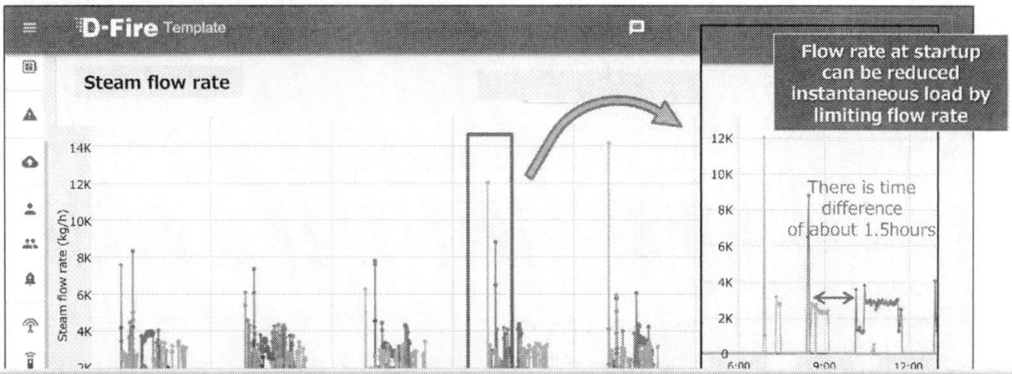

- ✓ By monitoring instantaneous values instead of integrated values, detailed steam usage can be grasped
- ✓ Possible to notice improvement point

Introduction example 2 - Food industry-

Expected effect

		Issue	Function	Effect
D-Fire original functions	Energy management	Steam usage status is not known precisely	Steam load management	-Reduction of equipment maintenance and renewal costs -Energy saving

14

Conclusion of presentation

IGRC 2020 MUSCAT

Development of IoT service named **D-Fire**

- ✓ For manufacturing factories
- ✓ Service characteristics
  - (1) Energy and equipment management functions
    - ➢ Develop various applications
  - (2) Low price for easy introduction
    - ➢ Lowering cost of service
    - ➢ Reducing initial investment
- ✓ Promoting introduction of IoT services for many manufacturing factories to help solving issues for entire factory

15

IGRC 2020 MUSCAT

THANK YOU

FOR YOUR ATTENTION

Service details of "D-Fire"

IGRC 2020 MUSCAT

Functions of platform

1. Store

- Function that can basically store anything
- Flexibly define the capture format according to the data to be stored

2. Visualization

- Function to visualize stored data in chart
- Specify the type of graph, the type of data you want to see, and the range

3. Cooperation

- Function to link data from various sensors and devices
- Release API * for coordination

* API is a rule for using the function from outside

4. Alert

- Function to send an alert email based on the value of the captured data
- By setting the threshold in advance, an alert is sent when the data is imported

Service details of "D-Fire"

IGRC 2020 MUSCAT

	Before use the platform	After use the platform
Collection	·Don't know how to get information ·Leaks occur with handwriting	·Get more information ·Can capture any information ·Record accurately and without omission
Accumu lation	·Can't accumulate because can't get it ·Remains only with analog paper ·May be incorrectly transcribed	·Automatically accumulated ·Easy to process because it remains digital ·Always record accurate values
Inflection	·Collected but not reviewed ·Can't be linked to other information ·Do not use information, rely on intuition	·Can look back the accumulated information ·Composite with other information ·Make informed decisions

Realization of easy introduction price

Consideration	Comparison item	Conventional service	D-Fire
Lowering cost of service itself	Service features	Original system for users	Has versatile application and can be deployed
Reducing initial investment	Data storage	User's server	Osaka Gas IoT Platform
	Contract method	Sale	Multi-year contract
introduction costs		**Some hundred thousand dollars**	**From 40 thousand dollars**

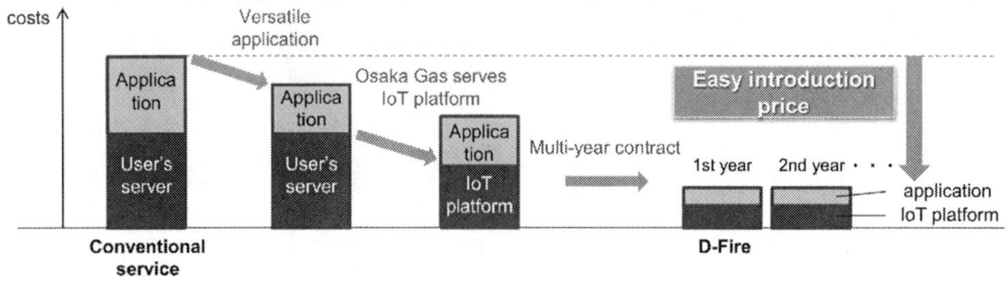

Why data storage difference reduce costs

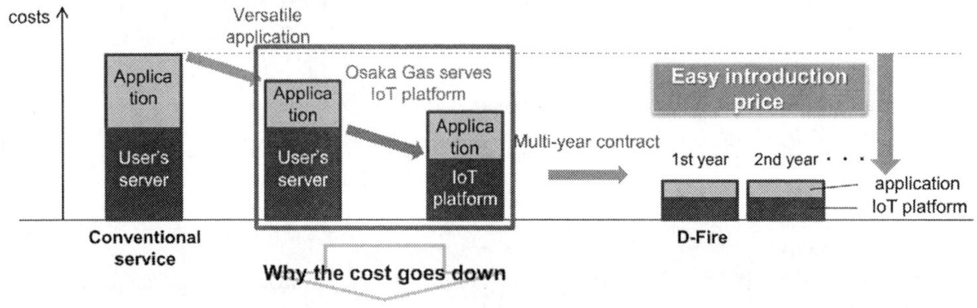

- ✓ User does not need to construct a data server with excessive capacity, and can pay Osaka Gas a platform usage price according to the required data capacity.
- ✓ It is possible to lower the price per capacity because Osaka Gas has a large capacity data storage.

Introduction example 1 - Metal industry-

		Function	Effect	Monetary value
Common IoT service functions	Automated data collection	-Automatic data acquisition	-Labor saving	$1,600/year
	Data linkage	-Linkage with existing systems	-Labor saving	$11,700 ~ 23,300/year
D-Fire original functions	Energy management	-Energy intensity unit management for each condition -Cross analysis of energy data and quality data	-Quick PDCA cycle -Energy saving -Quality stabilization	$32,900 ~ 65,900/year
	Equipment management	-Trend management of combustion conditions	-Reduction of lost profits -Lower maintenance costs	$26,500 ~ 194,000/year

Total merit : $72,700~284,800/year

Introduction example 2 - Food industry-

		Function	Effect	Monetary value
D-Fire original functions	Energy management	Steam load management	-Reduction of equipment maintenance costs	$125,000
			-Reduction of equipment renewal costs	$3,600/year
			-Energy saving	$32,000/year

Total merit : $125,000
$35,600/year

ABSORPTION OF CO_2 IN BLENDED AQUEOUS SOLUTION OF 2-AMINO-2-HYDROXYMETHYL-1, 3-PROPANEDIOL (AHPD) AND PIPERAZINE (PZ): MODELING OF CO_2 SOLUBILITY DATA USING ARTIFICIAL NEURAL NETWORK APPROACH

Dr. Ghulam Murshid
Department of Petroleum and Chemical Engineering
Sultan Qaboos University, Muscat. Oman

25th Feb 2020

IGRC 2020, Muscat, OMAN

About Sultanate of Oman....

Oman is a country situated in Southwest Asia, bordering the Arabian Sea, Gulf of Oman, and Persian Gulf, between Yemen and the United Arab Emirates with the area of 309,501 km².
Capital: Muscat

Sultan Qaboos University (SQU)

Oman's first Government University founded in 1986.
SQU is ranked at number 379 in the world's top universities ranking by the London-based Quacquarelli Symonds (QS)

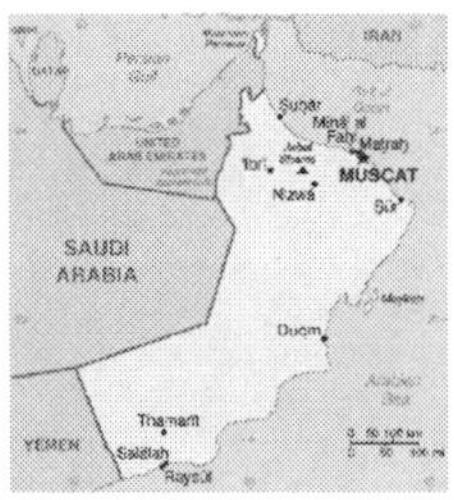

Motivation

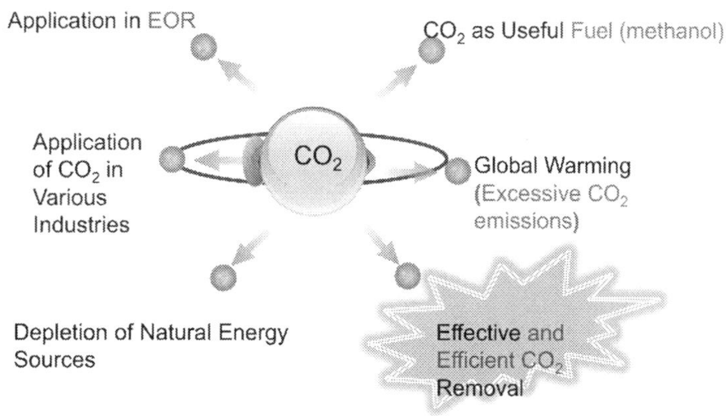

4

Sources of CO_2 Emission by Sector

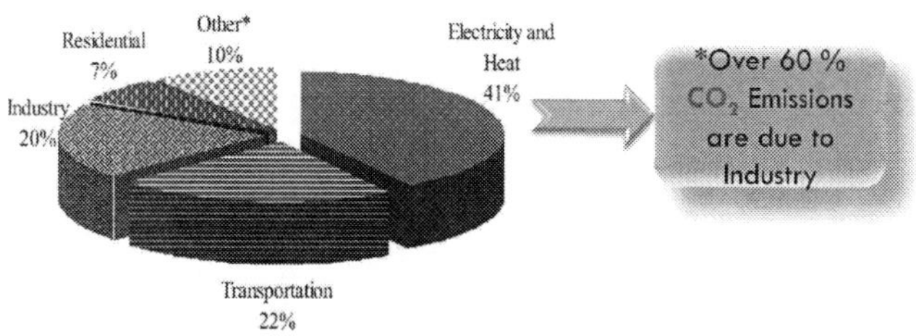

* World Energy Outlook (2019)

Contribution of CO_2 in Global Warming

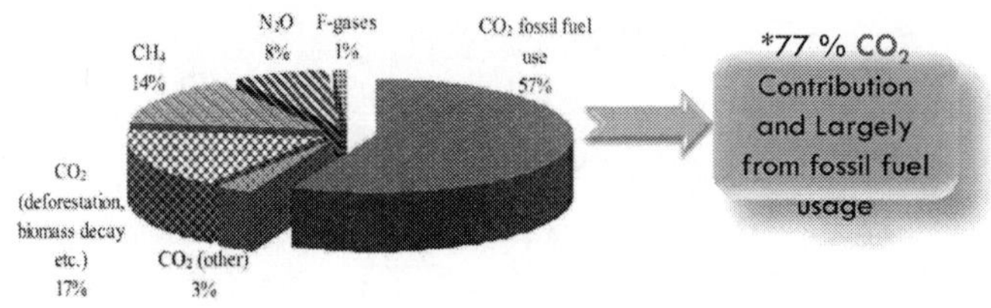

* Contribution of different green house gases to the global warming in 2004 (Source IPPC, 2007)

CO_2 Removal Technologies

- Current Technologies

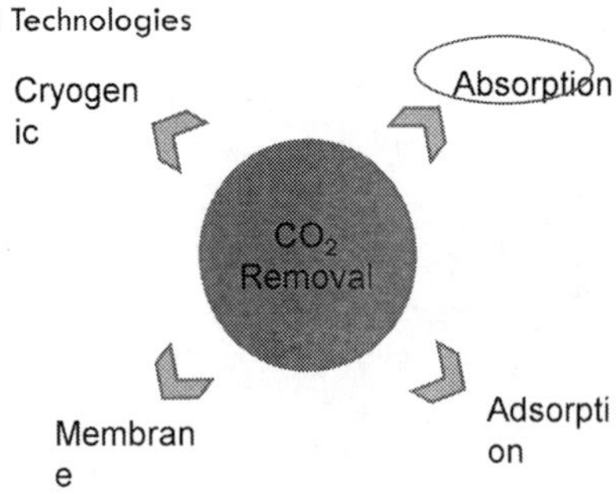

7

CO_2 Absorption Process

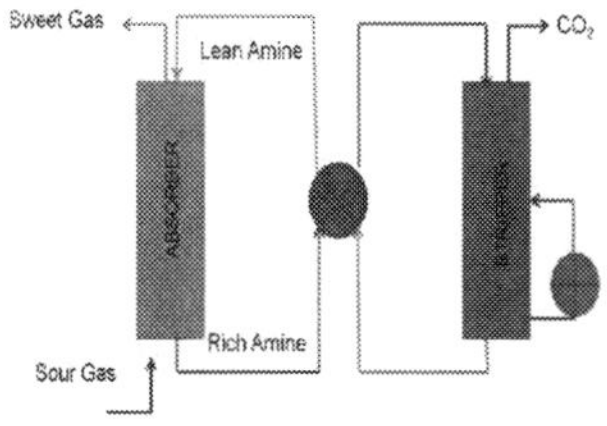

Flow Diagram of Amine Absorption Process

- Selection of Solvent
- Addition of Activator
- High CO_2 Loadings
- Higher Rate of Reaction

8

Selection of Solvents

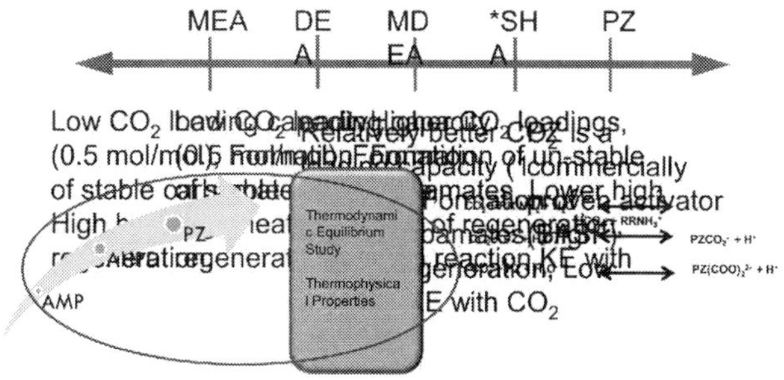

* Sterically hindered amines (SHA)

Literature Review

❑What is Steric hindrance

A SHA is defined structurally as a primary amine in which the amino group is attached to a tertiary carbon atom, or a secondary amine in which the amino group is [attached] to a secondary or tertiary carb[on]

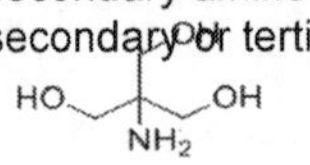

2-amino-2-hydroxymethyl-1, 3-propanediol
AHPD

Monoethanol amine
MEA

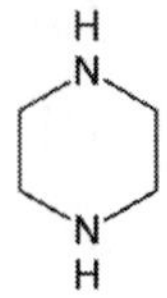

Piperazine (di-amine)

Literature Review

❑Thermodynamic Equilibrium Study

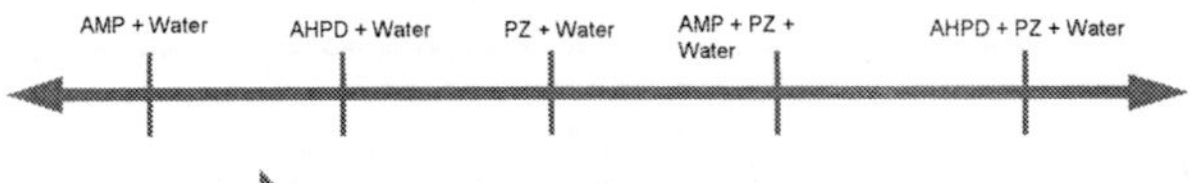

Limited to Atmospheric Pressure

11

Objectives

1. To measure the thermodynamic solubility of CO_2 in aqueous solutions of AMP, AHPD and PZ from lower to high pressure range
2. To study the effect of activator PZ on solubility of CO_2 in aqueous solutions of AMP and AHPD
3. To study the effect of temperature, pressure and concentration of solvents on the solubility of CO_2 in order to evaluate the solvent performance
4. To represent the solubility results through thermodynamic model
5. To measure the thermophysical properties including density, viscosity, surface tension, refractive index and thermal stability of solvents and to develop empirical correlations from the measured properties data.

METHODOLOGY FOR SOLUBILITY EXPERIMENTS

Experimental Setup and Methodology

13

Experimental Set Up

- High Pressure Gas Solubility Cell.
- Capable of measuring solubilities of gases and mixtures for various solvents

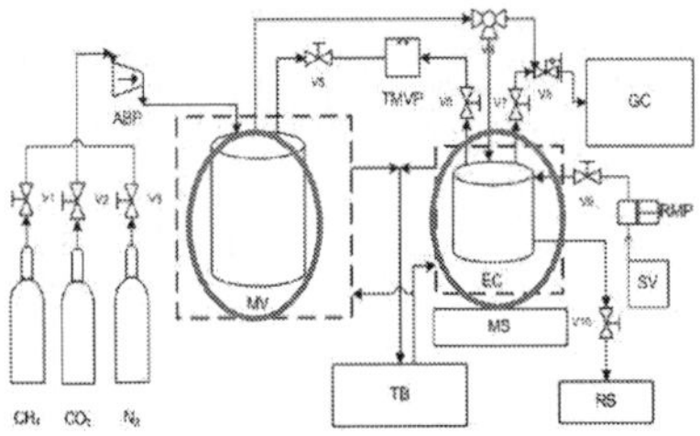

High Pressure Gas Solubility Unit

14

Experimental Set Up

Digital Picture of High Pressure Gas Solubility Unit

Methodology

- **Solubility Measurements**

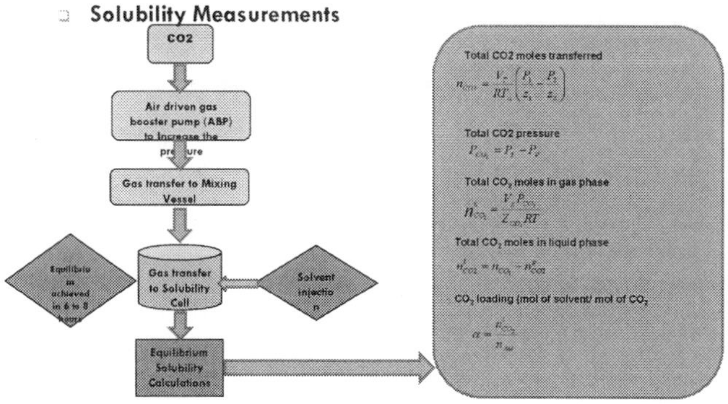

METHODOLOGY FOR MODELLING

Experimental Setup and Methodology

17

Methodology

- **Modeling**

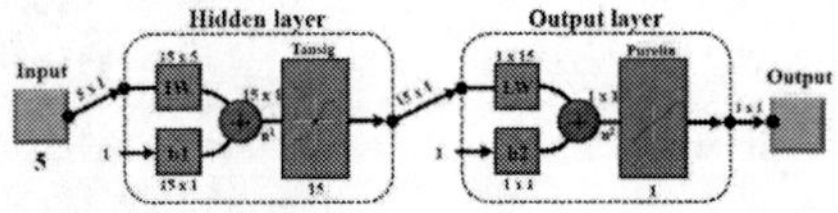

ANN architecture to predict the CO_2 solubility data in aqueous blended solutions of AHPD and PZ

Mean square error was applied to minimize the error

$$MSE = \frac{1}{N}\sum_{N=1}^{N}(x_{pred} - x_{exp})^2$$

18

Methodology

- **Modeling**

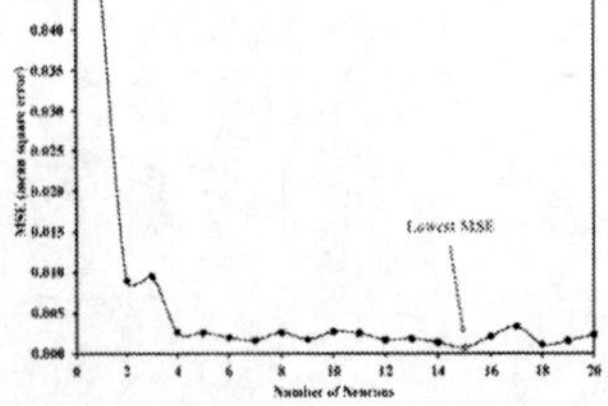

MSE vs Number of Neurons in the hidden layer

EFFECT OF ACTIVATOR (PZ) ON SOLUBILITY OF CO_2 ON AQUEOUS SOLUTIONS OF AHPD (AHPD/PZ)

20

Results and Discussion

Solubility of CO_2 in aqueous solutions of (AHPD/PZ) wt% at 303.15 K

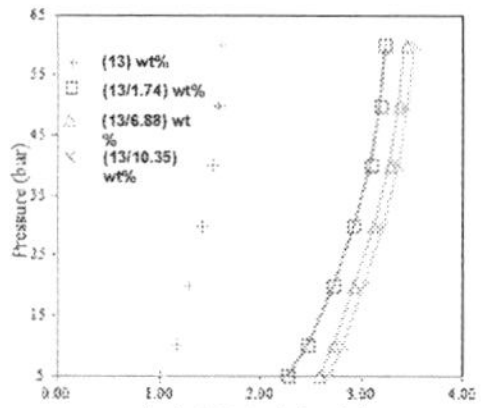

Solubility of CO_2 in aqueous solutions of (AHPD/PZ) wt% at 313.15 K

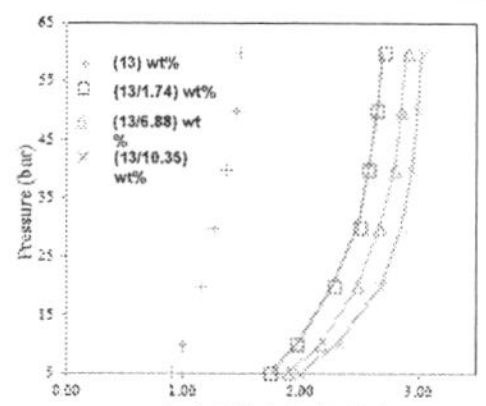

At a fixed temperature and solvent composition solubility increase with increasing pressure
At a fixed partial pressure and (AHPD/PZ) concentration, solubility decrease by increasing temperature
At a fixed AHPD concentration and temperature, solubility increase by increasing PZ concentration which means the increasing the (AHPD/PZ) ratio helps to increase the solubility

21

Results and Discussion

Solubility of CO_2 in aqueous solutions of (AHPD/PZ) wt% at 333.15 K

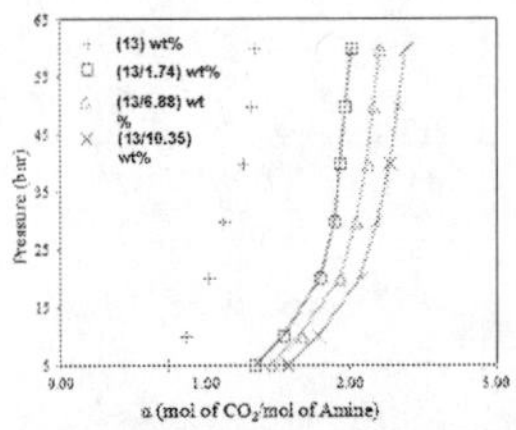

Effect of Temperature on Solubility of CO_2 in aqueous solutions of (AMP/PZ) at Various Pressures

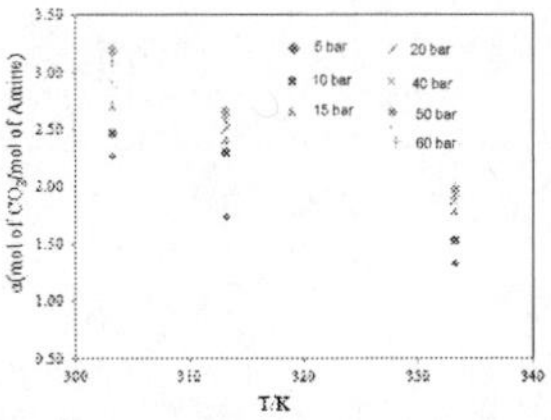

Similar approach of Modeling was adopted as for ternary solution of (AMP + PZ + water) system
The other six systems of AHPD with PZ with concentrations (13/1.74, 13/6.88, 13/10.35, 25/1.74, 25/6.88, 25/10.35) wt% show higher loadings with higher (AHPD/PZ) ratio at any fixed temperature and AHPD concentration

22

Results and Discussion

Solubility of CO_2 in aqueous solutions of (AHPD/PZ) wt% at 303.15 K

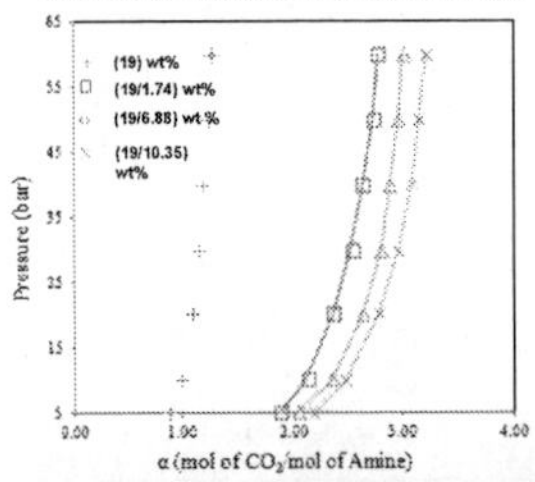

Solubility of CO_2 in aqueous solutions of (AHPD/PZ) wt% at 313.15 K

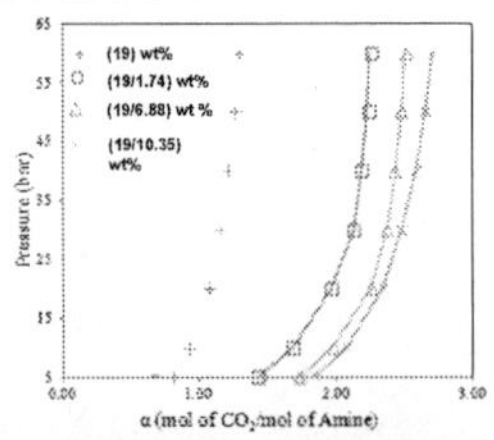

At a fixed temperature and solvent composition solubility increase with increasing pressure
At a fixed partial pressure and (AHPD/PZ) concentration, solubility decrease by increasing temperature
At a fixed AHPD concentration and temperature, solubility increase by increasing PZ concentration which means the increasing the (AHPD/PZ) ratio helps to increase the solubility

23

Results and Discussion

Solubility of CO_2 in aqueous solutions of (AHPD/PZ) wt% at 333.15 K

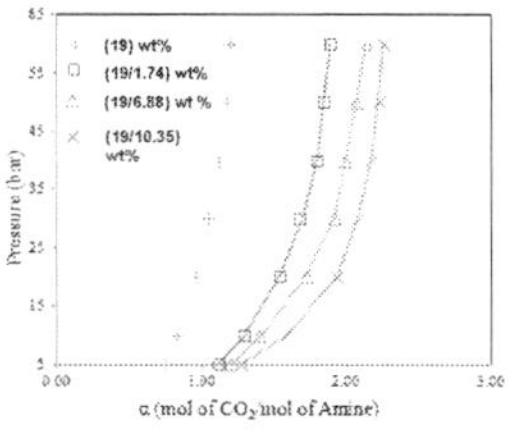

- At a fixed temperature and solvent composition solubility increase with increasing pressure
- At a fixed partial pressure and (AHPD/PZ) concentration, solubility decrease by increasing temperature
- At a fixed AHPD concentration and temperature, solubility increase by increasing PZ concentration which means the increasing the (AHPD/PZ) ratio helps to increase the solubility

24

Results and Discussion

Solubility of CO_2 in aqueous solutions of (AHPD/PZ) wt% at 303.15 K

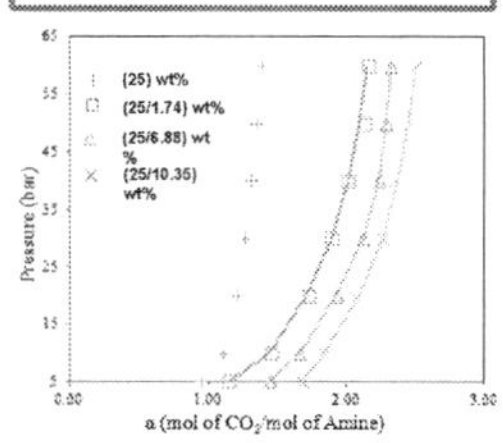

Solubility of CO_2 in aqueous solutions of (AHPD/PZ) wt% at 313.15 K

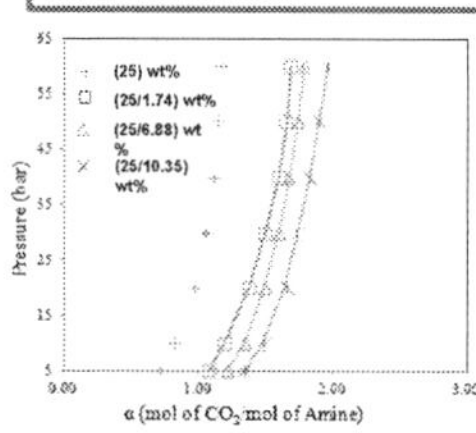

At a fixed temperature and solvent composition solubility increase with increasing pressure
At a fixed partial pressure and (AHPD/PZ) concentration, solubility decrease by increasing temperature
At a fixed AHPD concentration and temperature, solubility increase by increasing PZ concentration which means the increasing the (AHPD/PZ) ratio helps to increase the solubility

25

Results and Discussion

Solubility of CO_2 in aqueous solutions of (AHPD/PZ) wt% at 333.15 K

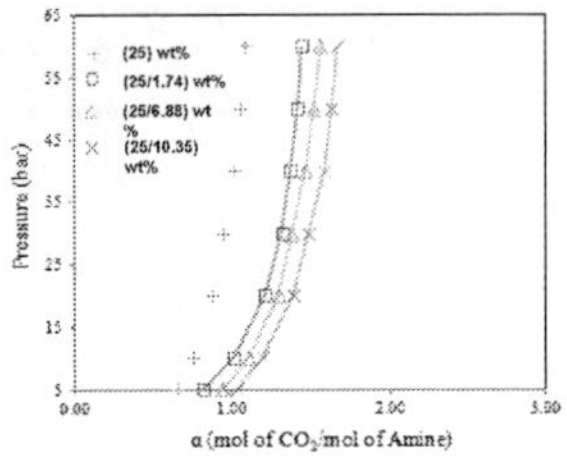

26

Results and Discussion

Comparison plot of experimental versus predicted CO2 loading data from ANN modeling

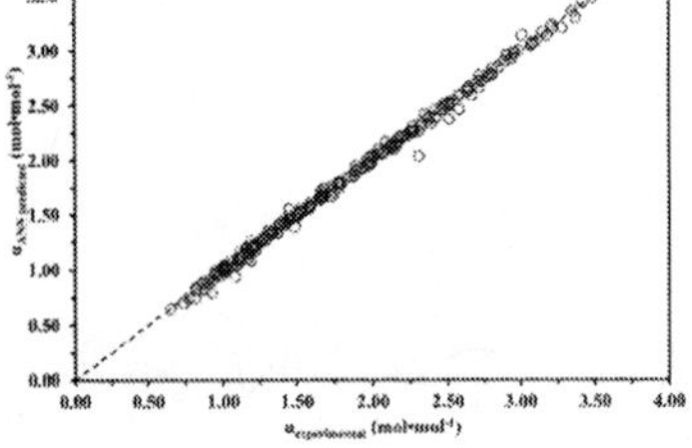

27

Conclusions

All the selected solvent mixtures perform better CO_2 loadings at high pressure conditions and offer higher CO_2 loading

Loading capacities of solvents mixtures, decrease at high temperatures due to higher vapor pressure. Therefore lower temperature and higher pressure is recommended for higher CO_2 loadings

The effect of activator PZ, is found to be positive on binary mixtures of AHPD

ANN Model fits well to the solubility data and therefore can be used to predict the CO2 loading at other operating conditions in order to design the absorber

28

References

Anderson, C.J., Tao, W., Jiang, J., Sandler, S.I., Stevens, G.W., Kentish, S.E.. 2011. An experimental evaluation and molecular simulation of high temperature gas adsorption on nanoporous carbon. Carbon 49, 117-125.

Aronu, U.E., Svendsen, H.F., Hoff, K.A.. 2010. Investigation of amine amino acid salts for carbon dioxide absorption. Int. J. Greenhouse Gas Control 4, 771-775.

Bishnoi, S., Rochelle, G.T., 2000. Absorption of carbon dioxide into aqueous piperazine: reaction kinetics, mass transfer and solubility. Chem. Eng. Sci. 55, 5531-5543.

Bougie, F., Iliuta, M.C., 2009a. CO_2 Absorption into Mixed Aqueous Solutions of 2-Amino-2-hydroxymethyl-1,3-propanediol and Piperazine. Ind. Eng. Chem. Res. 49, 1150-1159.

Bougie, F., Iliuta, M.C., 2009b. Kinetics of absorption of carbon dioxide into aqueous solutions of 2-amino-2-hydroxymethyl-1,3-propanediol. Chem. Eng. Sci. 64, 153-162.

Bougie, F., Lauzon-Gauthier, J., Iliuta, M.C., 2009. Acceleration of the reaction of carbon dioxide into aqueous 2-amino-2-hydroxymethyl-1,3-propanediol solutions by piperazine addition. Chem. Eng. Sci. 64, 2011-2019.

Chen, C.-C., Evans, L.B., 1986. A local composition model for the excess Gibbs energy of aqueous electrolyte systems. AIChE J. 32, 444-454.

Danckwerts, P.V., McNeil, K.M., 1967. The absorption of carbon dioxide into aqueous amine solutions and effects of catalysis. Transactions of the institution of chemical engineers and the chemical engineer 45, T32-T49.

Daneshvar, N., Zaafarani Moattar, M.T., Abedinzadegan Abdi, M., Aber, S., 2004. Carbon dioxide equilibrium absorption in the multi-component systems of CO_2 + TIPA + MEA + H_2O, CO_2 + TIPA + PZ + H_2O and CO_2 + TIPA + H_2O at low CO_2 partial pressures: experimental solubility data, corrosion study and modeling with artificial neural network. Separation and Purification Technology 37, 135-147.

Derks, P.W.J., 2006. Carbon dioxide absorption in piperazine activated N-methyldiethanolamine. University of Twente, p. 244.

Deshmukh, R.D., Mather, A.E., 1981. A mathematical model for equilibrium solubility of hydrogen sulfide and carbon dioxide in aqueous alkanolamine solutions. Chem. Eng. Sci. 36, 355-362.

Fausett, L., 1994. Fundamentals of neural networks: architectures, algorithms, and applications. Prentice-Hall, Inc.

Garg, S., Shariff, A.M., Shaikh, M.S., Lal, B., Aftab, A., Faiqa, N., 2016a. Measurement and prediction of physical properties of aqueous sodium salt of L-phenylalanine. J. Serb. Chem. Soc. 81.

Garg, S., Shariff, A.M., Shaikh, M.S., Lal, B., Aftab, A., Faiqa, N., 2016b. Selected physical properties of aqueous potassium salt of l-phenylalanine as a solvent for CO_2 capture. Chem. Eng. Res. Des. 113, 169-181.

Garg, S., Shariff, A.M., Shaikh, M.S., Lal, B., Aftab, A., Faiqa, N., 2016c. VLE of CO_2 in aqueous potassium salt of L-phenylalanine: Experimental data and modeling using modified Kent-Eisenberg model. J. Nat. Gas Sci. Eng. 34, 864-872.

29

References

Harris, F., Kurnia, K.A., Mutalib, M.I.A., Thanapalan. M., 2009. Solubilities of Carbon Dioxide and Densities of Aqueous Sodium Glycinate Solutions before and after CO_2 Absorption. J. Chem. Eng. Data 54, 144-147.

Heaton, J., 2008. Introduction to neural networks with Java. Heaton Research, Inc.

Holst, J.v., Kersten, S.R.A., Hogendoorn, K.J.A., 2008. Physiochemical Properties of Several Aqueous Potassium Amino Acid Salts. J. Chem. Eng. Data 53, 1286-1291.

Hook, R.J., 1997. An Investigation of Some Sterically Hindered Amines as Potential Carbon Dioxide Scrubbing Compounds. Ind. Eng. Chem. Res. 36, 1779-1790.

Hosseini Jenab, M., Abedinzadegan Abdi, M., Najibi, S.H., Vahidi, M., Matin. N.S., 2005. Solubility of Carbon Dioxide in Aqueous Mixtures of N-Methyldiethanolamine + Piperazine + Sulfolane. J. Chem. Eng. Data 50, 583-586.

Hu, W., Chakma, A., 1990. Modelling of equilibrium solubility of CO_2 and H_2S in aqueous amino methyl propanol (AMP) solutions. Chem. Eng. Commun. 94, 53-61.

Hussain, T., 2009. Checking the integrity of Global Positioning Recommended Minimum (GPRMC) sentences using Artificial Neural Network (ANN). Department of Technology and Built Environment. University of Gävle, p. 61

Jane, I.S., Li, M.-H., 1997. Solubilities of Mixtures of Carbon Dioxide and Hydrogen Sulfide in Water + Diethanolamine + 2-Amino-2-methyl-1-propanol. J. Chem. Eng. Data 42, 98-105.

Kent, R.L., Eisenberg, B., 1976. Better data for amine treating. Hydrocarbon process 55, 87-90.

Kohl, A.L., Nielsen, R., 1997. Gas purification. Gulf Professional Publishing.

Li, M.-H., Chang, B.-C., 1994. Solubilities of Carbon Dioxide in Water + Monoethanolamine + 2-Amino-2-methyl-1-propanol. J. Chem. Eng Data 39, 448-452.

Li, Y.-G., Mather, A.E., 1998. Correlation and Prediction of the Solubility of CO_2 and H_2S in an Aqueous Solution of 2-Piperidineethanol and Sulfolane. Ind. Eng. Chem. Res. 37, 3098-3104.

Lu, Wang. Sun, Li, Liu, 2005. Absorption of CO_2 into Aqueous Solutions of Methyldiethanolamine and Activated Methyldiethanolamine from a Gas Mixture in a Hollow Fiber Contactor. Ind. Eng. Chem. Res. 44, 9230-9238.

Lu, J.-G., Gao, L., Wang, J.-T., Lu, Z.-Y., Wang, C.-X., Wu, Q., Su, M., Liang, T.-Y., Zhao, Y.-X., 2015. Prediction and validation of physical property for a CO_2 capture agent of aqueous (potassium citrate + 2-amino-2-methyl-1-propanol). J. Nat. Gas Sci. Eng. 27, Part 3, 1574-1584.

Majchrowicz, M.E., Brilman, D.W.F., 2012. Solubility of CO_2 in aqueous potassium l-prolinate solutions—absorber conditions. Chem. Eng. Sci. 72. 35-44.

Mandal, B.P., Biswas, A.K., Bandyopadhyay, S.S., 2003. Absorption of carbon dioxide into aqueous blends of 2-amino-2-methyl-1-propanol and diethanolamine. Chem. Eng. Sci. 58, 4137-4144.

30

THANK YOU

COMTHERM s.r.o. & HELORO s.r.o.

INTEKOM Group of Companies

Use of „ Renewable Heat“ to increase „Intelligent 3-GenerationTechnology´s“ efficiency.

IGU IGRC 2020 MUSCAT

Imrich DISCANTINY /Jozef KONCZER / Tomas POTASCH / Ernest SZABO

COMTHERM s.r.o. & HELORO s.r.o.

INTEKOM Group of Companies

Transforming waste heat into „renewable, green heat“

Simultaneous energy production from paralell heat sources with optimized Absorption Chiller operation

IGU IGRC 2020 MUSCAT

Imrich DISCANTINY /Jozef KONCZER / Tomas POTASCH / Ernest SZABO

COMTHERM s.r.o. & HELORO s.r.o.

INTEKOM Group of Companies

„Intelligent technology“strategy:

- **Creating partnership between gas and renewable energy**
- **Making gas greener by lower CO_2 production**

IGU IGRC 2020 MUSCAT Imrich DISCANTINY /Jozef KONCZER / Tomas POTASCH / Ernest SZABO

COM-therm´s Power Plant Production / Specifications

City of KOMÁRNO / South Slovakia – 30 000 residents

- 30 km – total lenght of heat network for :
- **51,100 MWhr/year** – heat production & distribution for :
- 10.000 apartments + schools + hospitals + administrative buildings
- **10,700 MWhr/year** - power production
- **8,000 MWhr/year** – cooling production

Problem :

- The demands for production and delivery of cooling output grows constantly
- **for efficient air-conditioning of:**
  - technological spaces / server rooms, schwitchboards
  - residental buildings,etc.

Result :

The current worldwide electric power & heat production has a negative impact on the environment by :

- emissions
- leakage of forms of waste /condensation heat and **low potential heat** from various heat sources

Aero-thermical Energy – Renewables : heat emissions from surrounded facilities, geothermal well, solar collectors

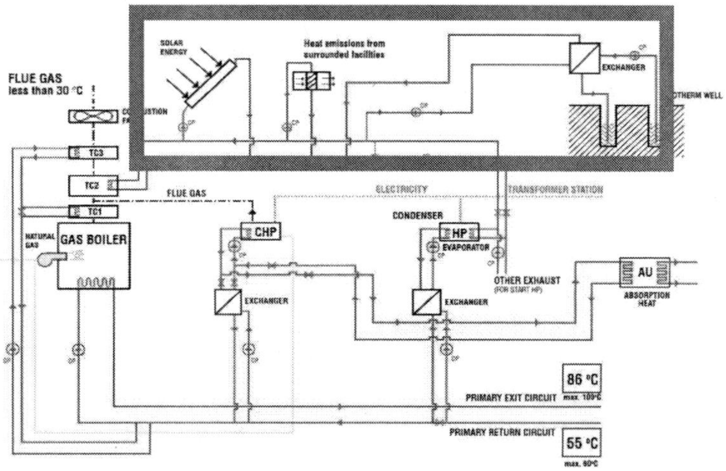

Thermal Condensators/Heat Exchangers Set

Flue Gases T≤ 30°C

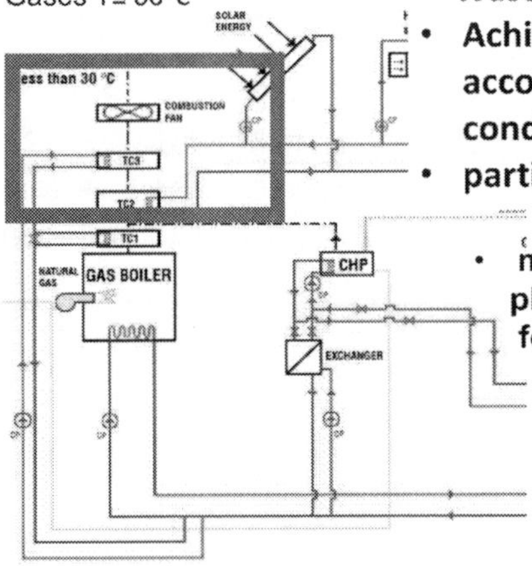

Waste Heat Recovery Technology :

- **Achieved higher degree of flue gases cooling is accompained with high level of water vapour condensation**
- **partial drying of the flue gases**

- **makes for central heating & power plants optimal physical conditions for : - essential reduction of Nox generated emisson volumes**
  - CO2 separation technologies

Intelligent 3-Generation Boiler Room Connection Scheme
Solar Energy / Water Storage Tank

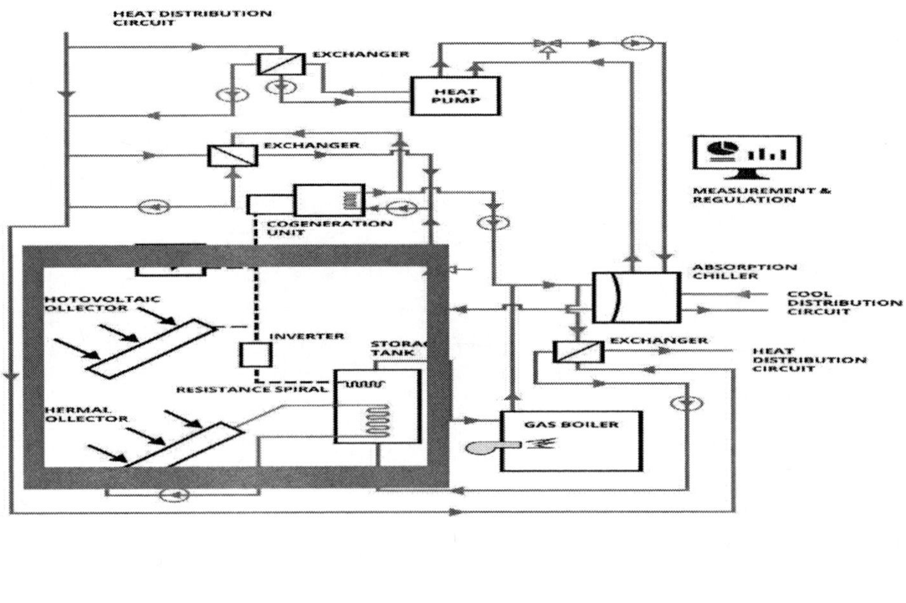

Current deployment / shortcomings of 3-generation operations

absorption chiller unit which :

- transforms thermal energy between different thermal levels in 3 heat exchange circuits :
  - hot circulating water from the CHP unit output
  - cooling water » removes the heat released by internal exchange » performed in cooling towers » results in heat loss
  - cold water circuit to the consumer

Current deployment / shortcomings of 3-generation operations

- CHP units work with temperature gradient of hot circulating water 90/70°C
- Production of cold : - with standard compressor cooling » energy inefficient, additional means / space requirements

  - absorption chiller unit

The technical solution of the new method of cooling – addressing the shortcomings of existing systems

AU cooling water circuit (30 ° C) is connected to the HP input and to the return line of the heating system (55°C):

- ensures an optimal gradient of 25/30 ° C
- uses low-potential waste heat from the AU´s circulation & secondary line » supports system´s heat output

Heat & storage water tank- connected to hot water circuit:

- enables **simultaneous** cooling/heating from multiple heat sources
  which so far was not known but required in practice
- stabilizes temperature conditions during various system fluctuations
- absorbs heat flow from other sources (solar energy)

BENEFITS of "INTELLIGENT 3-GENERATION" technology

- by connecting of heat pump to heat & cold production, all generated low-potential heat sources are used, marked as "Renewable" resp. "Green" heat"

Result: elimiated / reduced cooling towers installation

Power Plant´s 2019 Total Renewable Heat Production

/ generated by Heat Pumps section/: **10.7 MWhr/year**

Renewable Heat Production Gain : **20.9 %**

COP average achieved : **3.95**

Usability of 3 - generation technology

- Central heating sources (Heat treatment plants)
- Waste water treatment plants
- Industry facilities with constant heat/cold consumption (e.g. glass works, iron works, etc.).
- Wellness, swimming pools, geothermal springs
- Schools, hospitals, hotels, office buildings, institutions, etc.
- The technology is designed for areas with favorable / partially favorable climatic conditions

„ RENEWABLE GREEN HEAT " definition

- means low-potential thermal energy, obtained as a by-product from several energy sources of ***the energy sector / other industries*** *:*

- the waste heat from the flue gases of heat sources

- from the circulation line / secondary circuit of Absorbtion Chiller Unit -

- aerothermal energy of the surrounding outdoor/ internal air created by the operation of technology processes, which :

- *has so far been considered unusable under industrial conditions and has therefore been inefficiently escaped in various forms into the air with negative consequences for the environment of cities, municipalities and region*

„ RENEWABLE GREEN HEAT ” definition

- *If this low-potential energy is not used, unnecessary additional costs for its removal in the energy and industry production processes arise*
- *by defining of “renewable heat” was created a new, parallel category of RES research to address global environmental protection and climate change issues.*

Thank you for your kind attention !

Autonomous Integrity Recognition (AIR)

Haye-Jan Tholen
Product Owner AIR

Definitions & cautionary note

The companies in which Royal Dutch Shell plc directly and indirectly owns investments are separate legal entities. In this presentation "Shell", "Shell group" and "Royal Dutch Shell" are sometimes used for convenience where references are made to Royal Dutch Shell plc and its subsidiaries in general. Likewise, the words "we", "us" and "our" are also used to refer to Royal Dutch Shell plc and its subsidiaries in general or to those who work for them. These terms are also used where no useful purpose is served by identifying the particular entity or entities. "Subsidiaries", "Shell subsidiaries" and "Shell companies" as used in this presentation refer to entities over which Royal Dutch Shell plc either directly or indirectly has control. Entities and unincorporated arrangements over which Shell has joint control are generally referred to as "joint ventures" and "joint operations", respectively. Entities over which Shell has significant influence but neither control nor joint control are referred to as "associates". The term "Shell interest" is used for convenience to indicate the direct and/or indirect ownership interest held by Shell in an entity or unincorporated joint arrangement, after exclusion of all third-party interest.

This presentation contains forward-looking statements (within the meaning of the U.S. Private Securities Litigation Reform Act of 1995) concerning the financial condition, results of operations and businesses of Royal Dutch Shell. All statements other than statements of historical fact are, or may be deemed to be, forward-looking statements. Forward-looking statements are statements of future expectations that are based on management's current expectations and assumptions and involve known and unknown risks and uncertainties that could cause actual results, performance or events to differ materially from those expressed or implied in these statements. Forward-looking statements include, among other things, statements concerning the potential exposure of Royal Dutch Shell to market risks and statements expressing management's expectations, beliefs, estimates, forecasts, projections and assumptions. These forward-looking statements are identified by their use of terms and phrases such as "aim", "ambition", "anticipate", "believe", "could", "estimate", "expect", "goals", "intend", "may", "objectives", "outlook", "plan", "probably", "project", "risks", "schedule", "seek", "should", "target", "will" and similar terms and phrases. There are a number of factors that could affect the future operations of Royal Dutch Shell and could cause those results to differ materially from those expressed in the forward-looking statements included in this presentation, including (without limitation): (a) price fluctuations in crude oil and natural gas; (b) changes in demand for Shell's products; (c) currency fluctuations; (d) drilling and production results; (e) reserves estimates; (f) loss of market share and industry competition; (g) environmental and physical risks; (h) risks associated with the identification of suitable potential acquisition properties and targets, and successful negotiation and completion of such transactions; (i) the risk of doing business in developing countries and countries subject to international sanctions; (j) legislative, fiscal and regulatory developments including regulatory measures addressing climate change; (k) economic and financial market conditions in various countries and regions; (l) political risks, including the risks of expropriation and renegotiation of the terms of contracts with governmental entities, delays or advancements in the approval of projects and delays in the reimbursement for shared costs; and (m) changes in trading conditions. No assurance is provided that future dividend payments will match or exceed previous dividend payments. All forward-looking statements contained in this presentation are expressly qualified in their entirety by the cautionary statements contained or referred to in this section. Readers should not place undue reliance on forward-looking statements. Additional risk factors that may affect future results are contained in Royal Dutch Shell's Form 20-F for the year ended December 31, 2018 (available at www.shell.com/investor and www.sec.gov). These risk factors also expressly qualify all forward looking statements contained in this presentation and should be considered by the reader. Each forward-looking statement speaks only as of the date of this presentation, 25 February 2020. Neither Royal Dutch Shell plc nor any of its subsidiaries undertake any obligation to publicly update or revise any forward-looking statement as a result of new information, future events or other information. In light of these risks, results could differ materially from those stated, implied or inferred from the forward-looking statements contained in this presentation.

We may have used certain terms, such as resources, in this presentation that the United States Securities and Exchange Commission (SEC) strictly prohibits us from including in our filings with the SEC. U.S. Investors are urged to consider closely the disclosure in our Form 20-F, File No 1-32575, available on the SEC website www.sec.gov.

Introduction

HAYE THOLEN

- 10 years + in the gas industry
- Experience in process technology, production management
- Passionate digital product owner of AIR
- Based in The Netherlands

TEAM

Data Scientists	4
Data Engineering	1
Full Stack Developers	2
Inspectors	4
Scrum Master	1

3

AIR

WHY

Pernis Refinery estimated equipment

>150,000,000 meter piping

>10,000,000 flanges

We inspect to prevent

- Leakages → Safety & environment
- Production losses → $$
- Maintenance cost → $

Business drivers

- Consistent & standardized visual inspections
- Reduce HSE exposure, accelerate robotics
- Improved maintenance planning

WHAT

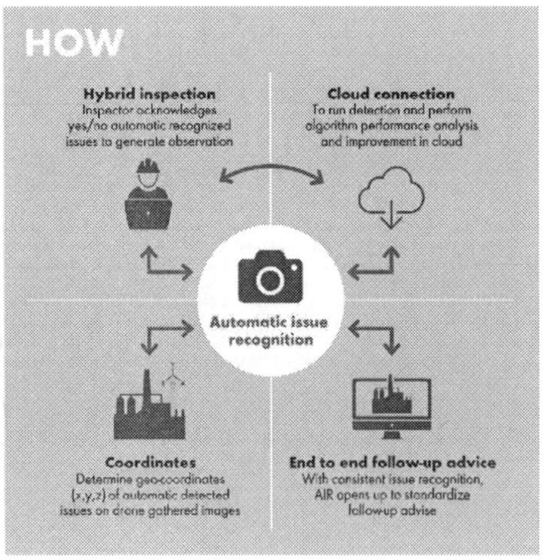

4

WHAT CAN AIR DO

Now

- Detect and classify a percentage of the following integrity issues on images send to the AIR cloud application;
  1. Corrosion on piping
  2. Corrosion studs and bolts
  3. Corrosion structural steel
  4. Insulation matrix exposed
  5. Insulation open seams
  6. Insulation damaged & open
  7. Insulation missing
  8. Insulation blanket
  9. Missing studs and bolts
- Distinguish Medium corrosion vs Light corrosion
- Integrated with existing integrity management software

WHAT CAN AIR CAN'T DO

- AIR can only visually detect Corrosion Under Insulation risk areas, not corrosion under insulation itself
- AIR analyzes visual data, generated by digital sensors. The integrity issues that can be detected are limited to a percentage of those that are captured on a digital image on a digital camera.
- AIR algorithm inference is currently build on the cloud and needs online connectivity (WiFi, 4G) to work. Currently, we cannot provide offline detection and classification.

Vision

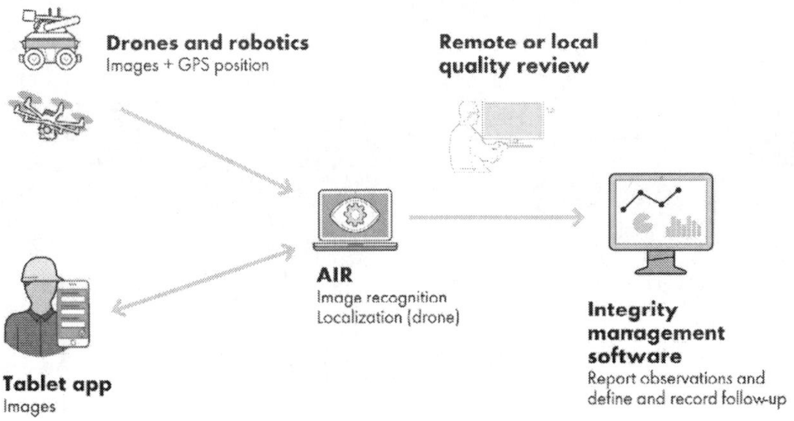

AIR has to ambition to improve the quality, consistency and efficiency of visual external inspections.

Based on image data, we have built a detection algorithm and integrated it with existing integrity management software.

AIR Roadmap

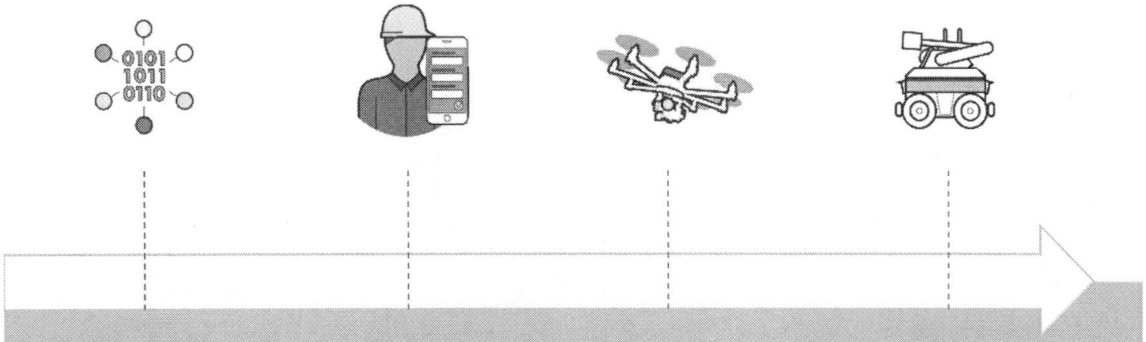

Machine Vision "Cycle" of AIR

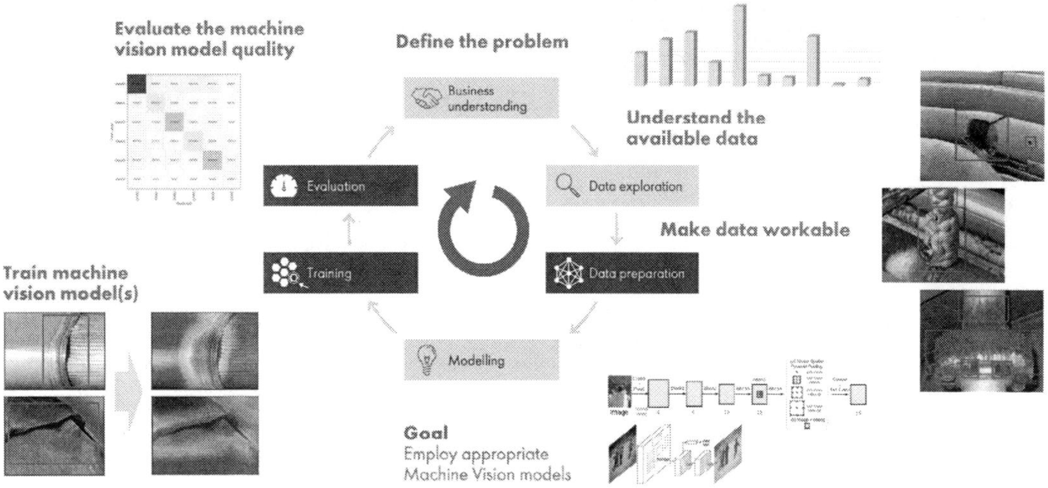

9

AIR annotation engine

We use different machine vision techniques to create labelled data

10

Way of working

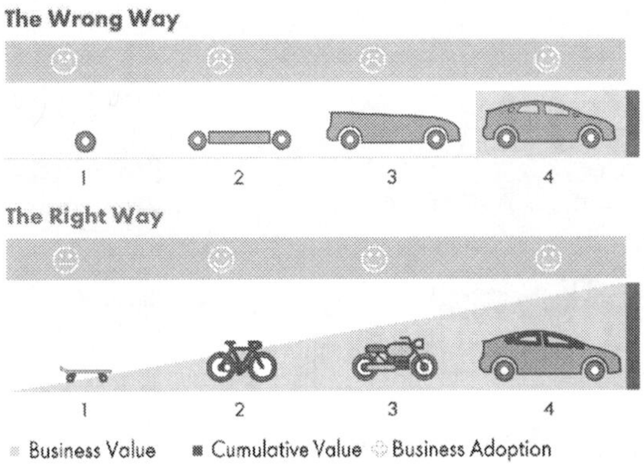

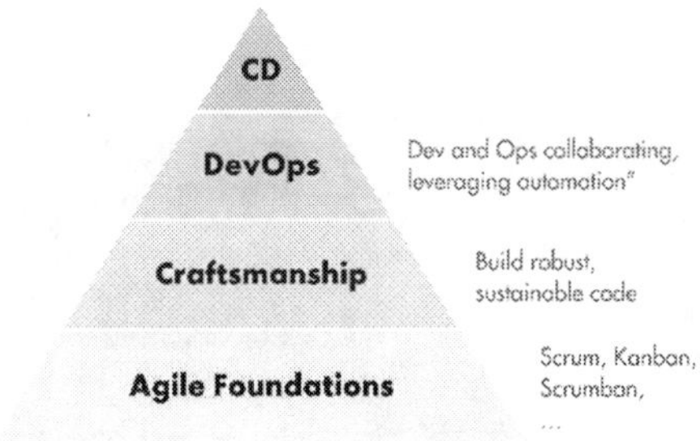

MVP is a product with **just enough features** to gather **validated learning** about the product and its continued development.

 31

Integration of Gas and Electricity with Residential Fuel Cell

Kentaro Ito
February, 2020
Residential Fuel Cell Business Development Dept.
TOKYO GAS

1. Product Concept of *ENE-FARM*
2. Sales Details
3. Product Line-up and *ENE-FARM mini*
4. Going Forward
5. Summary

1

ENE·FARM

What is *ENE-FARM*?

- *ENE-FARM* is a residential fuel cell micro-CHP (Combined Heat and Power) system that provides home with electricity, hot water, and heat with high energy efficiency.
- *ENE-FARM* enables power generation at home and covers 70% of power demand of a typical household.

The Product name *ENE-FARM* is a coined word for energy and farm.

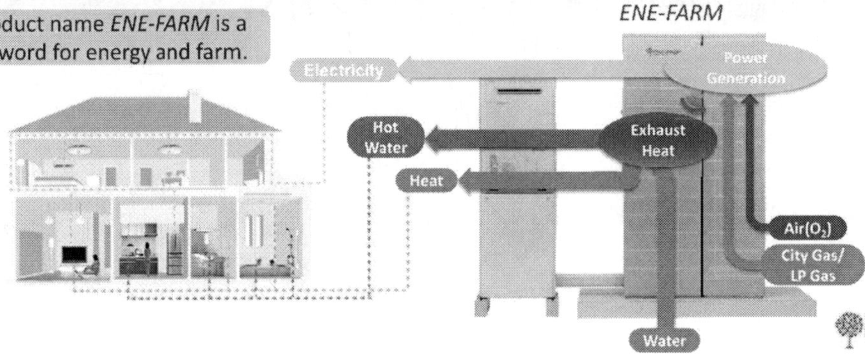

How does it work?

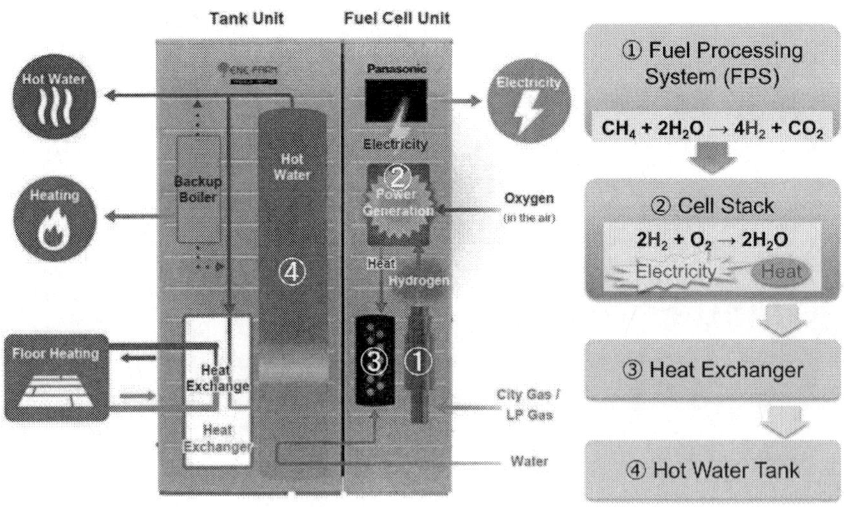

Benefits of *ENE-FARM*

TOKYO GAS

For Society

- Lowers carbon emission (reduces by 1.2 ton-CO_2/ year) *1

For Customers

- Lowers total energy bill (reduces by USD 600/ year) *1

*1: In case of a family of four in 120m² house.

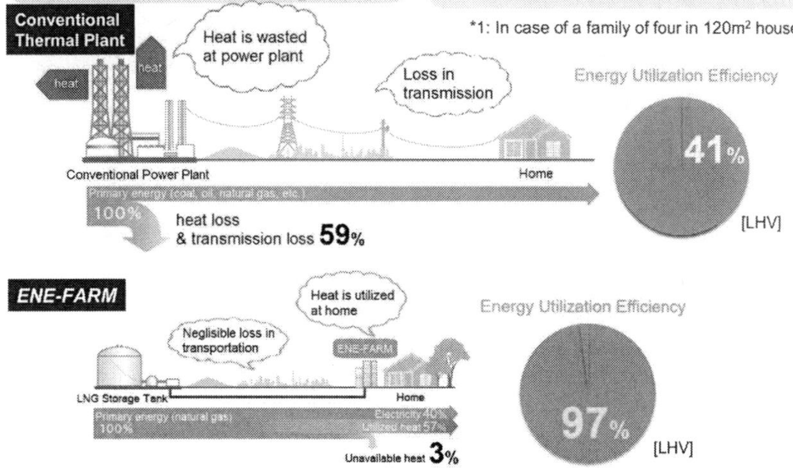

4

ENE·FARM

Continuous Power Generation during blackout

TOKYO GAS

- *ENE-FARM* can provide electricity, hot water, and heat even in case of power outages.
- **<u>In recent years, a number of typhoons caused large-scale power outages in Japan.</u>**
- **<u>The importance of preparation for blackout has been drawing more attention.</u>**

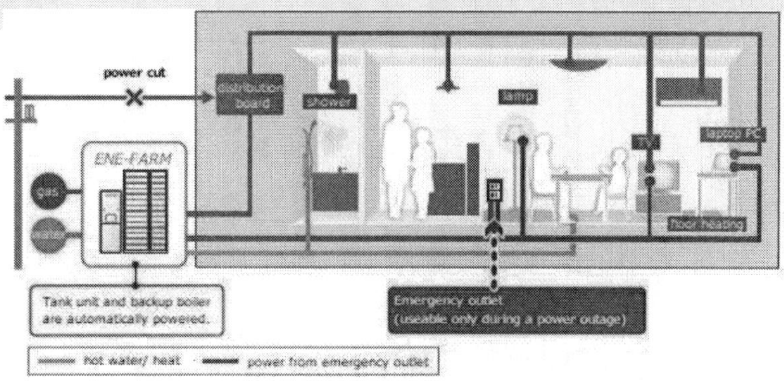

5

NE·FARM

The Purpose of the *ENE-FARM* Development

TOKYO GAS

Contribution to energy/environmental policies

ENE-FARM is a residential distributed power generation (CHP) system that contributes to energy conservation.
(The Japanese government plans to achieve 5.3 million *ENE-FARMs* by 2030.)

ENE-FARM is expected to responds to a social request to realize energy security and a low-carbon society.

Customers' merits

Reduce Expenses for electricity and gas by USD 600 a year.

ENE-FARM covers 70% of a typical household electricity consumption.

Reduce CO_2 emissions by 1.2 ton a year.

Emergency power supply in case of blackout.

Industrial promotion

World-first market entry of the residential fuel cell system that use Japanese state of the art technology.

Development of a new Industry in Japan by global exportation of *ENE-FARM*.

6

ENE·FARM

Number of Units Sold by Tokyo Gas

TOKYO GAS

- The unit sales of *ENE-FARM* has been steadily increasing.
- In Tokyo Gas area, total unit sales reached 120,000 in August 2019.

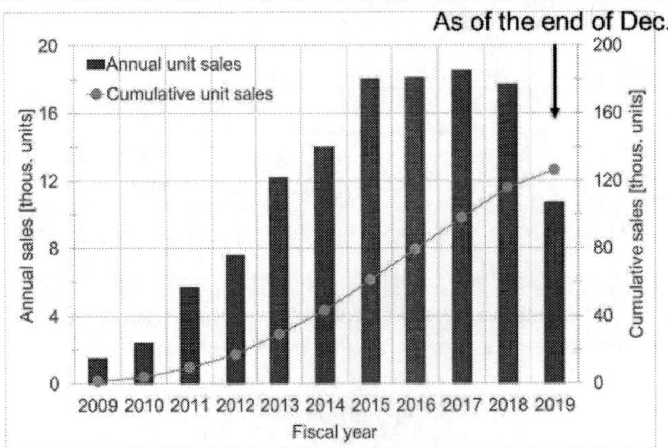

2009	- Commercial sales of *ENE-FARM* started.
2011	- The Great East Japan Earthquake occurred.
2014	- SOFC-type *ENE-FARM* was released. - *ENE-FARM* for condominium was released. - Continuous power generation during blackout became optionally available.
2019	- Continuous power generation during blackout became standardly available.

✓ Total unit sales in Japan reached 300,000 in November 2019.

7

ENE·FARM

Current Challnges

TOKYO GAS

- For further expansion of market, reduction of size and cost has been required.

- ENE-FARM has been installed not only to new-built houses but also to existing houses.
- The challenge is to increase the number of ENE-FARM installed to new-built condominiums.

■ *ENE-FARM* sold in Tokyo Gas Area (until March 2019)

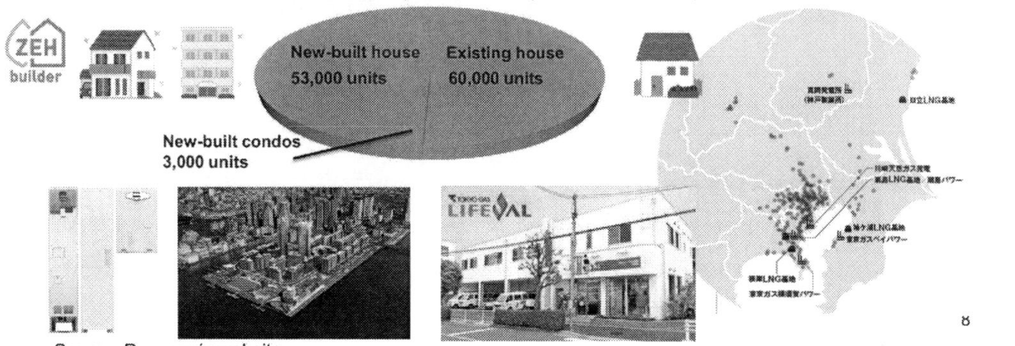

Source: Panasonic website

ENE·FARM

Product Line-up

TOKYO GAS

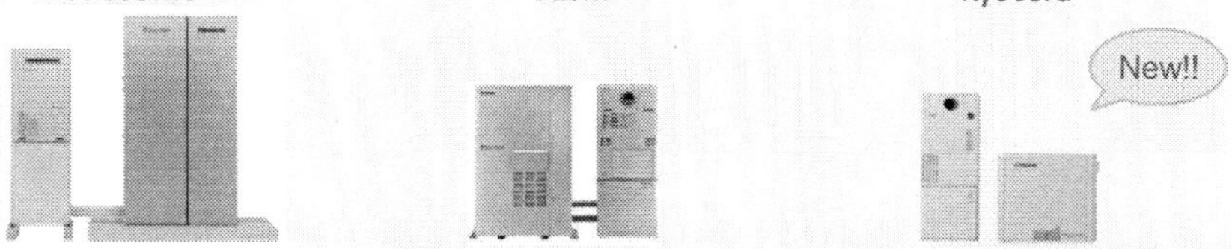

	Panasonic	Aisin	Kyocera
Type	PEFC*1	SOFC*2	SOFC*2
Rated power	700 W	700 W	400 W
Size	910(W) x 1,650(H) x 350(D) mm	780(W) x 1,220(H) x 330(D) mm	800(W) x 700(H) x 350(D) mm
Weight	95kg	106kg	80kg
Efficiency [LHV]	Power generation : 40% Overall : 97%	Power generation : 52% Overall : 87%	Power generation : 47% Overall : 80%
Hot water tank size	100 L	28 L	20 L
Continuous power generation during blackout	✓	✓	✓

*1: Polymer Electrolyte Fuel Cell *2: Solid Oxide Fuel Cell

TOKYO GAS

ENE-FARM mini

- Tokyo Gas announced on October 2019, the launch of the world's smallest residential fuel cell "*ENE-FARM mini*".

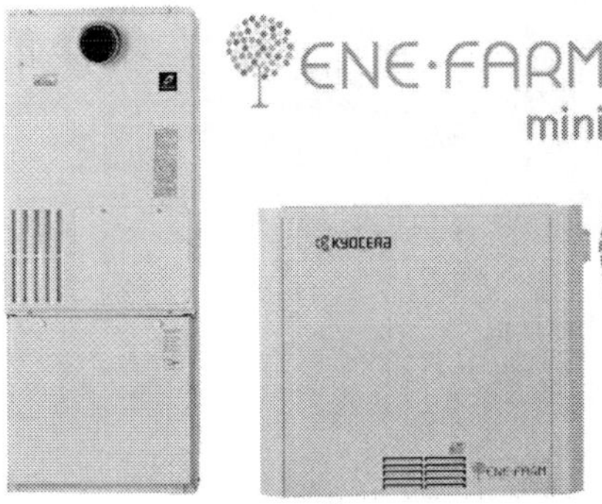

< Development Concept >

- **Expand potential markets**
- **Lower equipment costs**
- **Reduce commercial distribution and logistics costs**
- **Improve installation work**

10

ENE·FARM

TOKYO GAS

ENE-FARM mini -Installation Space-

- Since *ENE-FARM mini* can be installed to 500mm-wide alleyway surrounding house, it has potential to be installed in most detached houses in Japan.

- The size is as small as air conditioner condensing unit.
- Better installation flexibility is realized by reduced height (lowered exhaust gas port enables installation close to wall/window).
- Smaller foundation is available due to lightened weight and lowered center of gravity.

■ System size

Size	H700 x W800 x D350mm
Exhaust gas port	H415mm (Center of gas port)
Weight (body)	80kg (without water) , 103kg (with water)
Weight (foundation)	27kg (13.5kg x 2)

■ Installation image

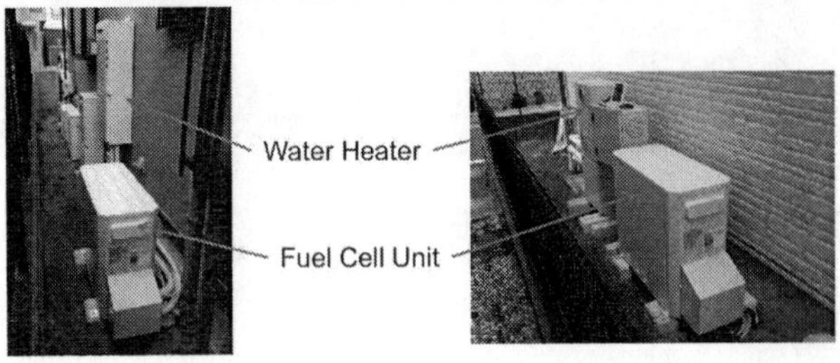

Achieved installation to 500mm-wide alleyway

11

ENE·FARM

ENE-FARM mini -Resilience for Emergency-

TOKYO GAS

- *ENE-FARM mini* can provide electricity, hot water, and heat even in case of power outages.

- Electricity is available as long as city gas is supplied to *ENE-FARM mini.* (max. 400W)
- Water in storage tank can also be utilized (e.g. for toilet flushing). (20L)

■Appliances powered by 400W *ENE-FARM mini* (example)

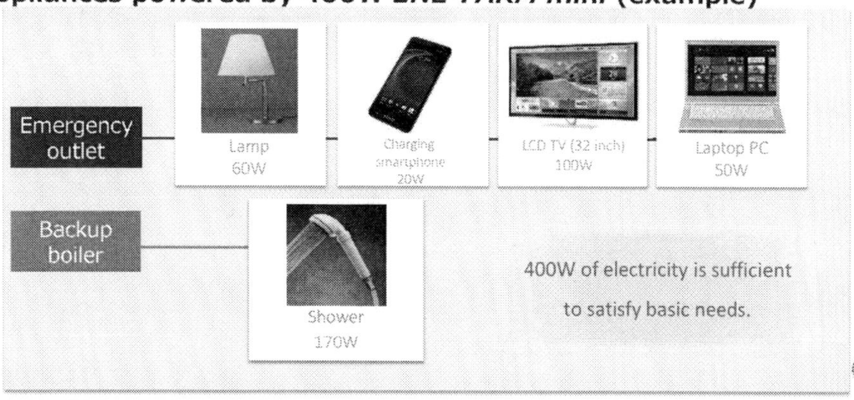

12

ENE·FARM

ENE-FARM mini -IoT Capabilities-

TOKYO GAS

- *ENE-FARM mini* can be remotely operated by smartphone or smart speaker.

- *ENE-FARM mini* can be connected to Purpose (boiler manufacturer) cloud server.
- All the services provided by Purpose Connect (such as remote bath filling and floor heating operation) are available. Besides, fuel cell operation status can be checked.

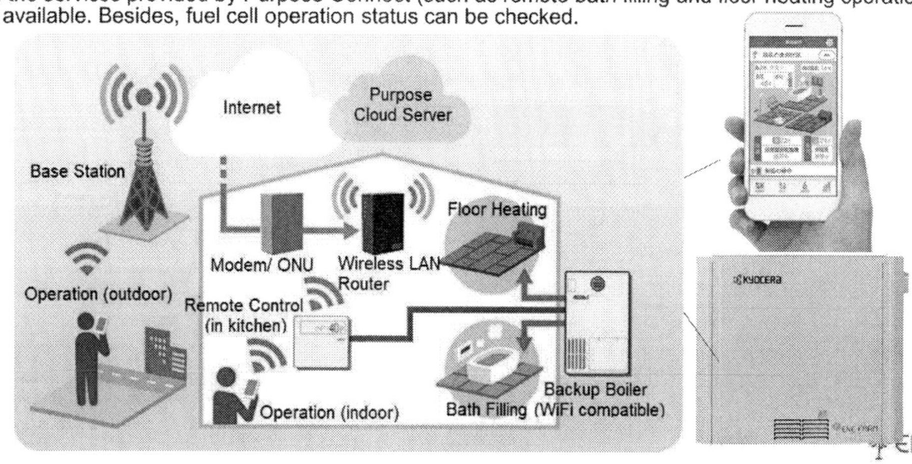

13

ENE·FARM

ENE-FARM mini -Economic and Environmental Performance-

- High power generation efficiency in low load range enables economic and environmental performance equivalent to previous models.

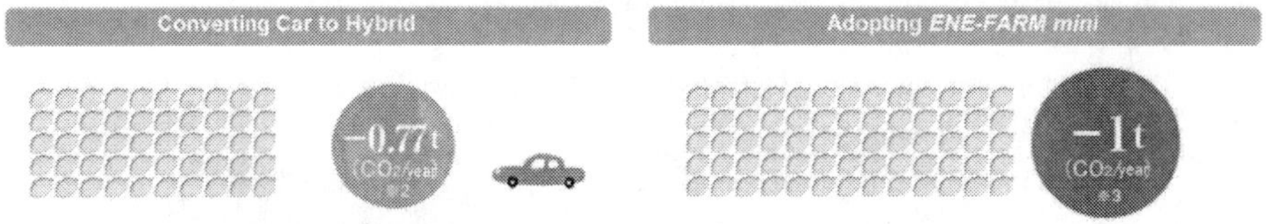

※2 試算条件：年間1万km走行時の場合のガソリン車とハイブリッドカーのCO2排出量を比較。ガソリン車の排出量を2.32kg-CO2/L（資源エネルギー庁家庭の省エネ徹底ガイド春夏秋冬（2017年8月より））燃費：ハイブリッド車37.2km/L、ガソリン車16.6km/L（国土交通省「自動車燃費一覧」（平成30年3月）データ）より京セラにて算出。
※3 試算条件：CO2排出係数：電気0.587kg-CO2/kWh、ガス2.277kg-CO2/㎥（資源エネルギー庁家庭の省エネ徹底ガイド春夏秋冬（2017年8月より）電力負荷：JIS C 8851 中間期　給湯負荷：JIS S 2075 中間期の条件にて京セラにて算出。使用環境により数値は異なります。

Going Forward -condominium install-

- More installation to condominium (in particular, existing condominium) is the key to further spread of *ENE-FARM*.
- Installation flexibility and power generation efficiency need to be improved.

- Installation space for solar panel is limited; *ENE-FARM* is promising in condominium market.
- Special product design is required to overcome installation restrictions in condominium.

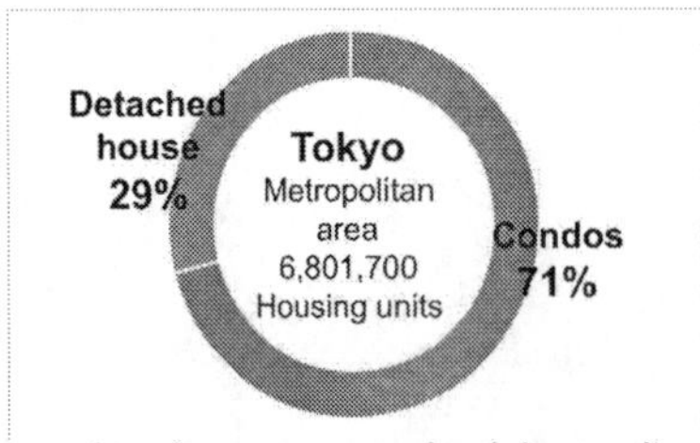

In urban area, condominium unit account for a large percentage.

Created based on 2018 housing and land statistics survey results

Installation to open corridor

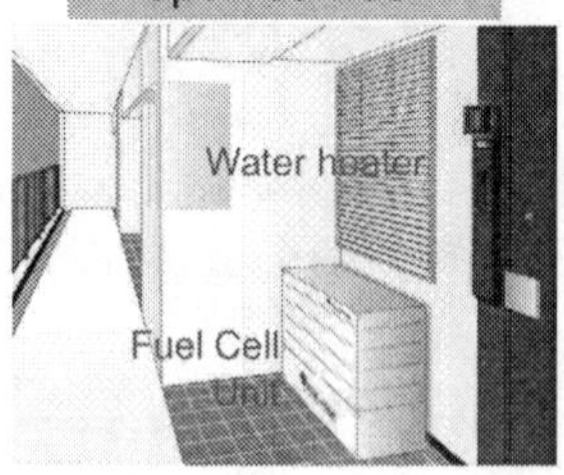

TOKYO GAS

Going Forward -VPP-

- It is expected that *ENE-FARMs*, storage batteries, renewable energies, etc. will be integrated into a virtual power plant (VPP) by utilizing IoT technology, contributing to a stable and inexpensive energy supply.

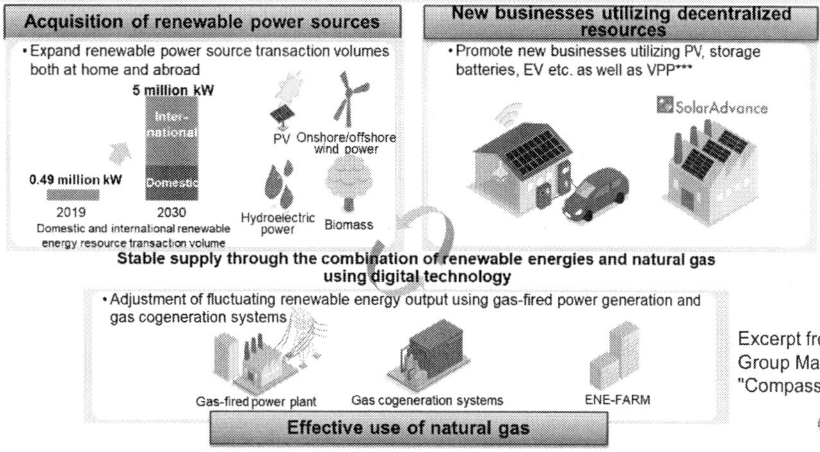

Excerpt from Tokyo Gas Group Management Vision "Compass 2030"

16

ENE·FARM

TOKYO GAS

Summary

- *ENE-FARM* is a residential fuel cell micro-CHP system that contributes to reducing CO_2 emission and saving energy bills.
- The unit sales of *ENE-FARM* has been steadily increasing. The total unit sales in Japan reached over 300,000.
- *ENE-FARM* has been gaining attention as disaster-resilient infrastructure since it can provide electricity, hot water, and heat even in case of power outages.
- Tokyo Gas developed and launched the world's smallest residential fuel cell "*ENE-FARM mini*" in October 2019.
- For further spread, the market will be expanded by improving installability in condominiums.
- The integration of *ENE-FARMs*, renewable energy, and batteries build a resilient and smart society.

17

ENE·FARM

TOKYO GAS

Thank you for your attention

TOKYO GAS

18

ENE·FARM

Evaluating the Resilience of Integrated Energy Systems Using Dynamic Simulation

Carsten Bode, Anne Senkel, Gerhard Schmitz

Institute of Engineering Thermodynamics
Hamburg University of Technology

International Gas Union Research Conference 2020, Muscat, Oman
February 25th, 2020

TUHH

ENGINEERING THERMODYNAMICS

Central Questions

TUHH
ENGINEERING THERMODYNAMICS

- Why are we looking at resilience?
- What is resilience?
- How do we quantify resilience?
- How do we model resilience?
- Example of an integrated energy system

Why are we looking at resilience?

TUHH ENGINEERING THERMODYNAMICS

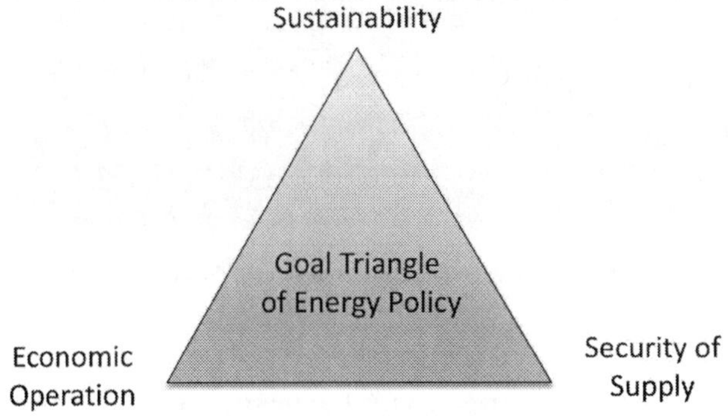

Why are we looking at resilience?

TUHH ENGINEERING THERMODYNAMICS

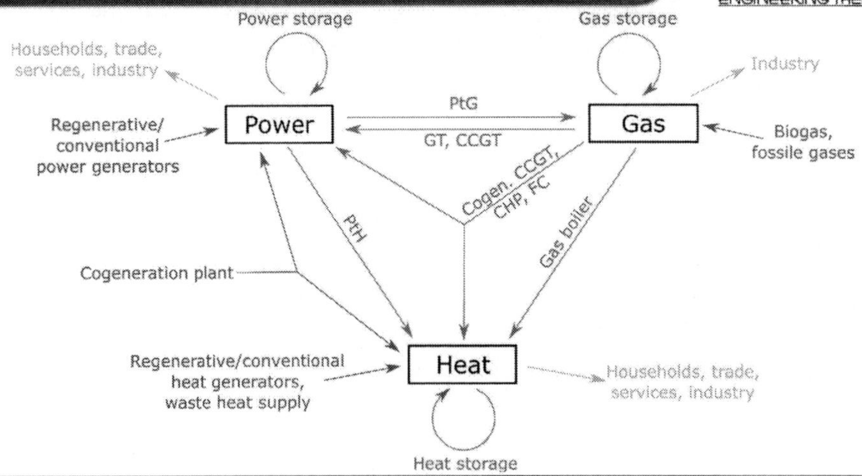

What is resilience?

So far resilience has mostly been defined in a qualitative way:

"a measure of the persistence of systems and of their ability to absorb change and disturbance"

C. S. Holling, 1973

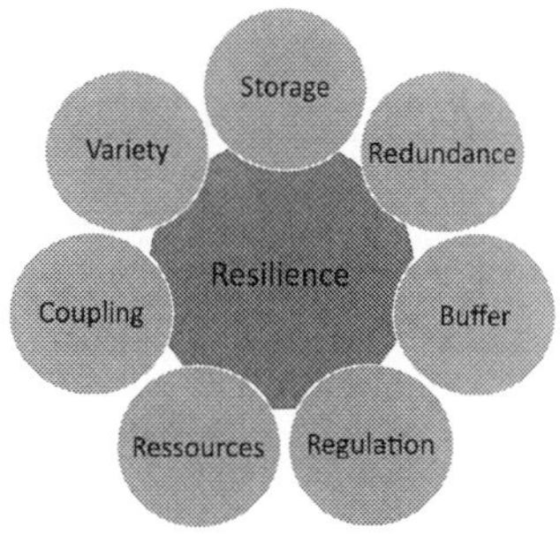

Translated from U. Brandt *et al.*, 2017

→ How can we quantify resilience?

How can we quantify resilience?

TUHH ENGINEERING THERMODYNAMICS

How fast does the system recover?

→ Recovery Time: $RT = \frac{t_r - t_d}{\Delta t_{norm}}$

How resistant is the system?

→ Maximum Deviation: $MD = \frac{\Delta x_{max}}{\Delta x_{norm}}$

How does the system react?

→ Performance Loss: $PL = \frac{\int_{t_d}^{t_r} \Delta x dt}{A_{norm}}$

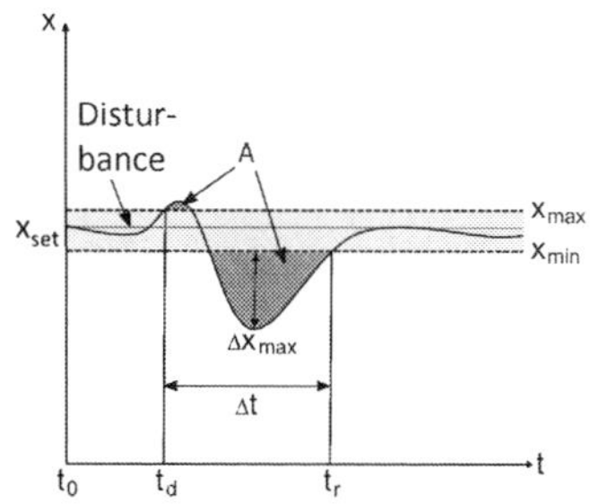

Irresilience Index: $IRI = RT \cdot MD \cdot PL$ **Resilience Index:** $\boldsymbol{RI = \frac{1}{1+IRI}} \in [0,1]$

How do we model resilience?

TUHH ENGINEERING THERMODYNAMICS

- TransiEnt
  - UsersGuide
  - Basics
  - Components
  - Producer
  - Consumer
  - Grid
  - Storage
  - Examples
  - SimCenter
  - ModelStatistics

TransiEnt Library:

- open-source library
- physically-based
- multi-domain
- dynamic simulation of integrated energy systems

MODELICA

www.tuhh.de/transient-ee/en/

Example

TUHH ENGINEERING THERMODYNAMICS

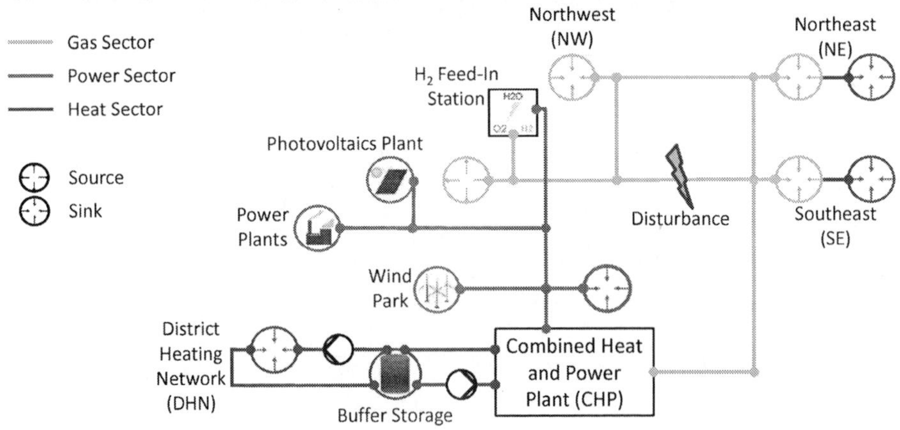

Results – Gas Sector

TUHH ENGINEERING THERMODYNAMICS

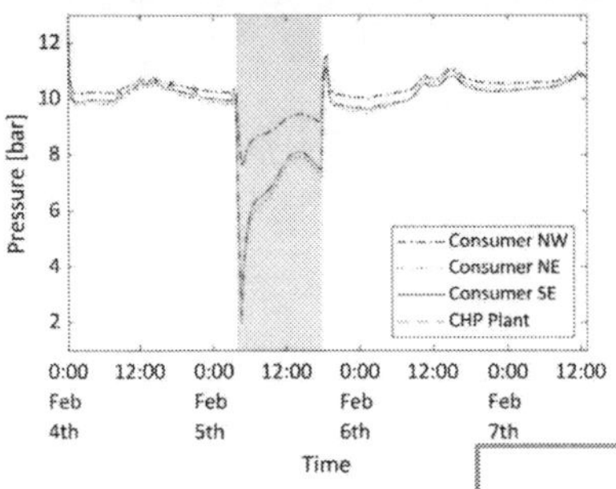

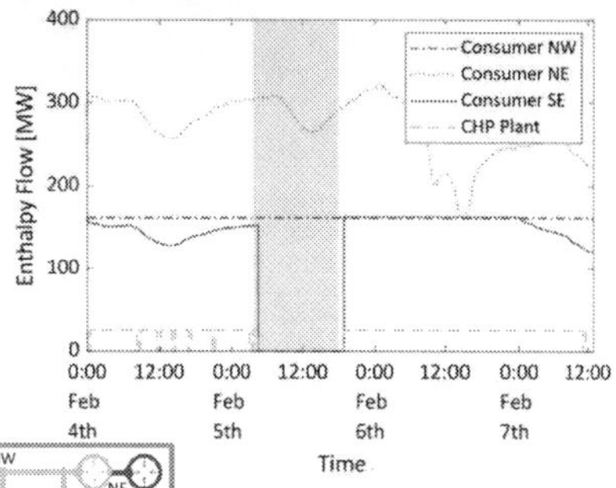

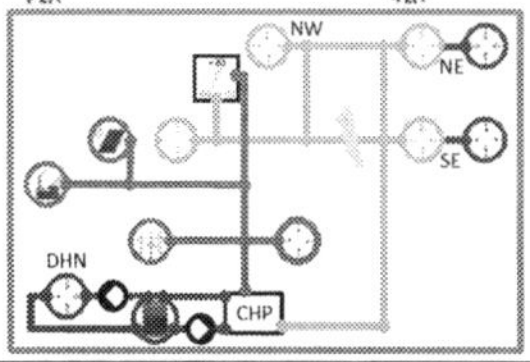

Results – Heat Sector

TUHH ENGINEERING THERMODYNAMICS

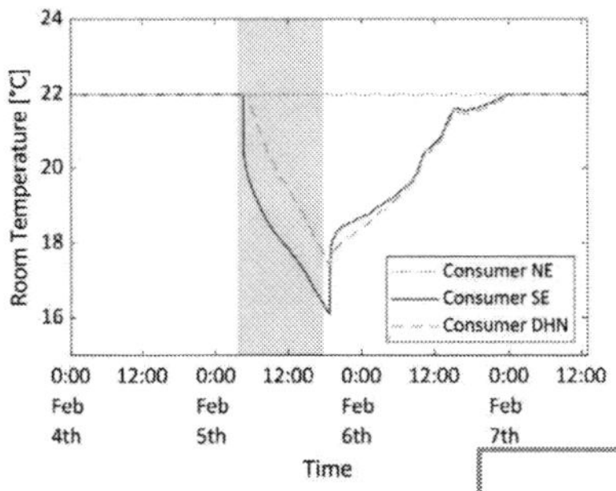

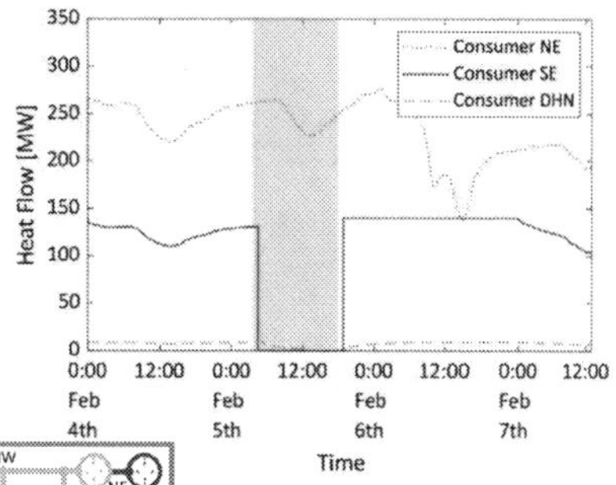

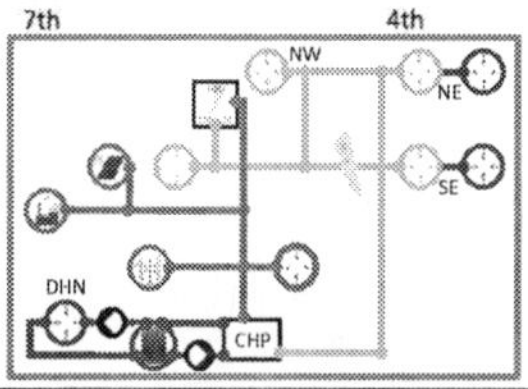

Results – Power Sector

TUHH ENGINEERING THERMODYNAMICS

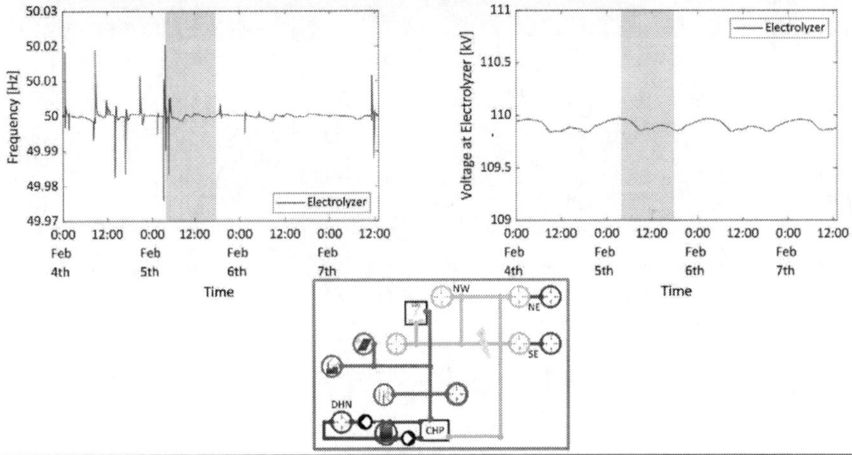

Results – Resilience Indices

TUHH ENGINEERING THERMODYNAMICS

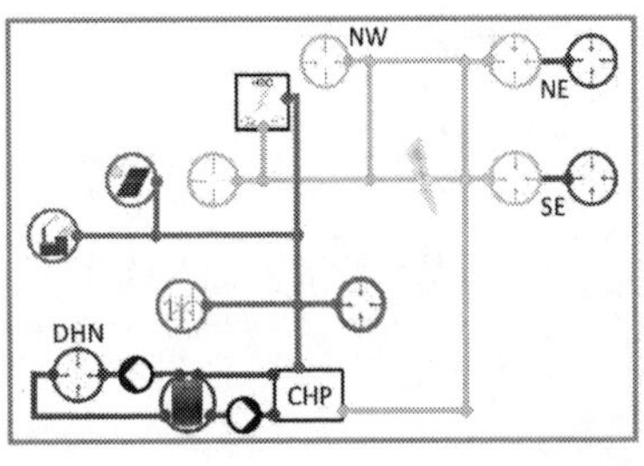

Sector	RI_{NW}	RI_{NE}	RI_{SE}	$RI_{CHP/DHN}$
Gas	1.00	1.00	0.20	0.17
Heat	-	1.00	0.21	0.24
Power	-	-	-	-

$$RI_{sys} = 0.78$$

Conclusion

TUHH ENGINEERING THERMODYNAMICS

- Resilience Index reflects the system's behavior
- Changes in the system's architecture can be assessed quantitatively
  - ➢ especially higher shares of renewables
- Quantitative evaluation and comparison of CO_2 emissions, costs and resilience is possible

Thank you for your attention!

TUHH ENGINEERING THERMODYNAMICS

TransiEnt Library

www.tuhh.de/transient-ee/

Carsten Bode, M.Sc.
c.bode@tuhh.de

Anne Senkel, M.Sc.
anne.senkel@tuhh.de

Prof. Dr.-Ing.
Gerhard Schmitz
schmitz@tuhh.de

Supported by:
Federal Ministry for Economic Affairs and Energy
on the basis of a decision by the German Bundestag

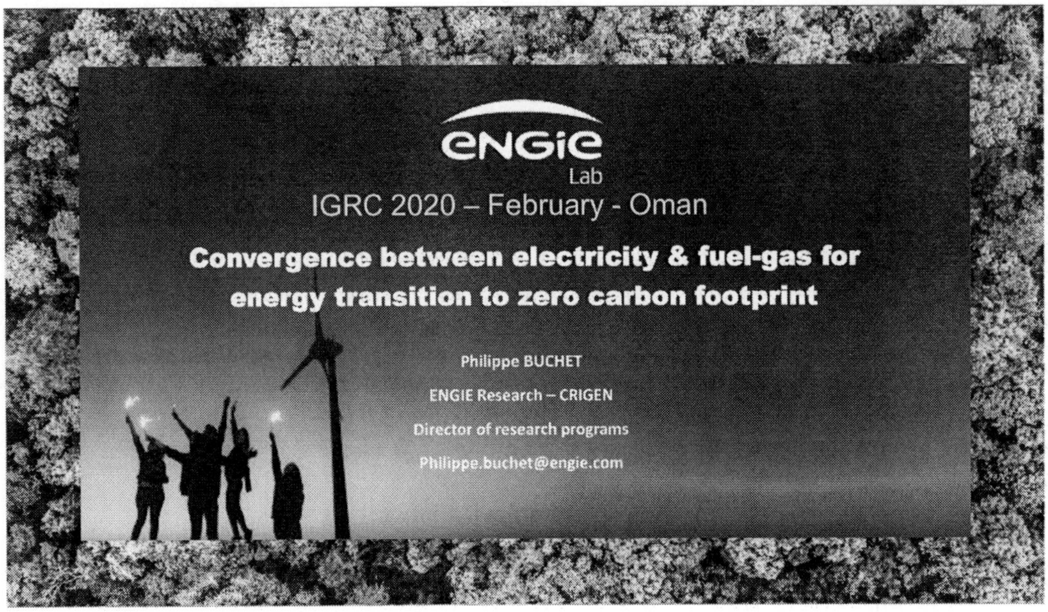
engie
Lab
IGRC 2020 – February - Oman
Convergence between electricity & fuel-gas for energy transition to zero carbon footprint
Philippe BUCHET
ENGIE Research – CRIGEN
Director of research programs
Philippe.buchet@engie.com

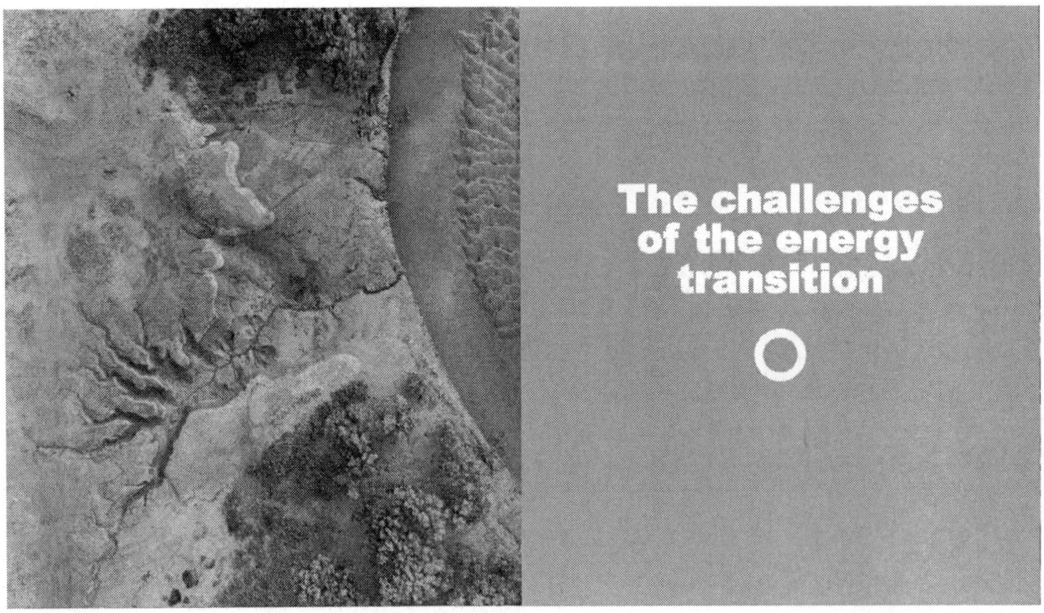
The challenges of the energy transition

The climate has never changed so quickly ...

and this trend should accelerate

- **2016:** the hottest year since official surveys (1.2 ° C higher than the average of the 20th century)
- **Top 5 of the hottest years: 2015, 2016, 2017, 2018, 2019**
- Acceleration of warming since the 1980s '

This infographic shows temperature anomalies in # countries

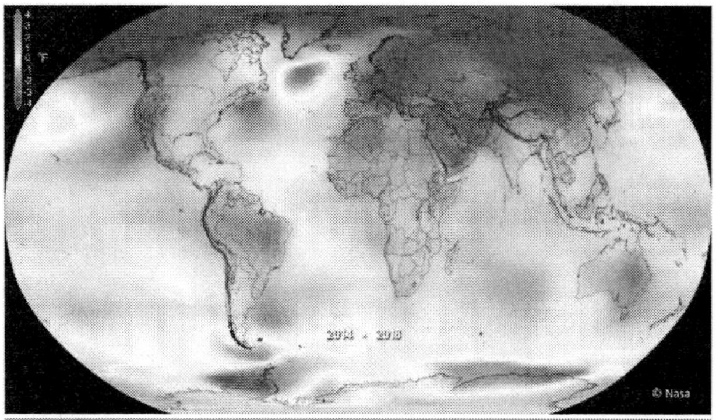

Temperature anomalies for the period 2014-2018.
The redder the areas, the stronger the anomaly.

An increase in CO2 emissions as global warming is increasingly felt on a global scale

- The day of the overshoot: in 2019, July 29. was In 1970, December 29.
- At least 15 natural disasters linked to climate change caused destruction costing more than $ 1 billion in 2019 and 7 of them at least $ 10 billion (*source Christian Aid*): floods (United States , India...), fires (California, Australia), exceptional heat waves in Europe...
- By 2030, 40% of water demand will not be met. Climate change could also lead to additional migration (10 to 20 million peoples).
- Above 4 ° C, the world is no longer insurable (*source: Axa insurance Company*)

The equation of the energy system to be solved is unprecedented

- 80% of the world's primary energy is carbon-based
- 1.2 billion people still lack access to electricity (16% of the world's population)
- A 1.5 ° C scenario requires drastic changes & imagine a world without CO2 in the long term

Primary energy world consumption

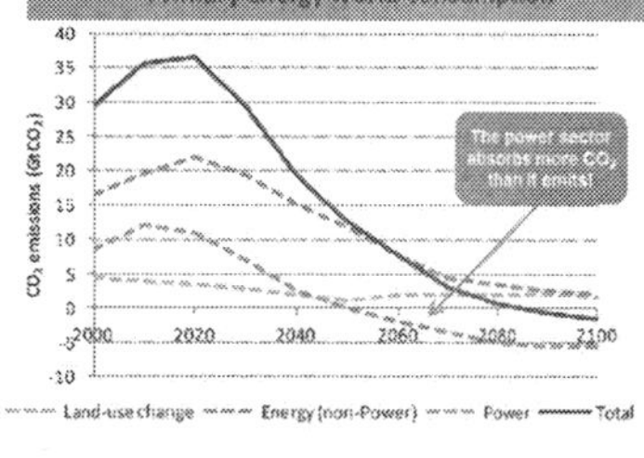

However, energy demand & CO2 emissions reached a record in 2018

85% of the increase in emissions in 2018 is caused by China, India and the United States

Global CO2 emissions from energy combustion reached a historic level of 33.1 GtCO2 in 2018, (+ 1.7% compared to 2017), due to:

- **Growth in primary energy demand (+ 2.3%,** twice the average growth since 2010), the result of sustained economic growth.
- **A slowdown in the implementation of energy efficiency policies.**
- **And this despite the growth of renewable energies** (+ 7% of renewable electricity production in 2018 - average growth of 6% since 2010). 40% of this growth is driven by China.

A rapid Scale-up of energy efficiency investments is needed

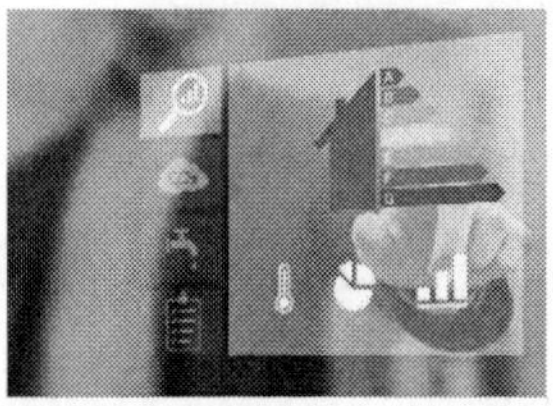

- **Spending on energy efficiency stagnated in 2018. $ 240 billion was invested** in energy efficiency in buildings, transportation and industry. *It would cost $ 1,300 billion a year between 2026 and 2040.*
- **In France, the surface of buildings was renovated at a rate of 1% per year over 2010-2018.** *It should be multiplied by 4 by 2030.*
- **Europe remains in first place for investment in energy efficiency,** followed by China and North America.

Faced with this challenges, State commitments & regulations are insufficient

- **Global CO2 emissions must now be halved by 2030 to keep global warming below 1.5 ° C in 2100.**
- **Despite some encouraging announcements**: exit of coal in Germany (2038) and Chile (2040), carbon neutrality by 2050 under discussion in Europe (Green pact), the commitments made to date by the states are insufficient.
- **Brazil** goes from **14th largest** emitter of CO_2 **to 6th** place in the world (deforestation exceeds 10,000 km² in 2019). The efforts to be made are greatly reduced by the new government.
- **China** accounted for almost a quarter of global emissions in 2017.
  - It remains the leading investor in renewable energy and will reduce its emissions from 2030
  - But its current fleet is 59% coal-fired and it will add 25% more power plants by 2030
- **The United States** has withdrawn from the Paris Agreement COP21. However, the country is expected to reach 2/3 of its initial commitment by 2030 thanks to state policies.

« I oppose irresponsible and job-destroying demands to limit the coal industry, Scott Morrison, Australian statesman»

ENGIE LAB CRIGEN - Electricity & fuel-gas convergence

Energy transition was pushed by authorities & regulation

...A second trend is growing

...Pull by local initiatives, by Customers requests, industrial companies

EV 100

RE 100

C40 CITIES

SCIENCE BASED TARGETS

GREENHOUSE GAS PROTOCOL

CDP A LIST 2018

CARBON TRUST

7

ENGIE LAB CRIGEN - Electricity & fuel-gas convergence

The civil society changes. Under pressure many innovations and business changes are emerging

- **Global protest calls** for accelerated fight against climate change
- **Other demonstrations** arise spontaneously from influential people who spread the message on social networks (eg Greta Thunberg for the strike of high school students).
- **On August 19, 2019, the Business Roundtable, a lobby grouping 188 American CEOs,** published a forum questioning the priority given to shareholders, for the benefit of all business stakeholders. **(200 CEOs** in Europe called for a European New Deal**)**
- **Climate Action 100+ (€ 33,000 billion in assets under management) obtained from Shell and BP that they set themselves CO2 reduction targets.**
- **7,000 companies share their environmental data as part of the CDP approach (x2 since 2010).**
- **The Science Base Target SBT label becomes the criterion for assessing a "2 ° C compatible" development strategy. (9 in 2015, 237 in 2020, and 359 to have signed up).**

SCIENCE BASED TARGETS

8

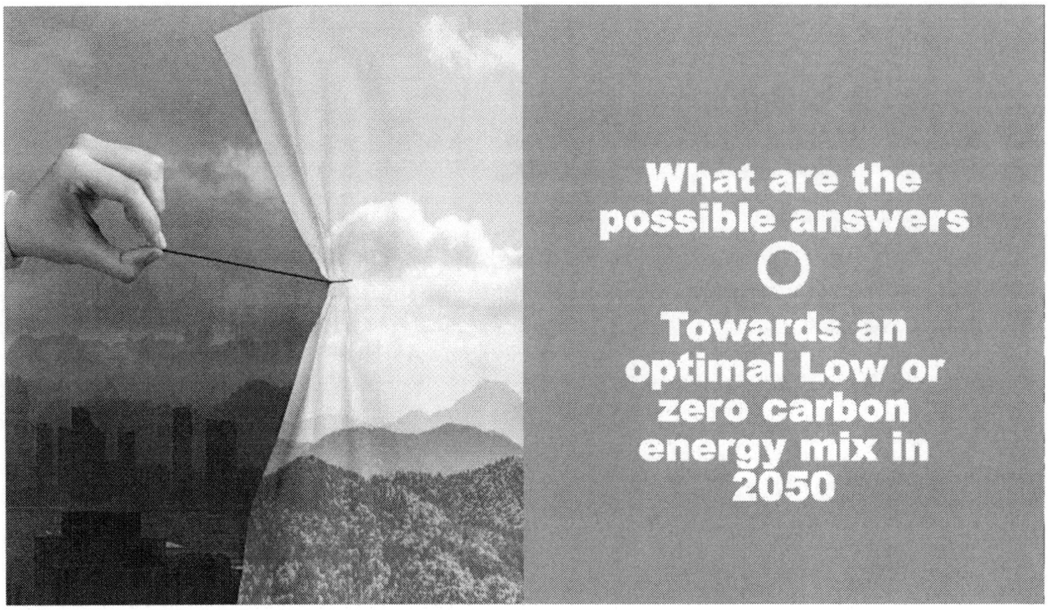

Towards an zero CO_2 Energy master plan with large convergence of electricity and Fuel-gas

- **Industrial, Buildings & Mobility sectors** will continue to account for the largest share of energy consumption in the world **(ie. CO2 emissions) through 2050**
- **For ENGIE the "optimal path" for a decarbonization trajectory is a Multi energy mix,** based on three key levers:
  - High level of **Energy efficiency through strong support of customer** ;
  - **Targeted electrification** of certain uses such as industrial processes, Mobilty in cities, based on a **massive development of electrical renewable energies;**
  - Development in parallel of the full potential of the **low carbon Fuel-gas** vectors: **Mainly renewable gases (Biogas, biomethane, H2...) but also Natural Gas during the transition period**
- **These scenarii include :**
  - **Drastic revisions of energy master plans** (mostly defined in 70's) to use more & more Renewable energies
  - **Development & implementation of disruptive technologies** & solutions for lowering carbon footprint + increase of industrial **automation (ex: industry 4.0)**

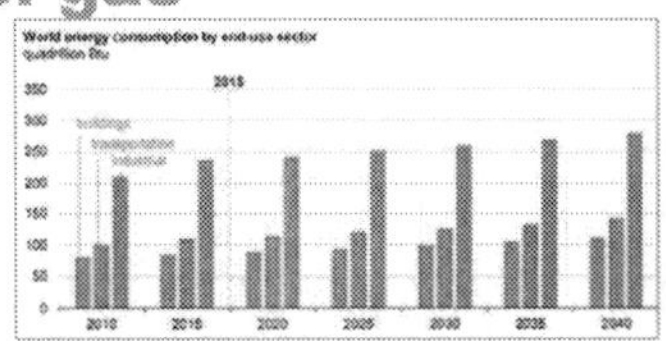

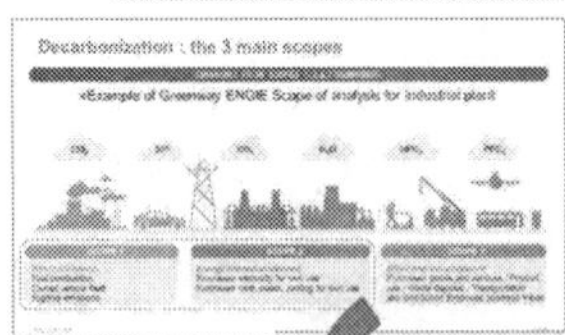

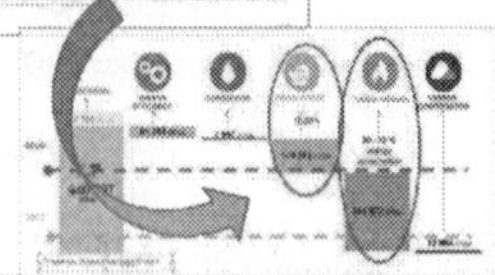

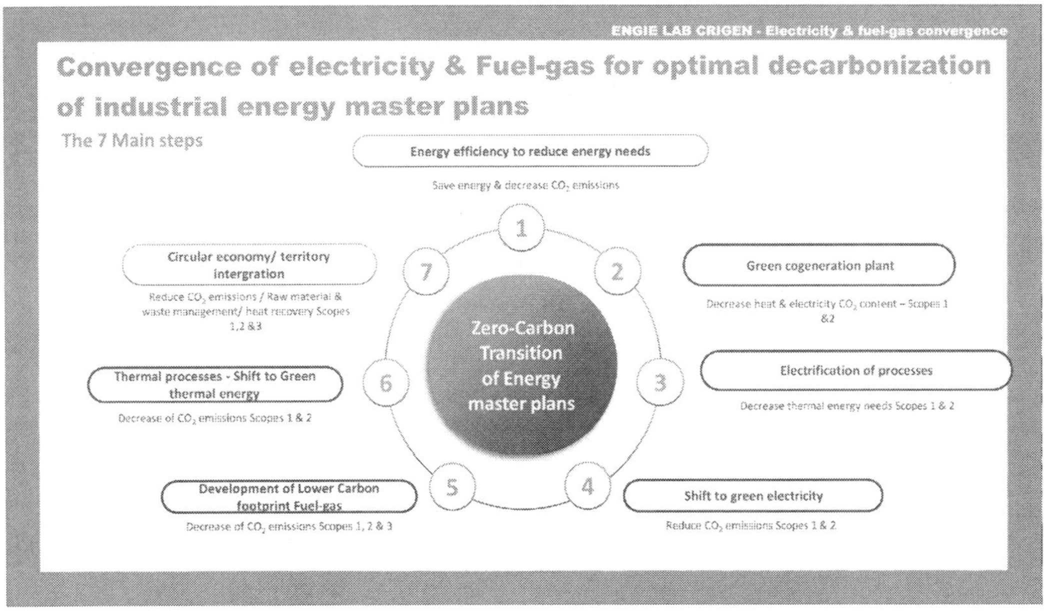
ENGIE LAB CRIGEN - Electricity & fuel-gas convergence
Convergence of electricity & Fuel-gas for optimal decarbonization of industrial energy master plans
The 7 Main steps
1
Energy efficiency to reduce energy needs
Save energy & decrease CO2 emissions
2
Green cogeneration plant
Decrease heat & electricity CO2 content – Scopes 1 &2
3
Electrification of processes
Decrease thermal energy needs Scopes 1 & 2
4
Shift to green electricity
Reduce CO2 emissions Scopes 1 & 2
5
Development of Lower Carbon footprint Fuel-gas
Decrease of CO2 emissions Scopes 1, 2 & 3
6
Thermal processes - Shift to Green thermal energy
Decrease of CO2 emissions Scopes 1 & 2
7
Circular economy/ territory intergration
Reduce CO2 emissions / Raw material & waste management/ heat recovery Scopes 1,2 &3
Zero-Carbon Transition of Energy master plans

Towards an optimal Low or zero carbon mixte energy master plan in 2050
Some examples

Ex.: Decarbonizing the Energy mix with fuel-gas technologies

To reduce CO_2 Footprint of Countries or territories many scenarios are studied. Nearly all of them include:

- Massive development of **renewables energies production** (**electricity** and **green Fuel-gas** & biomass..)
- **But, Development of (RE) renewables electricity** (**Wind , Solar;...**) generate 2 main issues:
  - RE are mainly intermittent sources (→ energy Storage)
  - Capacity will not be enough to reduce GHG emissions.

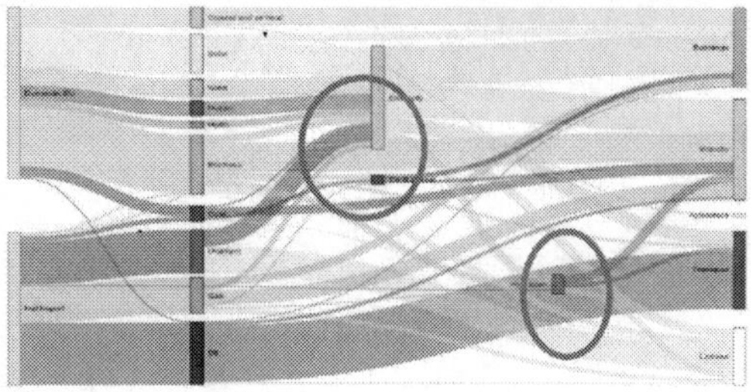

Sankey diagram - illustrative scenario of a low carbon EU energy system in 2050

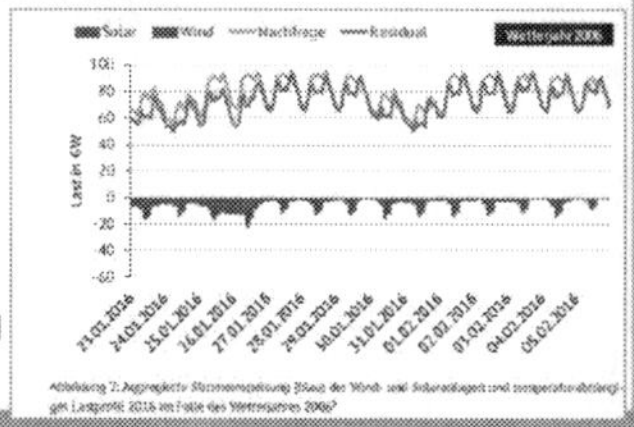

For example, In the past 10 years:

- The USA had CO_2 emissions decrease mainly due to a shift in power generation from coal to natural gas, not due to RE development
- Germany has maintained its CO_2 emissions while there has been strong wind and solar development (more than 100 GW of installed capacity) in its energy mix

ENGIE LAB CRIGEN - Electricity & fuel-gas convergence

Ex.: Decarbonising the electric mix with fuel-gas technologies

The electrification of the economy and the transition to a zero-carbon electricity system will account for 40% of the decarbonization effort by 2050.

This strategy implies an increase in electricity demand of about 40% by 2050, and therefore even more sustained efforts to decarbonize this sector.

Today, to answer to this request sustainable solutions are emerging or available & most of them **include multi-energy mix** :

- Development of **green electrical central assets** (Windmill , PV , Hydro, ...) plants **combined with decentralized** NG/ Biogas/biomass/hydrogen **Cogeneration** & management of energy demand (**Smart Grid**)
- Development of **Combined fuel-Gas to power** assets (CCGT, Biogas/biomethane /Hydrogen cogeneration / NG &/or H2 CCGT, ...)
- Development of **Hybrid power plant** (ex: Florida –USA)
- **Micro CHP** based on NG or H2 fuel cell or Hybrid technologies (Ex: Japan)
- ...

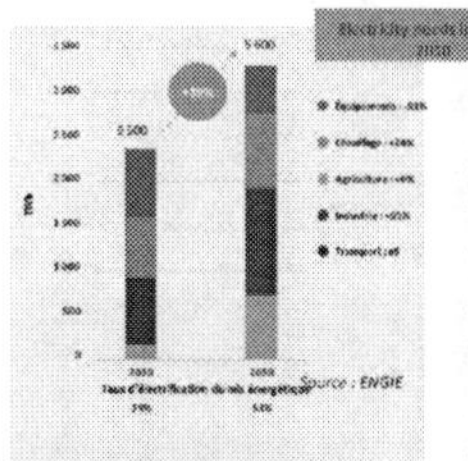

Source : ENGIE

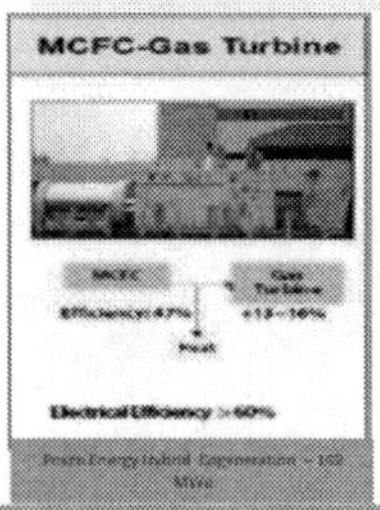

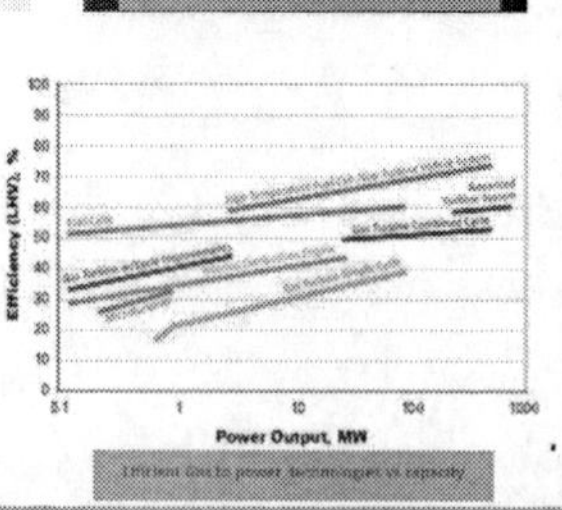

14

Flexiblity of smart grid with fuel-gas solutions

Sources : IRENA Report published - November 2018 &
Study DVGW / BEW Germany

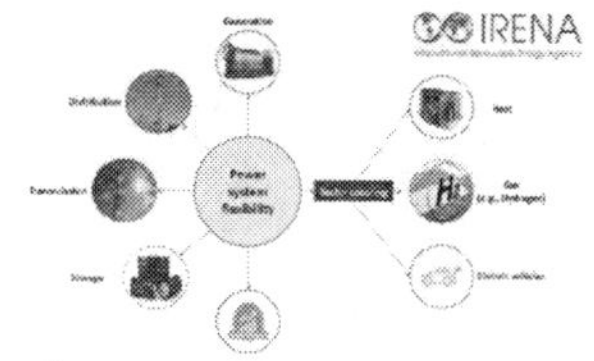

About "Power system flexibility for the energy transition"

Key elements of electrical grid flexibility is sector-coupling:

- **Decentralized Biofuel-gas generation to Power** (Bio cogeneration, renewable gas to power…).
- **The coupling of energy demand for heat**, Buildings & cities by using power to heat (e.g. heat pumps, micro **cogeneration** …)
- **Power to X** (source Hydrogen Europe):
  - Power to H_2/gas (e.g. hydrogen from renewable electricity) and power to mobility
  - Power to Synthetic methane (Methanation)
  - Power to H_2/gas for direct uses
- **Fuel-gas storage** vs battery storage (H_2/biomethane injection into gas grid - Gas storage capacity ~ 350 X electricity storage capacity in western Europe Countries)

DVGW/BEW source

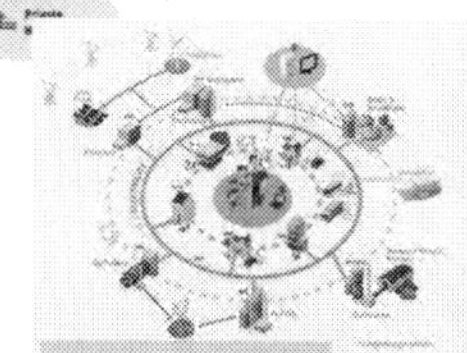

MARKOGAS/EUROGAS source

Power to gas & smart grid projects :

Examples of Industrial projects of Low Co2 footprint or zero carbon grids

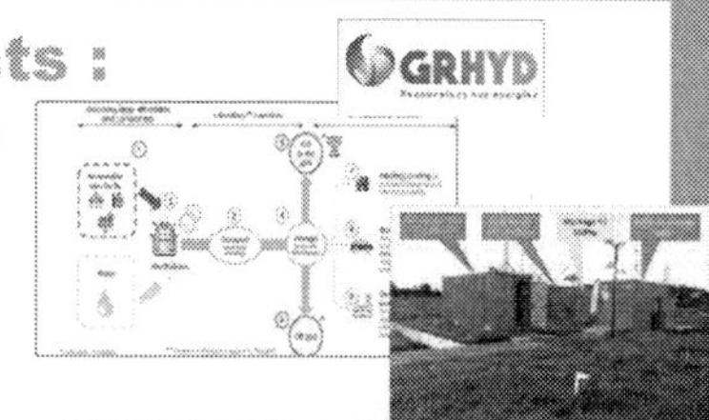

- **GRHYD (ENGIE – FR Dunkerque)**: Valorization of excess Green power to H2 & injection into NG grid (NG & H2 blend up to 20%)
- **Renewable Energy Integration Demonstrator (REIDS - Singapore) :** Testing and demonstrating the integration of solar, wind, power-to-gas technologies, storage as well as waste-to-energy and as well as other production to manage Smart Grid
  - **Leader** : NTU -Nanyang Technological University of Singapore
  - **Partners** : Accenture, EDF, Emerson, ENGIE, GE, Rolls-Royce, Schneider Electric, Vestas, …

- **BIG HIT**- **B**uilding **I**nnovative **G**reen **H**ydrogen Systems in **I**solated **T**erritories (**Scotland** - **Orkney Islands**): Power to Hydrogen solution, by electrolysis of water with green elec. . This hydrogen is then stored as high-pressure gas, transported and used at mainland Orkney.
  - 2 PEM electrolysers : 1 MW + 0,5 MW, 5 H2 tube-trailers & 75 KW H2 Fuel Cell -> heat and power + 5 H2 Fuel Cell road vehicles

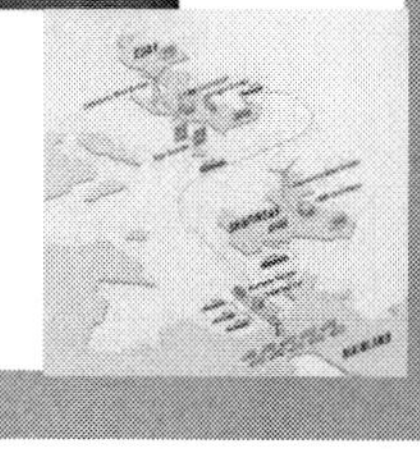

Decarbonization of industrial thermal processes - Ex: Glass melting furnaces (1/2)

Glass is a material that has multiple applications:

- Flat glass for the building and automotive industries,
- Container glass for food and drink....
- Special glasses for thermal insulation, electronics and lighting ...

- If the glass can be easily recycled after use (~ 50% recycled glass), **manufacturing glass needs lot of energy**) due to high temperature requirement
- **Annual Energy consumption is (Europe) ~ 300 PJ for the European glass industry** - ~ 0.64 Ton of CO2 / ton of glass) - 80% comes from fossil fuels (mainly NG) and 20% from electricity.

As important contributors to industrial CO2 emissions, glassmakers are today increasingly constrained by their customers on the "carbon" content of glass products and therefore on the reduction of CO2 emissions, particularly related to share thermal energy.

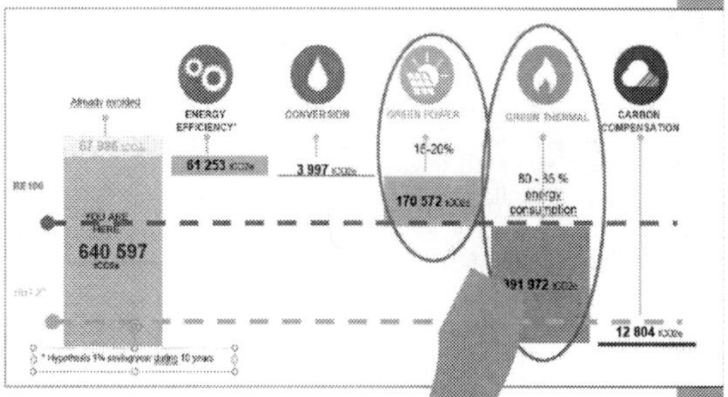

Decarbonization of thermal processes - Glass melting furnaces (2/2)

Electricity & Fuel-gas convergence for decarbonization of glass melting plant:

- Electrification + Green thermal heat :
  - ❖ New design of furnace with high ratio of electricity use or only electrical process
  - ❖ Combined Green electricity boosting & Biogas, Biomethane or Bio-syngas
- Use of lower carbon footprint fuel-gas: Green Power to H2 & Use of Natural Gas & H2 blend

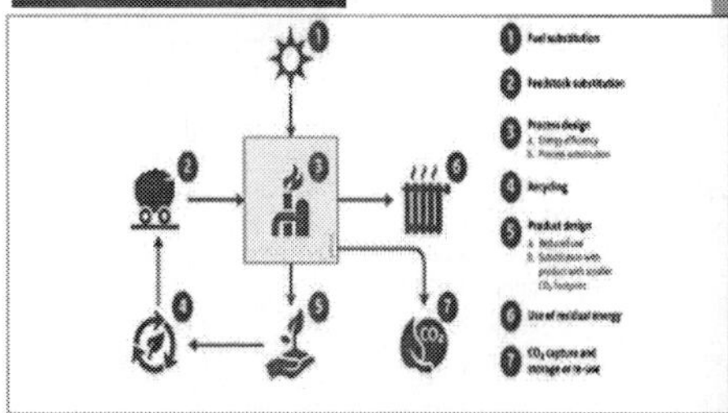

Energy transition roadmaps for glass plant – ref. TNO/ Owens – Illinois Study

Electricity & fuel-gas convergence : Buildings & Cities

Even if energy demand will be highly reduced, in domestic & commercial markets

- Scenario with **almost complete electrification** will imply very high increase of the electricity demand by 2050 and **will not necessarily lead to a drastic drop in CO2.** Furthermore European studies shows that would generate an additional cost of around € 300 / year on household bills probably nonacceptable).
- For commercial & domestic markets, **the mix scenario with limited electrification,** completed by gas appliances, seems to be better. Furthermore natural gas is considered as versatile energy carrier with robust, efficient appliances and low emission on its use (low NOx, no PM, low CO).
- In many Countries (Europe), **Future energy master plans will include renewable** gases (biomethane, RNG green methane from Biomass anaerobic digestion or gasification, NG +H2 blends...), **which will be easily accessible through the existing gas grids and easily used into NG appliances.**
- Finally, Smart electricity grids recommend a mix between central power production and decentralized power production or Optimize Micro grids.

✓ **Many opportunities for Gas appliances to contribute to Decarbonization of Buildings and cities :**
  - ✓ Gas driven heat pumps (GHP),
  - ✓ Stirling appliances,
  - ✓ Residential mini-cogeneration units
  - ✓ fuel cell heating systems.,
  - ✓ Hybrid boilers
  - ✓ Mini or micro-cogeneration (Europe , Japan, US...)

✓ **Several Proof of concept or demo are on going in the world**

19

zero carbon footprint districts - 2 demo projects

Atlantech low carbon district

Building types : dwellings, offices, data center, new and refurbished NZEB
Technologies : PV & BIPV, microgrid, DHC and geothermal, H2 production and storage, electric batteries, EV, H2 vehicles

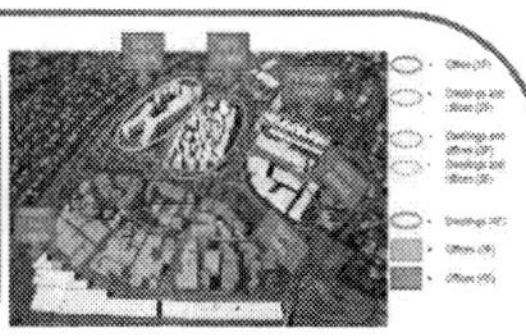

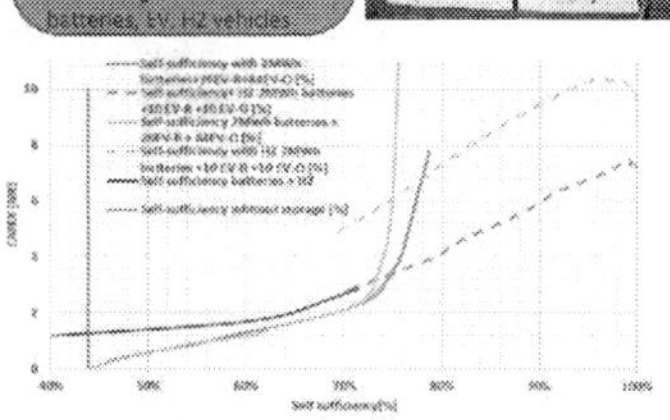

Main results
H2 cost-effective for 100% autonomous microgrid

Next steps :
- *Merit order of various energy sources and CO_2 calculation approaches for district and for the energy infrastructures*
- *Recommendations for microgrid operation and blockchain*

Châteauneuf

Building types : dwellings, offices, swimming pools, schools
Technologies : PV & BIPV, microgrid, gas cogeneration, electric batteries

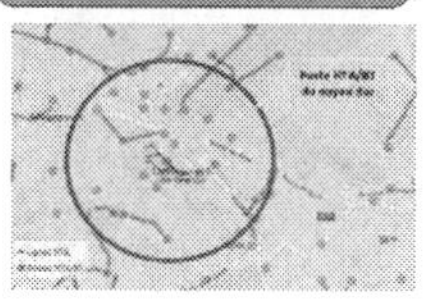

Main results :
- *Zero carbon footprint : Min Payback time 7 years*
- *The self-consumption rate increased if the number of clients is increasing => collective self-consumption thanks to the microgrid*

Next steps :
- *Improve the sizing tool*
- *New configurations and scenarios (increased perimeter) to study looking for CO_2 indicators*
- *Scale up the approach*

20

Vers une transition zéro-carbone

Electricity & fuel-gas convergence : Decarbonization of Mobility

- **Electric mobility** - The penetration of electric vehicles (EV) is accelerating (+ 68% in 2018) - at the end of 2019, 8 million EVs were in circulation at the end of 2018, 45% of which in China. **However, the share of EVs in the global passenger vehicle fleet remains below 1% & mainly focused on light mobility.**

- **Charging infrastructure remains the big challenge.** Today there are 630,000 public charging stations installed worldwide, but it will take much more to serve the growing fleet of VE.

- **CNG, LNG & Hydrogen mobilities are no doubt complements for Bus fleet, Heavy mobility and Urban light vehicles. Nevertheless, It remains:**
  - ✓ In development for Natural gas CNG & LNG mobility, but rapidly growing for Heavy mobility (long distance Bus fleets, heavy duty trucks , Ferry's ...) & also Urban mobility (Taxis, LV's, Bus fleets, Garbage trucks ...)
  - ✓ Very limited today for Hydrogen (Few Taxis or LV's,..), but Toyota and Hyundai both have ambitious development goals.

Multi-energy energy mix of low carbon mobility

21

Vers une transition zéro-carbone

Electricity & fuel-gas convergence : Examples- Decarbonization of Mobility

Power to Fuel-gas : Because of RE fluctuating and intermittent characteristics, Power produced from REs cannot match electrical demand to 100%. Valorization of production excess To fuel-gas for mobility is be profitable solution. Several project are ongoing.

- GRHYD project (France) Green power to **Hythane** fuel-gas for bus fleet (SNG blend - 80% NG +20% H2)

- **Wertle project E-gas/ AUDI** - Power to SNG via methanation process

- Power to H2 project. Green electricity to Hydrogen direct use (H2 Bus fleets) – **GNVert ENGIE** projects (Cities of PAU and Artois Gohelle – France),
  - Electrolyser
  - Compress & delivery station
  - New hydrogen buses

22

THANKS
© ENGIE 2019
ENGIE

RICE GRTgaz
Research & Innovation Center for Energy

Special session VI

Investigation of biomethane trace elements compositions: data science study to improve the knowledge

25 February 2020

Renewable Natural Gas in France and Europe

RICE GRTgaz Investigation of biomethane trace elements compositions: data science study to improve the knowledges_IGRC 2020 2

Renewable Natural Gas in Europe

- **European goals = 10% of renewable gases in 2030**
- **Biomethane injection well established in some countries**
  - 621 units in Europe (end of 2018)
  - Annual production = 22 TWh (2018)
  - Significant growth: +9% number of injection units
    - Germany: 216 units, 10 TWh/year
    - Denmark: 32 units, 1.75 TWh/year
    - Netherlands: 40 units, 1.1 TWh/year
    - UK: 88 units, 4.13 TWh/year

Data: Biomethane European Benchmark - France Biomethane with Sia Partners – December 2019 https://france-biomethane.fr/

Renewable Natural Gas in France

- Situation at the end of 2019

→ **102 RNG injection sites on the distribution network (97 for GRDF, 3 for RGDS, 1 for Regaz, 1 for Veolia);**
→ **13 RNG injection installations on the transmission grid (11 for GRTgaz, 2 for Teréga)**

- Expectations for 2020

→ **≈ +25/30 RNG on the distribution network**
→ **≈ + 12 RNG on the transmission network**

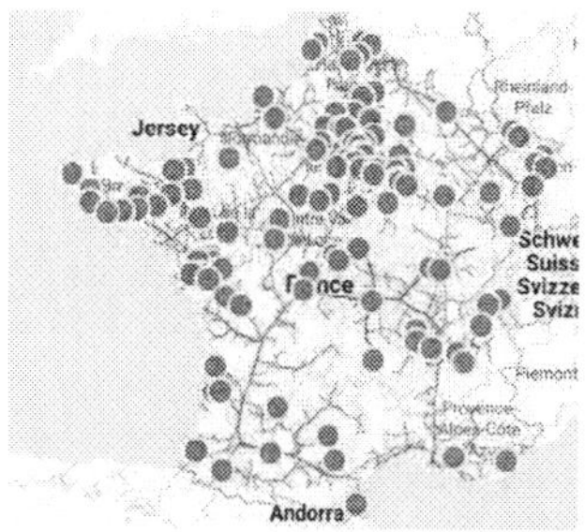

Data: https://opendata.reseaux-energies.fr/

Expectations of ADEME (French Environmental Agency)
→ **to set between 500 and 1400 RNG installation by 2030 (12 to 30 TWh)**
→ **New French policy (under discussion) around 8-10%**

The CARABIO project

- **CARABIO for ChARActerisation of BIOmethanes injected on the French transmission and distribution network**
- **Observations: limited data available on the literature for biomethane**
- **Consequences: limited knowledge on the trace compounds of the biomethane (vs. natural gas)**
- **Biomethane quality based on European standard EN 16723-1**
    - For some parameters = no threshold value
    - Threshold values may not be alligned with reality?
- **What about the compounds outside the specifications and standards?**
- **Need to increase our knowledge → inform the users of gas**

The CARABIO project

Several companies involved

storengy

GRDF

RICE GRTgaz Research & Innovation Center for Energy

GRTgaz

TERÉGA

The CARABIO project

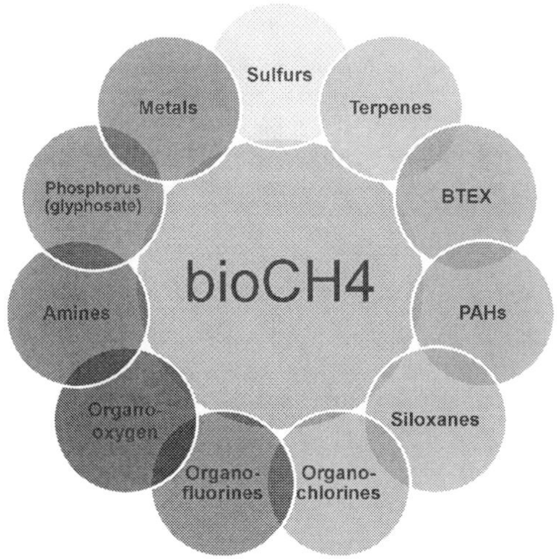

11 chemical classes
More than 450 compounds identifiable
Three-years projects
58 campaigns
5 different types of feedstocks:

- Agricultural
- WWTP
- Domestic waste
- Industrial waste
- Landfill

Data science study

- **Ranking the data by type of feedstocks and by type of upgrading process**
- **Total = 58 sets of data**

	Cryogenic	Pressure Swing Adsoprtion	Water scrubbing	Amines scrubbing	Membranes
Agricultural	0	5	4	1	15
Domestic waste	0	0	7*	0	1
Landfill	10*	0	0	0	0
Industrial waste	0	0	0	0	1
WWTP	0	0	0	0	14*

** Including one campaign dedicated to metals analysis*

- **A minimum of 10-15 sets of data is required to have consistence**
- **Some data shown will just give a tendency**

Stats on the number of compounds

Distribution of the number of compounds per site

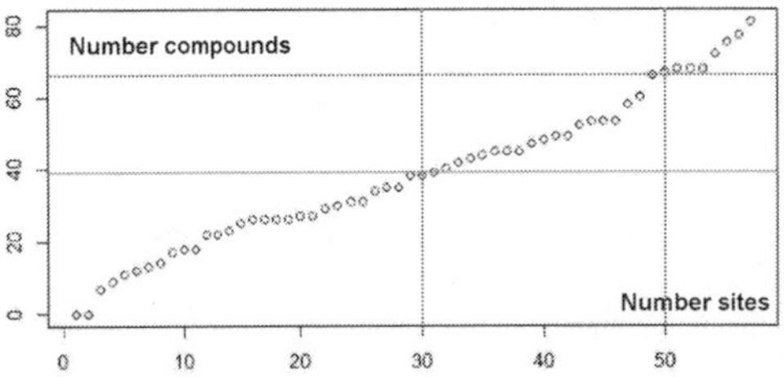

58 sites (including 3 for metals measurements)
30 sites at least 39 compounds
50 sites at least 66 compounds
At most, a site has 78 compounds (17% of the total)

Distribution of the number of sites per compounds

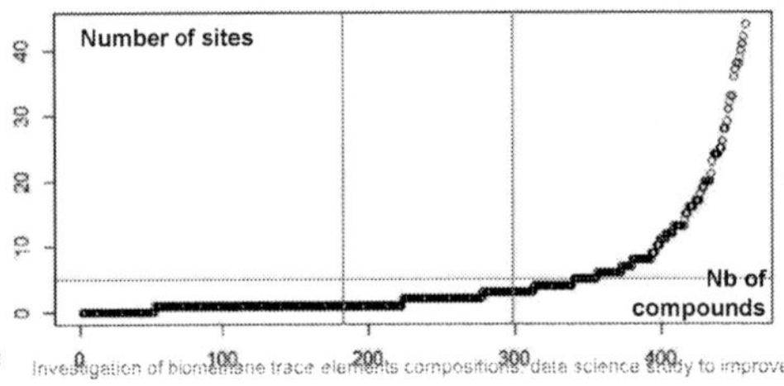

457 compounds
50 compounds are never presented (mix of standards used for the calibration)
173 compounds are just presented once
133 compounds are presented 2 to 5 times
0 compound isn't present on all sites
At most, a compound is presented on 77% of the sites

Occurrence of the compounds – main results

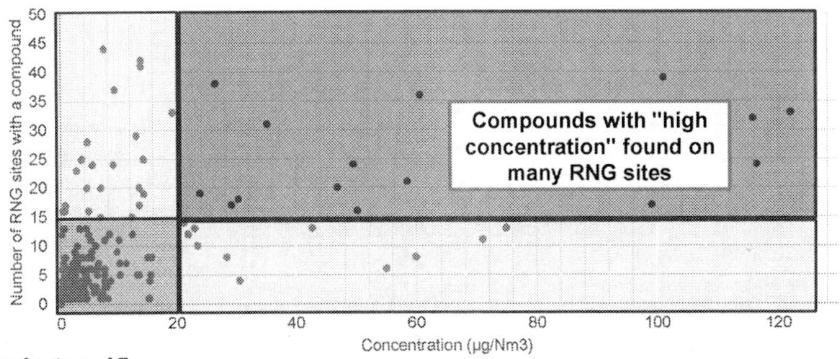

- **Compounds top 15:**
  - Ketones: 2-butanone, 2-pentanone
  - Terpenes: Limonene, pinene, camphene
  - BTEX: benzene, toluene,
  - Hydrocarbons: methylhexane, methylcyclopentane, cyclohexane, methylcyclohexane, n-heptane, n-octane, n-decane
  - Siloxanes: D4

Occurrence of the compounds – main conclusions

- **Compounds in the TOP 15**
- **In adequacy with the results expected**
- **Main detected compounds are:**
  - Ketones
  - Siloxanes
  - BTEX
  - Terpenes
  - Some hydrocarbons

- **Consistence with the current R&D efforts:**
  - Metrology for biomethane – EURAMET project – 16ENG05 → one of the goals is to develop standardized analytical tools to measure siloxanes, terpenes and halogenated VOCs (on-going) [end of the project June 2020]
  - GERG biomethane → working on siloxanes impact and improve knowledges on biomethane (on-going) [2019-2022]

Association rules – Data mining

- **Definition:**
  - Association rules = if-then statements that help to show the probability of relationships between data items within large data sets
- **Results for the upgrading process:**
  - *Rules observed at least 3 times*
  - *Confidence range = 95%*
  - PSA process → D3 and D4
  - Cryogenic process → 2-butanone, xylene, ethylbenzene
  - Trimethylsilanol, 3-propylfuran, cymene → membranes process

Association rules – Data mining

- **Results for the feedstocks & upgrading process:**
  - *Rules observed at least 3 times*
  - *Confidence range = 95%*
  - Agricultural & water scrubbing → pinene
  - Agricultural & PSA process → D3, D4
  - Landfill & cryogenic → 2-butanone, BTEX
  - Domestic waste & water scrubbing → 2-butanone, camphene, 3-methylhexane
  - n-propylacetate, 2-methyl-3-petanone, 2-propylfurane → Agricultural & membrane

→ Some rules are obvious, e.g. all the landfill facilities have cryogenic process, ditto domestic waste & water scrubbing

Association rules – Data mining

- **Results for the compounds: (A induces B)**

A	B
α-pinene	β-pinene
limonene	β-pinene
3-ethyltoluene	ethylbenzene
trimethylbenzene	p-xylene
trimethylbenzene	ethylbenzene
n-decane	m-xylene
n-decane	p-xylene
m-xylene	p-xylene
p-xylene	m-xylene
m-xylene	ethylbenzene
p-xylene	ethylbenzene

Hypothesis:
Rules observed at least 20 times
Confidence range = 95%

Data science study – main results

- **The landfill have about 40 compounds**
- **The sites using cryogenic and PSA process have less compounds than the ones using membranes**
- **A lot of compounds are just observed with Agricultural waste and/or membrane process**
- **Compounds top 15: terpenes (limonene, pinene, camphene), BTEX (benzene, toluene), ketones (2-butanone) → consistence with current R&D efforts**

- **This study enables us focus our analysis on specific sites like:**
  - Agricultural waste with upgrading unit (except membrane)
  - Continue to increase data on landfill, WWTP and domestic waste to have more consistence data

Next steps

- **Increase the knowledge on the RNG composition trace compounds and its impacts**

- **Reduce the cost of the gas quality and especially the biomethane injection plant**
  - One of the solution is to challenge the way RNG is controlled
  - → Find new tools/sensors to analyze biomethane composition
  - → Needs to be « Best cost »

- **RICE actions:**
  - **Continue CARABIO projects and to be involved in the European initiatives**
  - **Perform benchmark of potential sensors & analyzer, meet new suppliers, explain the biomethane needs and evaluate the analytical performances of the analyzers in our labs**

Contacts

Christophe Tastard
R&D Program Manager "Energy transition"
christophe.tastard@grtgaz.com

Dairo Ballestas Castro
R&D Program Coordinator "New methane"
dairo.ballestas@grtgaz.com

Amelie Louvat
R&D Program Coordinator "Gas impact on the safety of people & assets"
amelie.louvat@grtgaz.com

GRTgaz
Connecter les énergies d'avenir

grtgaz.com

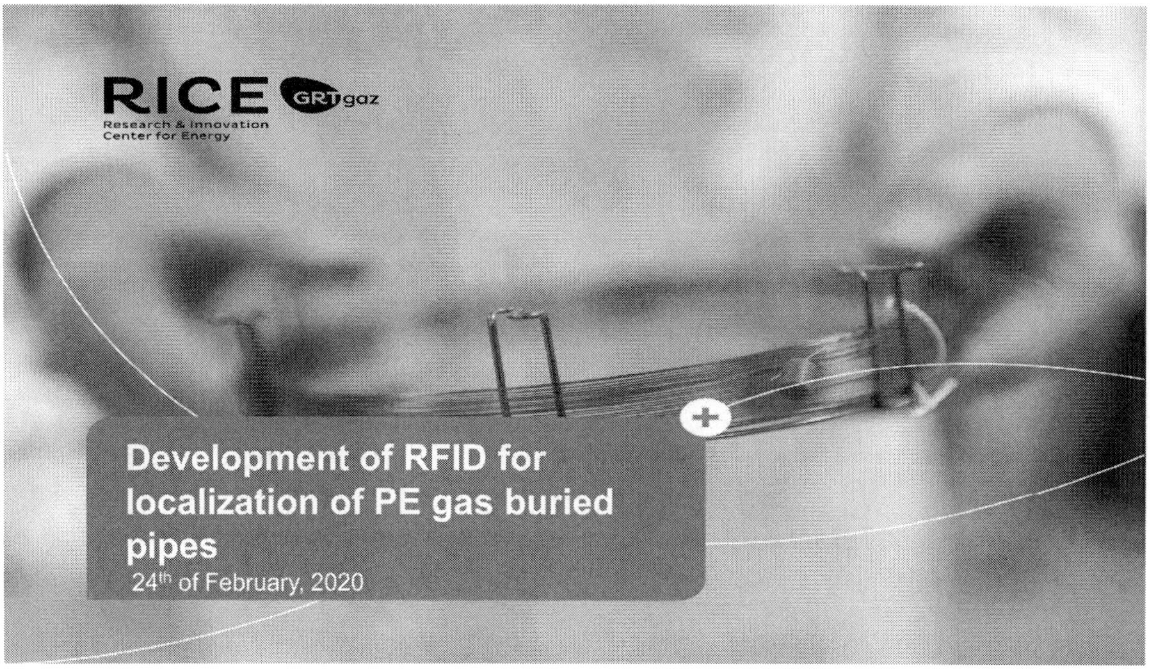

Context : Reducing Third Part Damage

- In France : 100 000 damages / year
- In US : an underground utility line damaged every six minutes
- A reinforced reglementation : One Call System in US and an equivalent one in France
- In US and in France : marking undergroung network is mandatory before digging

A big issue with the existing tools, especially for the small PE pipes

- In France: accurate cartography is mandatory in urban area since the 1st of January 2020
- New standards in US to have a better knowledge of each buried component especially for gas

What is commercialy available on the market today

- The solution commercially available consists of a standalone locator and the passive tags RFID
- Passive RFID tags are installed up *to 1,5 m on the top part of the pipe.*
- *The locator consists of a terminal, a* transmission/reception antenna to identify and geolocalize RFID passive markers
- For pipelines laid in an open trench
- After backfilling, with the portable Eliot® locator, the buried network can be located and tracked with a high accuracy and identified

MAIN REQUIREMENTS

- The aim of the 1st project developed with NYSEARCH and Eliot ® is to develop a passive tag for the pipelines installed by Horizontal Directional Drilling (HDD)
  - Designed for compliance with French and U.S standards
  - Minimum 1.52m (5ft) in depth
  - Compatible with all soil types
  - Provides information per ASTM F2897-15a (Tracking and Traceability Encoding System for Distribution Components)
  - Allocates space to include work order number
  - Geo-tagged data
- At the end of the PE pipes manufacturing proccess

↳ *New machine to design to lay the tags at the end of the extrusion line*

First results

- Developping a prototype which taking account that coiled pipes can twists and turns during installation with an **unpredictable** final RFID tag position
- A RFID **omnidirectional** and capability has to be developped

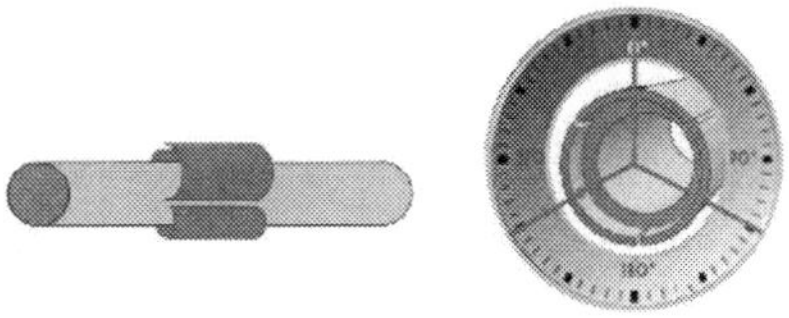

- The best prototype is a **macrotag** with 3 individual tags which individually contains a RFID chip

Integration of the macrotag on the extrusion line

- The prototype has been integrated manually to the extrusion of PE tubing, covered by a coating layer

Prototype tests

- The purpose of these tests was to verify the accurate detection whatever:

  - the pipe angle
  - the pipe depth
  - the soil in which the pipe is buried

- Different measures were taken for different types of soil (Gravel, sand, limestone, topsoil, clay

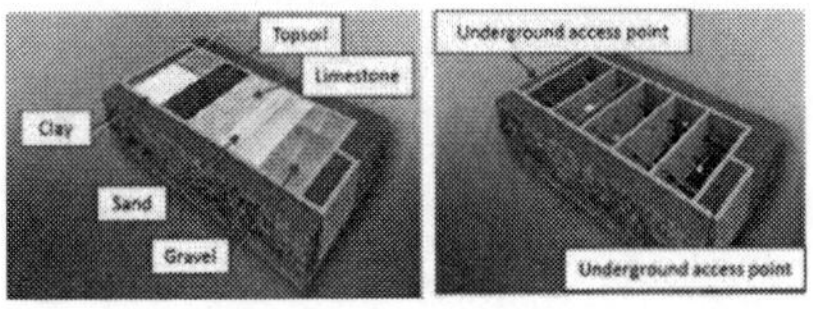

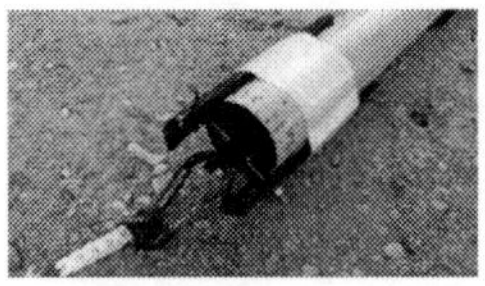

Industrial Phase : under progress

- Main steps :

  - Update specifications
  - Tag laying machine specifications
  - Laying machine design
  - Laying machine prototype
  - Laboratory tests
  - Locator software modification
  - Field tests in real conditions in France and in US

Universal locator : under development

- Based ont the VLOC PRO3 from Vivax Metreotech
- Will combine different functionalities :

  - Electromagnetic detection (VLOC PRO3 specifications)
  - EMS markers detection
  - RFID markers detection

- Main steps :
  - Feasibility study
  - Sensors/antenna development
  - Mechanical Integration
  - Tests in laboratory
  - Tests in real conditions

International Partnership

NYSEARCH Natural Gas RD&D

- NYSEARCH
  - voluntary research, development, and demonstration organization
- RICE GRTgaz – Research and Inovation Center for Energy
  - R&D center for gas infrastructures

GRTgaz

- ELIOT® division of the RYB Group ELYDAN
  - focused on research in innovative technological solutions dedicated to public works contracts.

ELYDAN
VOS PROJETS NOUS INSPIRENT

- ELYDAN
  - design and supply polymer products and solutions for the infrastructures, networks and buildings of tomorrow.

International Partnership

- Chevron Phillips Chemical Company LLC
  - Chevron Phillips Chemical is one of the world's top producers of olefins and polyolefins and a leading supplier of aromatics, alpha olefins, styrenics, specialty chemicals, piping, and proprietary plastics.

- Vivax-Metrotech
  - leading manufacturer of utility line locators, fault finders, marker locators, pipeline defect mappers, metal detectors and CCTV inspection systems

Perspectives: Augmented Reality

- Add some functions to the universal locator such as the visualization of the buried network in Augmented Reality
- A first POC in RICE : the locator allows the operator to visualize the route of the pipelines constructed from the detection points.

GRTgaz

Connecter les énergies d'avenir

grtgaz.com

A 100% green gas scenario in France by 2050?

France 2030: 10% of green gas

- State in the Law No. 2015-992 of 17 Aug 2015 relating to energy transition for green growth: 10% of gas consumed in 2030 should be from a renewable source = Objective set by the
- Biomethane from methanisation sector is well established in France, injecting biomethane transmission and distribution grid, with high growing projections for the next years

2018	2019
76 biomethane injection sites	107 biomethane injection stations at 30th Sept 2019

A theoretical potential of 460 TWh of renewable gas

ADEME, GRDF and GRTgaz report "*A 100% renewable gas mix in 2050? - Technical/economic feasibility study*" (2018) => conditions of a technical/economic feasibility of a gas system by 2050 based on a 100% renewable low-carbon gas

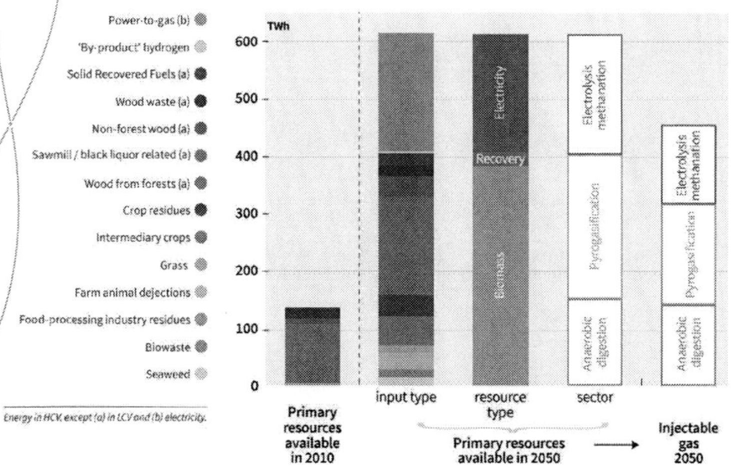

- All the renewable gas sectors required to fully decarbonise the gas
- High potential of biomass in France: need of a national transport network between French regions.
- Cost to replace the fossil gas by the renewable gases < the shadow price of carbon (250 €/tCO2 from 2030)

Example of potential synergies between RNG sectors

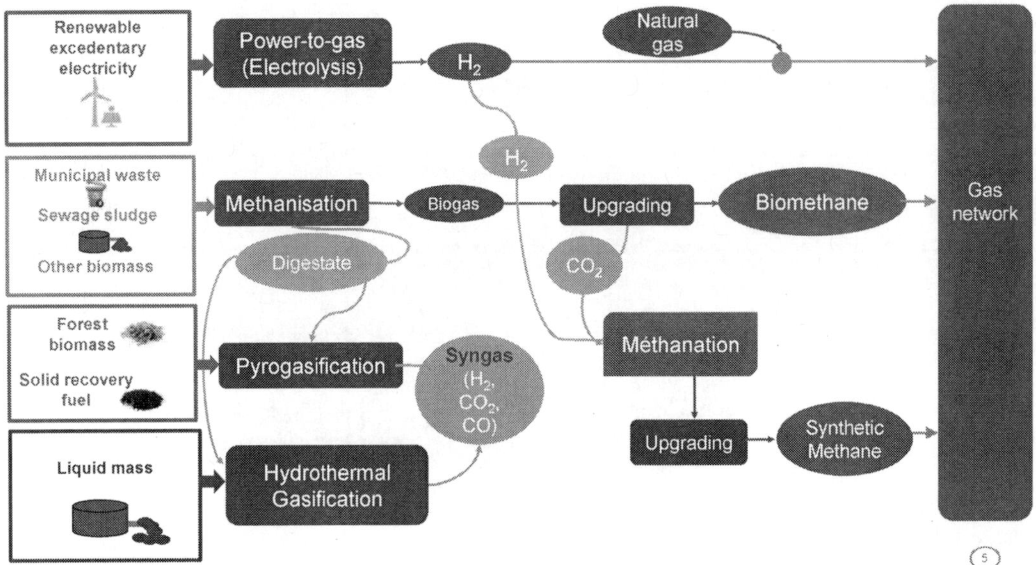

Classification GRTgaz : Public [ ] Interne [ ] Restreint [X] Secret [ ]

5

The Renewable Natural Gas R&D Roadmap

6

Renewable gases challenging keys

2021 1st potential revision of the European standard **EN 16723-1** about biomethane quality and publication of the 1st standard concerning the biomethane injection station (commission CEN/TC 234/WG-6).

2021/2023 GRTgaz
- **Hydrothermal gasification: 1st potential injection into the gas grid from an R&D pilot (GRTgaz target, under revision).**

2023/2028
- Reduction of the cost of the injected biomethane at a pace compatible with the decrease of the methanisation feed-in-tariff level (French Ministry for the Ecological and Solidary Transition)

2023/2025 TSO/DSO
- **Pyrogasification: 1st potential injections into the gas grids from low-power industrial unit (French TSO/DSO target)..**

2026 2nd potential revision of the European standard **EN 16723-1** about biomethane quality

2028 TSO/DSO
- Pyrogasification: reach 1TWh of injected gas (French TSO/DSO, target under revision).

2030
FR: 10% of green gas in the grid.
EUR :**10% of green in the energy consumption**
EUR : **reduction to 55% of the Greenhouse gas emissions compared to 1990**

Classification GRTgaz : Public [ ] Interne [ ] Restreint [X] Secret [ ]

7

2020-2030 R&D Roadmap: 3 axes

Issue: To reduce the costs of the RNG injeciton while supporting the development of the RNG production chain and better managing the possible impacts on the gas chain.

1 Optimization of the gas infrastructures (injection station) to reduce the cost of gas transit and to manage the impact of RNG

2 **Support of the development of RNG production processes in order to reach an industrial and flexible stage of the technical solutions *(R&D pilots and POC's)***

3 Accompanying the development of prospective « new grids »(*)

(*) New-grids such as the biogas "self-injection" and the centralized upgrading and injection of biomethane from raw biogas collected from several production sites

Classification GRTgaz : Public [ ] Interne [ ] Restreint [X] Secret [ ]

8

2020-2030 R&D Roadmap: areas of expertise

Complementary expertise are needed to all along the 2020-2030 roadmap. They are part of RICE core competencies.

Biomethane and Synthetic methane injection station

- Analysis and odorization at best cost
- Reliability of metering, analysis and odorization methods
- Global optimization of injection stations

Gas transmission

- Gas quality : deep characterization and development of analytical methods and impacts on the gas grid. (Ex : CARABIO, EMPIR)
- Network integrity: corrosion studies.
- Odorization : odor masking effects

Classification GRTgaz : Public [ ] Interne [ ] Restreint [X] Secret [ ]

9

2020-2030 R&D Roadmap: areas of expertise

Complementary expertise are needed to all along the 2020-2030 roadmap. They are part of RICE core competencies.

Production processes

- Reactor, separation and upgrading units
- Set-up of R&D pilots and demonstrators

"Self-injection" and Centralized upgrading/injection

- Technical and economic feasibility
- Raw gas quality and metering
- Leaks detection
- Raw gas compression and upgrading
- Raw gas gathering network: integrity and materials.

Transversal

- Work on European and International Standardization groups.
- External communication.

Classification GRTgaz : Public [ ] Interne [ ] Restreint [X] Secret [ ]

10

R&D Pilot Units for RNG

11

New R&D pilot and demonstrator projects have been started in the last years to support the technology development for the production of RNG from gasification, methanation (including Power-to-gas) and methanisation processes.

12

Methanisation Projects (1/1)

- **West Grid Synergy**
  - ✓ 1st experiment of the SMART GAS GRID for renewable gas in the region.
  - ✓ Combines digital and energy technologies to maximize biomethane production, develop new efficient uses of gas and optimize overall management of the energy system
  - ✓ Key example : 'reverse flow' installations at the interface between the distribution grid and the transport grid allowing bidirectional flow of biomethane.

Partners: GRTgaz, GRDF, Soregies, Morbihan Énergies, SIEML (French Maine-et-Loire energy union), SYDEV (Vendée energy union), the Pays de la Loire region and the Brittany region.

- **Méthagrid**
  - ✓ Goal: to develop a membrane bioreactor technologies for biogas upgrading applications.
  - ✓ Specific development: a biological methanation process to upgrade in-situ a biogas, inside of the digester (methanisation of biomass), in order to increase the CH4 production for injection into the gas grid
  - ✓ Intensification of the reaction via a membrane contactor which increase the H_2 content in the liquid phase

Partners: Cristal Union, GRTgaz, Vivescia, ARD, TMA-process, the R&D unit ABI of AgroParisTech and the CentraleSupélec Chair in Biotechnology.

Pyrogasificaiton Projects (1/4)

- **Synthane**
  - ✓ Pilot plant to produce synthetic methane by gasification using a patented process called 'SYNTHANE®', developed by ETIA: a high temperature pyrolysis facility
  - ✓ Goal: to confirm that the injectable quality of the methane produced from several feedstocks.
  - ✓ Tests: started in 2018, will be held until the end of 2020. Specific deep analysis campaigns performed by RICE.

Partners: GRTgaz, ETIA

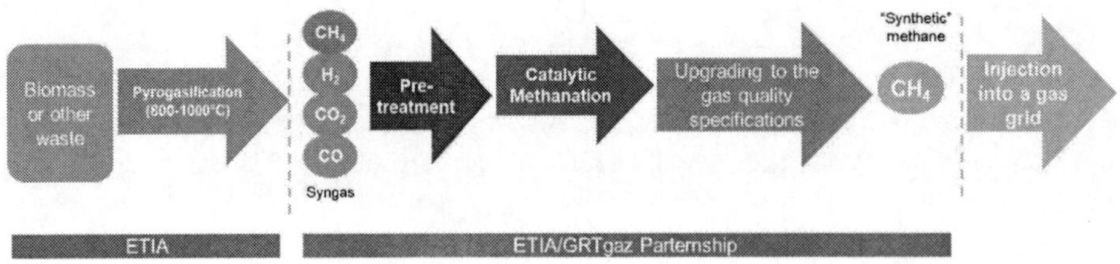

Pyrogasificaiton Projects (2/4)

- Titan 5
  - ✓ Pilot plant to produce synthetic methane by gasification (supplied by Leroux & Lotz process) followed by a biological methanation (supplied Terrawatt)
  - ✓ Tests: started in 2019, will be held until the end of 2020. Specific deep analysis campaigns performed by RICE.

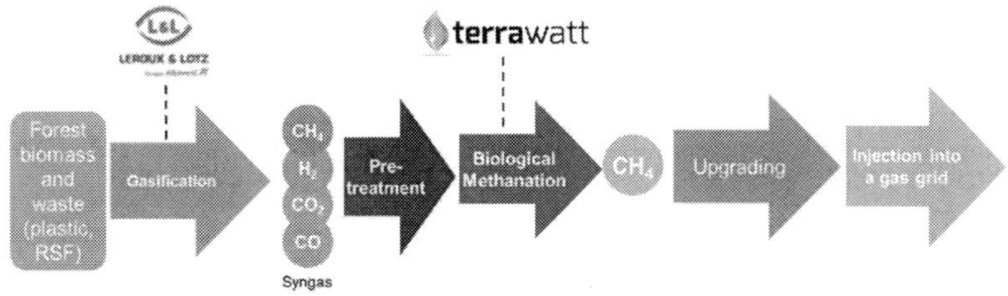

Pyrogasificaiton Projects (3/4)

- **Plainénergie**
  - ✓ Experimental installation to treat and convert into energy (via gas injection into the grids) a wide range of non-recyclable wastes collected at the Plaine de l'Ain region in France
  - ✓ Wastes include wood and the end of its life, plastic waste, bulky waste, agricultural waste, among others.
  - ✓ Tests: started in 2019, will be held until 2021. Specific deep analysis campaigns performed by RICE.

Partners: GRTgaz, CCPA, PIPA, Enosis, Provademse, Séché Environnement, ENOSIS, PROVADEMSE, INSAVALOR Platform, and the laboratoires DEEP and TBI from the INSA Lyon and Toulouse

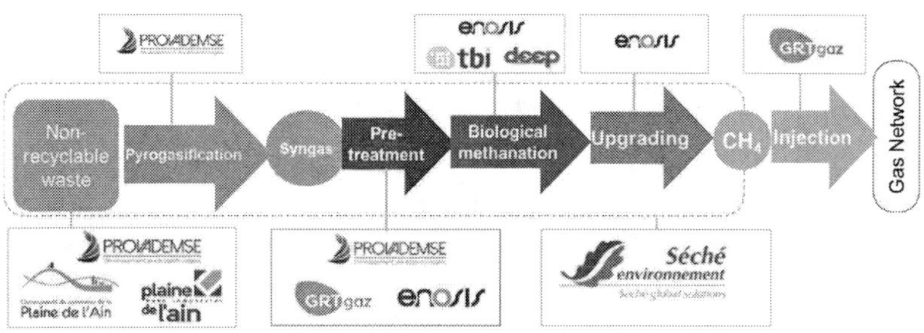

Pyrogasificaiton Projects (4/4)

- **Syctom et GRTgaz collaboration (framework: Cométha project)**
  - ✓ Partnership: share of expertise on gasification and methanation processes + gas quality measurements to be performed by RICE
  - ✓ Future goal: to recover and transform into injectable gas a part of the 7500 tons of municipal handled by Syctom, in a circular economy context.
  - ✓ Framework : Cométha project (the Syctom and the SIAPP) aiming to reduce the remaining waste from municipal waste, domestic/industrial waste water and rainwater effluents and to maximize the produced energy

Partners: GRTgaz and the Syctom (Intercommunal household waste treatment and recycling Syndicate)

Power-to-Methane and other Methanation Projects (1/2)

- **Jupiter 1000**
  - ✓ First demonstrator in France to recover surplus electricity from renewable sources and recycle CO_2
  - ✓ Power rating of 1 MWe for electrolysis (non-consumed renewable electricity converted into H_2) and a catalytic methanation process with carbon capture (CO_2 capture unit on the chimneys of a local industrialist).
  - ✓ The project reached its final phase of construction and launched the tests program

Partners: GRTgaz, McPhy Energy, Atmostat, CEA, Leroux et Lotz Technologies, CNR, GRTgaz, Teréga, Grand Port Maritime de Marseille

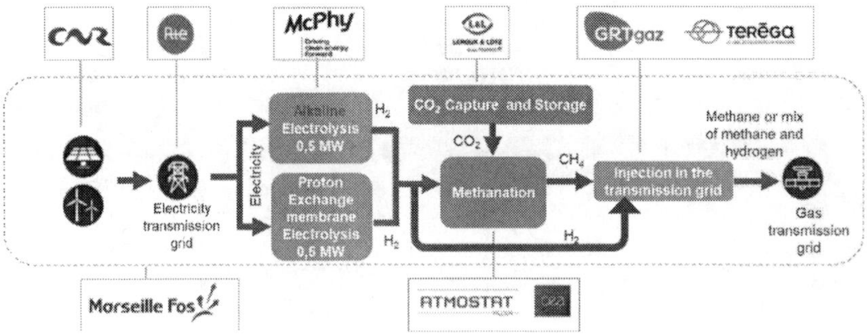

Power-to-Methane and other Methanation Projects (2/2)

- **MINERVE**
  - ✓ Low-power catalytic methanation demonstrator consisting of two in series fixed bed adiabatic reactors, placed adjacent to a biomass heating plant at the AFUL Chantrerie site
  - ✓ Global concept : recover the CO_2 contained in the combustion smoke from a heating plant + the H_2 from electrolysis (using the green electricity produced by solar panels and a wind turbine), then coupled to the innovative catalytic methanation reactor.
  - ✓ The produced methane feeds a CNG (Compressed natural gas) station.

Partners: GRTgaz, GRDF, Teréga and the Paris Diderot University

- **MARS** *(GERG Young Researchers' Awards in 2019)*
  - ✓ Renewable Methane Production by Catalytic reduction of CO_2 using sunlight
  - ✓ Project goal: contributing to the management of energy resources and environment by proposing a solution for recycling CO_2 and storing solar energy
  - ✓ Ongoing work: understanding the key parameters of the reaction, in order to be able to optimize the catalytic cycle and to scale up the process for future industrial applications.

Partners (main ones): AFUL Chantrerie, IMT Atlantique, Top Industrie and GRTgaz.

Classification GRTgaz : Public [ ] Interne [ ] Restreint [X] Secret [ ]

19

GRTgaz

Connecter les énergies d'avenir

grtgaz.com

DNV·GL

The future role of gas in a carbon constrained world

The power of joint innovation

Johan Knijp

February 25, 2020

IGRC 2020
MUSCAT
24-26 FEB | WWW.IGRC2020.COM

SAFER, SMARTER, GREENER

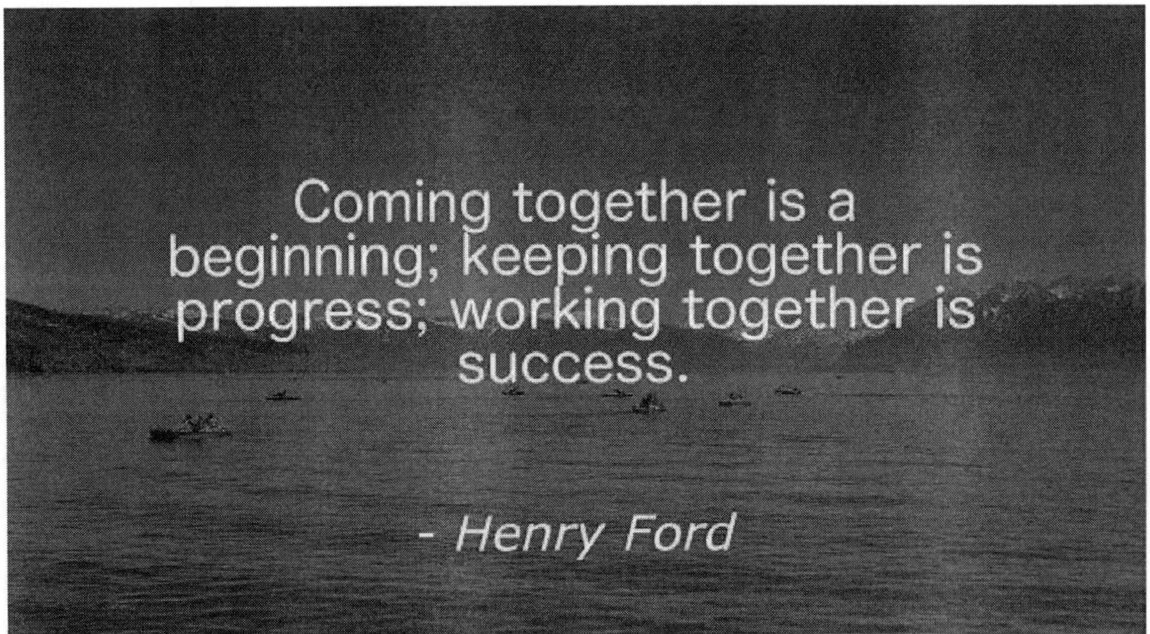

Decarbonization takes centre stage
Source: DNV GL – 2020 Industry Outlook

Percentage of respondents who expect their organization to increase investment in decarbonization this year

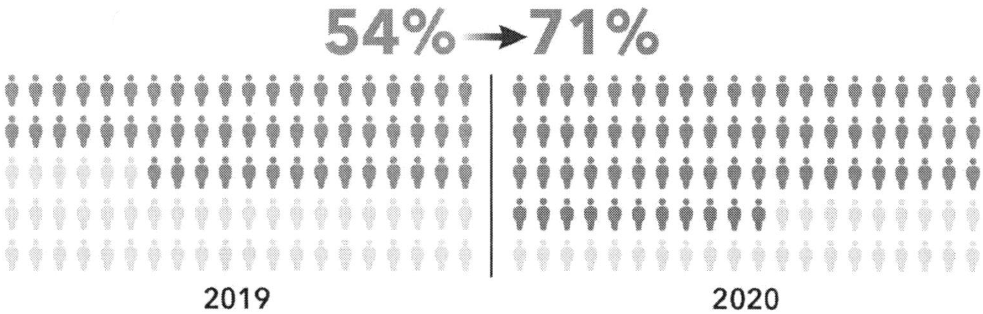

A decade of outlook research - DNV GL – 2020 Industry Outlook

- Yearly industry benchmark study on the outlook for the oil and gas industry
- Provides expert analysis of the key challenges facing the industry in the year ahead.
- Insights on industry sentiment, confidence and priorities.
- Launched in 2011 and is now in its 10th year.

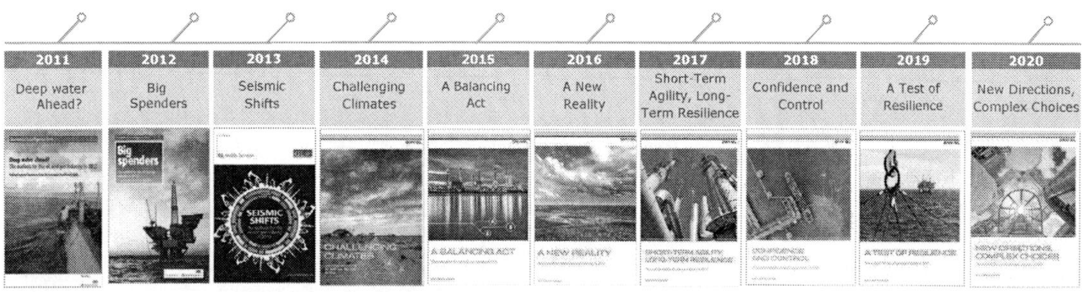

Content

1. The power of joint innovation

2. DNV GL's Energy Transition Outlook

3. The great Dutch gas transition

4. Examples of research in renewable gasses

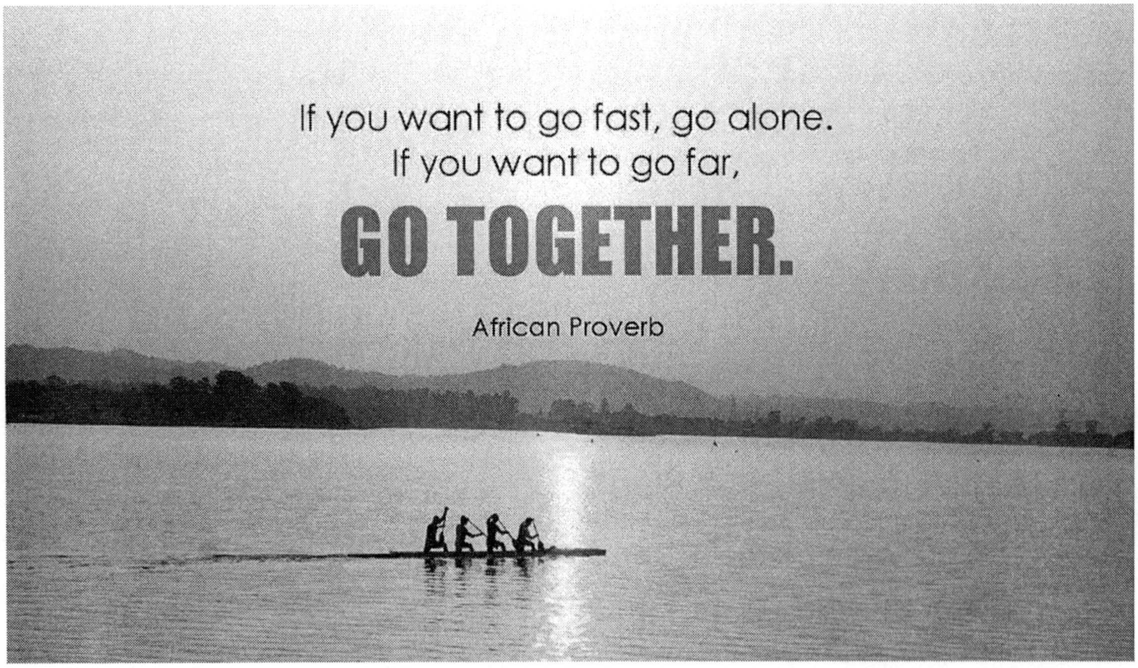

DNV GL - Global reach – local competence – including 50 years in natural gas

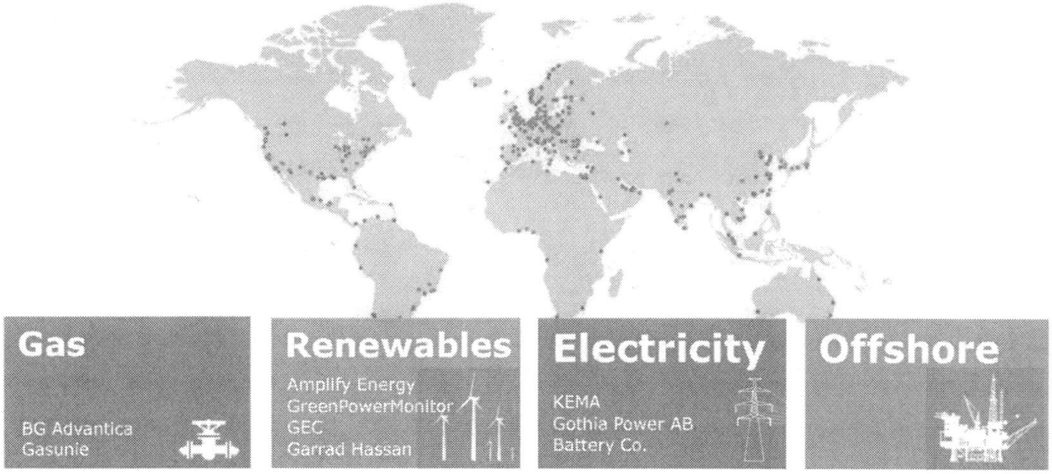

7 DNV GL © 2017

DNV·GL

A heritage of innovation and technology development in natural gas technology – a snapshot

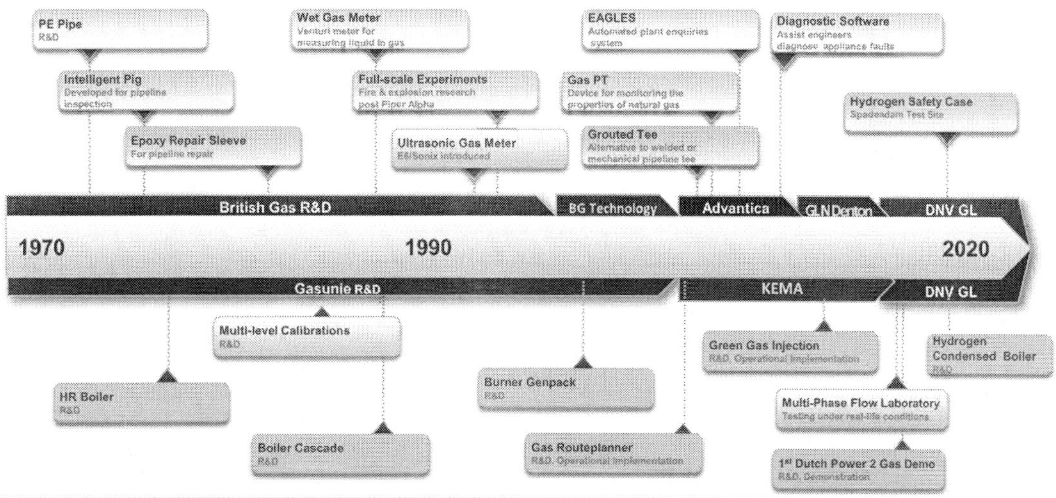

8 DNV GL © 2017

DNV·GL

Committed to innovation and cooperation

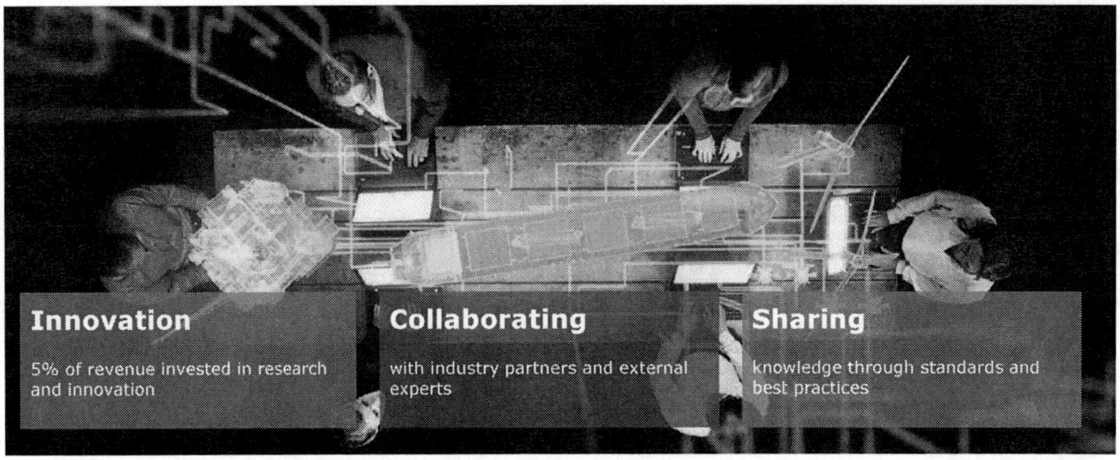

Committed to innovation – DNV GL's Joint Industry Project

- Our joint industry projects (JIPs) bring together the deep technical expertise of our people and our customers
- They help to tackle a variety of technical issues across the sector, and aim to develop new technologies, industry standards and recommended practices.
- Our project partners feed into the initial project scope of work, and have access to and input into all the research and development undertaken
- Many of the sector's best-known and most widely-adopted practices are the outcome of a joint industry project
- Yearly 50- 70 new JIP projects started

Putting our expertise to the test:
DNV GL's Research and Testing laboratory landscape for the gas industry

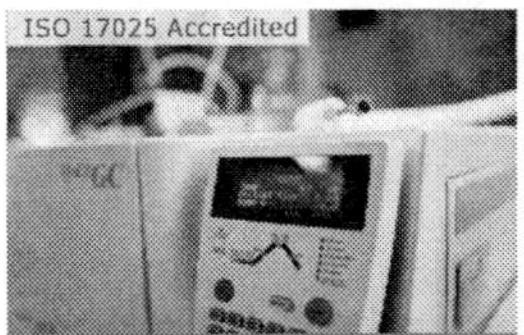

Gas analytical laboratory
Analyses of oils and gaseous fuels

Engines laboratory
Fuel combustion behaviour and performance testing

Technology Qualification
Multi purpose laboratory for sustainable technology testing

Fuel Research
Facilitating the introduction of new fuels

Multi-phase flow laboratory
Testing with different fluids (liquids, gases & solids)

Flow Calibration centres
Flow testing and calibration in the UK and NL

Hazardous testing
Full-scale hazardous trials in simulated real-world environments.

Hydrogen test facility
Development qualification of safe and reliable equipment

2. DNV GL's Energy Transition Outlook

Private and confidential

Energy Transition Outlook 2019 - A global and regional forecast to 2050

Close to equal split

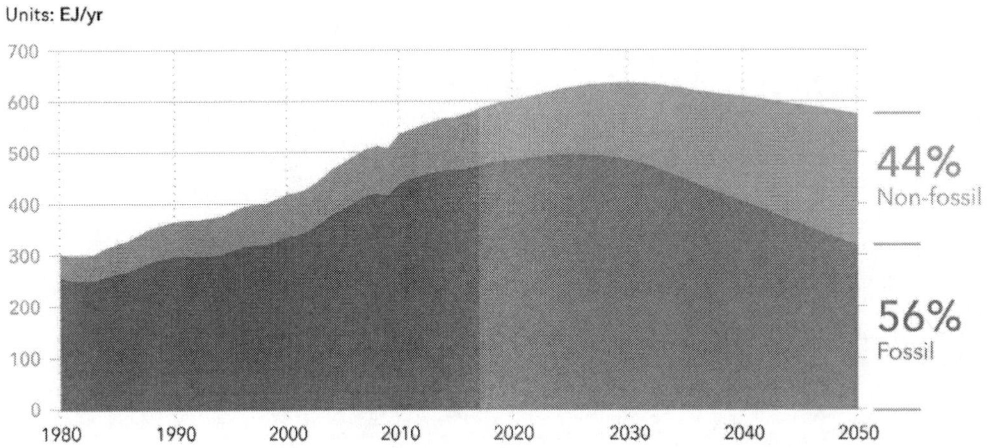

Electrification is a major theme of the ongoing transition

World electricity generation

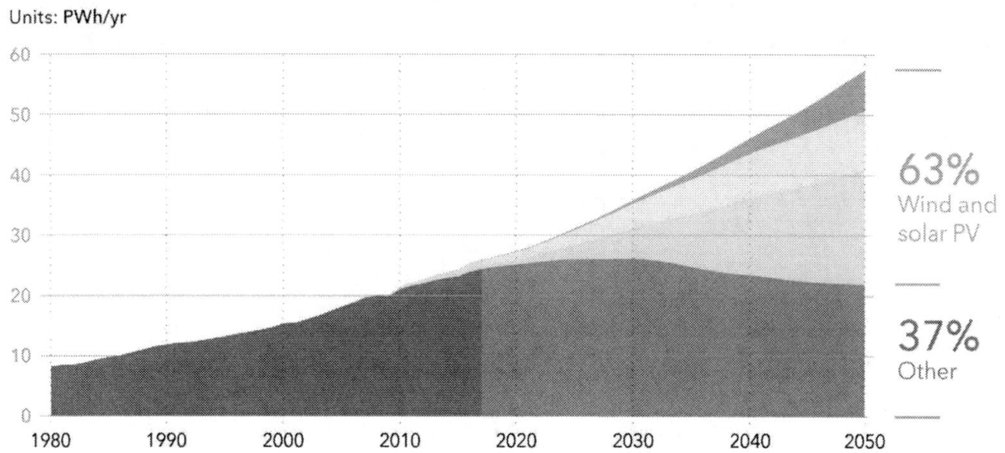

Gas takes over from oil as the largest energy carrier

Primary fossil fuel

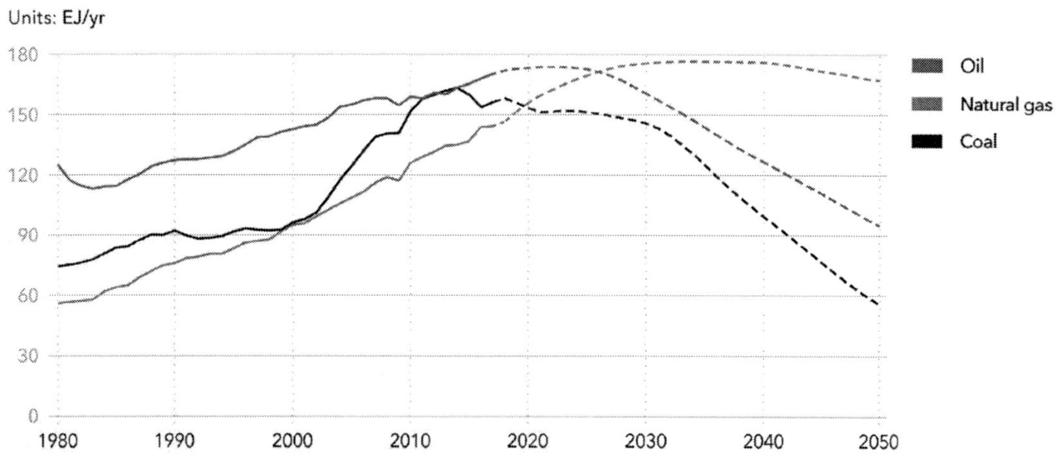

3. The great Dutch gas transition

The great Dutch gas transition

Embarking on an energy transition intended to lead to a complete phaseout of unabated natural gas consumption and production by 2050.

- The prospects for natural gas in the Netherlands changed dramatically between 2012 and 2018 due to rising concerns over climate change and induced earthquakes in the gas-producing province of Groningen
- In March 2018 the government announced that production from the giant Groningen field, for over 50 years the mainstay of Dutch gas production, will be phased out as quickly as possible and no later than 2030.
- In response to the political and public opposition to natural gas, the Dutch gas industry developed a strategy based on the continued use of 'molecules' in the energy system in the form of 'sustainable gases' (hydrogen, biogas, biomethane).

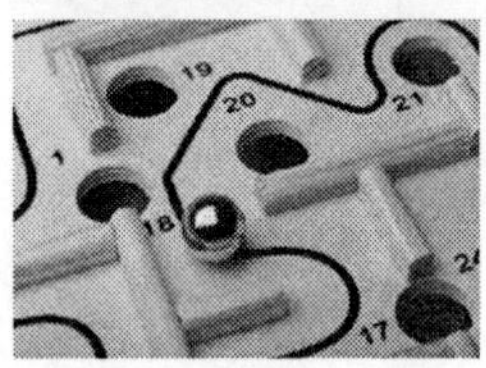

- Source: The Oxford institute for energy study – July 2019

New energy value chains

Energy System Framework – connecting sources with end users

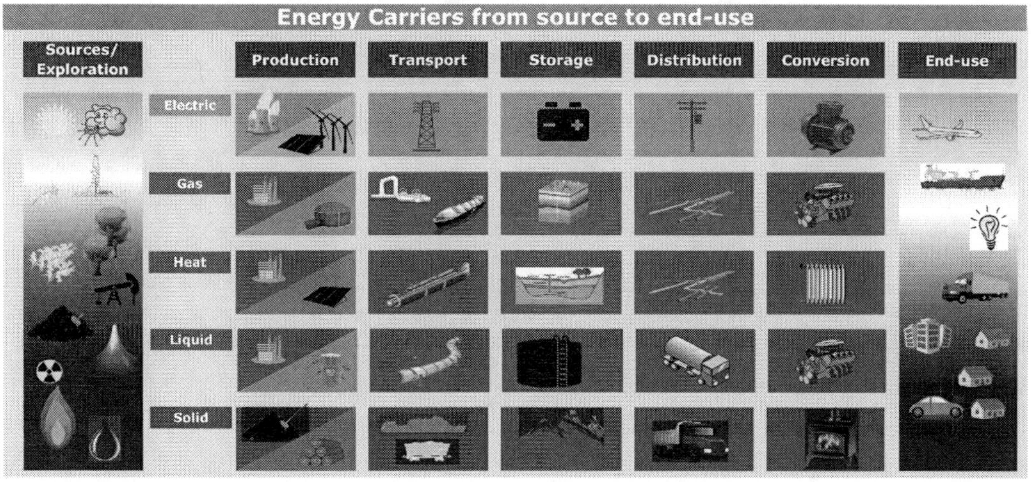

Example: Power and (bio) gas / hydrogen to mobility

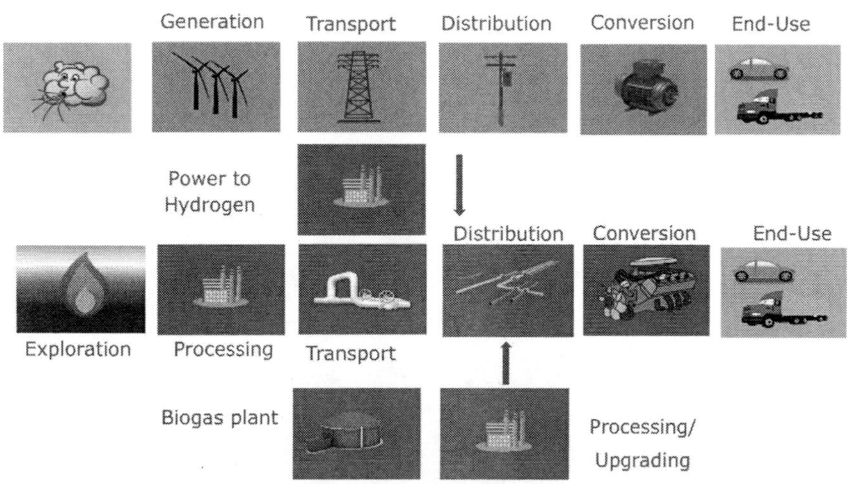

NORTHERN NETHERLANDS
HYDROGEN VALLEY
Welcome to Hydrogen Valley

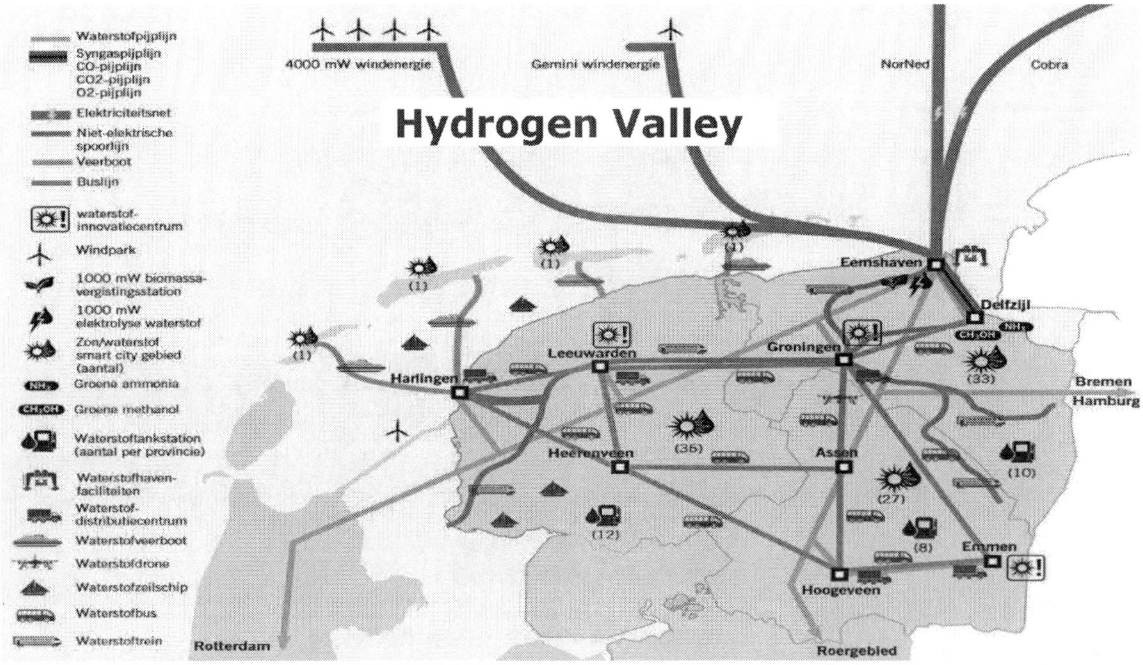
Hydrogen Valley
Waterstofpijplijn
Syngaspijplijn
CO-pijplijn
CO2-pijplijn
O2-pijplijn
Elektriciteitsnet
Niet-elektrische spoorlijn
Veerboot
Buslijn
waterstof-innovatiecentrum
Windpark
1000 mW biomassa-vergistingsstation
1000 mW elektrolyse waterstof
Zon/waterstof smart city gebied (aantal)
Groene ammonia
Groene methanol
Waterstoftankstation (aantal per provincie)
Waterstofhaven-faciliteiten
Waterstof-distributiecentrum
Waterstofveerboot
Waterstofdrone
Waterstofzeilschip
Waterstofbus
Waterstoftrein
4000 mW windenergie
Gemini windenergie
NorNed
Cobra
Eemshaven
Delfzijl
Leeuwarden
Groningen
Harlingen
Heerenveen
Assen
Emmen
Hoogeveen
Bremen
Hamburg
Rotterdam
Roergebied

4. Examples of research in renewable gasses

Welcome to HyStreet

Hydrogen Safety testing @ DNV GL's Spadeadem test facility (UK)

ROTTERDAM (NL) TO HEAT HOUSES WITH HYDROGEN
(Demonstration project)

- Hydrogen produced from green electricity via Power-to-Gas technology
- Transported via DSO Stedin's gas network
- Houses are heated directly with 100% hydrogen boilers.

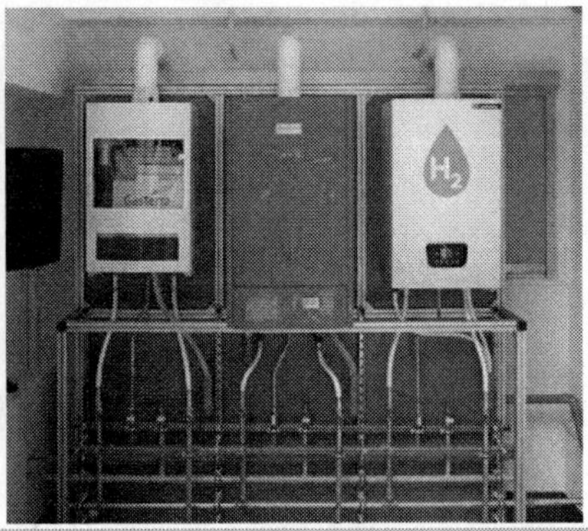

STEDIN
Gemeente Rotterdam
Ressort Wonen
BEKAERT
better together
remeha
DNV·GL
GasTerra

HyReady:
How to accommodate hydrogen blends in the natural gas grid?

Currently lack of guidelines for TSOs and DSOs.

Based on existing knowledge, HyReady is developing set of practical guidelines for adding hydrogen up to a 30% blend.

4 work packages ready:

- High-pressure gas networks and
- Low-pressure distribution networks.
- End user equipment
- Compressors

Results are easy available from a Wiki site

PG&E
FORTIS BC
ENBRIDGE
SoCalGas
enagas
gasNatural fenosa
Gas Networks Ireland
GRTgaz
GAZ system
TERĒGA
gasunie
DBI GUT
DNV·GL
GRDF

17 international partners from Europe, Asia, Canada and the US

Towards well-founded standards for siloxanes in bio-CNG

- An interesting and growing market for (bio)gas is the automotive market that uses **CNG as a transport fuel**
- Biogases can contain traces of **siloxanes**, which can cause mechanical problems to end-user appliances including gas engines
- DNV GL has performed several projects to test and study the performance of light- and heavy duty vehicle engines fuelled with gases containing realistic siloxane concentrations
- The results form the basis for well-founded siloxane **standards** to facilitate the smooth introduction of biogas in the European market

cen afnor GROUPE GERG

European Committee for Standardization

Modular bio-syngas pilot plant – technology qualification

Torrified biomass (bio coal) → Biomass transport → 2-step gasification → Syngas / Char

Two-step gasification is a promising technique on laboratory scale to produce syngas from biomass. In a two-step gasification process biomass is first pyrolysed and the pyrolysis vapours are then gasified.

The first step of this process will be investigated in a downer setup using torrefied biomass. In the gasification step no tar formation will take place.

torrgas

Torrgas is a Dutch, privately funded, clean technology company which has developed a cutting edge process for the production of Syngas from torrefied biomass.

Maturing methanation technology enables gas system decarbonization

Sharing comprehensive overview on chemical and biological methanation technologies and commercially available systems for grid injectable biomethane

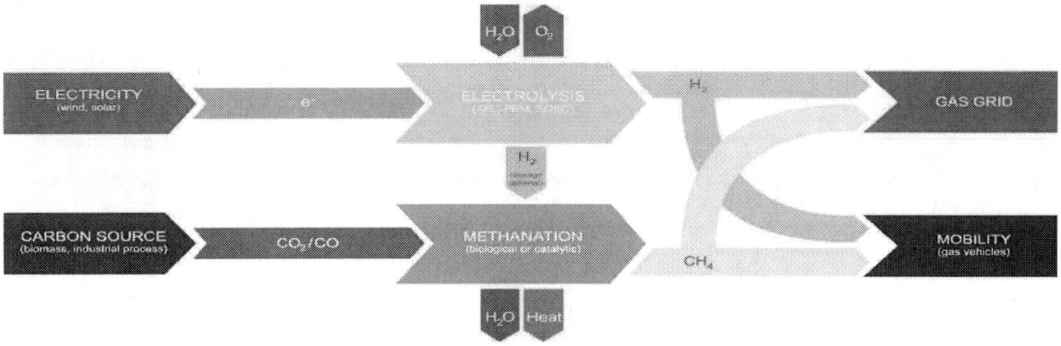

Source: Methanation: Technical fundamentals and market overview

gasunie new energy GasTerra

29 DNV GL © 2017 5 March 2020 DNV·GL

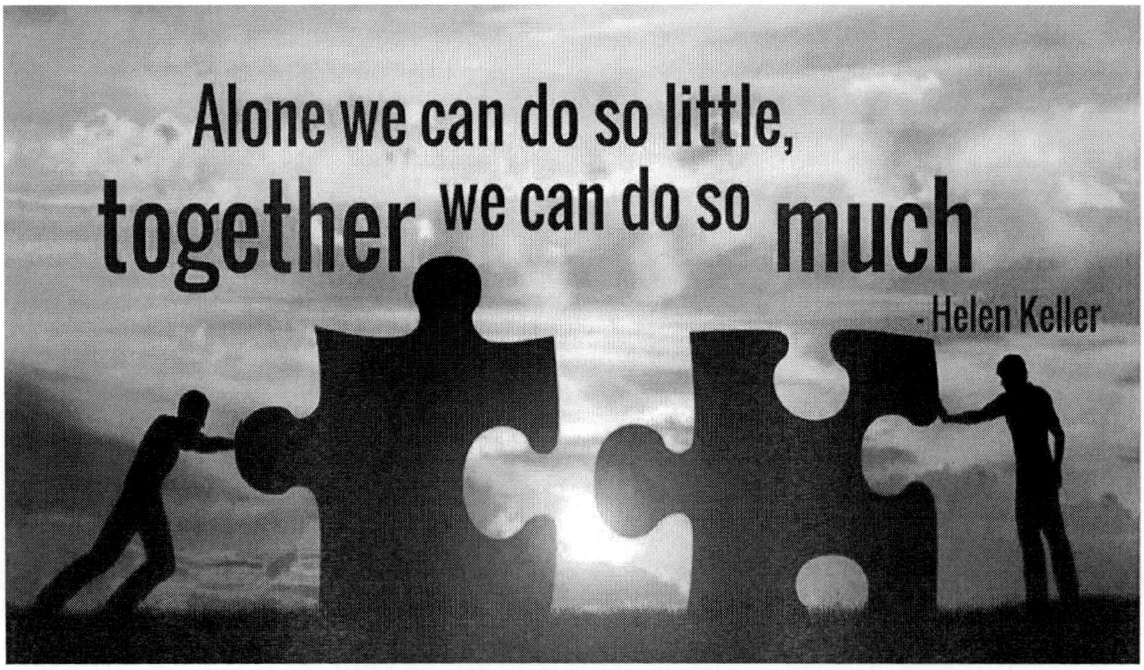

The power of joint innovation

Johan Knijp
Johan.Knijp@dnvgl.com
+31 650 26 24 10

www.dnvgl.com

SAFER, SMARTER, GREENER

OQ

Control Valve Maintenance Digitalization

Ahmed Al Mahrouqi
Head of Control & Automation, OQ

Care Share Dare

OQ

AGENDA

- About OQ
- OQ's Business Objectives
  - Control Valve Reliability for OQ's Operations
  - Condition Monitoring Implementation
  - Case Studies
  - OQ vision towards centralized monitoring
- Business Results Achieved
- Summary

Care Share Dare

OQ Gas Network

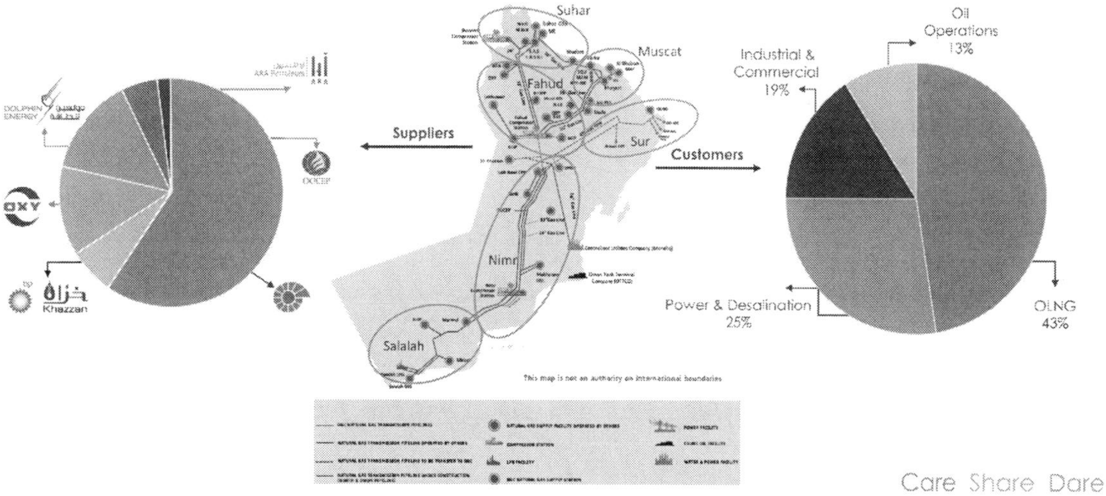

Care Share Dare

OQ'S OPERATIONAL RELIABILITY

- Enhance plant Availability and Pressure Control Valves (PCV) Reliability to acheive gas supply sustainability.
- Adopt technology to apply Predictive Maintenance to achieve Operation Excellence.
- Employ Digitalization to enable online remote monitoring for all critical valves from a Centralized Location.
- Reduce OPEX spending unnecessary.

Care Share Dare

TRADITIONAL MAINTENANCE PRACTICE

- Pressure Control Valve failure is a major contributor for gas outage and production losses.
- Around 500 PCVs scattered in 40 stations across 4000 Km pipeline to feed 100 customers.
- Conventional Time-based Maintenance lead to;
  - Good Valves dropped for repair.
  - Valves not planned for repair had issues.
  - Not able to plan the spare parts efficiently.
  - Potential delays for start-up.

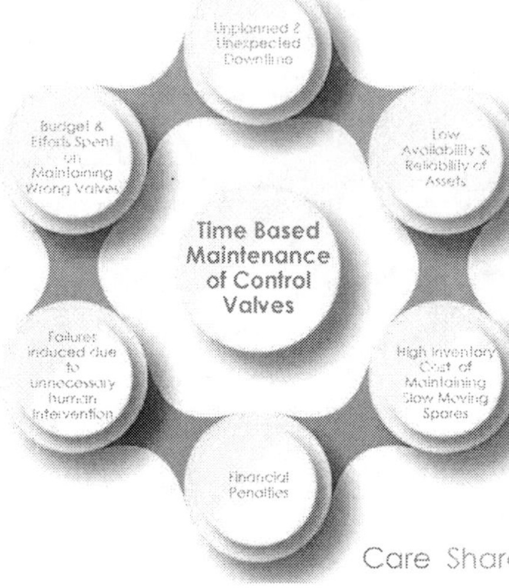

Care Share Dare

VALVE DIGITALIZATION ROADMAP

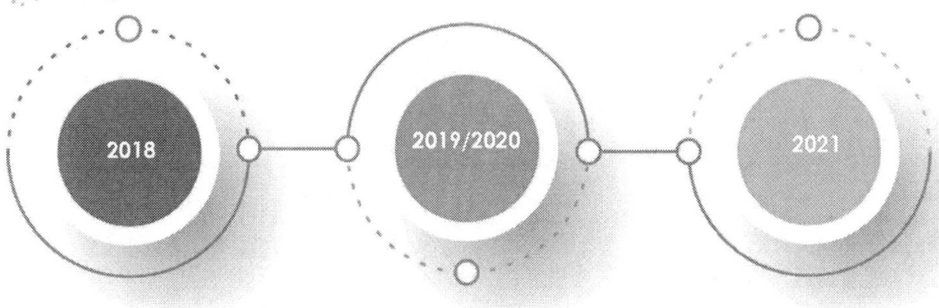

- Increasing Reliability issues on Control valves at Gas Network.
- Explore IIoT solutions and Plant Digitalization.
- Kick-Off Pilot project (8 tags - 4 months)

- 62 critical valves on 2019 was completed based on site readiness
- 52 critical valves on 2020
- Training and building in-house capabilities..

- Centralized monitoring system.
- Develop Process and procedures.
- KPI Dashboard.
- Further expansion across remaining sites.

Care Share Dare

SYSTEM ARCHITECTURE

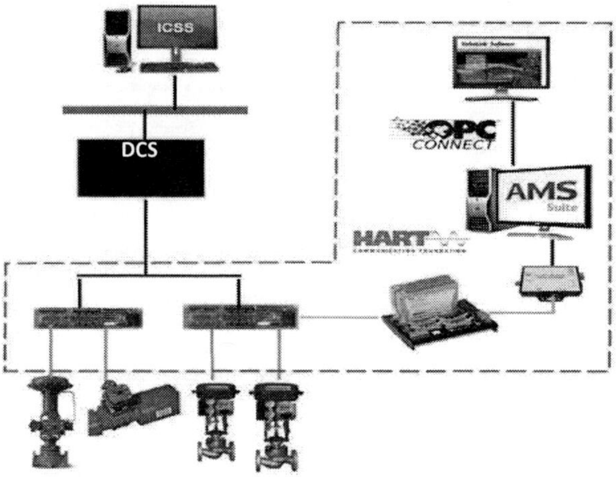

Technology Enabler

- FF /HART Platform
- Smart I/P Positioner
- Different Control Valves brands
- DCS/ICSS
- Assent Management Server.
- Valve Software.

Care Share Dare

WHAT CAN DIAGNOSTICS DETECT?

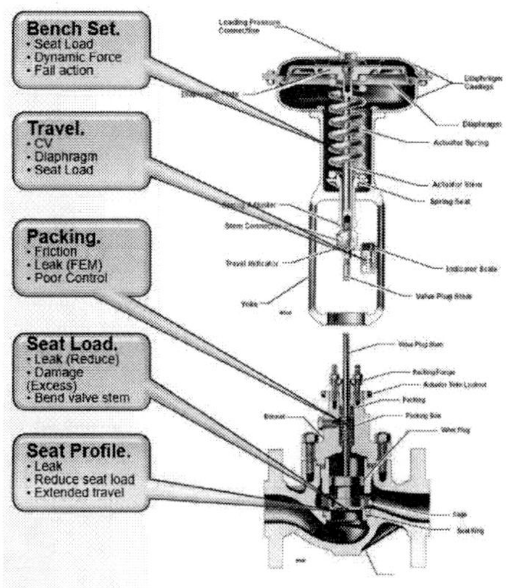

- Improper stroking speed
- Excess friction
- Insufficient air supply
- Poor throttling performance
- Seat integrity
- Erratic valve performance
- Pneumatic leaks
- Worn valve components
- Actuator problems

Care Share Dare

OPERATIONAL ACTIONS - ONLINE

01 ON-LINE: TRIGGER PROFILE
Captures data for 4 user defined variables when a defined threshold for the application and valve defined parameters is exceeded.

02 ON-LINE: AIR MASS FLOW
Detects both positive (supply) & negative(exhaust) air mass flow from the Relay. This diagnostic is useful for detecting leaks in the actuator or tubing to the actuator.

03 ON-LINE: I/P & RELAY INTEGRITY
Detect problems such as plugging in the I/P Primary, Plugging in the I/P nozzle, Instrument Diaphragm failures, I/P instrument O-ring failure & I/P calibration shifts.

04 ON-LINE: SUPPLY PRESSURE CHECK
Detects and Alerts in case of both low & high supply pressure readings. Used to detect and quantify drop in the air supply. This is helpful for identifying supply line restrictions.

05 ON-LINE: TRAVEL DEVIATION
Monitors actuator pressure and travel deviation from set point. This diagnostic is useful to identify if the valve is stuck, low supply pressure or travel calibration shifts.

Care Share Dare

MAINTENANCE ACTION – OFF-LINE

7 OFF-LINE: DYNAMIC VALVE SIGNATURE

- Evaluate valve friction, dead band, and shutoff capability.
- Calculate actuator spring rate and bench set. Identify potential packing problems.
- Compare current condition to previous baseline condition.
- Signature Analyzer provides pass/fail results for Max/Min Friction, Stem Integrity.

8 OFF-LINE: STEP RESPONSE TEST

- Evaluate how well the valve tracks an input change.
- Validate tuning parameters.
- Obtain a numerical analysis for overshoot, hysteresis, dead time.

Care Share Dare

CASE STUDY 1 – VALVE PASSING

Signature as found

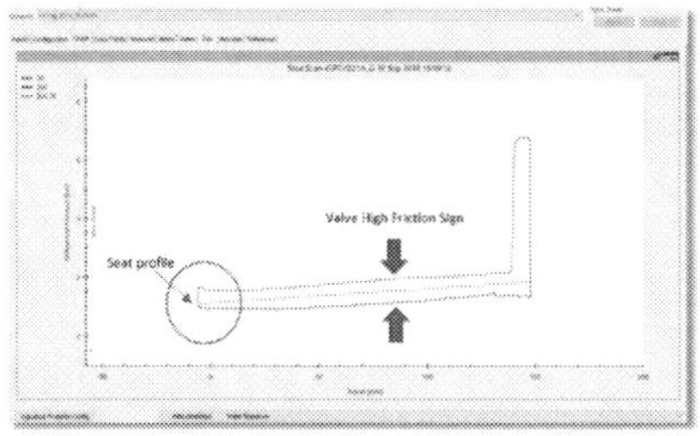

Signature as left

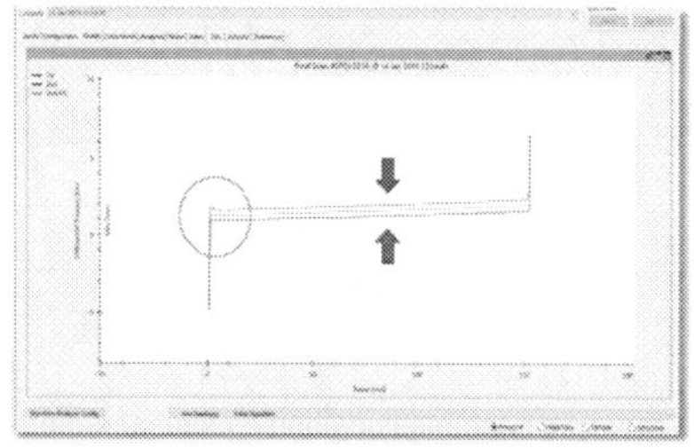

OBSERVATION (AS FOUND):

1. Abnormal Valve Close Profile
2. Class 5 Leak Class – Not achieved
3. Travel calibration is abnormal
4. Valve friction is High

IMPROVEMENT (AS LEFT)

1. Seat profile in good shape
2. Rated travel as per spec
3. Valve friction in good range

Care Share Dare

CASE STUDY 2 – VALVE STUCK / PERFORMANCE

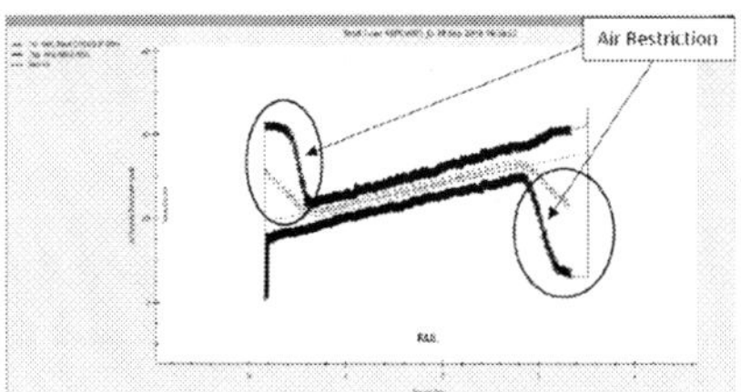

OBSERVATION:

- Valve stuck at 4.5% travel and unable to fully close.
- Air leaking from actuator stem seal.
- Packing leakage & positioner out of travel calibration.

RECOMMENDATION:

- Replace AFR.
- Replace Packing and seal ring
- Upgrade DVC positioner.
- Change output tubing size from 3/8" to 1/2".

Care Share Dare

PERFORMANCE DASHBOARD – SUR SITE

SUMMARY

VALVES THAT REQUIRE ATTENTION

12 — Valves Monitored

83% — Requires Attention

VALVE TAG#	STATUS
10-PCV-016A	External leak detected
10-PCV-016B	Friction out of bounds
10-PCV-021	Slight overtravel
10-PCV-025A	Friction out of bounds
10-PCV-025B	Slight overtravel
10-PCV-029	Valve is healthy
10-PCV-045A	Friction out of bounds. High Supply Pressure
10-PCV-045B	Valve is healthy
10-PCV-046	High Supply Pressure
10-PCV-047A	Friction out of bounds; External leak detected
10-PCV-047B	Possible Plug & Seat wear
10-PCV-048	External leak detected

MONTHLY VALVE STATUS

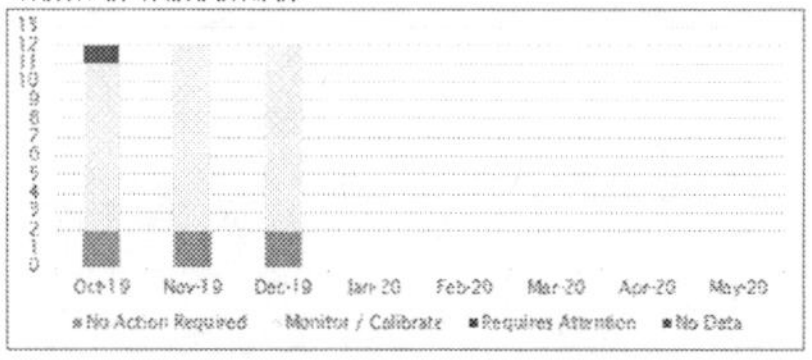

MONTHLY VALVE TREND

Tag Number	Oct-1	Nov-1	Dec-1	Jan-2	Feb-2	Mar-2	Apr-2	May-2
10-PCV-016A								
10-PCV-016B								
10-PCV-021								
10-PCV-025A								
10-PCV-025B								
10-PCV-029								
10-PCV-045A								
10-PCV-045B								
10-PCV-046								
10-PCV-047A								
10-PCV-047B								
10-PCV-048								

Care Share Dare

DETAILED SUMMARY

4 DETAILED STATUS & ACTIONS

TAG NO (AREA)	VALVE STATUS	STATUS	POTENTIAL IMPACT	RECOMMENDED ACTION(S)	PREVIOUS WEEK COMPLETED ACTIONS
10-PCV-016A		External Leak Detected.	a) Degraded valve performance or excessive seat load	a) Verify air leak from actuator & tubing (2.3)	
10-PCV-016B		Friction out of boundary limit.	a) Degraded valve performance	b) Overhaul valve (4.2) and replace packing and piston seals (3.7)	
10-PCV-021		Slight overtravel.	a) Calibration drift and loss of control accuracy.	b) Perform stroke adjustment (2.7) & Auto Calibrate the valve (1.2)	
10-PCV-025A		Friction out of boundary limit.	a) Degraded valve performance	a) Overhaul valve (4.2) and replace packing and piston seals (3.7)	
10-PCV-025B		Slight overtravel.	a) Calibration drift and loss of control accuracy.	a) Perform stroke adjustment (2.7) & Auto Calibrate the valve (1.2)	
10-PCV-029		Valve is healthy. Continue monitoring	a) No Potential Impact Identified.	b) Continue Monitoring	
10-PCV-045A		Friction out of boundary limit. High supply pressure detected.	a) Degraded valve performance or excessive seat load	a) Overhaul valve (4.2) and replace packing and piston seals (3.7)	
10-PCV-045B		Valve is healthy. Continue monitoring	a) No Potential Impact Identified.	a) Continue Monitoring	
10-PCV-046		High Supply Pressure.	a) Excessive seat load	a) Validate Supply pressure (1.1) and adjust according to datasheet value.	
10-PCV-047A		Friction out of boundary limit. External leak detected.	a) Degraded valve performance	a) Overhaul valve (4.2) and replace packing and piston seals (3.7)	

Care Share Dare

ACTION LIST CRITICALITY LEVEL

RECOMMENDED ACTION LIST – (Refer to Appendix for Level Classification)

	RECOMMENDED ACTION LIST		
	LEVEL 1		LEVEL 3
1.1	Validate supply pressure	3.1	Replace Air Set
1.2	Perform Auto Travel Calibration	3.2	Replace I/P Assembly
1.3	Run Performance Tuner or Manual Tuning	3.3	Replace Relay Assembly
1.4	Update valve spec sheet data	3.4	Replace Travel Sensor Assembly
1.5	Review & Clear Instrument Alert Record	3.5	Replace Positioner Assembly
1.6	Run Dynamic Scan	3.6	Replace PWB Assembly
1.7	Run Step Response Test	3.7	Replace Packing
	LEVEL 2	3.8	Replace Trim Set
2.1	Inspect I/P, O-ring & Screen Condition	3.9	Replace Valve Stem
2.2	Inspect Relay Seal & Relay Condition	3.10	Replace Actuator Stem
2.3	Verify Air Leak from Actuator	3.11	Replace Actuator Diagram & O-rings
2.4	Verify Valve Stem Condition		LEVEL 4
2.5	Inspect Positioner Feedback Arm Assembly	4.1	To do actuator assembly overhaul service in next available window.
2.6	Perform Relay Adjustment	4.2	To do valve body assembly overhaul service in next available window.
2.7	Perform Stroke Adjustment	4.3	To do Complete control valve assembly overhaul service in next available window.
2.8	Perform Travel or Pressure Sensor Adjustment		
2.9	Perform Analog Input Calibration		

Care Share Dare

PROJECT MILESTONES

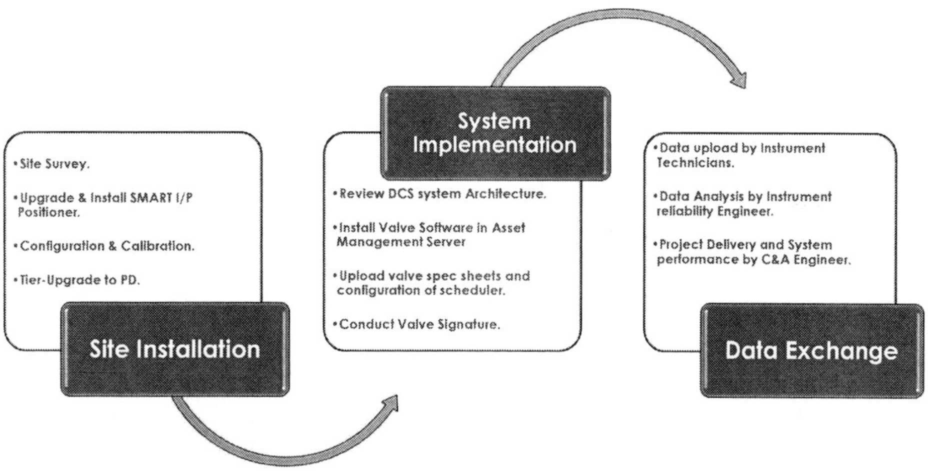

Care Share Dare

WORK PROCESS FLOW

- All 3 steps are CRITICAL to the success of the Project

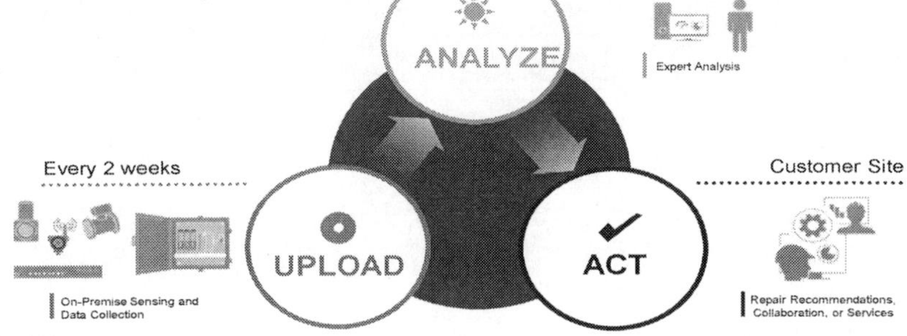

Care Share Dare

SUMMARY

- Estimated ROI from Valve Condition Monitoring at 40% of Overhauling Cost.
- Achieve Standardization for the requirement of control valve condition based maintenance.
- Monitoring the healthiness of the valves on-line 24X7 and predict abnormalities..
- Centralized Monitoring platform from Muscat HO to enable remote Access and Monitoring.
- Technology is enabler only, a focus is required on people development and process.
- Plan to expand the Online Condition Monitoring Program for the rest of Gas Network PCVs
- Plan to add Anti-Surge application and ESD (SIS) in condition monitoring roadmap.
- Explore Plant web Insight and Optics integration into Existing Systems Architecture

Care Share Dare

شكرا جزيلا لحضوركم

Thank you

Care Share Dare

International Gas Union Research Conference (IGRC-2020), 24-26 February 2020, Muscat, Oman

Tensile Performance Evaluation of PVC Pipes for Steam Injection Wells

Dr Zahid Qamar, Prof Tasneem Pervez, Dr Farooq al-Jahwari

Applied Mechanics and Advanced Materials Research Group $(AM)^2R$
Mechanical and Industrial Engineering Department
Sultan Qaboos University, Muscat, Oman

Engineering is the art of modeling materials we do not wholly understand, into shapes we cannot precisely analyze, so as to withstand forces we cannot properly assess, in such a way that the public has no reason to suspect the extent of our ignorance.

A R Dykes

MOTIVATION AND SIGNIFICANCE

Technology shapes society and society shapes technology.

Robert W. White

Polymer Pipes in Oil Drilling

- In some oil wells, viscosity of the crude is very high (heavy oil)
- Steam is injected to increase fluidity and mobility of the oil
  - Steam injection heats the crude oil in the formation, lowering its viscosity and vaporizing some of the oil to increase its mobility

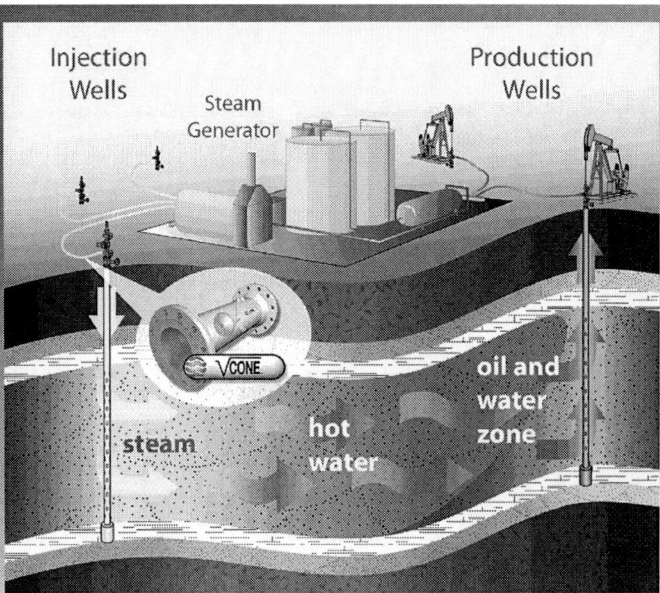

Schematic of steam injection process

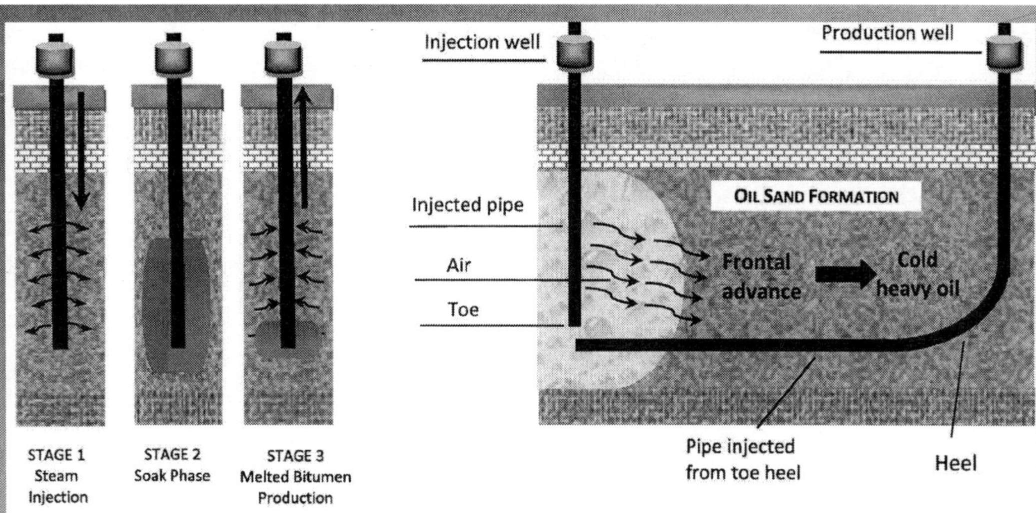

(a) CYCLIC STEAM STIMULATION PROCEDURE (b) TOE-TO- HEEL AIR INJECTION PROCEDURE

Cyclic Steam Stimulation (CSS) is a staged technique using Vertical Injection and Production

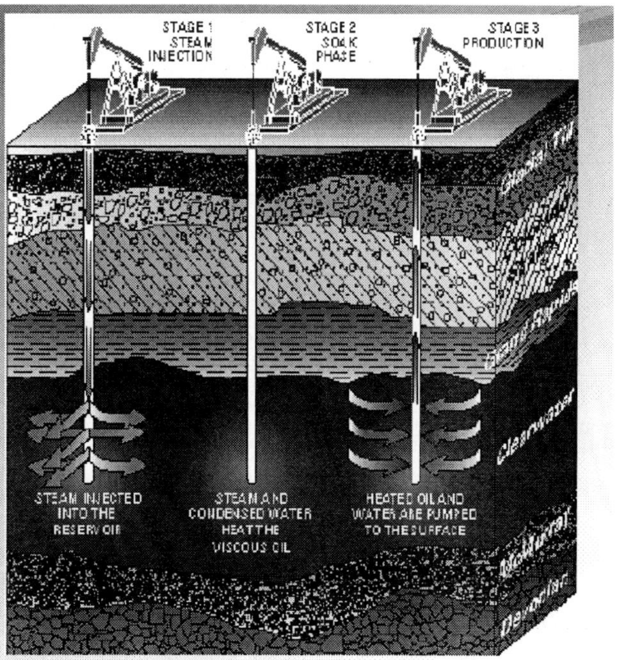

Polymer Pipes in Thermal Wells

- Steam injection requires large quantities of water, generally transported from nearby fields
- Some oilfields in the Gulf region, especially in Oman
  - Water-bearing sandstone is highly fragmented
  - Sand screens are required for sand control
- Sand screens made of carbon-steel are not reliable
  - Quickly undergo serious corrosion
- Some oilfields have started to try out non-metallic materials
  - Such as strengthened polymers (eg hard PVC)
  - Because of their non-corroding nature

Testing of Polymer Pipes

- Based on a hit-and-trial approach (in Oman)
  - Around 15% wells have reported failure due to screen collapse
- Petroleum Development Oman (PDO) wanted to obtain more information on strength of these pipes
- Experimental testing facility designed and developed at Sultan Qaboos University (SQU)
  - For integrity assessment of large-diameter hard polymeric pipes
  - Included design and construction of test setup and required jigs and fixtures
- Types of tests
  - **Pipe behavior under tensile loads**
  - Pipe behavior under compressive loads
  - Collapse behavior

Major Contributions

- Scientific understanding of different strength parameters of new pipe materials (high strength PVC)
- Prequalification of polymer pipes for appropriate field conditions
- Major savings in cost and time for thermally assisted oil wells
  - Once sufficient confidence is built with PVC concept, this well design can be introduced PDO wide

Financial Upside

- Currently PDO drills approximately 30 shallow water supply wells per year
  - Drill-site water, camp supply, R/O & frac unit supply, water-flood, steam generation
- Company-wide switch from carbon steel to PVC for water supply wells
  - Savings of approximately 150k-200k USED per well
  - Annual saving of 4.5-6.0 mil USD
  - Savings of 20-30 mil USD over a 5 year period

EXPERIMENTAL WORK

Testing Scheme and Parameters

- To satisfy requirements of international standards
  - Polymer pipes had to go through a 2-3 month ageing process before mechanical testing
  - In brine solution matching the salinity of the water field (1.02 sg, equivalent to 35 PPT)
- Testing under tensile loads
  - PVC pipe sections of 1m length and 280 mm dia
  - Three types of pipe sections
    - Plain
    - Built-up (pipe with threaded connection in the middle)
    - Slotted (screen)
  - 3 samples tested in each case

The three types of PVC pipes for tension testing: plain, built-up, and slotted

Stress-Strain Measurements

- Tension tests conducted on a Dartech universal testing machine (UTM)
  - Capacity of 2000 kN in tension mode and 4000 kN in compression mode
- Loads and deformations (elongation) directly recorded by machine
- Pipe dimensions measured before each test (length, outer diameter, thickness)
- For more precise and complete data recording, strain gages used
  - Another advantage: transverse (hoop) strains can also be recorded; measure of anisotropy behavior (Poisson's ratio)
- Machine readings converted into stress-strain data

Stress-Strain Measurements

- Expecting that deformations may be large (polymer pipe), very special high-deformation strain gages (TML company) procured
- All strain gages hooked up to Dartech machine thru a data logger and computer
- Two sets of strain gages (3 axial, 3 hoop) fixed on each test pipe, on top and bottom positions, on opposite sides of the pipe
  - Marked as TA1 (top axial-1), BH1 (bottom hoop-1), etc

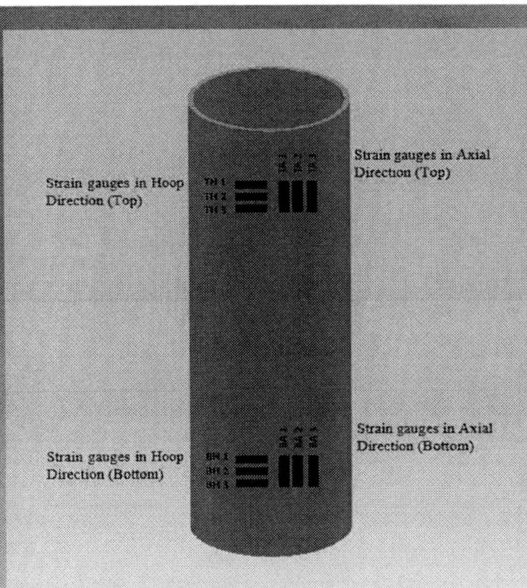

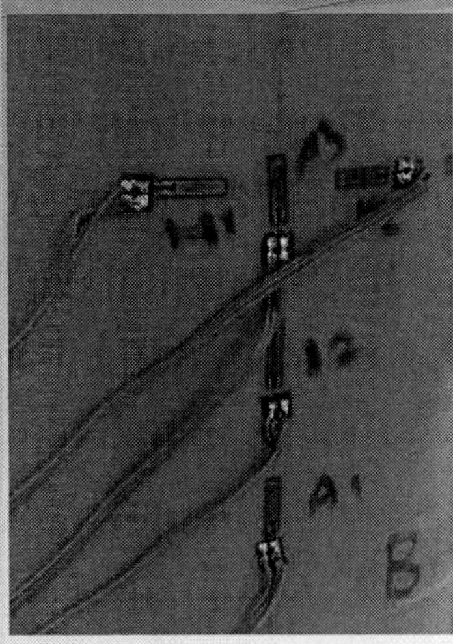

Sets of strain gauges in axial and hoop directions, near top and bottom ends of pipe

SALT WATER AGEING

Soaking in Salt Water

- Soaking to be done for
  - 3 PVC pipes of 7-m length (collapse)
  - 18 sections of 1-m length (tension + compression)
  - All of 280 mm diameter
- Required almost 5000 liters of saline water (roughly 1500 US gallons)
- Very difficult and time consuming to
  - Produce this amount of distilled water
  - Dissolve enough salt to get required salinity

Soaking in Salt Water

- Alternate (unique) solution
- SQU desalination research unit near the sea
- Seawater can be directly pumped through an underground tunnel and pumping system
- Judicious mixing of normal drinking water and high-salinity seawater
  - Required salt concentration

Soaking in Salt Water

- Challenges
  - Pumping of large amount of seawater (5000 liters)
  - Transportation from desalination plant to SQU
  - Special containers to be arranged to keep 18 pipes of 1-m length and 3 pipes of 7-m length under this saline water

Transportation and storage of saline water

Immersion of 18 sections of 1-m PVC pipe (plain, built-up, and slotted) in saline water for 2-3 months soaking

DESIGN AND CONSTRUCTION OF EXPERIMENTAL SETUP

The proper study of mankind is the science of design.

Herbert Simon

Design and Construction of Tensile Test Setup

- Tension tests; Dartech universal testing machine
- Main challenge; gripping of pipes
- Three major issues or design constraints
  - Pipe dia was too large to go inside the gripping system of the machine
  - PVC surface was too smooth to provide enough friction for good grip
  - Too much load could not be applied in the grip region, or the PVC material would fracture here rather than in the main pipe section
- Various gripping fixtures were designed, constructed, and tried out

Grout Design

- Idea based on application used in civil construction industry
- Grout: a chemical resin mixed with water that is used for extra high strength reinforcement in concrete structures
- A set of inside and outside metallic ring supports on each end of the PVC pipe
  - Prescribed gap between metal and PVC to pour in the grout mixture
  - After curing, grout would become a solid mass, exerting a large pressure and creating strong friction between metal and PVC
  - Hopefully, joint would not shear off under the maximum tensile load needed to fracture the main section of the pipe
- Grout could be single-sided or double sided
  - Single-grout design would be tried out first, being more efficient in terms of time, cost, material, and effort

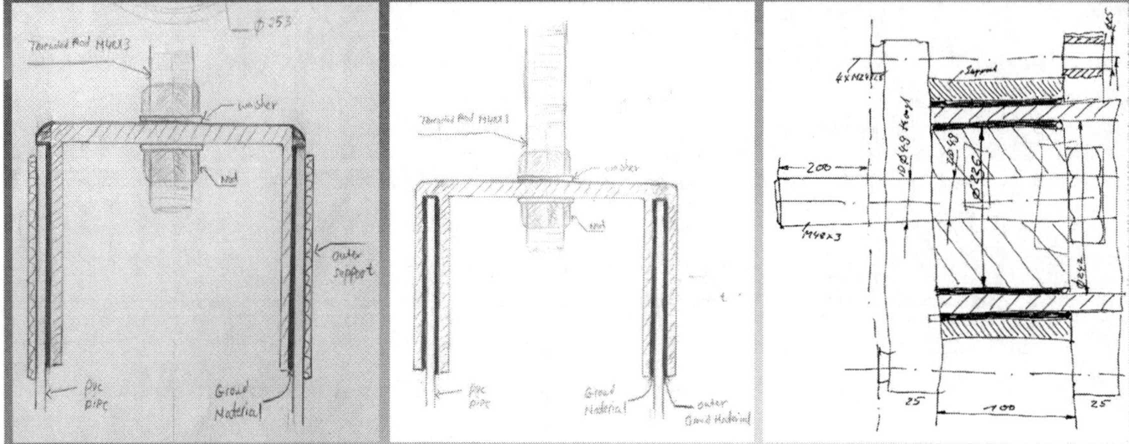

Design sketches for single-grout and double-grout designs

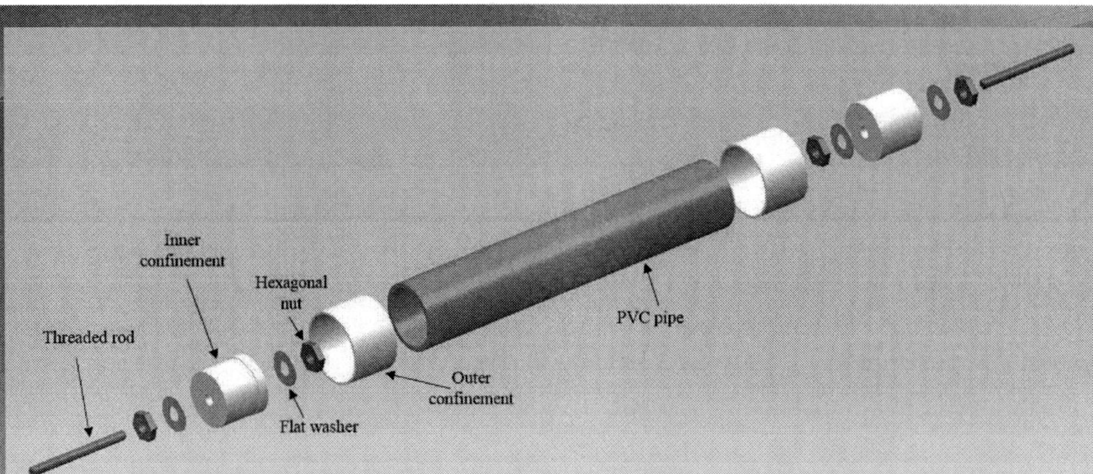

Exploded view of single-grout design for tensile test gripping fixture

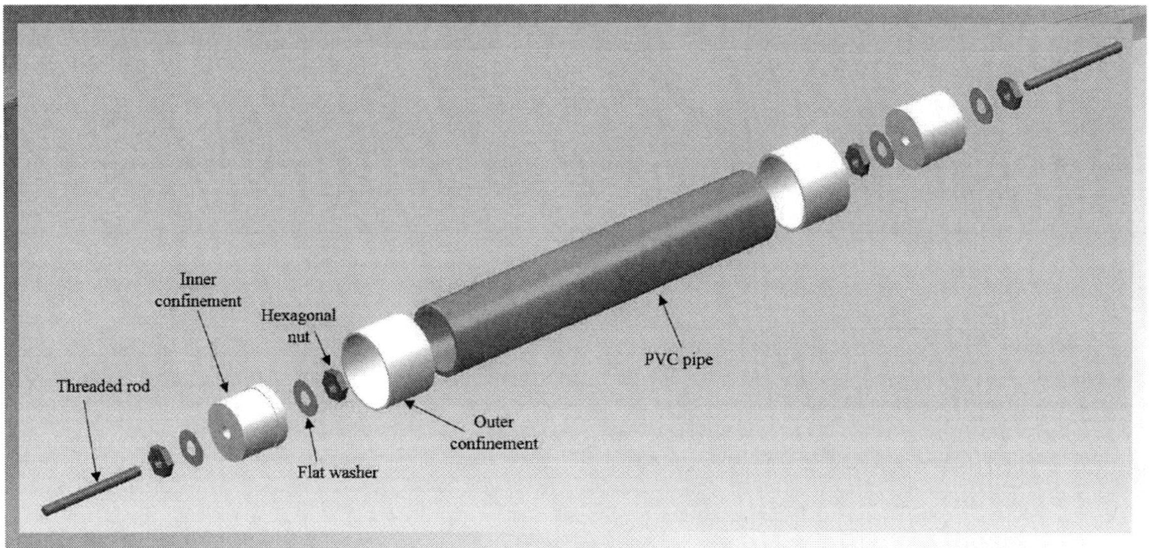

Exploded view of double-grout design for tensile test gripping fixture

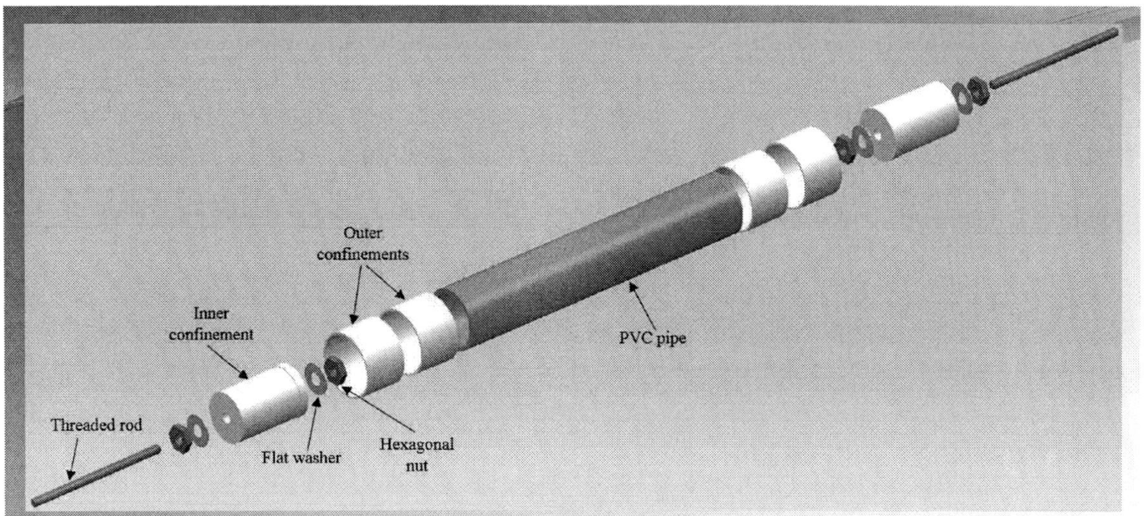

Exploded view of extended-length single-grout design for tensile test gripping fixture

Thread Design

- Grout design was successful for slotted PVC pipes
  - Lowest tensile strength
- Grout design failed in the case of plain and built-up pipes
- Another gripping fixture designed and constructed
  - Based on the idea of a threaded contact between PVC pipe and inner support ring

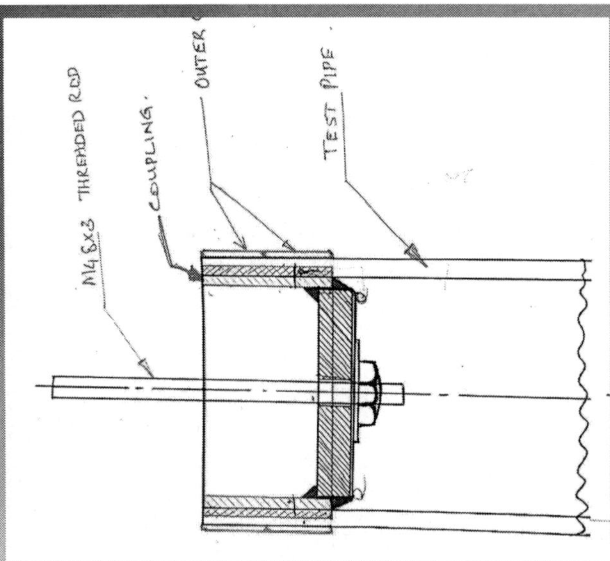

Sketch of threaded-contact design for tensile test gripping fixture

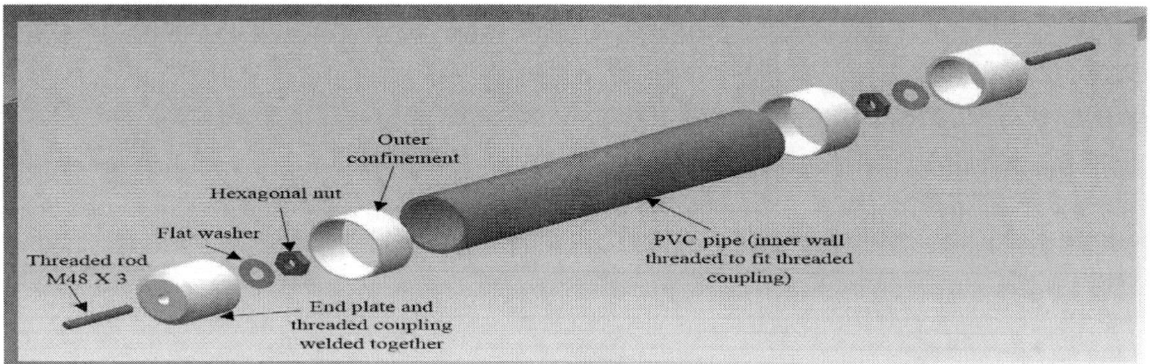

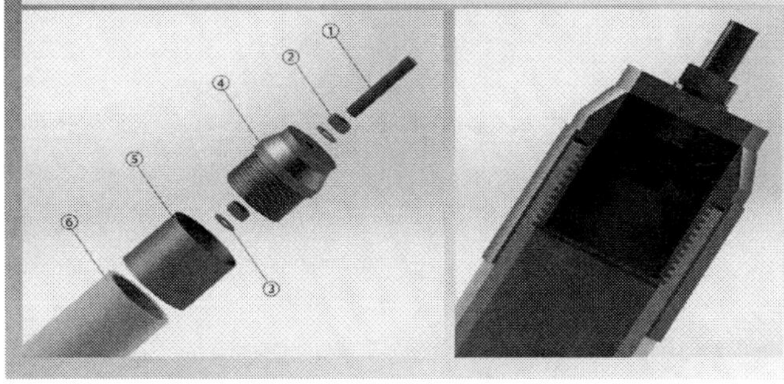

Exploded and assembled views of threaded-contact design for tensile test gripping fixture

TESTING UNDER TENSILE LOADS

Tension Testing

- Single-grout design used successfully for slotted PVC pipes
- Both single and double grout designs failed for plain and built-up pipes
  - Grout connection sheared and broke off before pipe could fracture
- Threaded connection also failed
  - PVC pipe fractured near the last thread of grip area rather than in the main pipe body

- Various stages in preparation of single-grout tensile test fixture
  - Assembly of tensile fixture components
  - Pouring of grout in the gap between metal ring and PVC pipe
  - Grout curing
  - Assembly after grouting, etc.

Single-grout tensile test setup; different stages of preparation and assembly

Testing of **plain** PVC pipe using single-grout fixture; grout assembly failed before pipe could fracture

Testing of **plain** PVC pipe using double-grout fixture; grout assembly failed before pipe could fracture

Components of threaded-contact tension fixture, and different stages of initial assembly

Various stages of final assembly of threaded-contact tension fixture

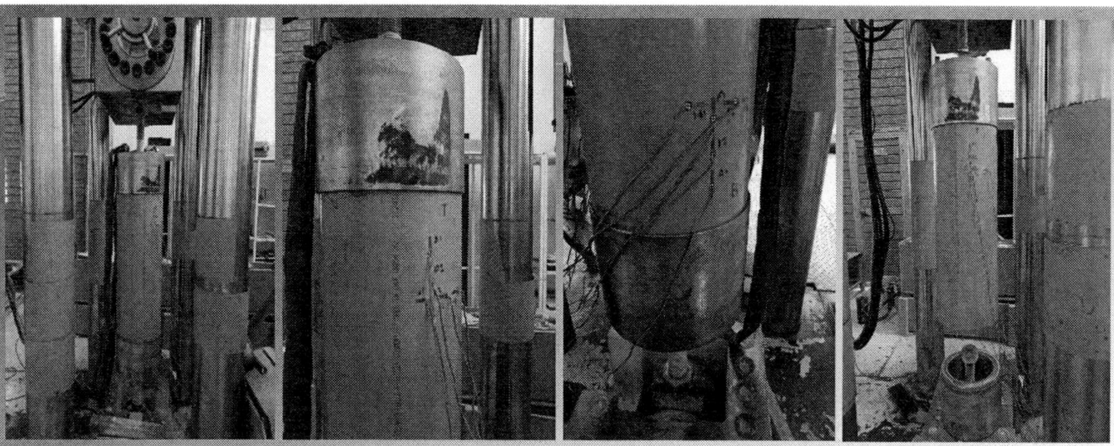

Testing of **plain** PVC pipe using threaded-contact fixture; pipe failed near last thread of grip area rather than in the main pipe body

RESULTS AND DISCUSSION

Tensile Test Results

- The system records force (kN), elongation/displacement (mm), and micro-strain data for each test
- Later; data converted into plots of stress vs strain
  - Using average values of pipe dimensions measured before each test
  - Length, outer diameter, thickness
- Pipe properties extracted from these graphs
  - Fracture stress
  - Fracture strain (ductility)
  - Elastic modulus
  - Poisson's ratio or anisotropy (ratio of hoop strain to axial strain)
- For elasticity value (Young's modulus)
  - Curve fitting done for roughly 10% (initial linear portion) of graph

Testing of slotted PVC pipe using single-grout fixture; pipe fractured in the middle section or near the end

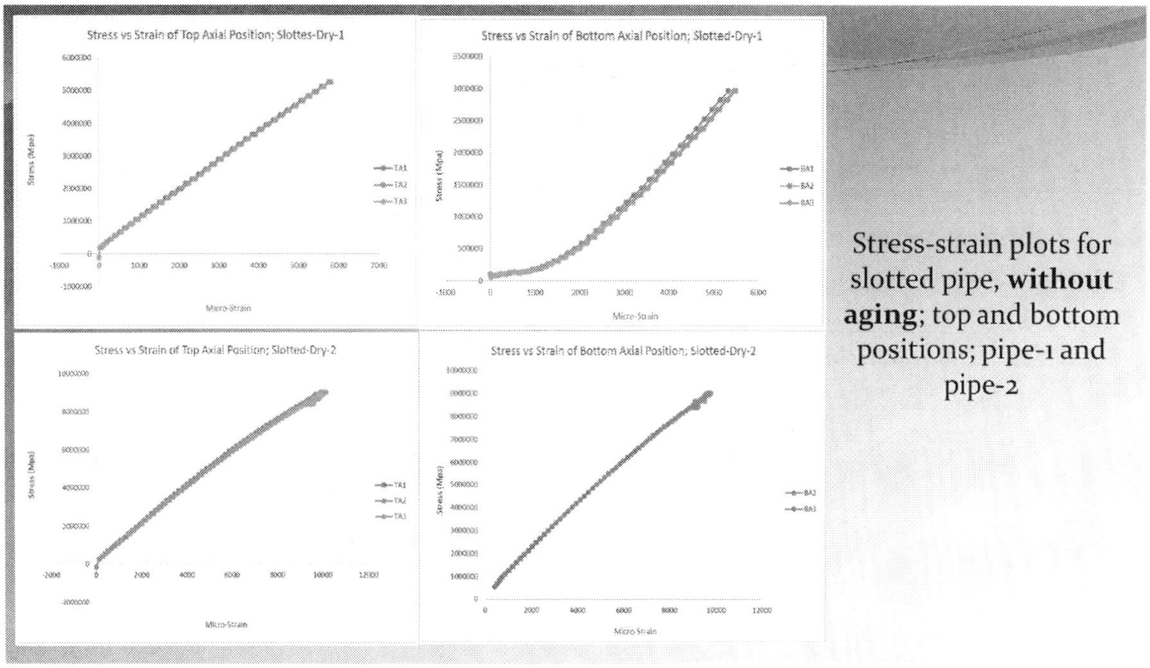
Stress vs Strain of Top Axial Position; Slottes-Dry-1
Stress (Mpa)
Micro-Strain
TA1
TA2
TA3
Stress vs Strain of Bottom Axial Position; Slotted-Dry-1
Stress (Mpa)
Micro-Strain
BA1
BA2
BA3
Stress vs Strain of Top Axial Position; Slotted-Dry-2
Stress (Mpa)
Micro-Strain
TA1
TA2
TA3
Stress vs Strain of Bottom Axial Position; Slotted-Dry-2
Stress (Mpa)
Micro-Strain
BA2
BA3
Stress-strain plots for slotted pipe, **without aging**; top and bottom positions; pipe-1 and pipe-2

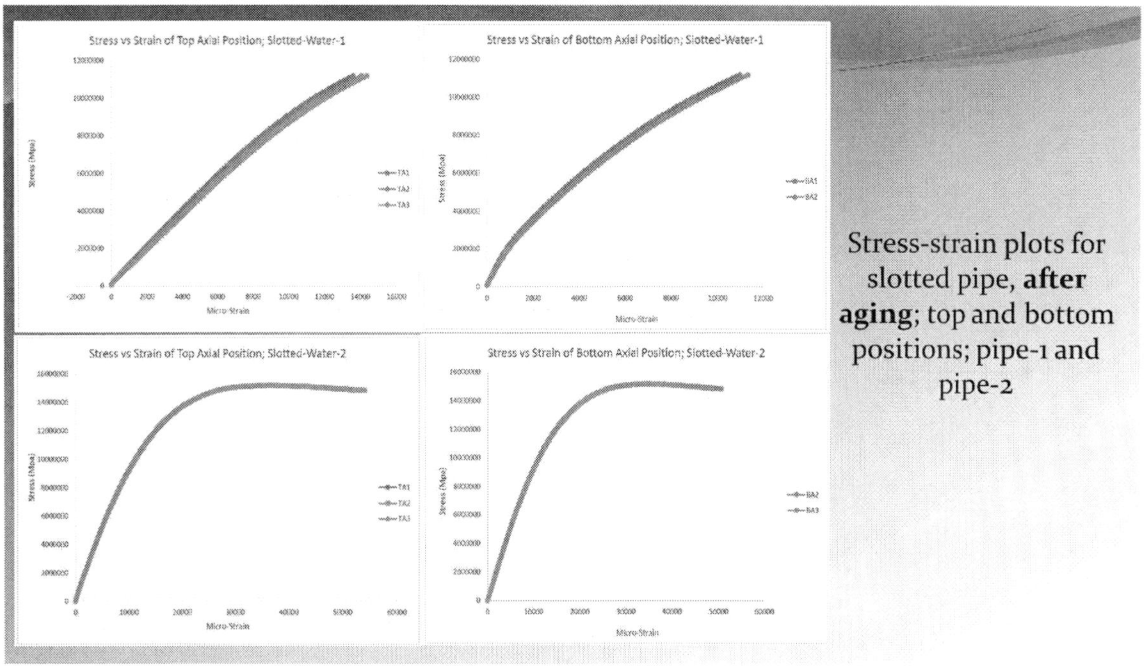
Stress vs Strain of Top Axial Position; Slotted-Water-1
Stress (Mpa)
Micro-Strain
TA1
TA2
TA3
Stress vs Strain of Bottom Axial Position; Slotted-Water-1
Stress (Mpa)
Micro-Strain
BA1
BA2
Stress vs Strain of Top Axial Position; Slotted-Water-2
Stress (Mpa)
Micro-Strain
TA1
TA2
TA3
Stress vs Strain of Bottom Axial Position; Slotted-Water-2
Stress (Mpa)
Micro-Strain
BA2
BA3
Stress-strain plots for slotted pipe, **after aging**; top and bottom positions; pipe-1 and pipe-2

Summary of Test Results

Slotted PVC Pipes			Modulus of Elasticity (Gpa)						
Test (Tension)			TA1	TA2	TA3	BA1	BA2	BA3	AVG
Pipe type	Wet / Dry	Trail							
Slotted	Wet	1	0.916	0.883	0.86	1.059	1.021	Fail	0.956
Slotted	Wet	2	0.815	0.805	0.811	Fail	0.834	0.826	0.818
Slotted	Dry	1	0.874	0.873	0.875	0.707	0.709	0.775	0.808
Slotted	Dry	2	0.984	0.967	0.957	0.972	0.928	0.946	0.959

Slotted PVC Pipes			Pisson's Ratio						
Test (Tension)			TH1/TA1	TH2/TA2	TH3/TA3	BH1/BA1	BH2/BA2	BH3/BA3	AVG
Pipe type	Wet / Dry	Trail							
Slotted	Wet	1	0.466	0.418	Fail	0.51	0.48	Fail	0.469
Slotted	Wet	2	0.394	Fail	Fail	Fail	0.379	0.392	0.388
Slotted	Dry	1	0.457	0.45	Fail	0.485	0.459	0.452	0.461
Slotted	Dry	2	0.426	0.417	0.425	0.429	0.416	0.411	0.421

Slotted PVC Pipes			Ultimate Strength (Pa)	Fracture Strength (Pa)	Percentage Elongation
Test (Tension)					
Pipe type	Wet / Dry	Trail			
Slotted	Wet	1	1.12E+07	1.12E+07	1.40%
Slotted	Wet	2	1.52E+07	1.52E+07	4.90%
Slotted	Dry	1	5.26E+06	5.26E+06	0.57%
Slotted	Dry	2	9.020E+06	9.020E+06	1.01%

Discussion of Results

- Good to observe that average values of stiffness of slotted PVC pipes are almost the same before and after aging
  - Value of elastic modulus determined by considering only linear portion of stress-strain curve, and fitting a straight line through these data points
  - Polymer pipes (not metallic); no portion of stress-strain diagram is really linear
  - So this straight-line fit is rather approximate

Discussion of Results

- Effect of salt-water soaking on anisotropy behavior is not significant
  - Poisson's ratio for slotted pipe (before and after ageing) remains in the 0.43-0.44 range
  - Decrease in pipe diameter (deformation in hoop direction) would be roughly 43-44% of increase in pipe length (deformation in the axial direction) under tensile loading

Discussion of Results

- Surprisingly, tensile strength (maximum stress that slotted PVC pipes can withstand before tensile failure) after aging is almost 46% higher
- This is of course good news for the field engineers

Acknowledgements

Authors acknowledge the support of Sultan Qaboos University (SQU) and Petroleum Development Oman (PDO) for this work.

Science never solves a problem without creating ten more.

Bernard Shaw

Many a thank!

IGRC 2020
MUSCAT
24-26 FEB | WWW.IGRC2020.COM
HOSTED BY
Oman LNG L.L.C.
PRESENTED BY
IGU
UNDER THE PATRONAGE OF

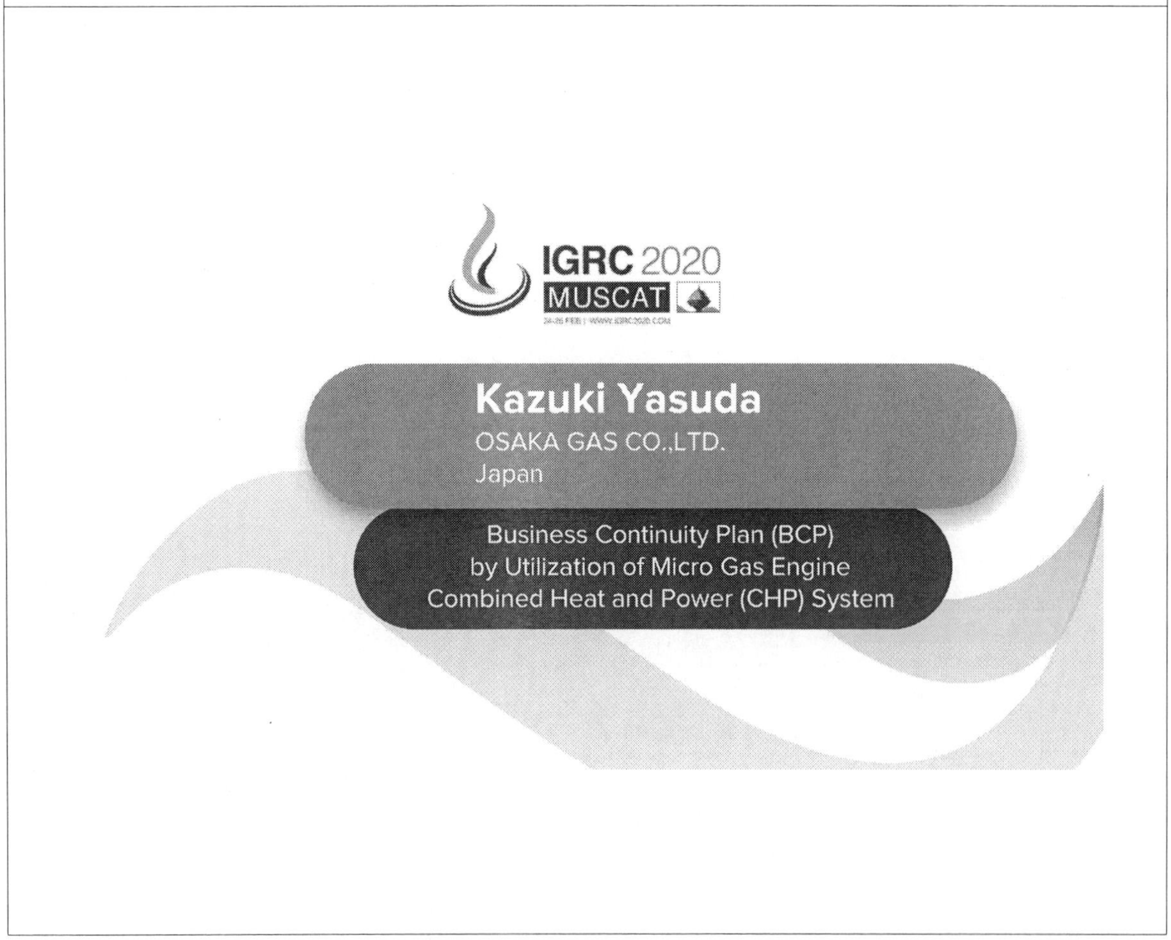
IGRC 2020
MUSCAT
Kazuki Yasuda
OSAKA GAS CO.,LTD.
Japan
Business Continuity Plan (BCP)
by Utilization of Micro Gas Engine
Combined Heat and Power (CHP) System

1. About Osaka Gas

- Started business operations in 1905
- Serving over 7million customers in Kansai region
- Headquarters in Osaka, Japan
  - Overseas affiliates -
    Europe(London), North America(New York, Houston)
    Oceania(Perth), ASEAN(Singapore, Bangkok)
- Imports about 8 MTPA of LNG
- Accounts for approx. 25% of the natural gas supply market in Japan
- Developing 3.1 GW of power business
- Investing/ operating in 13 countries for Upstream & Downstream businesses

2. Gas Engine Heat Pump (GHP) System

·GHP system is an air conditioning system driven by natural gas.

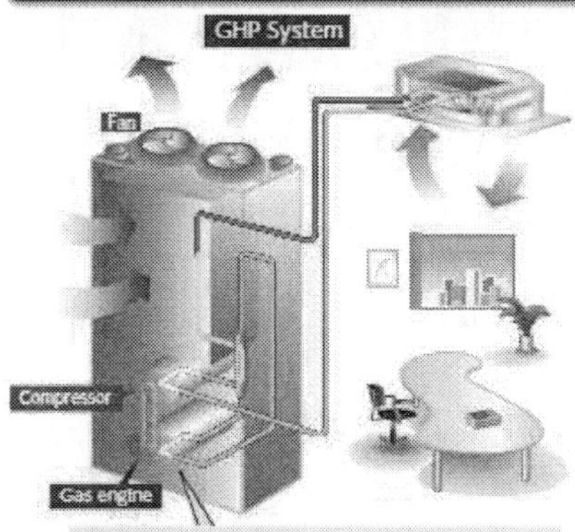

Drive the compressor with gas engine.

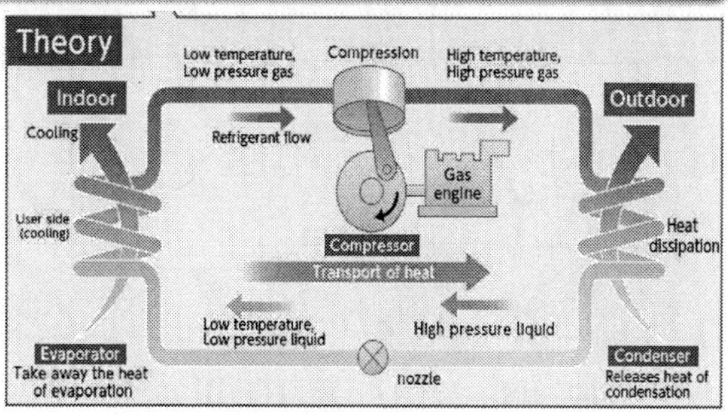

■when liquid evaporates, it takes away the surrounding heat, and when the gas condenses and liquefies, it generates heat.

■A heat pump cycle is a system that uses this property to cool and heat the refrigerant by compressing the refrigerant with a compressor and repeating liquefaction and vaporization.

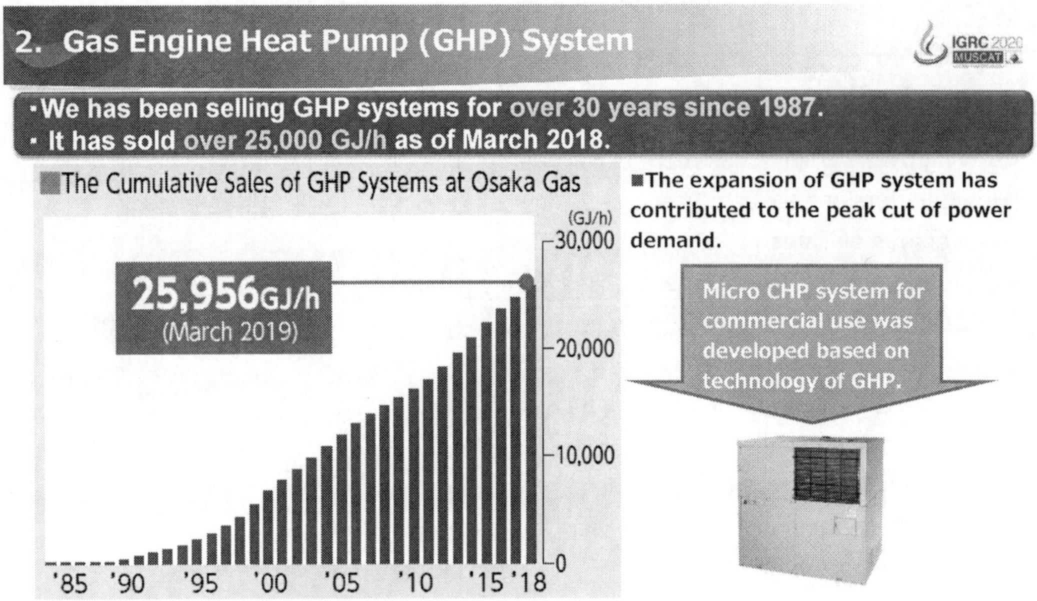
2. Gas Engine Heat Pump (GHP) System
IGRC 2020 MUSCAT
・We has been selling GHP systems for over 30 years since 1987.
・It has sold over 25,000 GJ/h as of March 2018.
■The Cumulative Sales of GHP Systems at Osaka Gas
(GJ/h)
30,000
20,000
10,000
0
25,956GJ/h
(March 2019)
'85 '90 '95 '00 '05 '10 '15 '18
■The expansion of GHP system has contributed to the peak cut of power demand.
Micro CHP system for commercial use was developed based on technology of GHP.

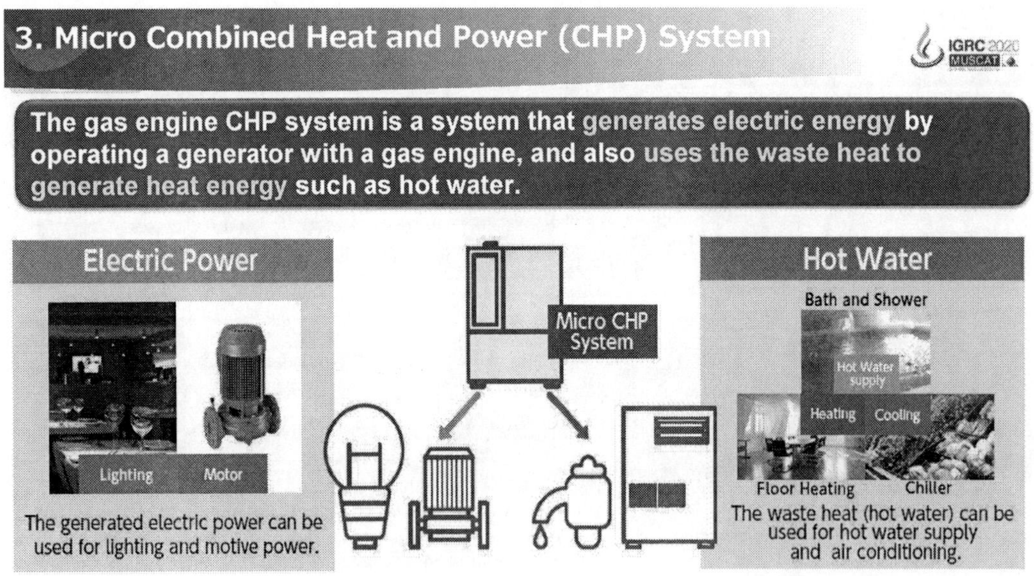
3. Micro Combined Heat and Power (CHP) System
IGRC 2020 MUSCAT
The gas engine CHP system is a system that generates electric energy by operating a generator with a gas engine, and also uses the waste heat to generate heat energy such as hot water.
Electric Power
Lighting
Motor
The generated electric power can be used for lighting and motive power.
Micro CHP System
Hot Water
Bath and Shower
Hot Water supply
Heating
Cooling
Floor Heating
Chiller
The waste heat (hot water) can be used for hot water supply and air conditioning.

3. Micro Combined Heat and Power (CHP) System

IGRC 2020 MUSCAT

The capacity of micro CHP systems is from 5 kW to 35 kW.

Capacity	5kW	9.9kW	25kW	30kW	35kW
Number of phase Voltage	One-phase three-wire system 100V / 200V		Three-phase three-wire system 200V		
Type	CP5D1(Z) -SNJG	CP10D1(Z) -SNJG	CP25D1(Z) -SNJG	CP35D2(Z) -SNJY	CP35D2(Z) -SNJG
Vender	YANMAR ENERGY SYSTEM CO.,LTD.				
Package appearance					
Generation efficiency	29.0%	31.5%	33.5%	32.4%	33.5%
Heat recovery ratio	56.5%	55.0%	52.0%	54.5%	54.5%
Overall efficiency	85.5%	86.5%	85.5%	86.9%	88.0%
Response to power shut down	●	●	●	●	●

■In the case of 35 kW, the power generation efficiency is 33.5% and the heat recovery ratio is 54.5 %.

The overall efficiency reaches 88.0%.

3. Micro Combined Heat and Power (CHP) System

IGRC 2020 MUSCAT

We have already sold 3,500 packages of Micro CHP system (over 50 MW) since the sales launched in1997.

■The Cumulative Sales of Micro CHP Systems at Osaka Gas

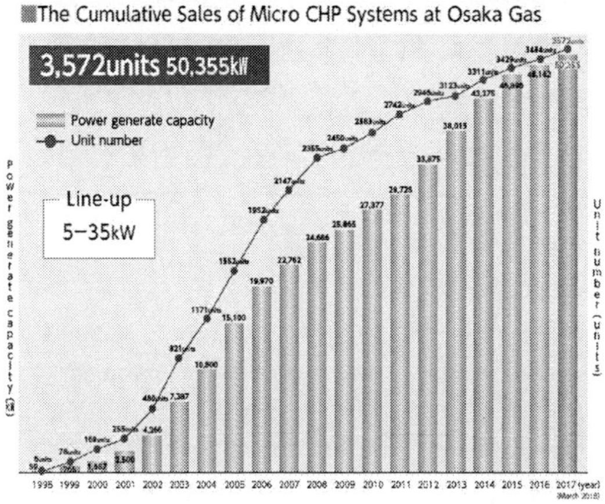

●Category of Customer Installed Micro CHP

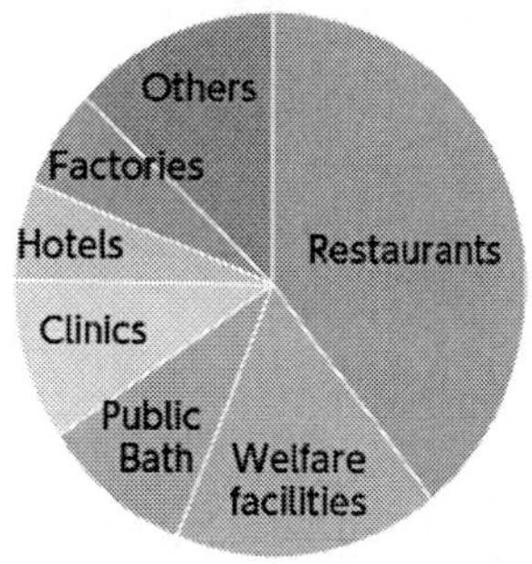

4. Back up Supply System of Micro CHP

IGRC 2020 MUSCAT

Some micro CHP systems can supply power even in the event of a grid power shut down.

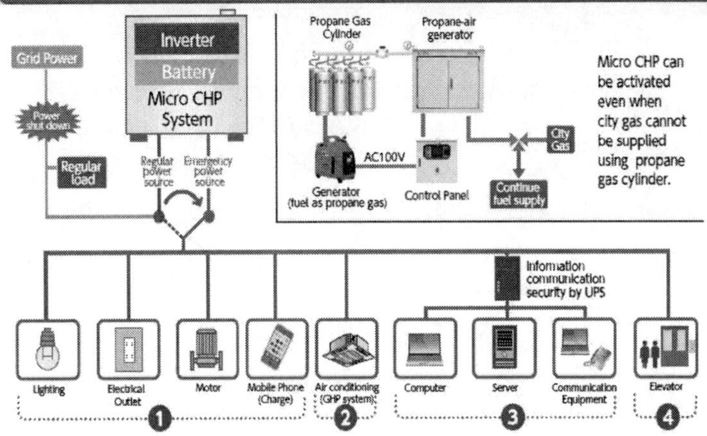

1. It can continue power supply during power shut down.
2. Also continue Air-conditioning in combination with GHP.
3. The combination with UPS can resolve instantaneous power interruption and realize continuous power supply. ※UPS :Uninterrupted Power Supply
4. It also allows using elevators in power failure.

5. Recent Major Disasters in Japan

IGRC 2020 MUSCAT

In recent years, severe disasters involving large-scale power shut down have frequently occurred in Japan.

Disaster Name	Accrual Date	Disaster Scale	Damage Summary
Great East Japan Earthquake	March 11, 2011	Magnitude 9.0	Power shut down in Tohoku area 4.66 million houses
Osaka Northern Earthquake	June 18, 2018	Magnitude 6.1	Power shut down in Kansai area 170 thousand houses
Hokkaido Iburi Eastern Earthquake	September 6, 2018	Magnitude 6.7	Power shut down in Hokkaido 2.95 million houses
Typhoon No. 21 in 2018 (Asian name "Jebi")	September 4, 2018	Super Typhoon Category 5	Power shut down in Kansai area 1.71 million houses
Typhoon No. 15 in 2019 (Asian name "Faxai")	September 9, 2019	Super Typhoon Category 4	Power shut down in Tokyo area 930 thousand houses
Typhoon No. 19 in 2019 (Asian name "Hagibis")	October 6, 2019	Super Typhoon Category 5	Power shut down in Japan 520 thousand houses

6. Typhoon No. 15 (Asian name "Faxai")

Typhoon No. 15 (Asian name "Faxai"), which occurred on September 5, 2019, landed in Japan on September 9, causing devastating damage mainly in the Tokyo area.

Typhoon No. 15 in 2019 (Asian name "Faxai")	September 9, 2019	Super Typhoon Category 4	Power shut down in Tokyo area 930 thousand houses

Collapse of power transmission towers

Road blockage by fallen trees

7. Mutsuzawa Smart Wellness Town

Micro CHP system supplied electricity and hot water (heat) at the "Mutsuzawa Smart Wellness Town" , and achieved results as a disaster prevention base.

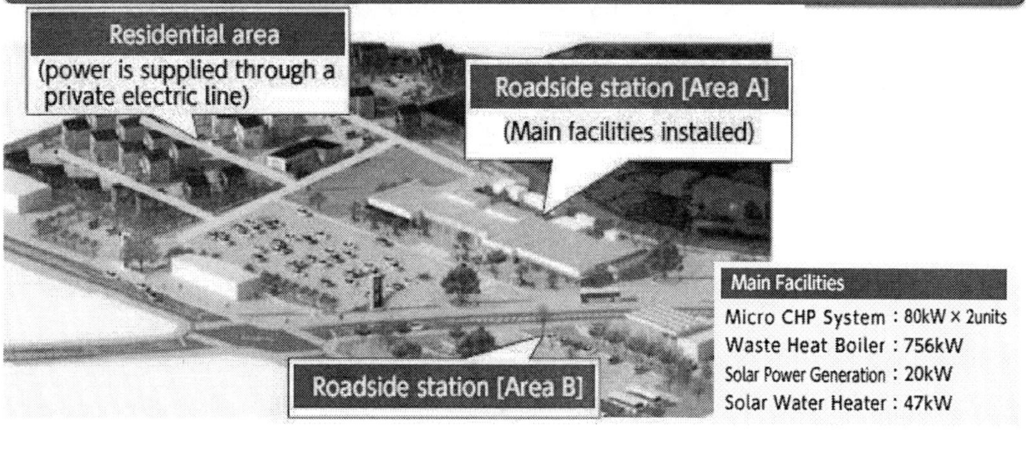

7. Actual Contribution Case of Micro CHP in Disaster

IGRC 2020 MUSCAT

Mutsuzawa Smart Wellness Town was operated as a disaster prevention base and the Micro CHP system supported this.

Supplies power during grid power shut down

Free access to hot water showers and toilets for local residents

8. Conclusion

IGRC 2020 MUSCAT

- In Japan, Micro CHP systems are widely introduced and used for commercial such as restaurants and clinics.
- We believe that the system will also contribute to power supply for houses as not only a regular power source but also an emergency power source in Europe, North America and Asia.
- We hope to work on overseas sales of the micro CHP system in the future.

IGRC 2020
MUSCAT
THANK YOU
FOR YOUR ATTENTION

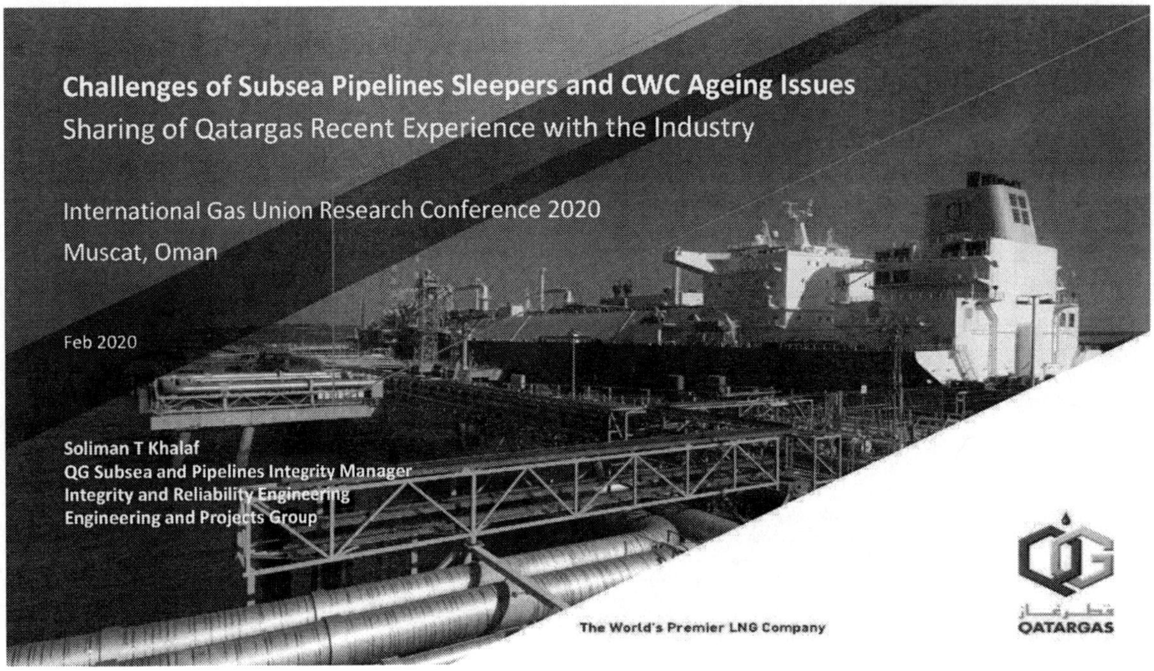

Agenda

1. Aim and Objectives
2. Background
3. Description of the Issue
4. Subsea Pipelines Sleepers Design and Sizing- Key Learning to Share
5. Sequence of Events
6. The Risk
7. The Response and Corrective Actions
8. Key Learning to Share
9. Suggestions and Recommendations
10. Q&A

Aim and Objectives

- ❑ Aim is **to share** the outline of **a finding** from a **recent** Subsea Pipeline Under Water Inspection **(UWI) activities**
- ❑ **Qatargas is willing** to share this **operational experience with the industry**
  - ✓ **To drive** early appreciation of **the impacts of design and construction aspects** on **pipelines full life integrity**
  - ✓ **To allow** the Industry **appreciate** the **challenges operating** Companies may encounter **in managing subsea sleepers and** Concrete Weight Coating **(CWC) issues** including **ageing,** for **live subsea pipelines**.
- ❑ This is a **critical topic** when dealing with **high pressure sour gas pipelines** and an area with **limited material available**
- ❑ **Lessons learnt** can be taken for **future projects planning and design works.**

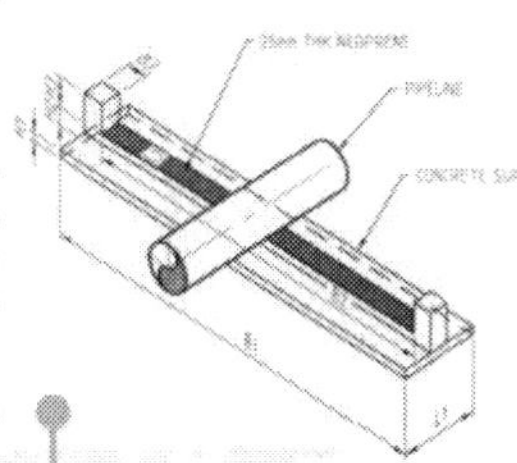

- ❑ **Visualize** the **potential damages,** caused over time to the pipelines, by inaccurate designed or installed concrete sleepers, used for elevating pipelines at subsea **crossings**, subsea **Tie-ins** arrangements, or **lateral buckling** management. This definitely **confirms the need to**:
  - ➢ **Enhance crossing designs** for the new projects.
  - ➢ **Carefully consider sleeper engineering, construction, installations and QA/QC.**
  - ➢ **Drive focus on details including the sleepers** while conducting subsea inspections
  - ➢ **Timely enhanced pipelines inspection, integrity and assessment** tasks.
  - ➢ **Importance of** subsea and pipelines emergency response systems **(EPRS)** for operating facilities.

QATARGAS

Background

- ❑ Subsea **sleepers** with retainers are **often used** on subsea pipelines at subsea tie-ins, pipelines crossing arrangements or where **movement is anticipated** as a result of thermal cycles, Lateral buckling, etc.
- ❑ The key concept is to **encourage the pipeline movement, in an engineered and controlled manner,** at **selected locations, over the sleepers,** rather than over the seabed.
- ❑ **Over time, different forms of damages may happen** to sleepers, the pipeline and/or its CWC.
- ❑ Damage may happen when the **pipe rubs against sleepers guide as the pipes moves,** if all aspects including the **max pipeline operating conditions and Installation tolerances** are not adequately considered during the project phases.
- ❑ This abrasive action may **compromise** the pipeline CWC and corrosion coating and **expose the pipe shell to time dependent failures mechanisms** (Dents, Deformation, Gouges or Fatigue, others).
- ❑ **NDT or repairs may drive pipeline outage** and/or pressure drops (a potential business impact)

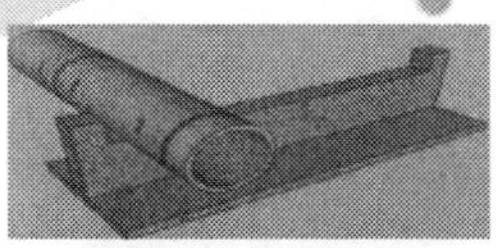

- ❑ In Qatargas, the recent observation of a pipeline crossing sleepers damage,
  - ✓ **The Case has been carefully managed and timely responded, safely.**
  - ✓ Now under detailed **RCFA and evaluation by a specialised team to identify the Lessons Learnt and drive permanent corrective actions, enhancements and improvements.**
  - ✓ As a **positive commitment to support the Industry,** the topic was **widely communicated to our Contractors and Development Team to consider** for **new facility developments.**
  - ✓ An active and effective **pipeline emergency response system (EPRS) in place,** managed by skilled subsea and pipelines integrity engineers, as part of **QP EPRS club,** to effectively respond to, and manage any situation while introducing permanent solutions.

QATARGAS

Description of the Issue

- ❑ The referred pipelines were built during the last ~10 years **utilizing concrete sleepers in managing all subsea crossings** and for managing the **pipeline lateral buckling.**
- ❑ During a recent **scheduled UWI,** the referred pipeline at a **subsea cable crossing,** were reported to have a **dislocation, combined with damage to 3 sleepers.**
- ❑ Following this **discovery, Company in-house EPRS procedures were implemented** including formal RA.
- ❑ **Detailed UWI** surveys were conducted **utilizing divers, for all crossings** and all sleepers were closed visual inspected **(CVI) and mitigations were introduced.**
- ❑ As all pipelines has **WYE arrangements resting on sleepers with anticipated** pipeline **frequent movement ,** the detailed **divers CVI** was extended to all WYEs.
- ❑ The **lateral friction force on the sleepers with** Neoprene plate is expected to be **below** the **pipeline to soil friction.**
- ❑ **Localized CWC damage reported** at some sleepers.
- ❑ Evaluation **identified few engineering** and the **execution** related **contributing factors.**
- ❑ **Ongoing efforts to evaluate all the findings.**

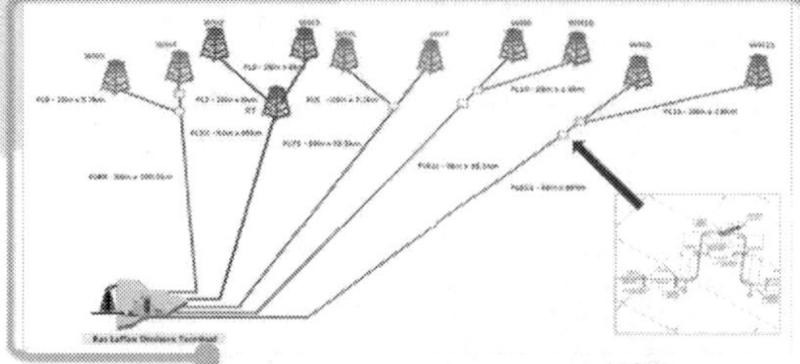

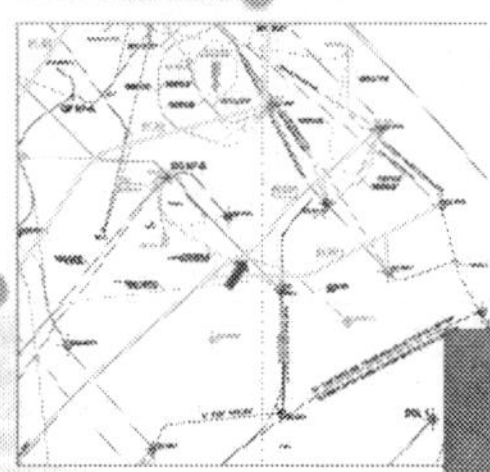

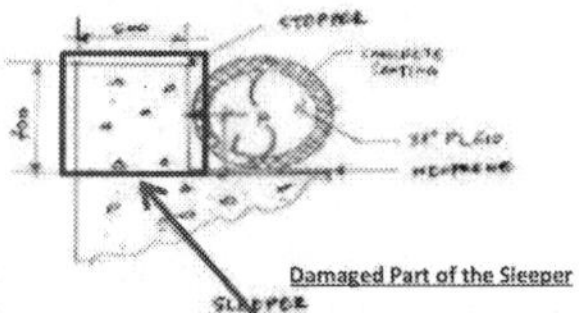

Damaged Part of the Sleeper

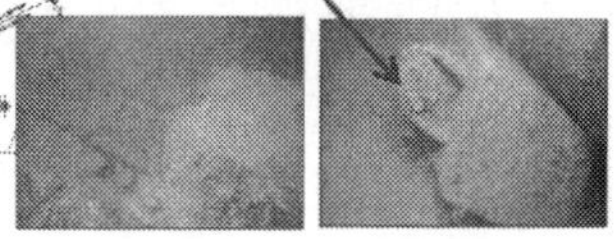

QATARGAS

Description of the Issue

The analysis concluded that:

- ❑ Progressive **seasonality displacement** of pipelines leading to **Neoprene plates dislocation** and damage.
- ❑ Sleepers and pipeline Interaction caused Localized **damages to** pipeline **CWC, sleeper Neoprene plates and sleeper retainer.**
- ❑ Sleepers **rebars are considered a potential** risk to the pipeline.
- ❑ Managing the sleepers issues, for live pipelines, is **an operational challenge** and may be of major impacts if left in place.

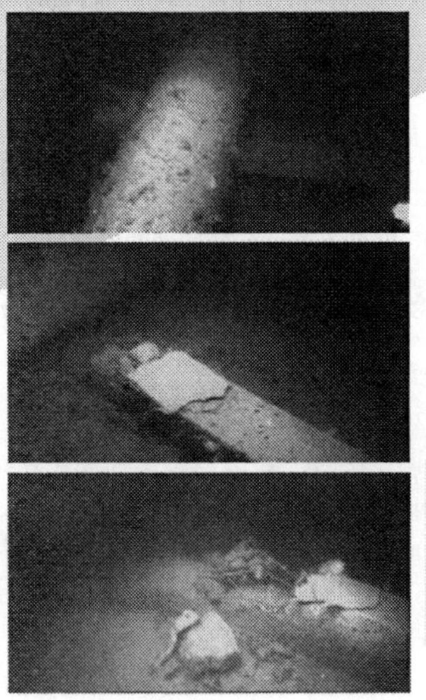

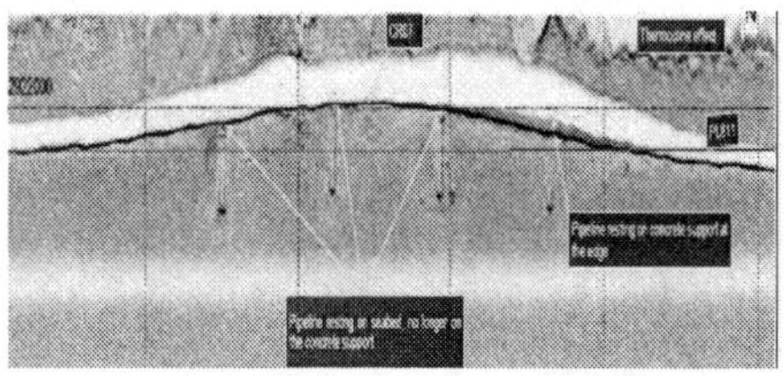

Initial RCFA findings : Indications of issues in the

- **Crossings and** Sleepers design
- Project **QA/QC (Engineering & Execution)**
- **Attention to details in critical zones**

QATARGAS

Subsea Pipelines Sleepers Design and Sizing- Key Learning to Share

- **Over time, cracks initiate** at sleepers with Sharp Corners with no trims for stress concentration enhancements.
- The sleeper design **didn't fully consider the buckling analysis predicted max lateral movement** of 11m, greater than the width of sleeper and Installation tolerances of 3m.

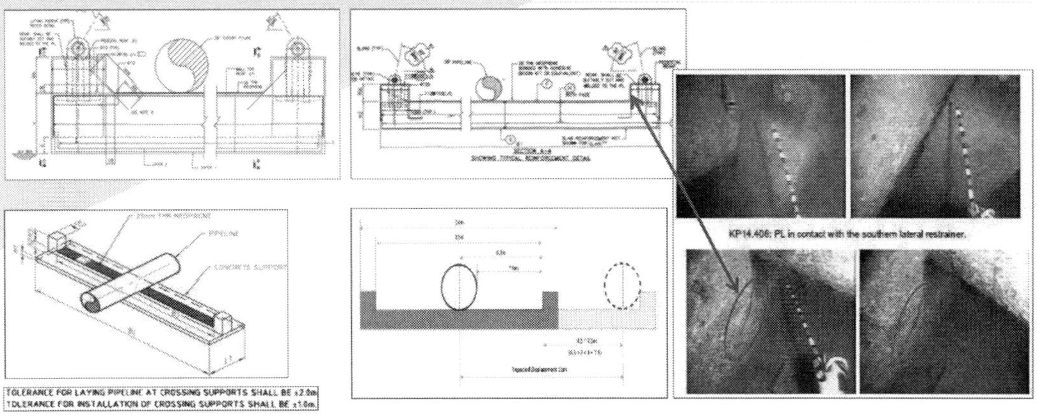

Sequence of Events

Potential Pipelines Failure due to CWC Coating Damage at Sleepers and Development of Gouges, Cracks or Dents Over Time.

- **Pipeline Movements** under **operational** and environmental effects
- **Ageing of the adhesive material** used to attached the Neoprene plate to the top of the sleeper
- **Neoprene plate dislocated** and get damaged
- Pipeline **CWC subject to friction** and **shear** under **compression**
- Micro cracks in the CWC getting enlarged and start to **degrade the CWC integrity and falls off**
- Pipeline **without CWC** starts to interact with the sleepers
- Pipeline **Corrosion Coating** starts interacting with the concrete sleeper and **gets damaged** under friction and compression
- **Over time** the pipeline **shell at 6 O'clock position** will be subjected to **deformation, dents and/ or gouges** which will lead to cracks
- Risk of **potential Pipeline failure and leaks.**

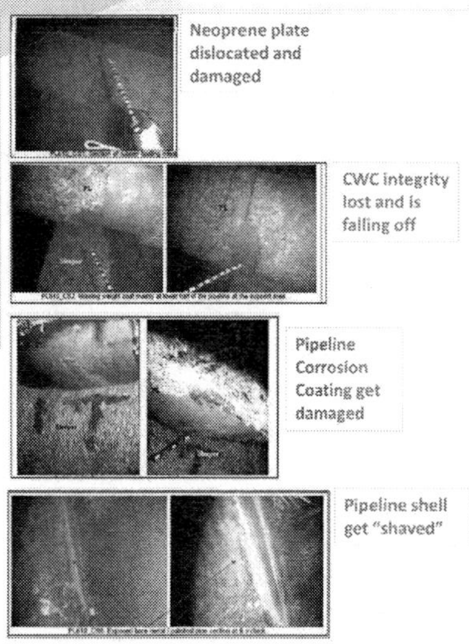

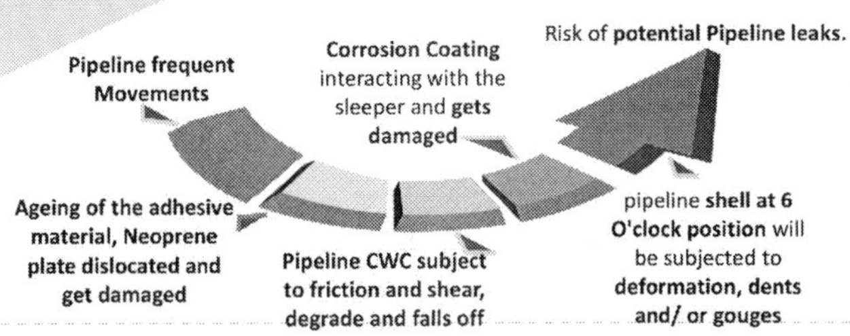

Sequence of Events

Potential Pipelines Failure due to Continued Interaction with Supports Restrainers and Development of Gouges, Cracks or Dents Over Time.

- Pipeline **Movements** under **operational** and environmental effects
- **Ageing of the adhesive material** used to attached the Neoprene plate to the top of the sleeper
- **Neoprene plate displaced** and gets damaged
- Pipeline side movement on the sleeper and **touching the retainer**
- Pipeline **CWC** at **3/9 O'clock** interacts with the sleeper and **gets damaged**
- Micro Cracks present in the sleeper retainer gets enlarged under bending and shear from the pipeline, **degrade the sleeper retainer integrity, causing damage and dislocation under excessive load and/or frequent pipe applied force**
- **Sleepers end retainers will fall down**, causing **pipeline dislocation** and **exposure to sleepers reinforcement steel bars**
- Pipeline shell has a **potential for deformation, dents and/or gouges** which will lead to development of cracks
- Risk of **Potential Pipeline failure** and leaks

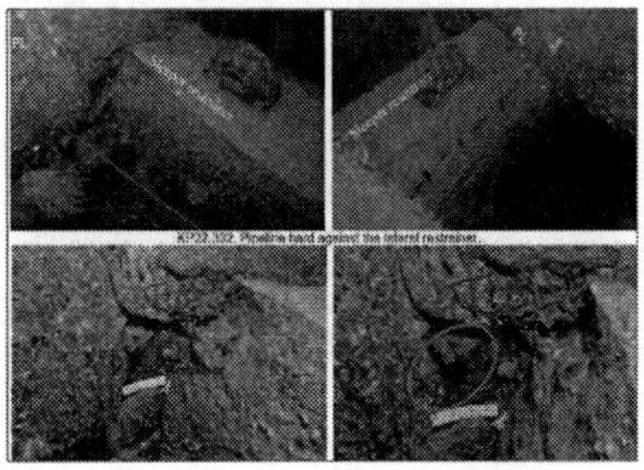

QATARGAS

The Risk

- ❑ **Risk screening and evaluation process** is a **key part of** Qatargas **established Integrity and reliability management system** for the operating facility.
- ❑ On the reporting of the discoveries, **a formal risk evaluation took place**, lead by the pipeline technical authority **(TA)** and supported by selected subject matter experts **(SME)**.
- ❑ The **risk scenarios identified were potential HSE and business impacts due to further degradation of the pipelines** at the areas of damaged CWC and sleepers as a result the pipeline and sleeper's interaction and friction **leading to potential pipeline loss of containment (HSE)** and pipeline **extended outage for repair (Business).**
- ❑ The **initiating event** maybe **one or a combination of deformation, gouging, cracks, Fatigue or mechanical impacts** for the pipeline body and/or welds as a result of the pipeline continuous movement over broken sleeper parts due to nature of pipeline operation and the impacts of surrounding environment.
- ❑ Key recommendation, in addition to **the engineering assessment**, were **to remove the broken sleepers, re-support the pipeline , steady pipeline operation** and implement more **frequent pipeline mentoring program.**

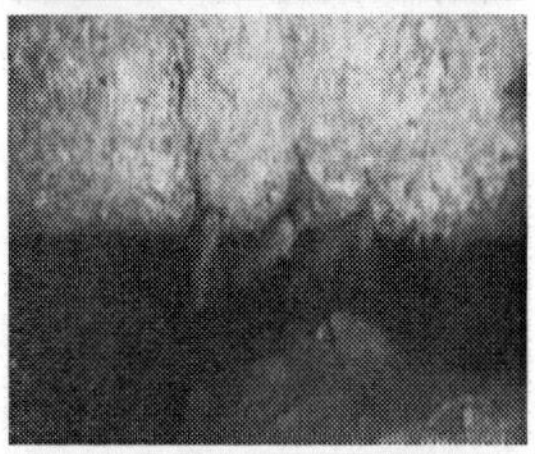

Remain part is potential to cause dent to the pipeline

QATARGAS

The Response and Corrective Actions

Key RA Recommendations

- Conduct Engineering Assessment,
- Remove the broken sleepers,
- Re-support the pipeline
- Conduct detail CVI and ILI
- Steady pipeline operation
- Implement more frequent pipeline mentoring program.

Damaged sleepers removed

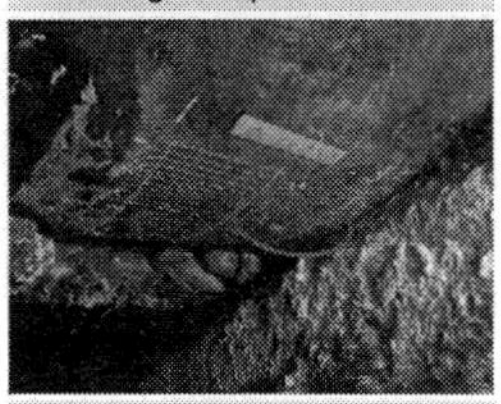

Detailed CVI and NDT Conducted

Implemented Recommendations

- Engineering assessment was conducted
- The damaged sleepers were removed.
- Developed free span was rectified.
- ILI was used to identify any localized deformation of internal / external corrosion
- External NDT was conducted
- 6 monthly frequent pipeline mentoring program initiated
- All recommendations were safely and timely implemented.
- Entire activities managed with no interruption to offshore gas supply.
- Ongoing planning for permanent solutions.

QATARGAS

Key Learning to Share

Ageing of the Neoprene plate and the integrity of its bonding material, the ageing of pipeline CWC and its potential damage under continues sliding/ lateral movements of the pipelines **all are of a negative impact on pipeline integrity at crossings sleepers.**

The designers shall take consideration that **subsea system effectiveness is the effectiveness of all its components**, including pipeline, concrete coatings, concrete sleepers, neoprene, etc. **Failures starts at the weakest part**

Integrity management of the subsea sour gas pipelines is a vital part of the Qatargas Integrity and reliability management. **Effective Integrity Management Starts at the Design Phase.**

The quality and effectiveness to response to pipelines integrity threats depends on the skills available to identify the risk and to act safely, timely and correctly.

The importance of the subsea and pipelines EPRS system and club

QATARGAS

Key Learning to Share

The need to **consider subsea sleeper full life cycle integrity and a Fit- For-Purpose design**, that is a **design** for **functionality;** for **safety** and for **maintainability,** considering anticipated pipeline movement, **construction tolerance and ageing in mind.**

All sleepers **components should be carefully designed, constructed and installed** to encourage the smooth pipeline movement in a control manner and mitigate potential damage of the CWC and the Neoprene plates.

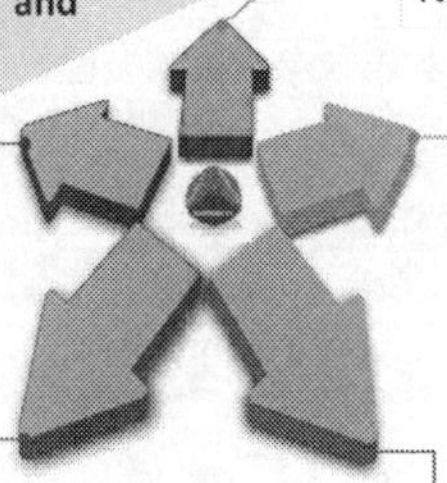

The impact of losing the CWC and **potential damage of the pipelines at the sleepers to be visualised at the design phase.**

Predicted pipelines cycling loading and operation mode in case of frequent offshore and onshore facilities shut down should be considered in the fatigue life all the pipelines system components.

The frequency and scope coverage of the UWI for critical zones and detailed inspection of sleepers need to be covered and considered in industry integrity management references as a good practice. Classically this is not an area of focus.

Suggestions and Recommendations

- **Consider** subsea sleeper **full life cycle integrity** and a **Fit- For-Purpose design**, for **functionality; safety** and **maintainability.**
- **All** sleepers **components** should be carefully **designed, constructed and installed** to encourage the **smooth pipeline movement in a control manner** and mitigate **potential damage** of the **CWC** and the **Neoprene plates.**
- Detailed engineering with **attention to details in critical zones** (Crossings, Tie-in, Buckling zones, etc.) should be considered
- The need for **elevated crossing arrangements** especially **surrounding subsea cables to be evaluated.**
- Carefully evaluate the **pipeline CWC design at sleeper locations**, its QA/QC

- Carefully **consider the Impacts of anticipated pipeline movement** within the subsea Tie-in arrangements and pipeline crossings **on the pipeline CWC, the sleepers and its Neoprene plate.**
- Consideration of careful **evaluation** of **actual pipeline configuration**, not simplified analysis.
- **Neoprene plate and its bonding material application**: Material selection and detailing among components (Adhesives), Fabrication QA/QC, **Prequalification for regional environmental effects** (Sea water quality effects), effects of **frequent pipeline sliding** , seasonality (and seabed temp.), CWC impact, etc.
- Focus on **projects engineering effective Interdisciplinary checks**.

- Concrete **Sleepers Fabrication, Cracking and Ageing**: Material selection (E.G Cement Type , Adhesives , Coating , etc.) **Cracks developed** during manufacturing, lifting, transporting etc., and its "long term" effects. Sea water quality and potential accelerating sleepers damages. **Reinforcing bars potential corrosion and aging due to water ingress from cracks.**
- **As-built** should consider reporting **sleeper positions** and its ends (E/N) **for future trending.**

- The **frequency and scope coverage of the UWI for critical zones and its sleepers** need to be considered in industry integrity management references **as a good practice.**

"Effective Integrity Management Starts at the Design Phase."

"Effective Integrity Management Starts at the Design Phase."

Thank You !!

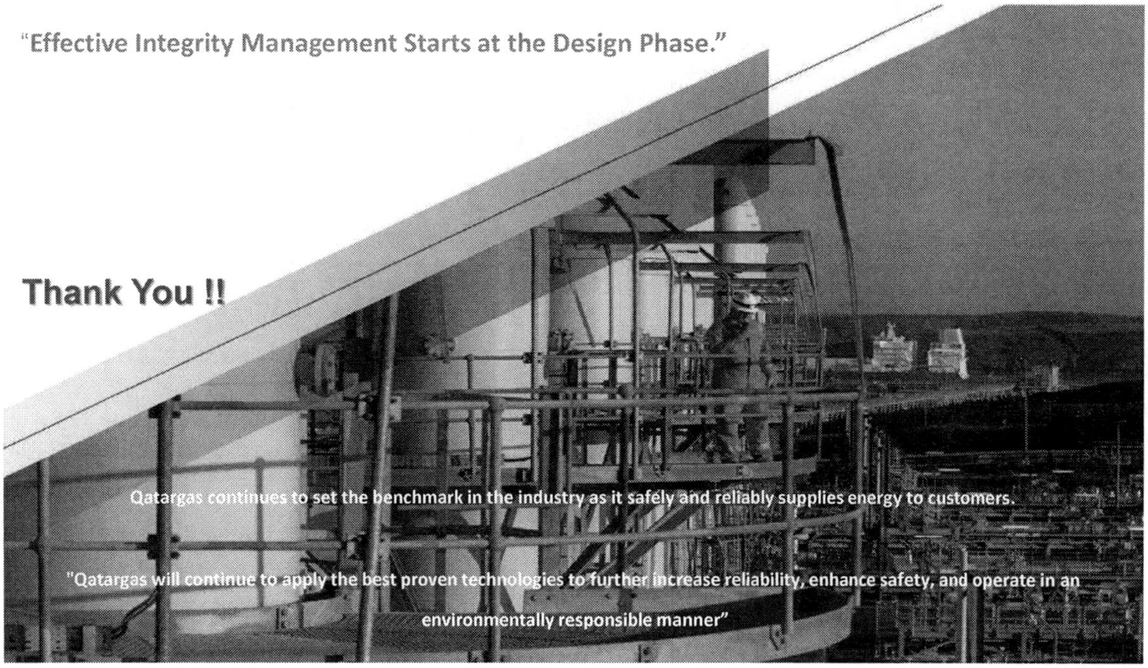

Any questions?

Sultan Qaboos University
College of Engineering
Department of Mechanical
nd Industrial Engineering

SQU

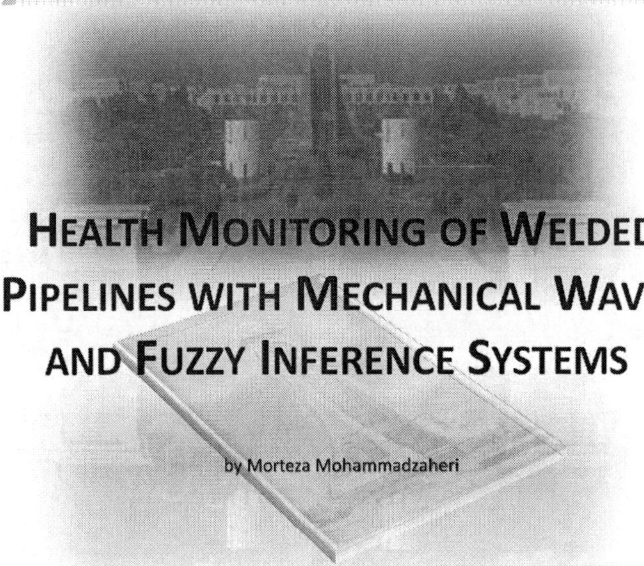

HEALTH MONITORING OF WELDED PIPELINES WITH MECHANICAL WAVES AND FUZZY INFERENCE SYSTEMS

by Morteza Mohammadzaheri

1

Sultan Qaboos University
College of Engineering
Department of Mechanical
nd Industrial Engineering

SQU

EXISTING METHODS

- test in the entire pipe
  - Electromagnetics (pigs)
  - Flame Detection
  - Optic Fibres
- test on some points
  - Ultrasonic Waves
  - Flow/Pressure Measurement

2

Sultan Qaboos University
College of Engineering
Department of Mechanical
and Industrial Engineering

SQU

WHAT ABOUT MECHNICAL WAVES?

Mechanical Waves (freq. below 5 kHz)
have further range and are cheaper to use than
Ultrasonic Waves (freq. over 100 kHz)

Problem: change in pipeline affects mechanical waves in a complicated way

Research Question: Can mechanical waves provide meaningful information about changes in pipelines?

3

Sultan Qaboos University
College of Engineering
Department of Mechanical
and Industrial Engineering

SQU

A MISSION: FAULT ISOLATION

Abaqus Model of a 50 meters 2" pipe with welding at every 5 meters
element types and mesh size are verified

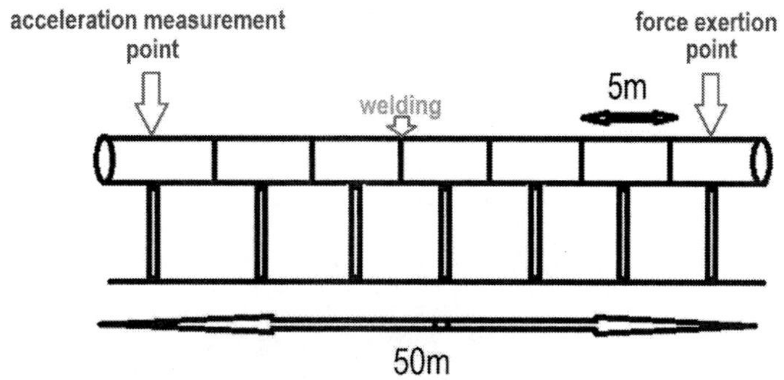

4

FEM METHOD VERIFICATION

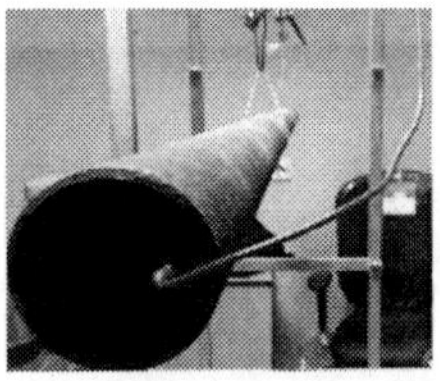

ABAQUS software
mesh size : 1 cm
element type : explicit shell
meshing technique : structured quad

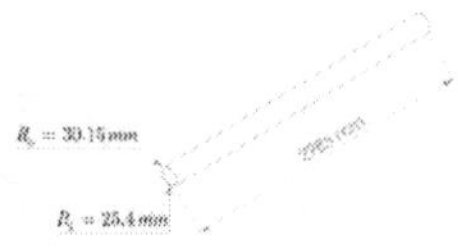

FEA RESULTS[Hz]	Modal Analysis [Hz]
40.101	40.020
110.03	110.95
214.26	216.55
351.05	356.018

FEM SIMULATIONS

Force =300 N in 0.001s, 15 fault locations,

fault is a 2mm diameter hole

no fault or with a fault at one of following locations from the end (in m)

3,6,9,12,15.5, 18,21,24,27,30.5,33,36,39,42,45.5

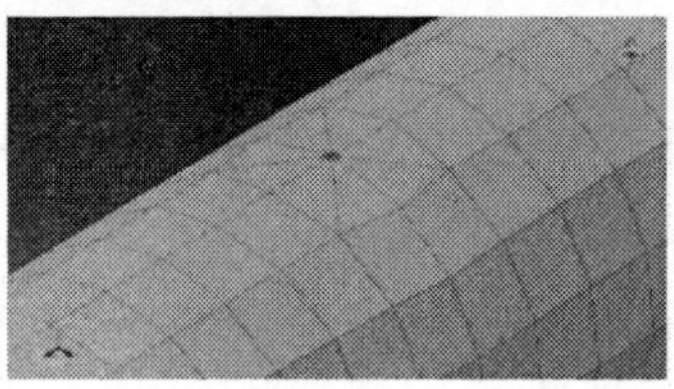

Sultan Qaboos University
College of Engineering
Department of Mechanical
nd Industrial Engineering

SQU

FREQUENCY RESPONSE

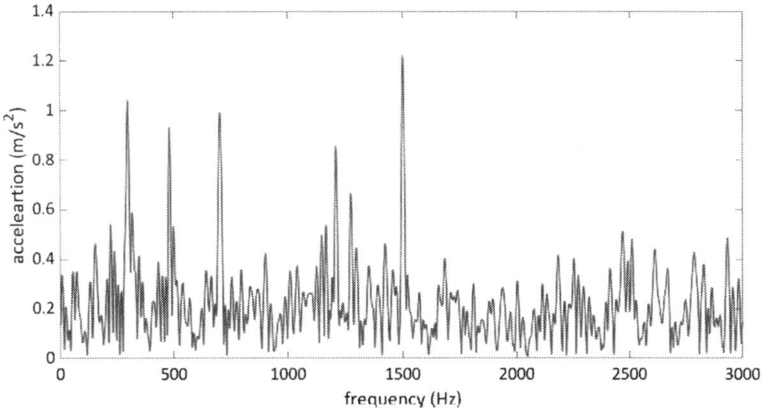

are peaks of the plot (resonance frequencies) the best points for fault detection?

Sultan Qaboos University
College of Engineering
Department of Mechanical
nd Industrial Engineering

SQU

FINDING MOST SENSITIVE FREQUENCY AREAS TO CHANGE

- **Partitioning** the frequency range (3kHz) into one hundred 30 Hz areas. Finding acceleration average in each partition.

Result: An array of **o**ne hundred numbers for each faulty and faultless pipe

- **Deducting** the array of average accelerations of each faulty pipe from the same array for the faultless pipe

Result: An array of one hundred number for each fault: Fault Signature

- The **variance** of acceleration within all faults is calculated for each frequency areas (out of 100 areas). The higher variance, the more sensitive to change of fault location.

Sultan Qaboos University
College of Engineering
Department of Mechanical
and Industrial Engineering

SQU

THE MOST SENSITIVE FREQ. AREAS

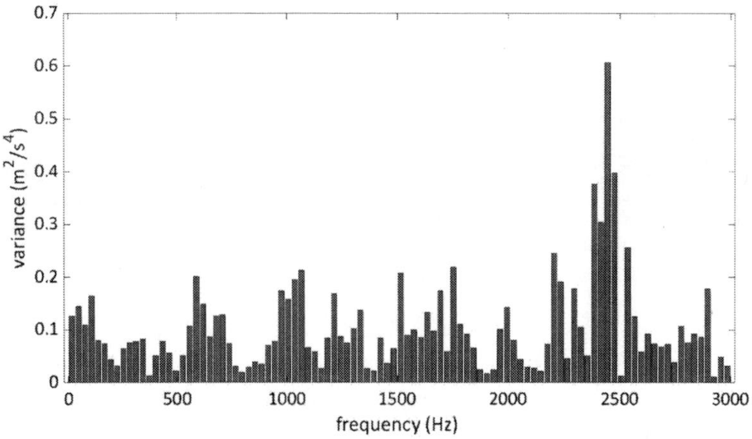

9

Sultan Qaboos University
College of Engineering
Department of Mechanical
and Industrial Engineering

SQU

RESONANCE IS NOT EVERYTHING

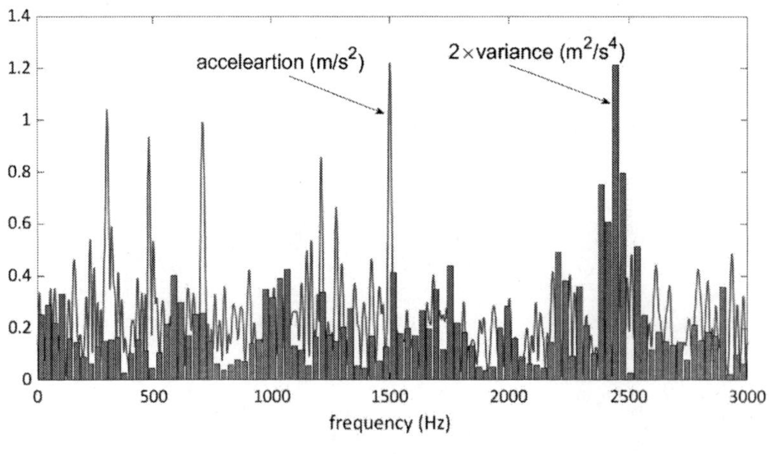

10

Sultan Qaboos University
College of Engineering
Department of Mechanical
nd Industrial Engineering

SQU

A FUZZY MODEL

$$\mu_{ij} = \exp\left(-\frac{(u_i - c_{ij})^2}{2\sigma_{ij}^{\;2}}\right). \qquad \hat{y} = \frac{\sum_{j=1}^{n}\left(\overbrace{\left(\sum_{i=1}^{3} a_{ij}u_i + a_j\right)}^{j^{th}\ rule\ output}\prod_{i=1}^{3}\mu_{ij}\right)}{\sum_{j=1}^{n}\underbrace{\prod_{i=1}^{3}\mu_{ij}}_{j^{th}\ rule\ weight}}.$$

$\hat{y} = estimated\ output$

$u_i = input, i = 1,\ldots, size\ of\ fault\ signature$

modelling data validation data test data

3,6,9,12,15.5,18,21,24,27,30.5,33,36,39,42,45.5

11

Sultan Qaboos University
College of Engineering
Department of Mechanical
nd Industrial Engineering

SQU

TEST RESULTS FOR FUZZY MODELS

Complete Model

- developed with the data of all, 100, frequency areas
- 1111 parameters
- 18 m is estimated 12.0 m
- 33 m is estimated 39.0 m
- σ(standard deviation)=6.0 m
- *b* (Estimation bias)= 0.0 m

real location =

Estimated location+0.0±12.0 m

Selective Model

- developed with 10 freq. areas (variance > 0.2)
- 341 parameters
- 18 m is estimated 17.1 m
- 33 m is estimated 24.5 m
- σ (standard deviation)=3.8 m
- *b* (estimation bias)=-4.7m

real location =

Estimated location-4.7±7.6 m

What if the fault location moved 30 cm per simulation rather than 3 m?

12

Sultan Qaboos University
College of Engineering
Department of Mechanical
and Industrial Engineering

SQU

DIFFICULTIES

Computing

Running numerical models for several (1000s) times to provide enough data for data-driven modelling (e.g. fuzzy modelling)

A Realistic Numerical Model (e.g. FEM)

Solution: **Adaptive Numerical Models**

numerical models fitting to the data collected from the site

13

Sultan Qaboos University
College of Engineering
Department of Mechanical
and Industrial Engineering

SQU

WHAT THE FINAL PRODUCT LOOK LIKE

1. Development of a rough numerical model
2. Adaptation of the numerical model to fit to the collected data from the site
3. Simulation of a variety of probable faults and collecting their signatures and outputs (existence, size, shape, location)
4. Training the results of step 3 to data-driven models
5. The algorithm now can start health monitoring using the received data

14

THANK YOU

Special Thanks to Medco Energi Oman

MEDCOENERGI

Any Questions?

MUSCAT UNIVERSITY
Cranfield University
Techno-Economic Assessment of CO_2 Monetization Towards Methanol Production
Lanah Al Sammarraie
MSc Process System Engineering
Contact
Ending
Results
Methods
Objective
About
1

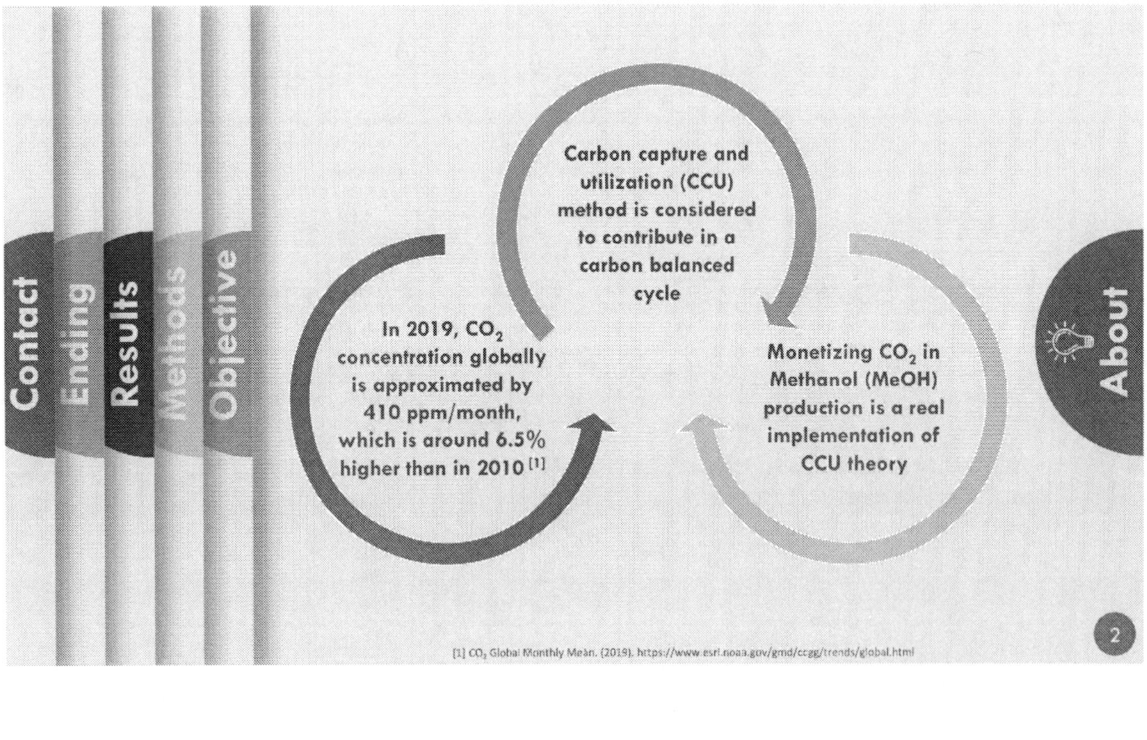
Carbon capture and utilization (CCU) method is considered to contribute in a carbon balanced cycle
In 2019, CO_2 concentration globally is approximated by 410 ppm/month, which is around 6.5% higher than in 2010 [1]
Monetizing CO_2 in Methanol (MeOH) production is a real implementation of CCU theory
Contact
Ending
Results
Methods
Objective
About
[1] CO_2 Global Monthly Mean. (2019). https://www.esrl.noaa.gov/gmd/ccgg/trends/global.html
2

Contact
Ending
Results
Methods
1
Monetizing 500 ton CO_2/day in MeOH production process.
2
Optimize the base model of Milani et. al [2]
3
Provide technical, environmental & economical evaluations of the processes.
Objective
about
[2] D. Milani, R. Khalilpour, G. Zahed, A. Abbas, A Model-Based Analysis of CO_2 Utilization in Methanol Synthesis Plant, 10, p. 12-22, Journal of CO_2 Utilization, Elsevier (2015).
3

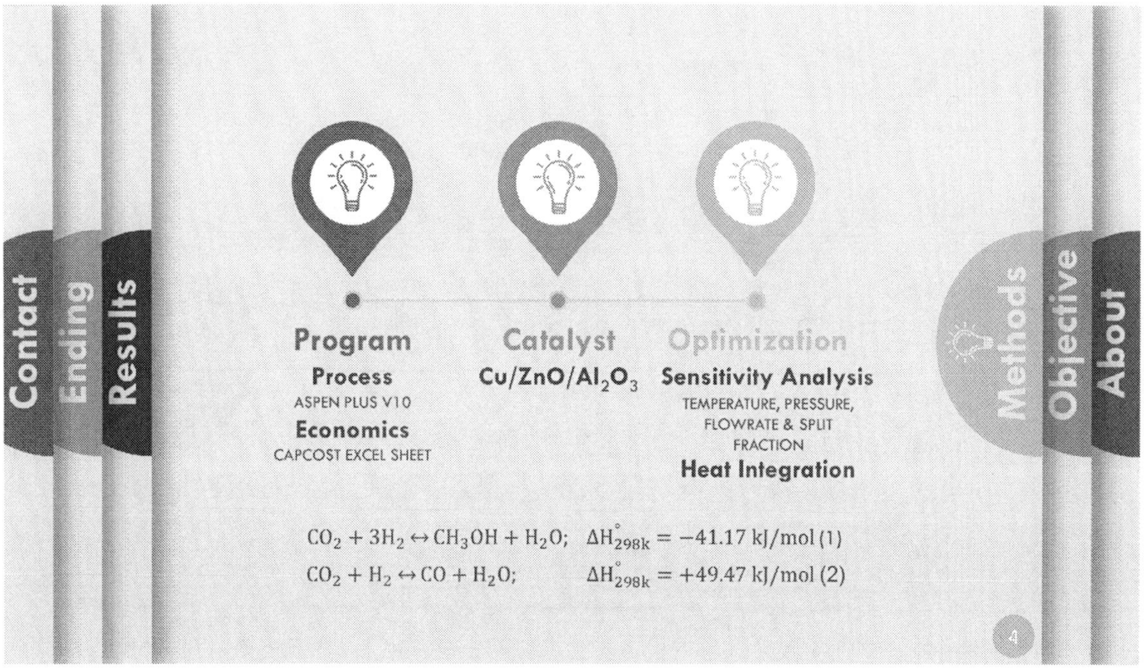

Contact
Ending
Results
Program
Process
ASPEN PLUS V10
Economics
CAPCOST EXCEL SHEET
Catalyst
$Cu/ZnO/Al_2O_3$
Optimization
Sensitivity Analysis
TEMPERATURE, PRESSURE, FLOWRATE & SPLIT FRACTION
Heat Integration
$CO_2 + 3H_2 \leftrightarrow CH_3OH + H_2O$; $\Delta H^\circ_{298k} = -41.17$ kJ/mol (1)
$CO_2 + H_2 \leftrightarrow CO + H_2O$; $\Delta H^\circ_{298k} = +49.47$ kJ/mol (2)
Methods
Objective
About
4

- Increasing H_2O flowrate
- Reducing flash units temperature
- Minimizing the purge
- Adding turbines & heat exchanger

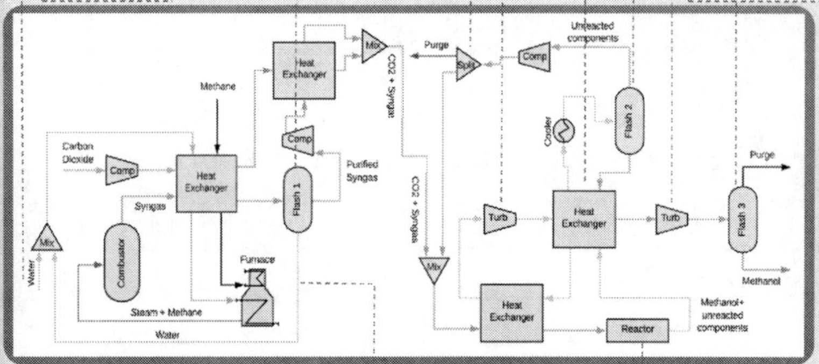

Figure 1: Developed model of syngas and MeOH production

- Full recycle of H_2O from "Flash 1"

- Increasing the inlet temperature
- Reducing the bed void
- Reducing L/D ratio

5

Contact Ending Results Methods Objective About

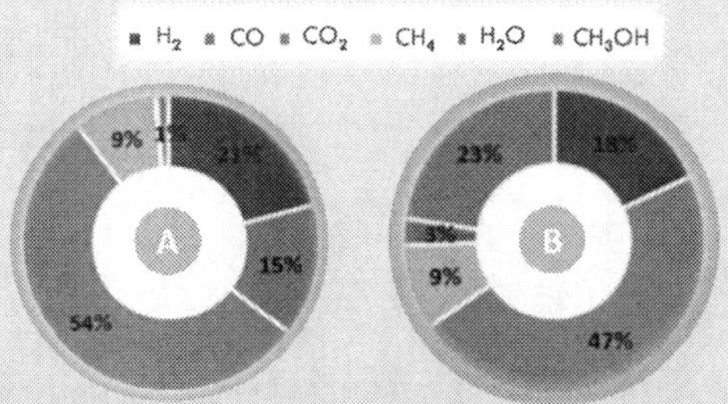

Figure 2: New reactor's components percentage (a) Input (b) Output

$$CO_2 + 3H_2 \leftrightarrow CH_3OH + H_2O$$

$$CO_2 + H_2 \leftrightarrow CO + H_2O$$

Table 1: Technical & environmental assessment of the base & optimized models.

Comparison	Base Model	New Model
Conversion [%]	2.24	14.38
Yield [%]	3.7	19.1
Net CO_2 Consumption [%]	-7.5	66.8

6

Contact
Ending
Results
Methods
Objective
About

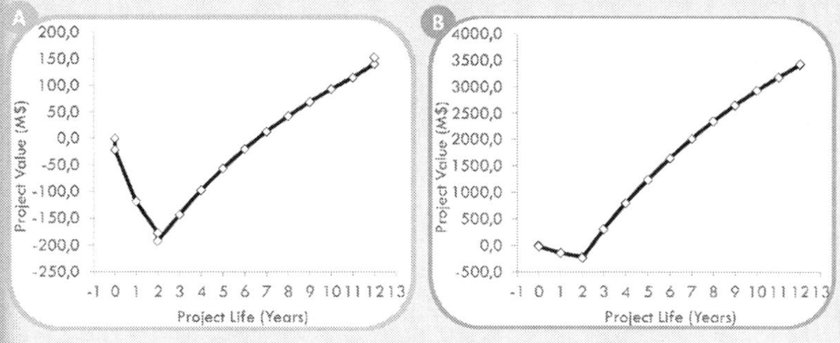

Figure 3: The cumulative cash flow diagram of (a) base model (b) developed model

Table 2: Economical assessment of the base & optimized model

Comparison	Base Model	New Model
Net Present Value [$]	152.94 M	3430.94 M
Discounted Payback Period [years]	3.6	0.4

7

Contact
Ending
Results
Methods
Objective
About

Conclusion

- Optimized process is promising to be established
- Technical assessment:
  - ➢ Conversion increased up to 14.38%
  - ➢ The yield increased by 5 times approximately
- Environmental assessment:
  - ➢ Process consumed 66.8% of CO_2
- Economical assessment:
  - ➢ Increase in net present value increased to $3.4B
  - ➢ Reduction in discounted payback period by 9 times

Future

- Burn the purge gasses using a flare stack.
- Add purification column to increase the purity of methanol.
- Investigate the effect of changing the feedstock price on the economical assessment of the project.

8

www.linkedin.com
/in/lanah-al-
sammarraie
lanah.sammarraie
@gmail.com
Muscat
Sultanate of
Oman
Contact
services
teams
Methods
Objective
About
Dr Nader Mosavat
Muscat University
Dr Dawid Hanak
Cranfield University
Dr Ali Nabavi
Cranfield University
9

IGRC2020
Muscat, Oman
February 24, 2020

Contribution of natural gas as mid- and long-term countermeasure for global warming and the road to the carbon neutrality

T. Nagata, F. Sano, K. Akimoto
Systems Analysis Group
Research Institute of Innovative Technology for the Earth (RITE)

About RITE

2

- Mission : R&D of industrial technologies that contribute to the conservation of the global environment and the progress of the world economy
- Established in July 1990 (Supported by MITI, local governments, academic circles and industries)
- Location : Kansai Science City (Kyoto, almost the boarder between Kyoto and Nara)
- Activities : Development of innovative environmental technologies for CO2 mitigation

Organization of RITE

3

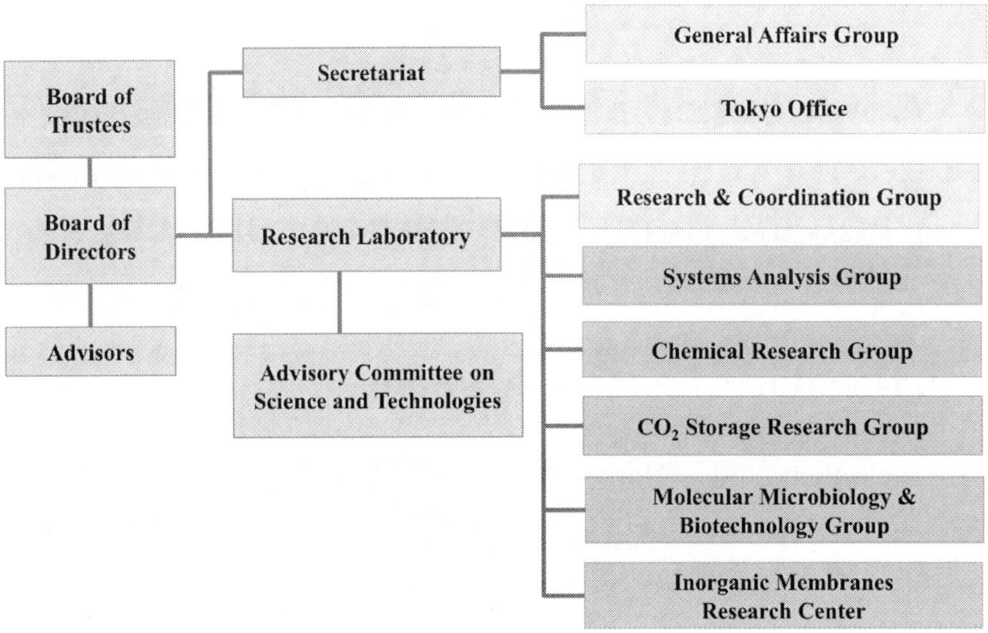

Contents

4

1. Background

2. Evaluation of natural gas systems

3. Evaluation of hydrogen and synthetic methane

Contents

1. Background

2. Evaluation of natural gas systems

3. Evaluation of hydrogen and synthetic methane

Basic Structure of Paris Agreement (PA)

Long-term Global Goal:
- Keeping temperature rises to well below 2℃ relative to pre-industrial levels, pursuing 1.5℃, and
- Achieving net zero GHG emissions (Carbon Neutrality) in the latter half of the 21st century.

Global Stocktake :
- NDCs are to be internationally and comparatively reviewed and evaluated from the viewpoint of meeting long-term targets.

NDC and Long-term Strategy:
- All of the member nations are required to submit their emission targets as the Nationally Determined Contribution (NDC),
- Pledge & Review
- NDCs are updated every five years
- Submit Long-term Strategy toward 2050 by 2020.

COP21(Dec. 2015、PA was adopted)、PA enforced (Nov. 2016)、
COP24(Implementation scheme excl. market mkm agreed)、US notified UN of Paris climate accord withdrawal (Nov. 2019)、COP25

Policy of Japan for Innovations in Energy and Environment

RIT 7

Paris Agreement : Adopt(2015)

5th Basic Plan for Science and Technology decided in Jan. 2016: proposing Super Smart Society (Society 5.0))

5th Strategic Energy Plan of Japan decided in July, 2018: Target for 2030 maintained、Vision for 2050 as an ambitious goal

IPCC 1.5°C Special Report (2018): net zero emission required in 2050

G20 in Japan (June 2019), Long-term Low Emission Development Strategy was submitted to UN.; G20 Innovation Action Plan in Karuizawa

Sept. 2019: International meetings for Hydrogen and Carbon Recycling

Oct. 2019: Green Innovation Week (TCFD Summit, ICEF, RD20)

Progressive Environment Innovation Strategy was decided (Jan. 2020)

Global CO2 emissions

RIT 8

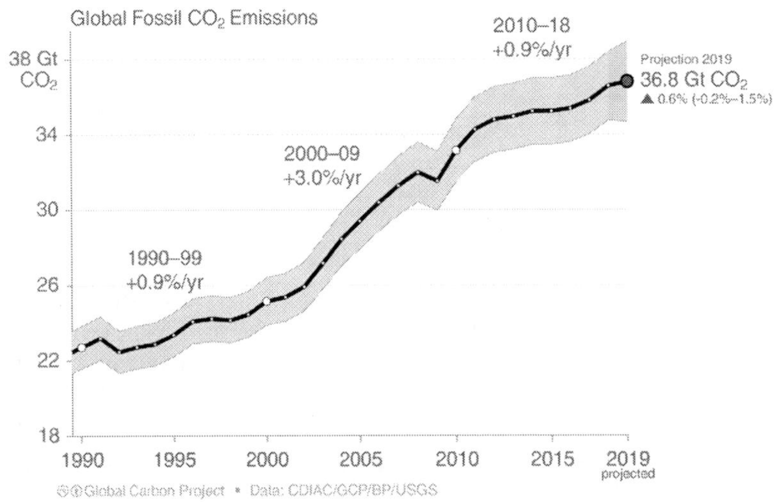

Source: Global Carbon Project

- Despite global and national efforts to mitigate global warming, global CO2 emissions have continued to rise.
- Along with long-term ambitious emission reduction targets, it is necessary to pursue effective emission reduction measures.

For Carbon Neutrality

9

- Final energy sources are required to be electricity and hydrogen in principle (+ bioenergy and direct heat use by solar heat etc.). Fuel cells by hydrogen provide also electricity for final use.
- Synthetic methane from CO2-free hydrogen and recovered CO2 can be used for final energy. (In this case, recovered CO2 plays only a carrier for hydrogen energy.)
- Productions of electricity and hydrogen are required to be decarbonized using renewables, nuclear power, CCS etc.
- However, the completely no use of hydrocarbon will be unrealistic, and therefore, even for carbon neutrality (net zero emissions), some gross emissions combined with negative emission technologies (NETs) of forestation, bioenergy with CCS (BECCS), DACS etc. can be accepted as real world response measures.
- Because the scenarios based on strong dependence on NETs will be weak in achievability and exert high negative impacts on biodiversity, (economically autonomous) low energy demands will be important for the achievement of decarbonized society.
- The transition to decarbonization is inevitable. Emission reductions should be implemented, considering total costs of both climate change damages and mitigation costs on the way to decarbonization.

Today's topics

10

The transition to decarbonization is inevitable. Emission reductions should be implemented, considering total costs of both climate change damages and mitigation costs on the way to decarbonization.

Natural gas is the energy source with the lowest CO2 intensity among fossil fuels. It also has a potential to utilize with intermittent renewable electricity, such as PV or wind power, since gas engine systems have quick response to load and can stabilize its output.

1. Quantitative analyses of natural gas systems
   (such as cogeneration and fuel switch)
2. Quantitative analyses of hydrogen and synthetic methane

using DNE21+, a linear programming model that enables systematic cost evaluation of various energy and CO2 reduction technologies.

Contents

11

1. Background

2. Evaluation of natural gas systems

3. Evaluation of hydrogen and synthetic methane

Features of DNE21+

12

World energy and global warming countermeasure evaluation model

- Linear programming model
- Model evaluation period: 2000-2100 years
  (Representative time: 2005, 10, 15, 20, 25, 30, 40, 50, 70, 2100)
- World region division: 54 region
- Interregional trade: coal, crude oil/oil products, natural gas/synthetic methane, synthetic oil, ethanol, hydrogen, electricity and CO2
- Bottom-up modeling (accumulating individual technologies):
  Energy supply sector (power generation sector, etc.)
  CO2 capture and storage
  Energy demand sectors
  (steel, cement, paper, chemical, aluminum, transportation, and consumer)
- Top-down modeling: considering welfare losses by using price elasticity
- Concretely model about 300 technologies

Modeling of cogeneration system (CGS)

Consider CGS options in industry, commercial and residential sectors

Assumption of CGS efficiency (LHV %)

		2015	2030	2050
Industry	power generation efficiency	49.0	51.0	54.5
	heat recover ratio	36.2	34.8	31.2
Commercial 1	power generation efficiency	42.3	47.5	50.7
	heat recover ratio	36.2	31.0	27.8
Commercial 2	power generation efficiency	41.0	44.0	47.0
	heat recover ratio	34.0	31.0	28.0
Residence	power generation efficiency	39.7	47.8	55.0
	heat recover ratio	55.2	45.0	37.8

Initial cost of CGS (thousand yen/kW)

	2015	2030	2050
Industry	150	150	150
Commercial 1	230	230	230
Commercial 2	300	300	300
Residence	1820	430	430

Assumed scenarios

Scenario	Remarks
Baseline	Not take into account any special measures against global warming
SCC	Internalization of external environmental costs equivalent to social cost of carbon (SCC) *47$/tCO2 in 2050
2DS	Below 2°C (>50%): Corresponding to IEA ETP [2DS]
B2DS	Well below 2°C (>66%): Corresponding to IEA ETP [B2DS]

Socioeconomic scenarios

- SSP2 ("Middle of the Road" scenario); **Global population: 9.2 billion in 2050, and global GDP growth: 2.4%/yr between 2000 and 2050.**
- SSP1 ("Sustainability" scenario); **Global population: 8.6 billion in 2050, and global GDP growth: 2.6%/yr between 2000 and 2050.**

CO_2 emissions

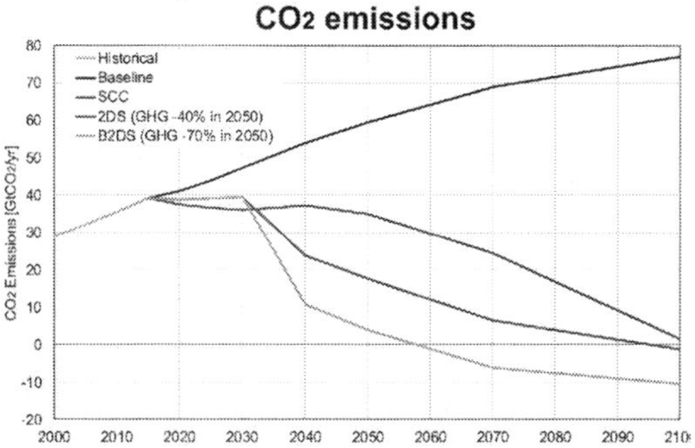

GHG emissions

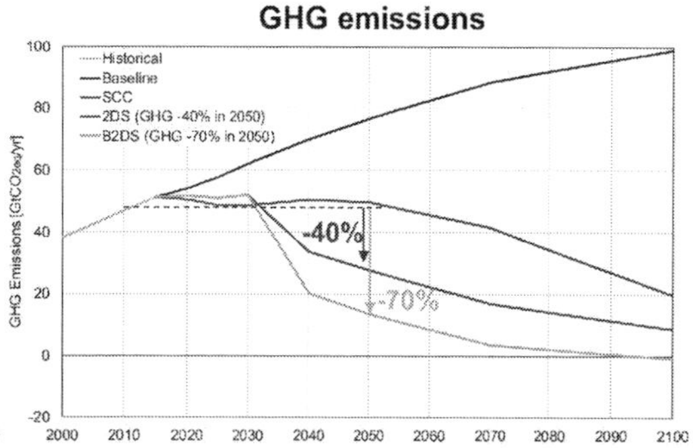

Note) Baseline emissions are not the assumed scenarios but are the resulting emissions by using the DNE21+ model. The figure shows the emissions of SSP2.

※ The emissions of 2DS and B2DS by 2030 were constrained by the submitted emission targets of NDCs of individual nations.

Primary energy supply (world, 2050)

RIT 15

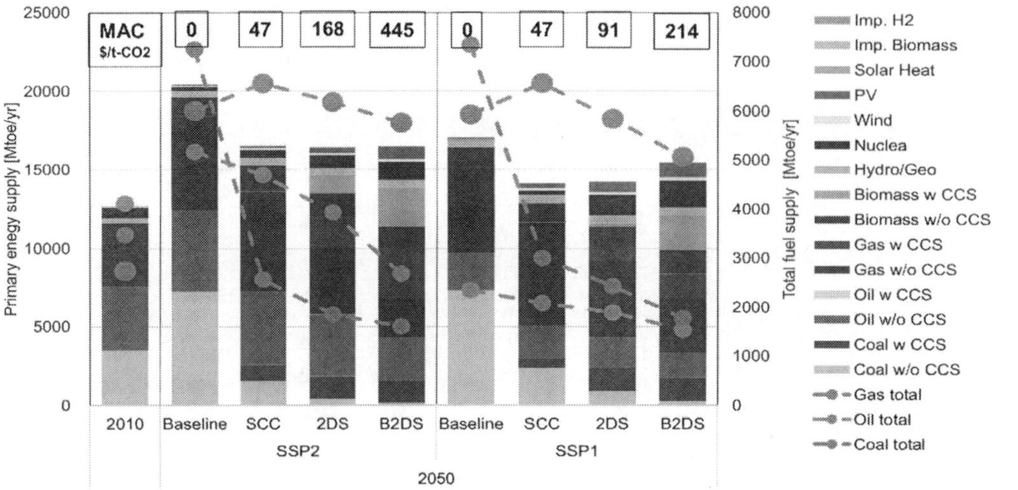

- Compared to the current situation, natural gas supply increase.
- Coal and oil will decrease significantly in SCC, 2DS, and B2DS scenarios, while natural gas will increase in SCC and stay almost same in 2DS scenario comparing baseline scenario.

Generated power (world, 2050)

RIT 16

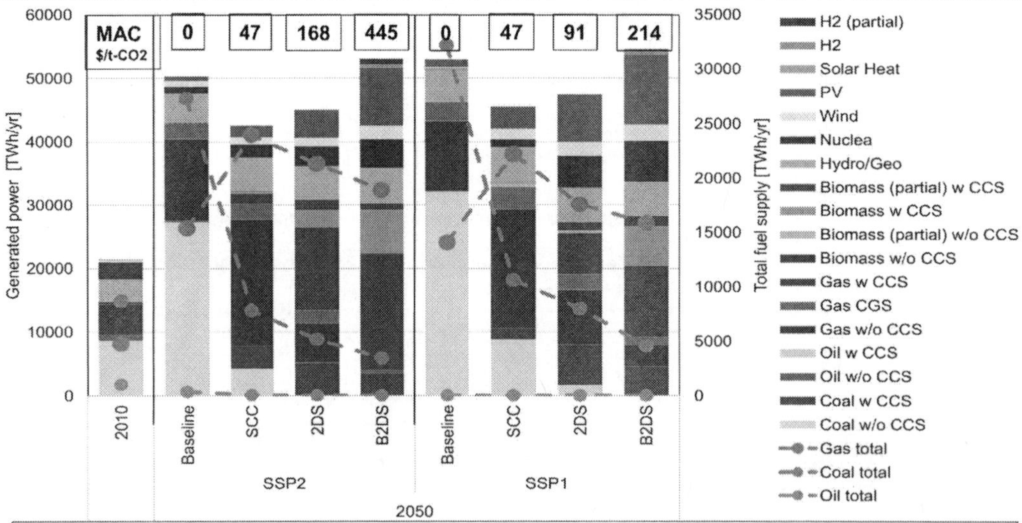

- Natural gas power including CGS increase greatly particularly in the SCC case, however need to convert to CCS, nuclear and renewable power at very high MAC (2DS, B2Ds).
- Compared to the baseline, natural gas power remains at a high level while coal-fired power generation is significantly reduced.

Energy consumption in industry sector (world, 2050)

17

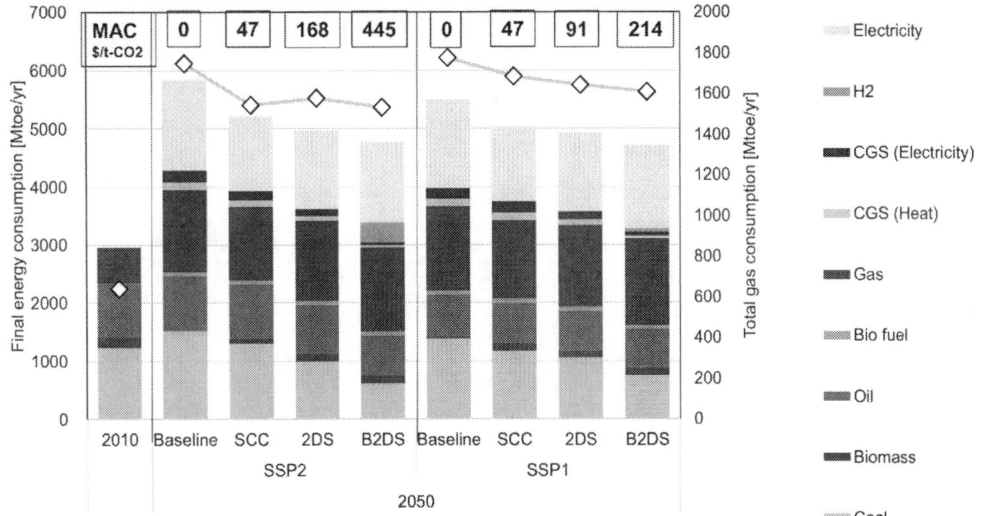

- **Demand for gas increases compared to the current situation.**
- **However, gas slightly decreases compared to baseline while coal decrease greatly.**

Results for evaluation of natural gas systems

18

- **Basically, expansion of natural gas use is economically rational until 2050 even in the 2 °C scenarios.**
- **However, when deep reductions of nearly net-zero CO2 emissions is required in 2050, CCS is required for natural gas power generations.**
- **There is a possibility of negative emission options such as BECCS, so that the economic rationality of using natural gas power without CCS still remains even for the requirement of nearly net-zero emission.**
- **With B2DS, the economics of using hydrogen in some of the power generation and industrial sectors also will be cost effective in 2050, but it is still limited.**

Contents

19

1. Background

2. Evaluation of natural gas systems

3. Evaluation of hydrogen and synthetic methane

Overall flow of hydrogen and synthetic methane

20

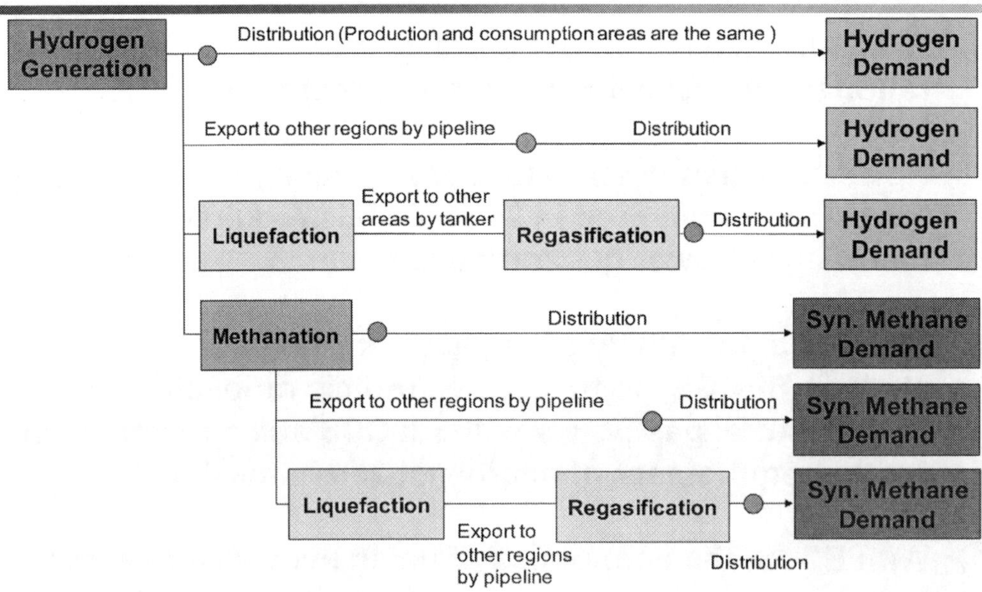

- **Hydrogen production technology: coal gasification, gas reforming, biomass gasification, water electrolysis, production in the petroleum refining sector**
- **Hydrogen distribution cost: assumed to be twice gas distribution cost per calorific value**

Assumption of PV generation

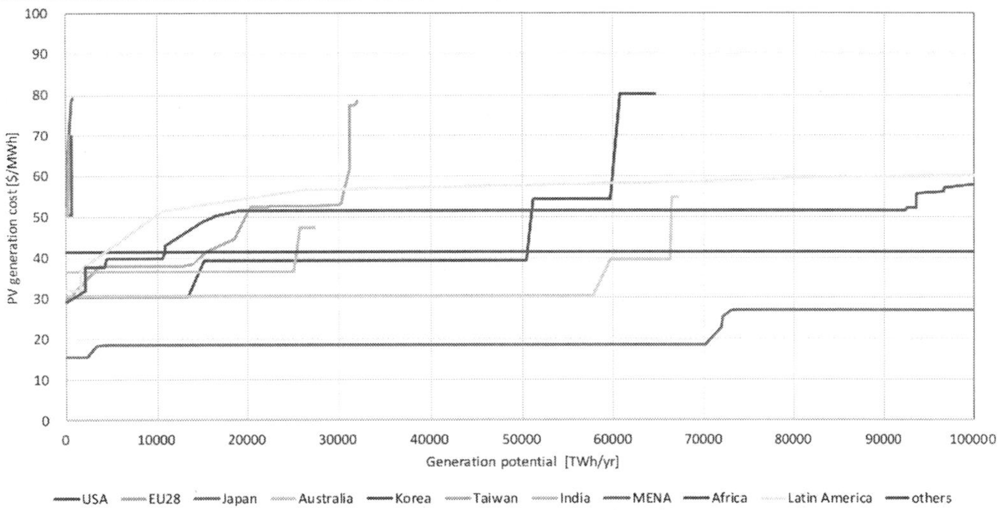

- PV equipment costs in the Middle East and North Africa (MENA) are assumed to be the same cost level as India.
- Assuming future facility costs for other countries and regions based on the SDS (Sustainable Development Scenario) in IEA World Energy Outlook 2018.
- Estimated PV generation (Wh) cost including grid-based solar radiation intensity data.

Cost assessment in 2050

Analyzing scenario: Corresponding to 2 ℃ target
GHGs of world are reduced by 40% in 2050 compared to 2010
(equivalent to 2DS of IEA ETP2017)
and G7 countries achieve 80% reduction

	CO2 MAC ($/t-CO2)	CO2 reduction cost (billion $/y)
USA	549	464 (1.72)
Canada	280	96 (5.05)
UK	289	36 (1.05)
France	167	25 (0.92)
Germany	279	57 (1.86)
Italy	291	22 (1.36)
Japan	512	82 (1.09)
Other countries	143	866 (1.35)
World		1649 (1.48)

Numbers in parentheses are percentages of GDP expected at baseline

- G7 countries have higher CO2 reduction costs than other countries
- Especially, the USA and Japan exceed 500 $/t-CO2

Hydrogen import / export

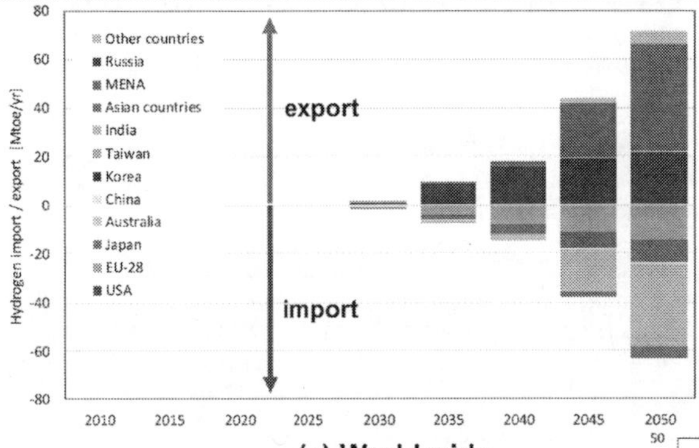

(a) World wide

· India imports mainly from the Middle East and North Africa, EU-28 imports mainly from others. (Switzerland and Norway), Japan imports mainly from the US

· Import mainly from regions with short transport distances.

· MENA and the United States are the major exporters.

· The main importing countries are India, EU-28 and Japan.

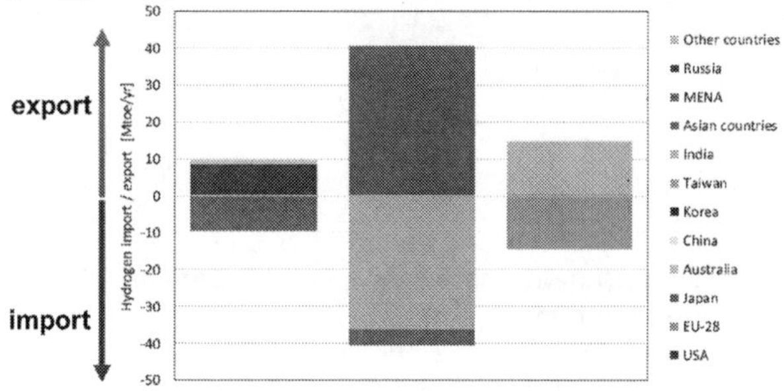

(b) Major importing countries

Usage of synthetic methane and production by region

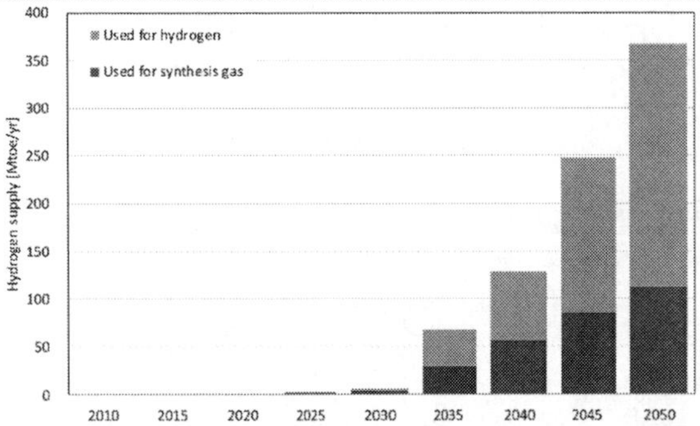

Used for hydrogen / synthetic methane

· MENA have the highest production of hydrogen, followed by Australia and EU-28.

· USA produce almost no synthetic methane. It is because CCS costs are low and CO2 is stored rather than use for methanation.

· Approximately 30% of hydrogen produced will be used as synthetic methane in 2050.

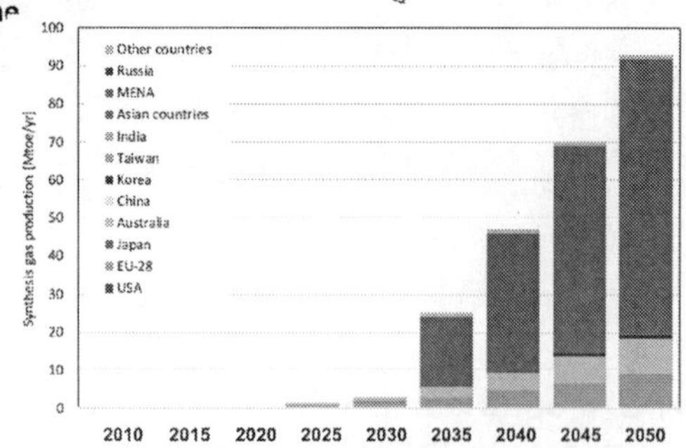

Synthetic methane production by region

Synthetic methane import / export

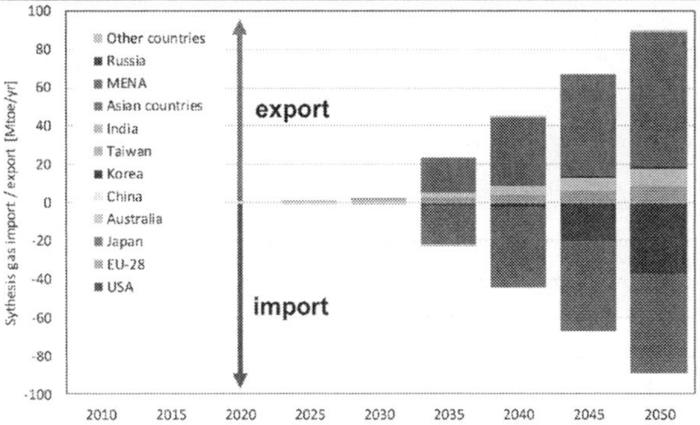

(a) World wide

- Japan and USA import synthetic methane since their high CO2 MAC.

- Japan import synthetic methane from MENA and Australia.
- USA import synthetic methane from MENA and EU-28.

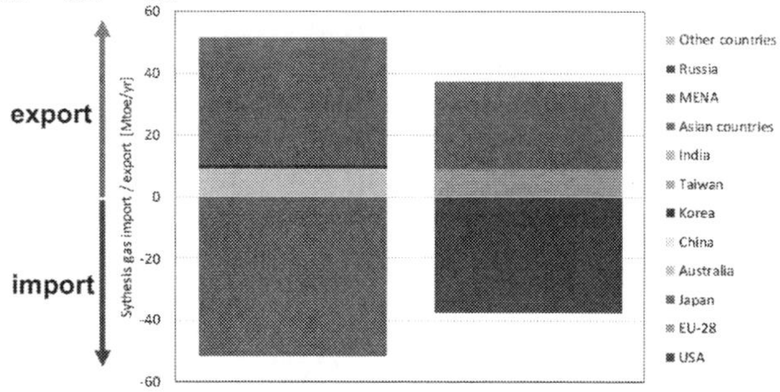

(b) Major importing countries

Results and discussions for evaluation of hydrogen and synthetic methane

- Synthetic methane take a certain role for the end use both industrial and building sectors.
- The role of synthetic methane is especially important when there are large difference in MAC between countries.

- Under high uncertainty for climate change future technological progress of various global warming countermeasures, synthetic methane and hydrogen can be effective options to achieve net zero emission.
- It is important to continue R&D for economic viability.

Appendix

Primary energy supply (world, 2030)

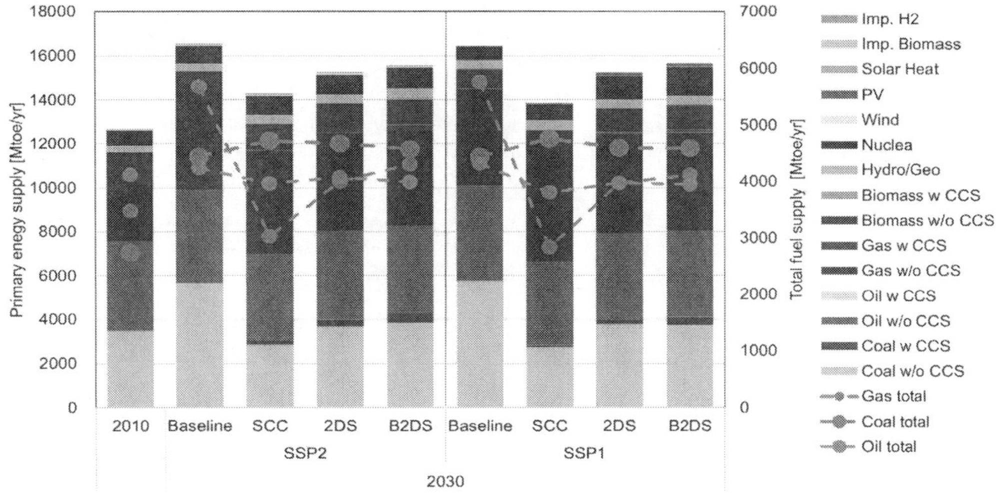

➢ **In 2030, demand of coal and oil will decrease, while demand of gas will increase in the scenario of internalizing external environmental costs, 2DS and B2DS compared to the baseline.**

Primary energy share (world, 2050)

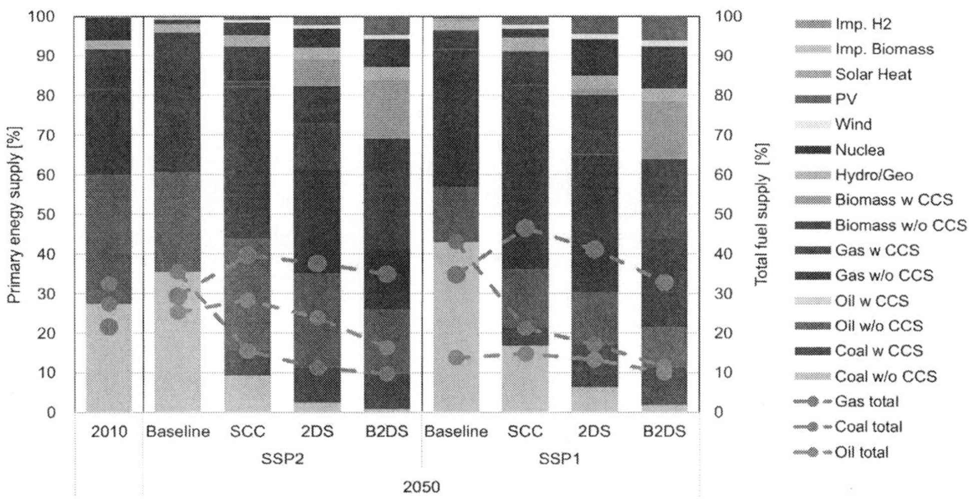

➢ **In terms of share, gas usage is higher than the baseline for SSP2, and remains close to the baseline for SSP1, even in the B2DS scenario, while coal and oil decrease.**

Generated power (world, 2030)

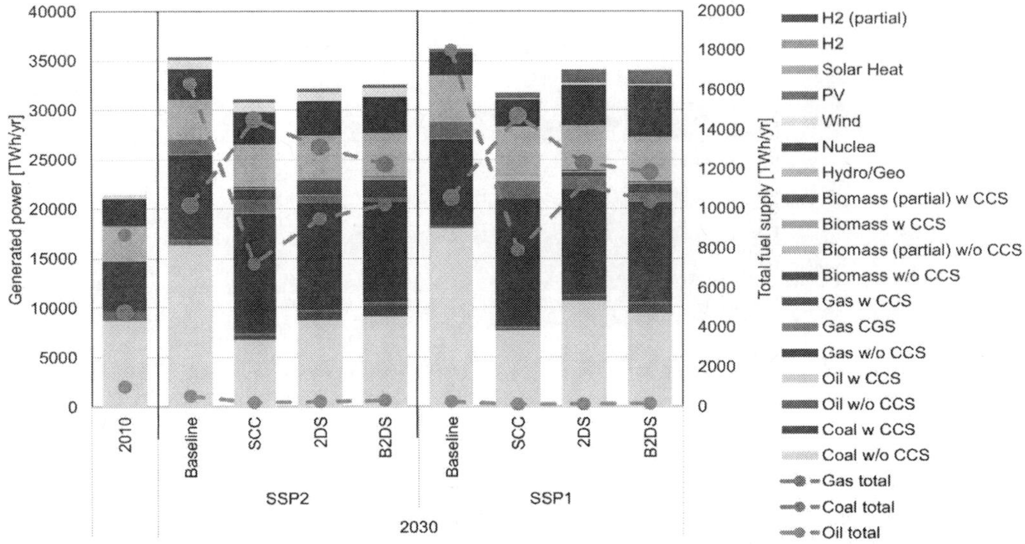

➢ **Coal-fired power increases significantly as the main power source for baseline scenario, however at the level of CO2 MAC of several tens of dollars/t-CO2, switch from coal to natural gas proceed.**

Primary energy supply (Japan, 2030)

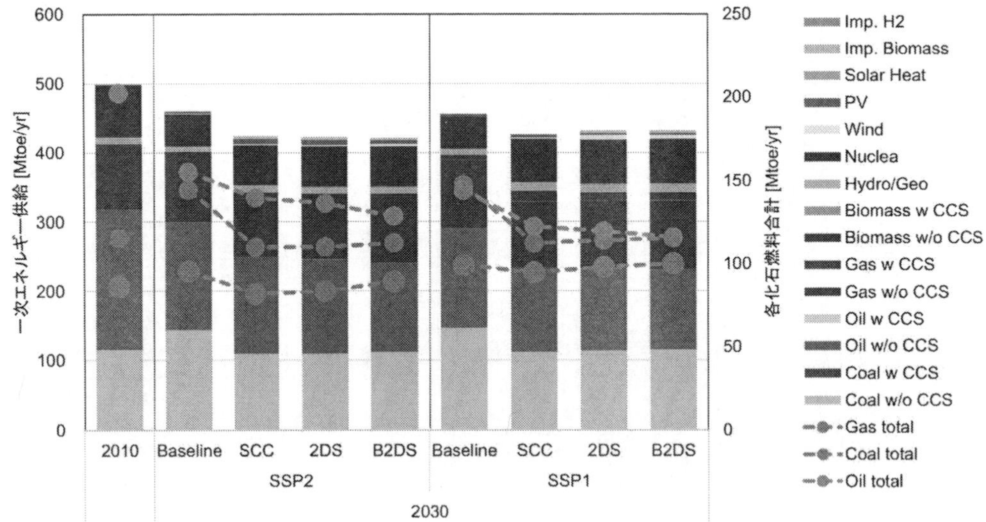

➢ **Coal and oil fall below baseline in emission reduction scenarios, but natural gas will be largely maintained**

Primary energy supply (Japan, 2050)

33

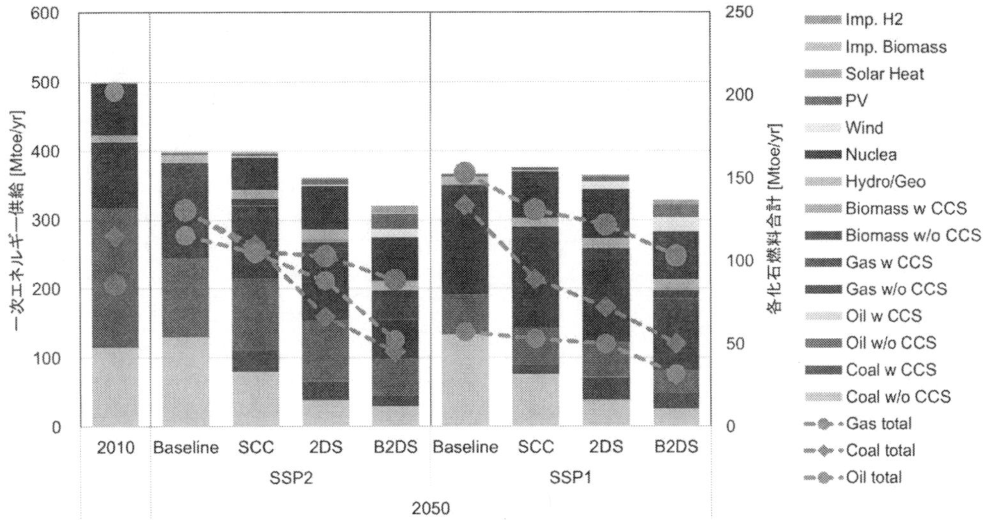

Natural gas decreases in emission reduction scenarios, but the extent of the decrease is smaller compared to other fossil fuels.

Generated power (Japan, 2050)

34

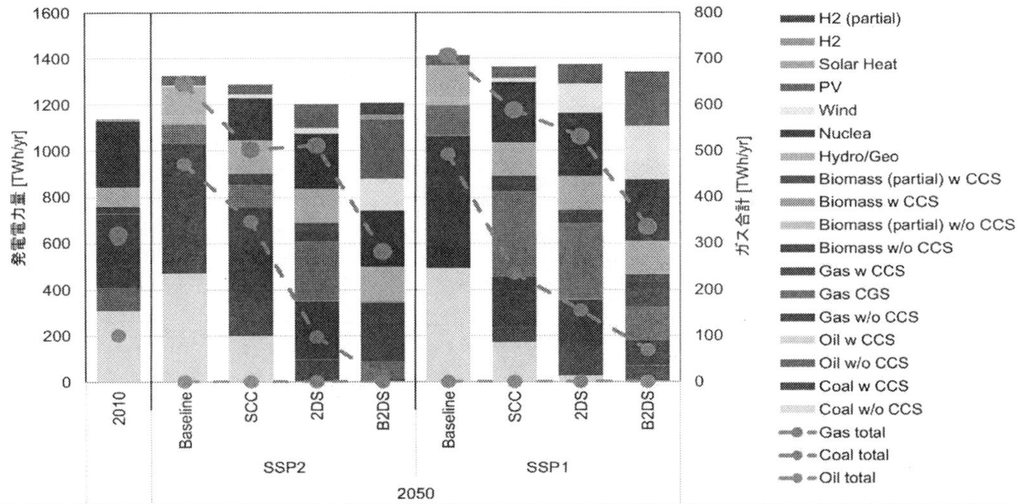

- **Power generated by gas and CGS increases at the baseline.**
- **In the CO2 reduction case, the power of the entire gas decreases due to the expansion of nuclear and renewable, but the reduction amount is smaller than that of coal.**
- **Oil-fired power generation is originally small.**
- **In the scenario of CCS and 2DS, CGS is dominant, but in the case of B2DS, the use of CGS becomes difficult.**

Categorization of deep emission reduction scenarios for below 1.5 °C

RITE 35

Fossil fuel and industry ● AFOLU ● BECCS

Source) IPCC SR15

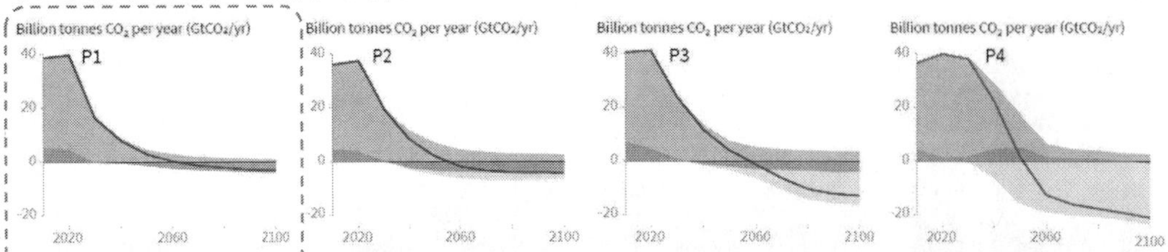

P1: A scenario in which social, business, and technological innovations result in lower energy demand up to 2050 while living standards rise, especially in the global South. A down-sized energy system enables rapid decarbonisation of energy supply. Afforestation is the only CDR option considered; neither fossil fuels with CCS nor BECCS are used.

P2: A scenario with a broad focus on sustainability including energy intensity, human development, economic convergence and international cooperation, as well as shifts towards sustainable and healthy consumption patterns, low-carbon technology innovation, and well-managed land systems with limited societal acceptability for BECCS.

P3: A middle-of-the-road scenario in which societal as well as technological development follows historical patterns. Emissions reductions are mainly achieved by changing the way in which energy and products are produced, and to a lesser degree by reductions in demand.

P4: A resource and energy-intensive scenario in which economic growth and globalization lead to widespread adoption of greenhouse-gas intensive lifestyles, including high demand for transportation fuels and livestock products. Emissions reductions are mainly achieved through technological means, making strong use of CDR through the deployment of BECCS.

Much lower energy demand scenarios than those of SSP1

SSP1

SSP2 (Middle scenario)

SSP5

Low ← Final energy demands → High

- Low energy demand is induced autonomously on economic principle through technological and social innovations
- Low carbon prices (business based measures even without strong climate polices)

✓ The total risk management is important, and various kinds of technologies play their own roles.
✓ On the other hand, opportunities of innovations in end-use technologies, induction of low energy demands, and their impacts on total climate change mitigation should be more focused. (P1)

- High carbon prices (harmonization among nations are required to avoid carbon leakage)
- Large-scale deployments of CDR, e.g., CCS, BECCS, DACS, are required.

Highlights of Japan's Long-term Strategy under PA

Proclaiming a "decarbonized society" as the ultimate goal and aiming to accomplish it ambitiously as early as possible in the second half of this century, while boldly taking measures towards the reduction of GHGs emissions by 80% by 2050

Chapter 3: Cross-sectoral Measures for Achieving a Virtuous Cycle of Environment and Growth

Section 1: Promotion of Innovation

・Promoting innovation for practical application and wide usage of cross-sectoral decarbonizing technologies leading to drastic reduction of GHG, achieving the cost to enable adoption in the society

(1) Progressive Environment Innovation Strategy

・Setting clear targets such as costs, maximizing provision of provision of the resources from both the public and private sectors, seeking out and creating potential technologies in Japan and abroad, setting challenges based on the needs, strengthening support for making it to the business case

・Challenging R&D, and enhancing alliances among R&D institutes with facilitation of international joint R&D activities [Research and Development 20 for clean energy technologies(RD20)]

・Target setting and visualizing of issues for commercialization
- Realizing hydrogen cost equivalent to existing energy: e.g. lowering manufacturing cost of CO_2-free hydrogen to 1/10
- Realizing products utilizing CCU/Carbon Recycling at the level equivalent to existing products in terms of cost, nuclear power(such as Reactor, Fusion)

(2) Innovation in Economic and Social Systems/lifestyle

Section 2: Promotion of Green Finance

・Appropriately "visualizing" corporate initiatives in innovation and mobilizing finance for innovation by financial institutions

(1) Mobilizing green finance through Disclosure including TCFD※ and dialogue ※Task Force on Climate-related Financial Disclosures

・Industry: Expanding on the TCFD Guidance/Scenario Analysis Guide / Financial sector: Formulating a guidance on green investment

・Creating a venue for dialogue between industries and financial sector (TCFD Consortium)

・Promoting discussion and sharing the above initiatives with the world (TCFD Summit)

(2) Promoting initiatives to expand ESG finance

・Promoting ESG finance (Support to the issuance of green bonds, encouraging local ESG finance), the development of ESG Dialogue Platform, enhancing ESG finance literacy, ESG Finance High-Level Panel

Section 3: Business-led Promotion of International Application, and International Cooperation

・Promoting international application of products and goods with high environmental performance/ promoting co-innovation benefiting participants from both countries

(1) Promoting international application of decarbonizing technologies together with policy and institutional support and rule-making

・Promoting wider application of decarbonizing technologies and reductions of GHG emissions through improving business environment by working for institutional development in partner countries and leading international rule-making (e.g. establishing public and private-sector initiatives in ASEAN, and creating appropriate international frameworks for using market-based mechanisms)

(2) Strengthening Development and Investment of infrastructure contributes to reduction of CO_2 emissions

・Promoting development and investment of energy and urban/transport infrastructure abroad in order to contribute to the global reduction of CO2 emissions consistent with the long-term goals stipulated in the Paris Agreement (e.g. renewable energy such as offshore wind power and geothermal power, hydrogen, CCS&CCU, Carbon Recycling, smart cities)

(3) Building basis for decarbonized society on a global scale

・Supporting partner countries in the formulation of NDCs and mitigation plan, enhancing transparency in the entire supply chain

●Innovations in Progressive Environment Innovation Strategy (Yamaji's Understanding for the System of Innovations)

Energy Conversion :
Renewables
Flexible Power NW
Hydrogen Supply Chain
Innovative Nuclear Tech.
Low Cost CO2 Capture ⇒CCUS

Industry :
Free from Fossil Resources
Electrification
CO_2 Recycling

Agriculture and Others :
Reduction of Methane・N_2O
Increase in Sink (Soil/Forest、Blue Carbon)
Smart Agriculture

Transport :
Diversified Green Mobility

DAC

Building :
More Energy Conservation
Smart City
Social Innovation (Sharing, Tele work, etc.)

General Purpose Tech. :
Digital Tech. (Big Data、AI、Blockchain、・・・)
Power Electronics、Material、Energy Storage、Sensing、Biotech.、etc.

37

Moonshot Goal No. 4:

Realization of Sustainable Resources Circulation to Recover the Global Environment by 2050

The mission of this Moonshot Goal Candidate is to develop technology for reducing the emissions of greenhouse gases and pollutants to contribute to the recovery from the on-going issues of global warming and environmental pollution. The concept of the this theme consists of two pillars of **"Cool Earth"** and **"Clean Earth"**

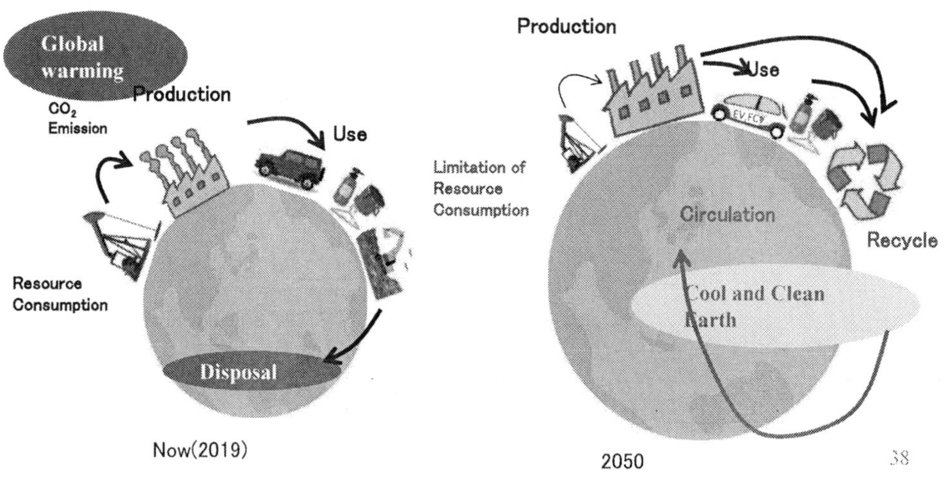

38

A Vision of Society 5.0 (“Super Smart Society) presented in the 5th Science and Technology Basic Plan in January 2016

Realizing “Society 5.0” (“Super Smart Society”)

Around the world, initiatives that use networks and the Internet of Things (IoT), centered on manufacturing fields, are now coming out. In Japan, the use of such networking will not be limited to manufacturing. Instead, it will be extended to various other fields in order to promote economic growth, the formation of a healthy and long-living society, and social transformation. In addition, it will help the fruits of science and technology to penetrate into all kinds of fields and spheres, and thereby lead to enhanced business capability and higher quality services.

We will share our vision of the future, which is characterized by the sophisticated integration of cyberspace with physical space (“the real world”) and work to enhance it, while further pursuing a series of measures aimed at its realization, under the concept of “Society 5.0”.*

* The history of humankind reveals that the evolution of human society has been fueled by technological advances, with key steps along the way as a hunter-gatherer society, agrarian society, industrial society, and, today, an information society. “Society 5.0” is next, and we'll achieve it by mobilizing the full power of STI.

It is necessary to aim at “systemization” of services and businesses, system advancement, and coordination between multiple systems. Therefore, we will promote the measures needed to develop a common platform for this goal (called “Society 5.0 Service Platform”), through collaboration between industry, academia, and government and the relevant government ministries.

39

Impacts of “Super Smart Society” (Society 5.0)

Society 5.0 is characterized by the sophisticated integration of cyberspace with physical space (“the real world”)

Society 5.0 is capable of providing the **necessary goods and services to the people who need them at the required time and in just the right amount**; a society that is able to respond precisely to a wide variety of social needs; a society in which all kinds of people can readily obtain high-quality services, overcome differences of age, gender, region, and language, and live vigorous and comfortable lives.

Beyond energy saving, Society 5.0 promotes sharing economy to shift the industry from manufacturing to service providers leading to a ultimate circular society. ⇒ **Information replaces Material**

⇓

Care for Rebound Effects

1) car/ride share → car ↓ → car productions ↓ → material needs ↓ → energy/CO_2 ↓
2) smart maintenance → component replace↓ → material needs ↓ → energy/CO_2 ↓
3) IC tag for all component → recycling rate ↑ → material needs ↓ → energy/CO_2 ↓

40

Image of standard scenarios by models and scenarios required for deep cuts in a real world

41

Model world:
Ordinary technology progress

Realistic world requirement:
Innovations stimulated & implemented

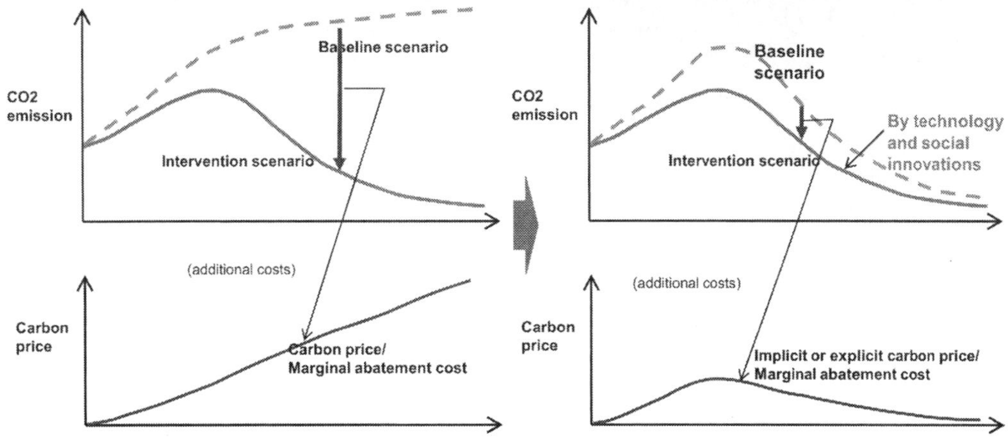

High carbon prices of over 100$/t$CO_2$ in real price are unlikely to be accepted globally in a real world. Technology and social innovations which will bring low (implicit or explicit) carbon prices (including coordination of secondary energy prices) are key to achieve deep emission cuts.

Innovations in end-use technologies through IT and AI, and the induced social changes

42

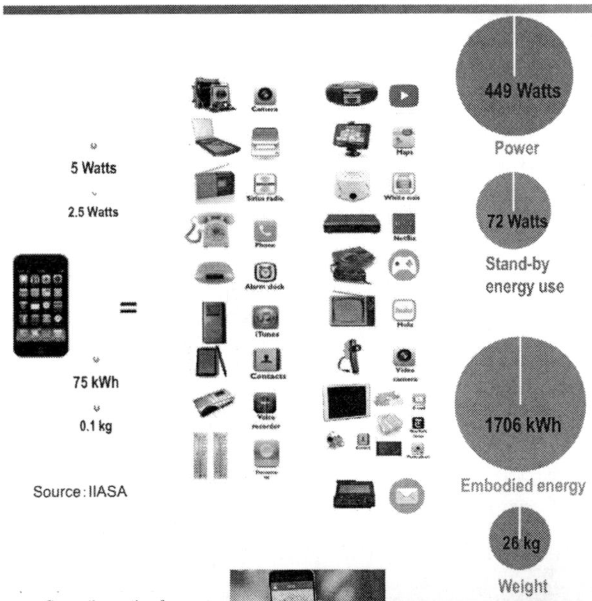

Source: IIASA

- Energy consumption is not our purpose, but is just a phenomenon accompanying with consumption of goods and services, which is conducted for our welfare increase. Energy embodied in goods and services must be taken into account.
- The end-use products and services will usually diffuse rapidly, and the embodied energy and CO_2 may decrease rapidly.

Operation ratio of automobiles is about 5%. The large room for the improvement exists by the achievement of fully autonomous cars.

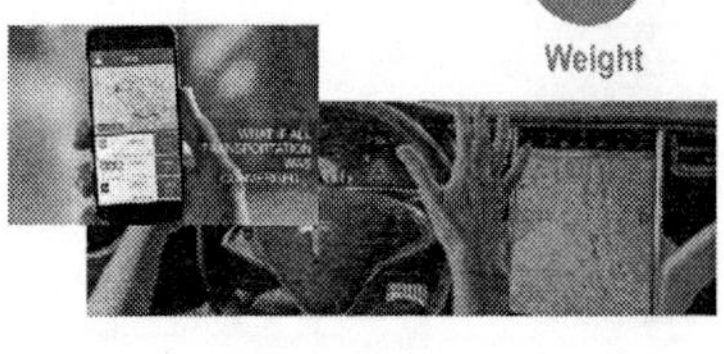

There are large opportunities to achieve social changes and to increase energy efficiency through fully autonomous cars, food systems, etc. which can be induced by innovations of IoT, AI etc.

Transport: CASE

Connected; Service & Shared

Autonomous; Electric

Operation ratio of automobiles is about 5%. The large room for the improvement exists by the achievement of fully autonomous cars.

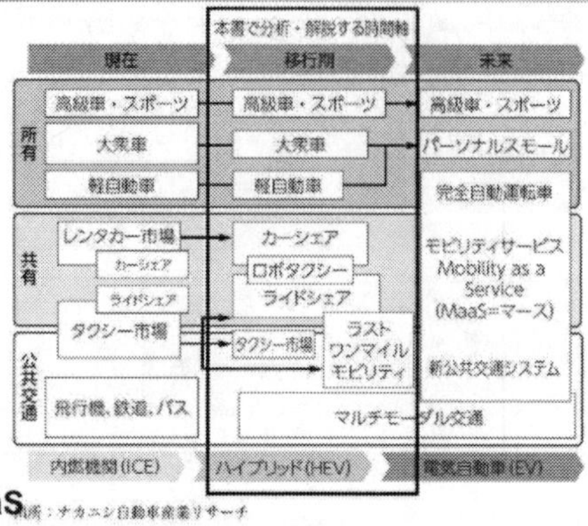

EV "e-Palette" only for Autono-MaaS

出所：ナカニシ自動車産業リサーチ

Source) Toyota

Changing the shape of cars

Possibility of integration of cars and near distance airplane

The sharing may reduce number of cars and the consumptions of materials, and change the f...

Airbus, Audi

Source) Jari Kauppila, ALPS International Symposium (2019)

Publication

- **Progresses of digitization**
- **Reduction of paper medium books and newspaper circulation (Paper reduction, Embodied energy reduction)**
- **Decrease of physical bookstores (Energy reduction for construction and maintenance, Energy reduction for access to physical bookstores)**
- **Newspaper delivery declining? (Reduction of transport energy)**

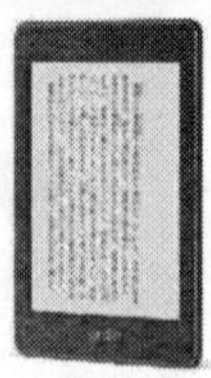

Food

- It is regarded that approx. 30% of GHG emission (even more depending on boundaries) comes from food system. Also food wastes and losses are about one third of total production globally (would be less in Japan).
- More accurate food demand forecast through AI / ICT could lead to decrease of food wastes and losses and to reduction of energy consumption and GHG emission accordingly.
- Consequently, reduction of plastic containers, store spaces in supermarkets, energy for refrigerators / freezers and transport energy could be triggered.

They could be huge contribution to the achievements of SDGs as well.

Food

- It is regarded that approx. 30% of GHG emission (even more depending on boundaries) comes from food system. Also food wastes and losses are about one third of total production globally (would be less in Japan).
- More accurate food demand forecast through AI / ICT could lead to decrease of food wastes and losses and to reduction of energy consumption and GHG emission accordingly.
- Consequently, reduction of plastic containers, store spaces in supermarkets, energy for refrigerators / freezers and transport energy could be triggered.

They could be huge contribution to the achievements of SDGs as well.

4 Challenge 1: Leadership in the effort to achieve Net-Zero CO_2

- In its overall business activities, the Tokyo Gas Group will **work to achieve Net-Zero CO_2 emissions** including customer emissions and **lead the transition to a decarbonized society**.
- We will use **technologies and expertise for the effective use of natural gas** to promote decarbonization in the electricity and heat sectors as well as for CO_2 capture technologies.
- We will **contribute to reduce carbon emissions on the scale of 10 million tons** by 2030 (which exceeds Japan's target ratio*) and lead the way to reducing CO_2 emissions on a global scale.

<Efforts to achieve Net-Zero CO_2 in the Tokyo Gas Group's style>

Efforts domestically and internationally

– 2030 | – around 2050

Emissions reductions not only in the Tokyo metropolitan area but on a global scale

1: Zero emissions

- Expanded use of renewable power sources
- Development of decarbonization technologies for gaseous energy

→ **Decarbonization in the electricity and heat sectors**

2: Zero emissions achieved through offsetting

- **Effective use of natural gas** (reduced energy consumption through efficient use of natural gas, coordination with the fluctuating output of renewable energy sources)
- Use of **CCUS**** technologies (capture of emitted CO_2)
- Incorporation of **successful reduction efforts overseas**

The combination of these activities will lead to Net-Zero CO_2

* Japan's reduction target ratio: The greenhouse gas reduction target in the Intended Nationally Determined Contribution submitted to the United Nations (26% reduction in FY 2030 as compared to FY 2013).
** CCUS: Carbon Capture, Utilization, and Storage

5

4 Action 2: Decarbonization technology innovations

- **We will promote innovations in essential core technologies that contribute to decarbonization** in the period leading up to 2030.
- After 2030, we will promote **hydrogen production and direct use** that make use of both domestic and global **renewable power sources** and the introduction and use of **methanation technology*, etc. to meet the demand for heat**. In addition, **these means will be combined and utilized** for achieving Net-Zero CO_2 through IoT, AI.

Roadmap to Net-Zero CO_2

Development of essential core technologies using open-innovation

Hydrogen production technology

High-efficiency fuel-cell technology

CO_2 separation and capture technologies

× Partner companies, universities and research institutions

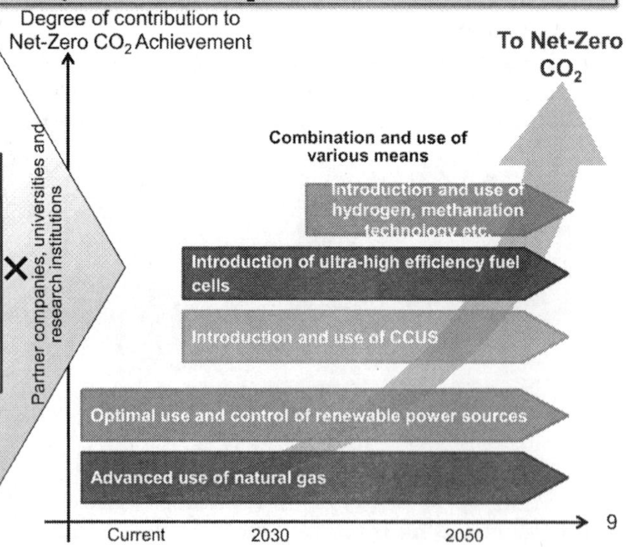

* Methanation technology: Producing carbon neutral methane by combining hydrogen and CO_2

9

Activities toward decarbonized society

Three Challenges	Five Actions

1: Leadership in the effort to achieve Net-Zero CO_2
2: Establishment of a value co-creation ecosystem
3: Transformation of the LNG value chain

① Coordination of renewable energies and natural gas
   - Smart energy network

② Decarbonization technology innovations
   - Carbon neutral LNG
   - Hydrogen station
   - High-efficiency fuel cell
   - CCUS
   - Hydrogen, Methanation

③ Resolving problems in daily life and businesses

④ Enhanced resilience functions through the use of natural gas

⑤ Overseas expansion

49

Coordination of renewable energies and natural gas: smart energy network

- Through the construction of smart energy using cogeneration (CGS), we will contribute to the creation of attractive towns that can save energy and CO_2, and continue to supply electricity and heat in the event of a disaster (S+3E).

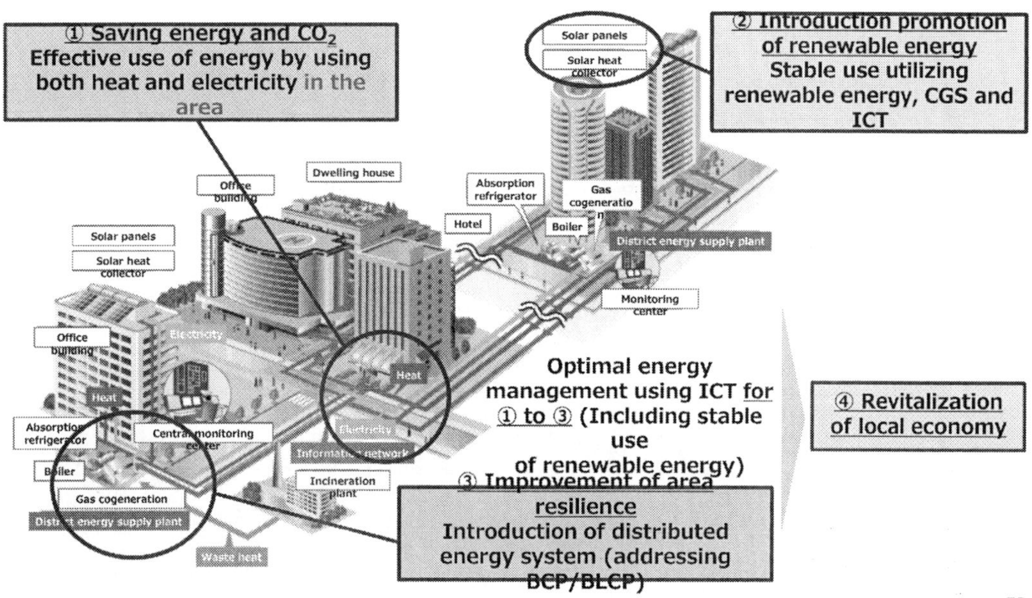

50

Note: Implementation examples of "City center type" regional revitalization

Tamachi smart energy network

I block

Ⅱ-2 block

Main equipment

Gas engine CGS	【I block】 370kW×2 【Ⅱ-2 block】 1,000kW×5	Solar heat collector	【I block】 288m² 【Ⅱ block】 82m² under construction
Fuel cell	【I block】 105kW×1	Solar panels	【I block】 73kW customer ownership

Note: In addition to the above, use renewable energy and unused energy from wind and underground tunnel water

Based on the city planning vision of Minato city, promote public-private partnerships to create a low-carbon disaster-resistant town. Optimal operation realized by supply and demand coordination control.
[Supply energy: electricity, hot water, cold water]

Toyosu smart energy network

Main equipment

Gas engine CGS	6,970W×1	Generation using gas pressure difference	Generation: approx. 650kW Cold: approx. 200RT
		Solar panels	2,000kW class customer ownership

Based on the "Toyosu Green Eco Island Concept" in Koto city, we contribute to the development of a low-carbon, self-reliant and safe town utilizing large CGS, unused energy, etc. The thermal/electrical network will be gradually expanded according to the progress of town development (planned).
[Supply energy: electricity, steam, cold water]

51

Decarbonization measures: Carbon neutral LNG

- Carbon neutral LNG is an LNG that is regarded as generating no CO_2 on a global scale by burning natural gas by offsetting the CO_2 generated in the process from mining to combustion of natural gas with CO_2 credits.
- We purchased first carbon neutral LNG from shell as Japanese LNG buyer and began providing it to customers.

Tokyo Gas purchases carbon neutral LNG ... Press release (Jan. 18, 2019)

Tokyo Gas Co., Ltd. (President: Takashi Uchida; "Tokyo Gas") and Shell Eastern Trading (Pte) Ltd. have been engaging in joint discussions and as a result, for the first time in Japan, Tokyo Gas has decided to receive the supply of carbon neutral liquefied natural gas (LNG).

Under this Master Agreement, with regard to this supply of carbon neutral LNG, Shell 's carbon credits will be used to compensate the full carbon dioxide (CO_2) emissions generated – from exploring for and producing the natural gas to use by the final consumer.
The carbon credits are purchased by Shell from a global portfolio of nature-based projects, and each carbon credit is subject to a third-party verification process.

By receiving the supply of carbon neutral LNG, Tokyo Gas will reinforce its effort toward low carbon society as well as providing a new low carbon product to our customers.

Tokyo Gas received Japan's first-ever LNG cargo on November 4th 1969 and this year marks the 50th anniversary. Tokyo Gas Group will continue to provide stable and safe source of energy to all our customers.

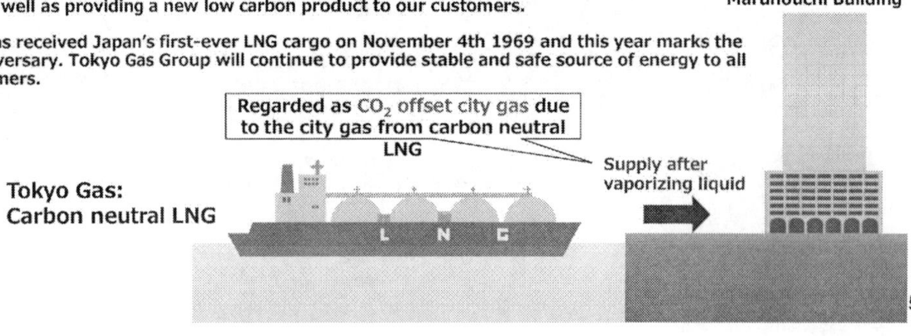

52

Decarbonization measures: Hydrogen station

- Tokyo Gas is contributing to the diffusion and expansion of fuel cell vehicles (FCV) through the construction and operation of hydrogen stations, utilizing the technology for producing hydrogen from city gas.

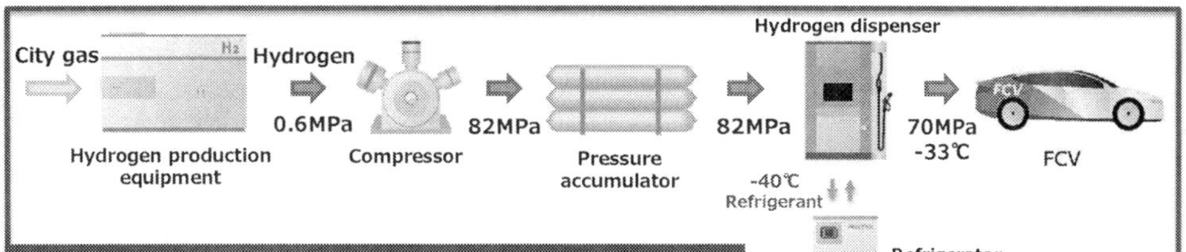

Configuration of city gas on-site hydrogen station

Nerima Hydrogen Sta.
- Opened in Dec. 2014 (first in Kanto)
- Medium scale (300m³/h)
- Off-site

Senju Hydrogen Sta.
- Opened in Jan. 2016
- Small scale (100m³/h)
- Off-site

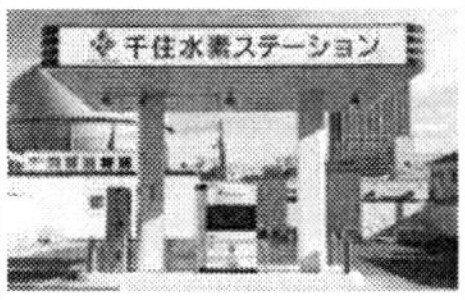

Urawa Hydrogen Sta.
- Opened in Feb. 2016
- Medium scale (300m³/h)
- Off-site

Toyosu Hydrogen Sta.
- Opened in Jan. 2020
- Large scale (also for bus)
- Off-site
- Supply carbon neutral hydrogen through carbon neutral LNG

Also for Fuel Cell Bus

53

Decarbonization measures: Ultra-high efficient fuel cell

- The power generation efficiency of gas cogeneration system (CGS) has improved dramatically. In recent years, the efficiency has been approaching 50%, which is comparable to large-scale thermal power on demand end basis.
- We are developing a distributed power generation, small fuel cells, that achieves high power generation efficiency, which surpasses the efficiency of next-generation thermal power plants.

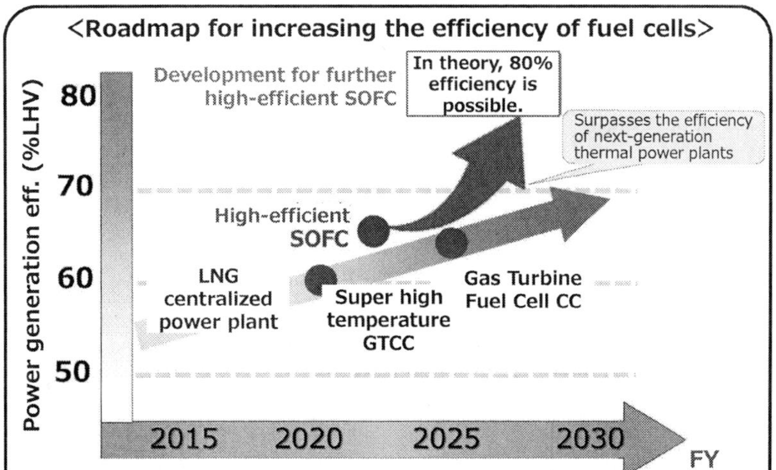

Based on "Technology roadmap for next-generation thermal power generation." Assumed transmission loss is 5%.

Prototype of high-efficient SOFC

54

Decarbonization measures: CCUS

- CCUS is an abbreviation for Carbon dioxide Capture, Utilization and Storage.
- In order to realize technology to capture CO_2 emitted from city gas equipment at customer and use it as a resource (dry ice, concrete products, carbonates, etc.), we have developed a system to efficiently capture CO_2 and to effectively use of captured CO_2.

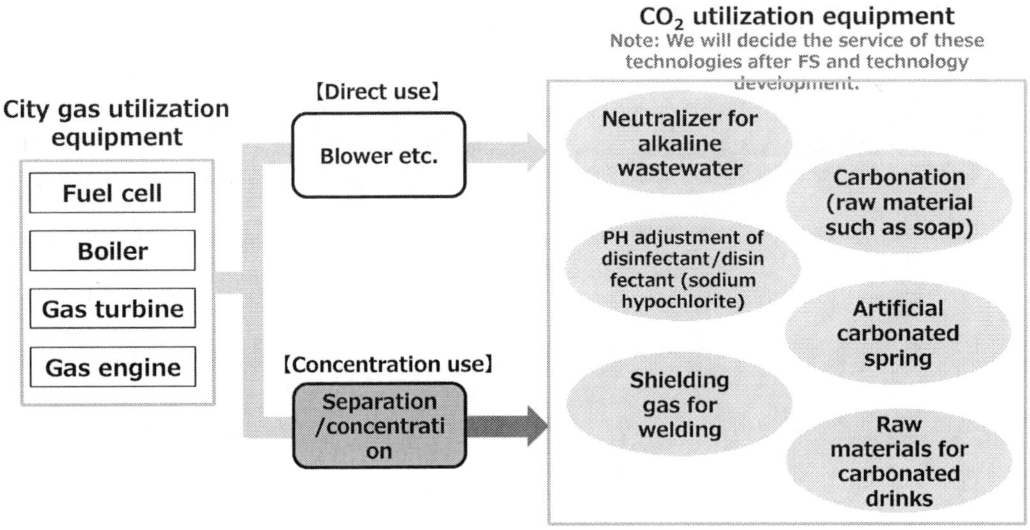

55

Note: CO_2 storage technology using microbubbles

- In order to realize Carbon Capture Storage (CCS) efficiently, we are working with RITE to commercialize a technology to convert CO_2 into microbubbles and store more CO_2.
- This technology makes it possible to store CO_2 in large quantities and efficiently, leading to a reduction in CCS costs. In addition, it is expected to be applied to Enhanced Oil Recovery (EOR), which increases oil production by injecting CO_2 into the underground to increase the fluidity of oil in the reservoir.

Microbubbles technology

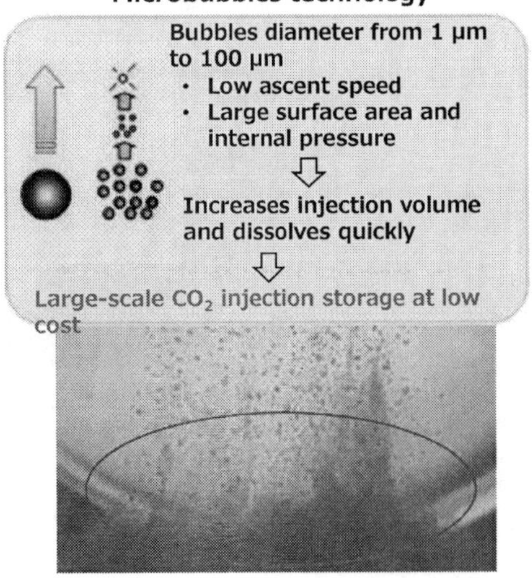

For realizing microbubbles

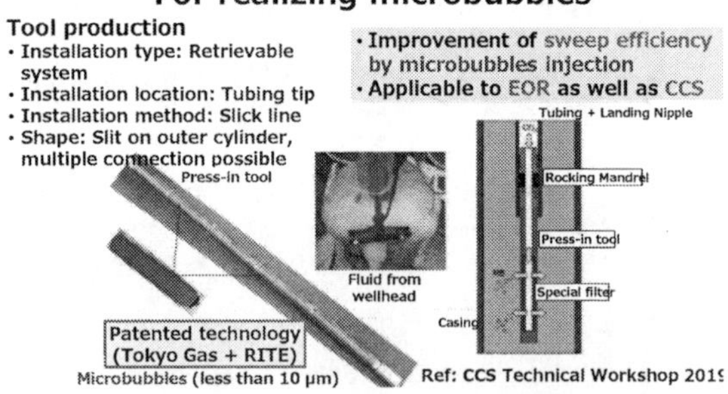

Ref: CCS Technical Workshop 2019

License agreement with Junlun Petroleum for patented technology (co-owned with RITE).

Takashi Honjo, Senior Managing Director of RITE
The 12th Japan-China Energy Conservation and Environment Forum in Nov., 2018

Ref: RITE press release

56

Decarbonization measures: Hydrogen, Methanation

- Methanation is a technology for synthesizing methane. In order to generate carbon free methane, methane is synthesized with CO_2 free hydrogen from renewable energy and recovered CO_2.

  $$4H_2 + CO_2 \rightarrow CH_4 + 2H_2O$$

- Methanation technology can effectively utilize existing city gas infrastructure such as pipelines and LNG receiving terminals (manufacturing sites), and gas systems and gas consuming equipment such as cogeneration, hot water supply, and air conditioning. Methanation technology is attracting attention as a decarbonization option for gaseous energy.
- We are working on the development of new hydrogen production technology to reduce the cost of hydrogen as a raw material for the practical use of metha

【Overview of utilization of hydrogen and methanation】

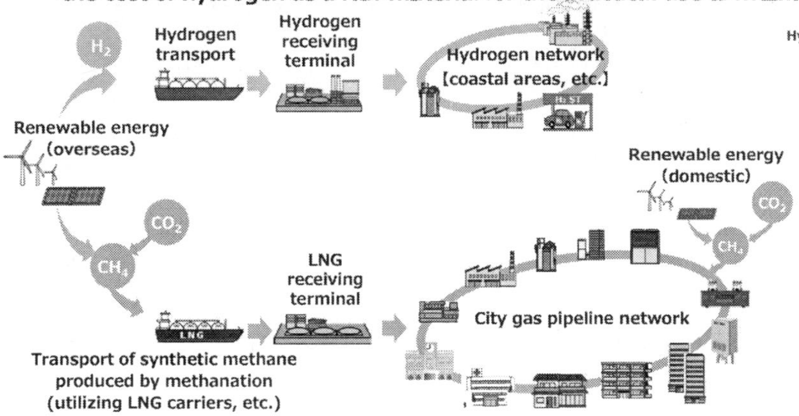

【Hydrogen production cell stack】

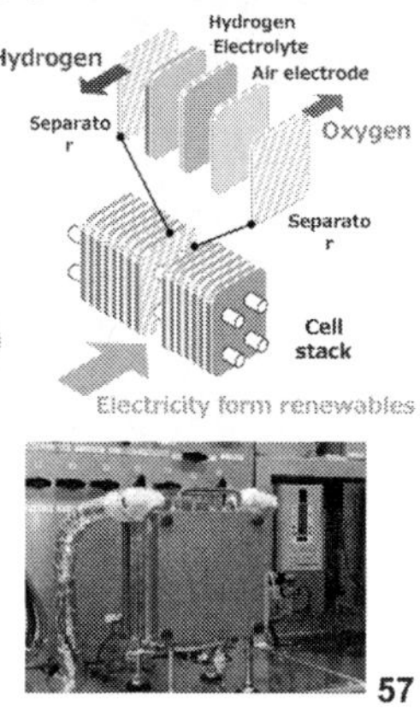

57

UCD ENERGY INSTITUTE

esipp
Energy Systems Integration Partnership Programme

Synthetic Natural Gas (SNG) Production: Optimal Configuration and Significant Variables

Devasanthini Devaraj, Eoin Syron & Philip Donnellan
School of Chemical & Bioprocess Engineering, University College Dublin, Ireland.

Outline

UCD ENERGY INSTITUTE

Renewable Energy
- Power-to-Gas (PtG)
- Large scale long-term renewable integration

Power-to-Methane (PtM)
- Synthetic Natural Gas (SNG)
- Multiple input sources

SNG Production Process
- Optimal Configuration
- Significant Variables

Motivation

UCD ENERGY INSTITUTE

1. Renewable energy targets
   a. Increased variability and grid stability
2. Power-to-Gas process chain
   a. Electricity → Hydrogen or Synthetic Natural Gas (SNG)
   b. Injection in existing gas networks
3. Recycling CO_2
   a. Complete decarbonisation
   b. Alternation in business of gas networks

Power-to-Gas Process Chain

UCD ENERGY INSTITUTE

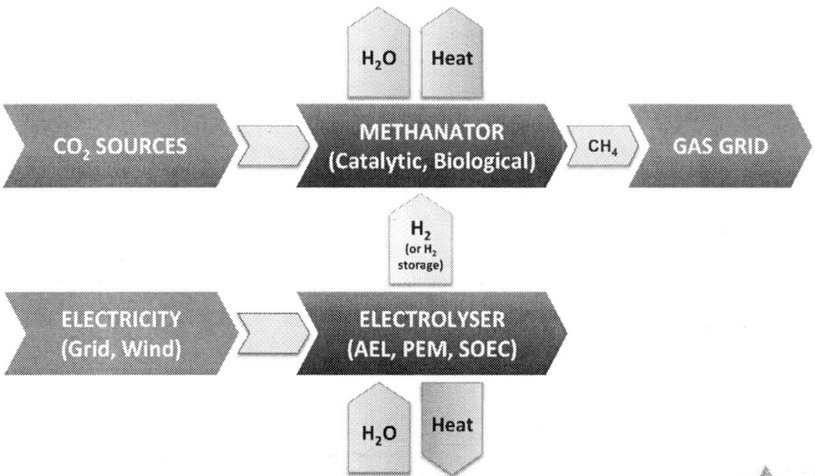

Methodology Development

UCD ENERGY INSTITUTE

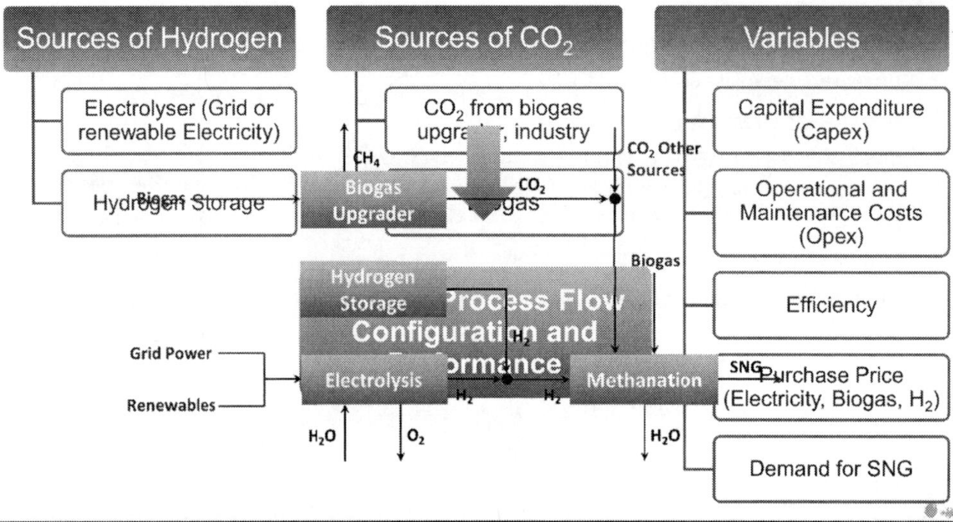

Step-wise Procedure

UCD ENERGY INSTITUTE

- Identify units in the Process Flow — Methanation, Electrolyser, Biogas upgrader, H_2Storage
- Define the Variables — Capex, Opex, Efficiency, Purchase and Selling Prices
- Define Equations and Constraints — Input/output, Demand, Mass balance, Capacity, Binary
- Objective function (Minimise Production Cost)

Objective Function

UCD ENERGY INSTITUTE

$$Production\ Cost\ =\ \frac{\sum_{t=1}^{n} Capex_t + Opex_t + Purchase\ Price_t \Big/ (1+r)^t}{SNG_t}$$

where,

$Capex_t$	*Capital expenditure*
$Opex_t$	*Operational costs*
$Purchase\ Price_t$	*Purchase price of input streams (biogas, CO_2, power)*
SNG_t	*Amount of SNG produced*
r	*Discount rate (6% for all costs)*
t	*Year*

Step-wise Procedure

UCD ENERGY INSTITUTE

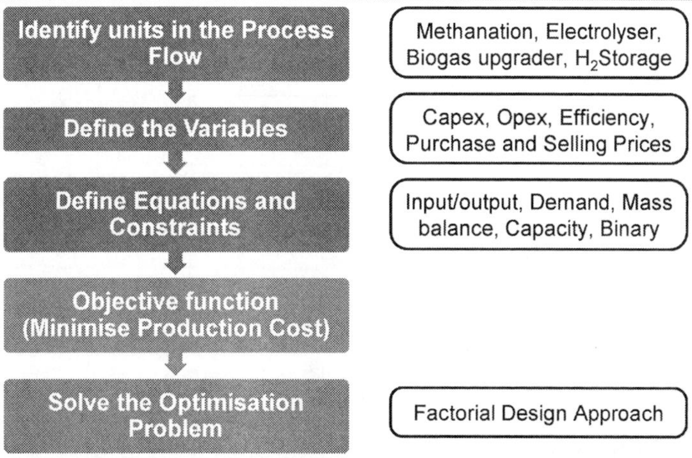

Factorial Design Approach

UCD ENERGY INSTITUTE

1. 10 Independent Variables (IV)
2. 1 Dependent Variable (DV) - Production Cost
3. Factorial design for the model
   a. 1 min and 1 max value for all IVs
   b. 2^{10} possible configurations
   c. Yearly approach (up to 20 years)

Capex of Methanator

Opex of H_2 Storage

Run	Year	Capex				Opex			
		M	E	BGU	H_2St	M	E	BGU	H_2St
1	1	300	220	15	3.84	4.5	3.3	0.002	0.058
2	1	300	220	15	3.84	4.5	3.3	0.002	0.16
4	1	300	220	15	3.84	4.5	3.3	0.004	0.16
:	:	:	:	:	:	:	:	:	:
:	20	1500	955	25	8	30	19.1	0.004	0.16

Year

Step-wise Procedure

UCD ENERGY INSTITUTE

- Identify units in the Process Flow — Methanation, Electrolyser, Biogas upgrader, H_2Storage
- Define the Variables — Capex, Opex, Efficiency, Purchase and Selling Prices
- Define Equations and Constraints — Input/output, Demand, Mass balance, Capacity, Binary
- Objective function (Minimise Production Cost)
- Solve the Optimisation Problem — Factorial Design Approach
- Output Analysis (ANOVA) — One-factor & Factorial

Results - Optimal Process Flow

UCD ENERGY INSTITUTE

Variables used in Objective function

Production Cost per kWh_{SNG}

Units (M - Methanator, E - Electrolyser, etc)

Run	Year	Capex				Opex				Cost	M	E	BGU	H_2St	CC
		M	E	BGU	H_2St	M	E	BGU	H_2St						
1	1	300	220	15	3.84	4.5	3.3	0.002	0.056	0.434	1	0	1	1	0
2	1	300	220	15	3.84	4.5	3.3	0.002	0.16	0.842	1	1	0	0	1
4	1	300	220	15	3.84	4.5	3.3	0.004	0.16	0.454	1	0	1	1	0
:	:	:	:	:	:	:	:	:	:	:	:	:	:	:	:
:	20	1500	955	25	8	30	19.1	0.004	0.16	1.711	1	0	1	1	0

Binary Values 0 & 1 – indicate the presence/absence of a unit in optimal flow

Optimal Process Flow

UCD ENERGY INSTITUTE

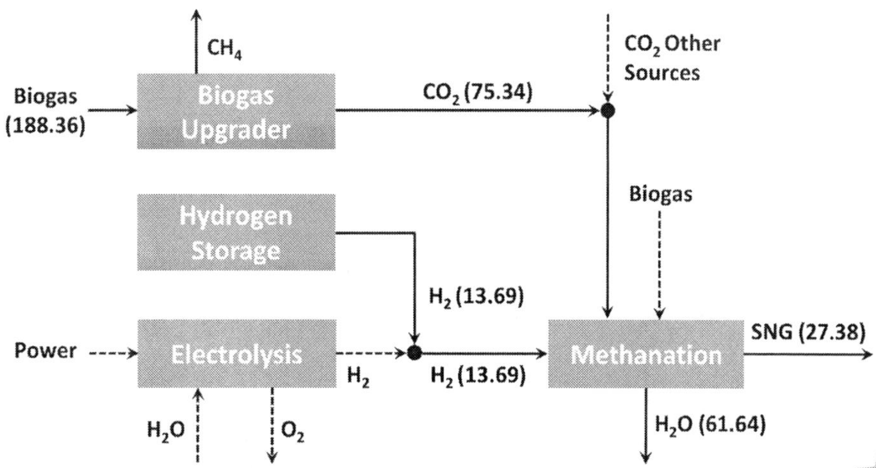

Discounted SNG Production Costs

UCD ENERGY INSTITUTE

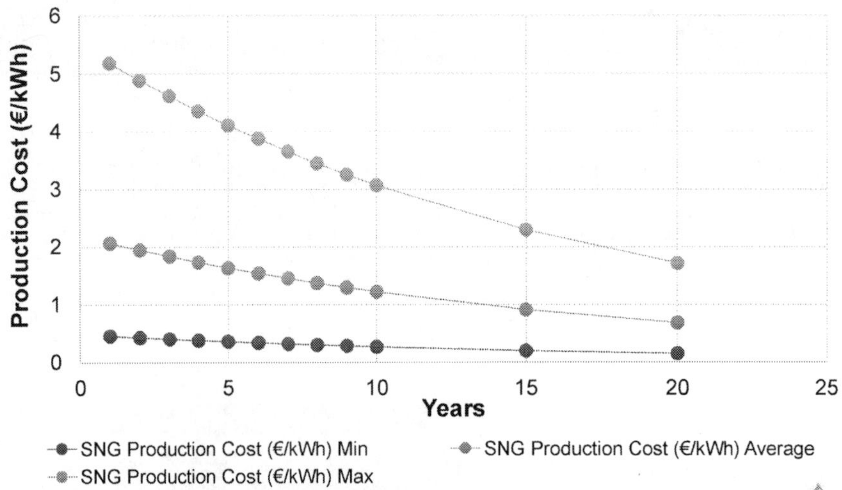

Single factor ANOVA Observation

UCD ENERGY INSTITUTE

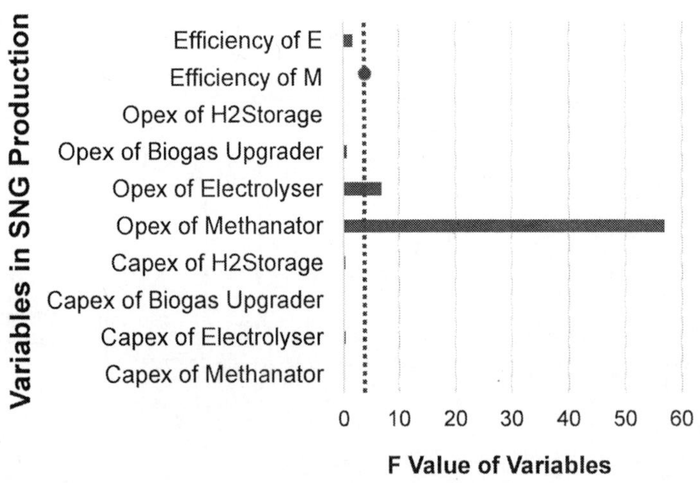

Conclusions

UCD ENERGY INSTITUTE

- ✓ Optimal process flow setup in SNG production
- ✓ Production cost per kWh_{SNG}
  - Year 1 - €0.45
  - Year 20 - €0.17
- ✓ Long-term production is preferable
- ✓ Oxygen income is beneficial
- ✓ Operating cost of Methanator and Electrolyser have most influence in determining the production cost

UCD ENERGY INSTITUTE

esipp
Energy Systems Integration Partnership Programme

Thank you

Email
devasanthini.devaraj@ucdconnect.ie

Funded by
Science Foundation Ireland sfi For what's next

PETRONAS

CO_2 RECYCLING: CO_2 FIXATION INTO METHANE USING RUTHENIUM BASE CATALYSTS

Nor Hafizah Berahim, Akbar Abu Seman,
Mohd Ghaddaffi Mohd Noh

Presentation to International Gas Union Research Conference
Muscat, Oman
25.02.20

Open

Content

01. INTRODUCTION

02. OBJECTIVES

03. CATALYST DESIGN AND SYNTHESIS

04. CHARACTERIZATION

05. CATALYST PERFORMANCE

06. CONCLUSION

PETRONAS

Open

INTRODUCTION

PETRONAS

Open

Background – Why Carbon Removal and Utilization

- Monetization of high CO_2 gas field will results in potential CO_2 emission after the acid gas separation process.
- Venting of CO_2 will impact the environment and results in carbon tax penalty, and thus CO_2 needs to managed/mitigated.

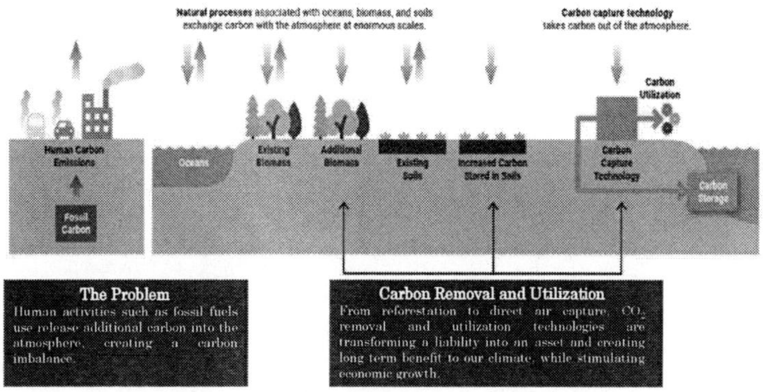

www.global2initiative.org.

PETRONAS

Open

CO_2 Utilization Pathways

Catalytic Conversion
- Formic Acid
- Syngas CO/H$_2$
- Methanol
- **Methane**
- Dimethyl Carbonate (DMC)
- Biopolymers

Disposal
- Mineralisation
- CO2 Hydrate Formation
- EGR/EOR
- Direct Usages

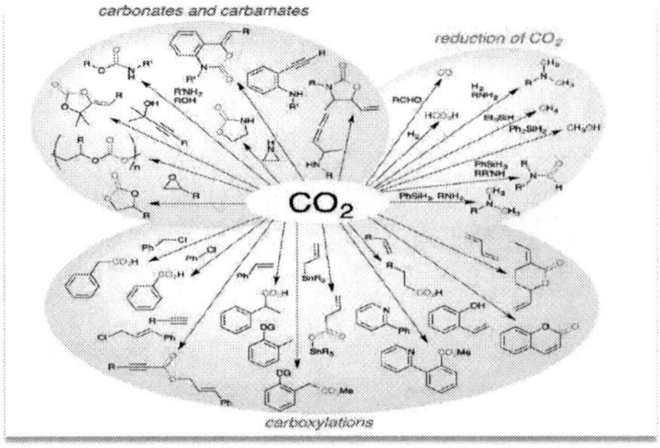

CO_2 conversion to methane was selected due to ease of integration with natural gas sales infrastructure

PETRONAS

Open

CO_2 Methanation

- The methanation of CO_2 is an exothermic reaction, typically operating between 200 – 450 °C, depending on catalyst and experimental conditions.

$$CO_2 + 4H_2 \leftrightarrow CH_4 + H_2O \qquad \Delta H = -164\ kJ/mol \quad (R1)$$

- The most widely accepted mechanism of the methanation reaction is the combination of an endothermic reversed water gas shift (RWGS) reaction (R2) and an exothermic CO methanation (R3).

$$H_2 + CO_2 \leftrightarrow CO + H_2O \qquad \Delta H = 41\ kJ/mol \quad (R2)$$

$$3H_2 + CO \leftrightarrow CH_4 + H_2O \qquad \Delta H = -206\ kJ/mol \quad (R3)$$

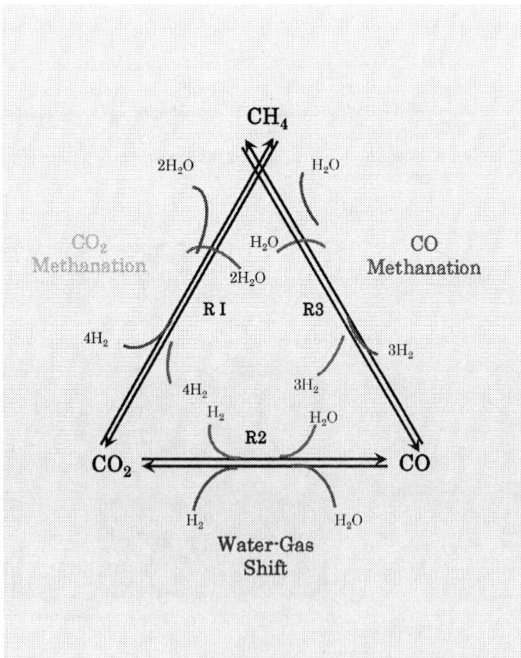

PETRONAS

Open

OBJECTIVES

PETRONAS

Open

01 To synthesis Ru based catalysts with different supports (Al_2O_3 and SiO_2)

02 To investigate the activation (reduction) of synthesized catalyst

03 To characterize synthesized catalysts for textural properties and phase identification analysis

04 To evaluate performance of synthesized catalysts for CO_2 Methanation

PETRONAS

Open

CATALYST DESIGN AND SYNTHESIS

PETRONAS

Open

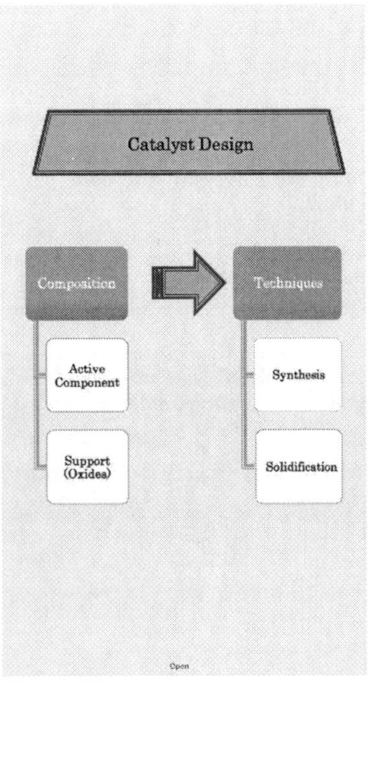

Wet Impregnation Method

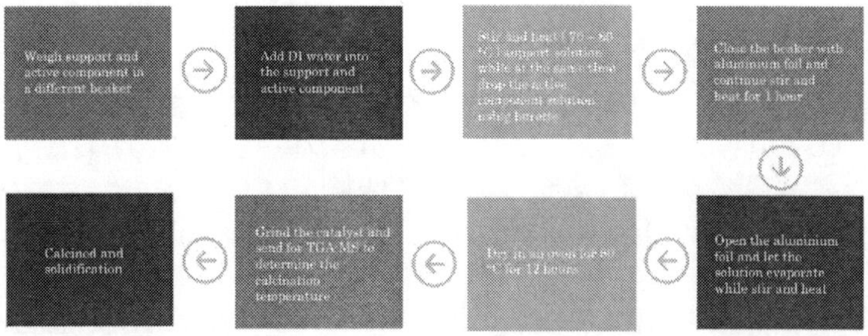

CHARACTERIZATION

PETRONAS

Open

Catalyst characterization

01 **TEMPERATURE PROGRAMMED REDUCTION**
Determination of reduction conditions

02 **BRUNAUER - EMMETT - TELLER**
Measurement of specific surface area

03 **X-RAY DIFFRACTION**
Phase identification

04 **FIELD EMISSION SCANNING ELECTRON MICROSCOPY**
Investigation of molecular surface structure

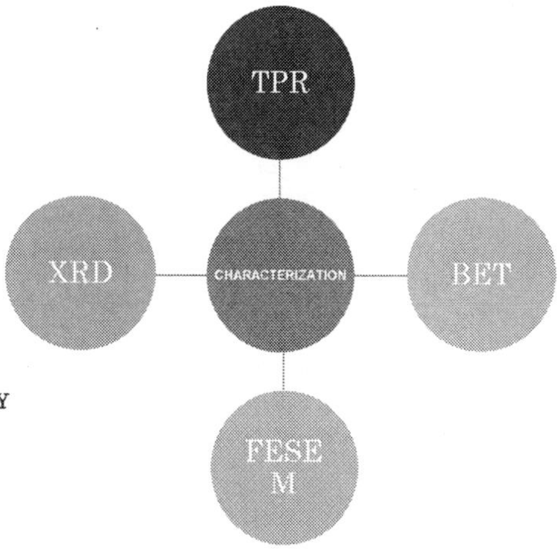

PETRONAS

Open

TPR Analysis

5% H_2 in Nitrogen; Temp: 40-900 °C; Analysis Time: 30min

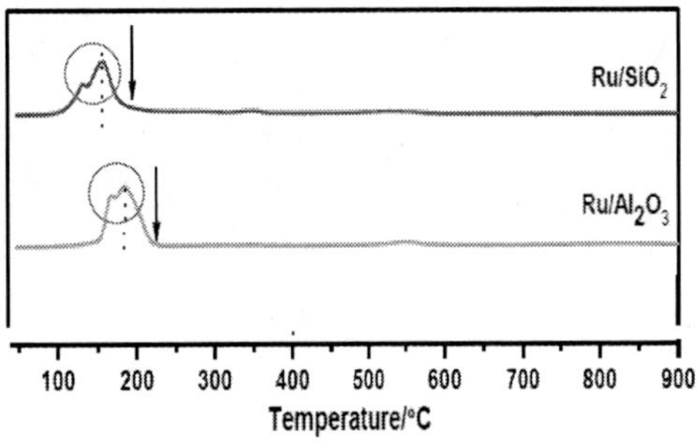

H_2-TPR profiles of Ru/metal oxide catalysts

$$Ru^{2+} \rightarrow Ru^{1+} \rightarrow Ru^{0}$$

Catalyst	Reducing Peak Temperature (°C)	Reducing End Temperature (°C)	H_2 Consumption (mmol)
Ru/Al_2O_3	180	231	1.022
Ru/SiO_2	176	220	1.003

PETRONAS

Open

BET Surface Area Analysis

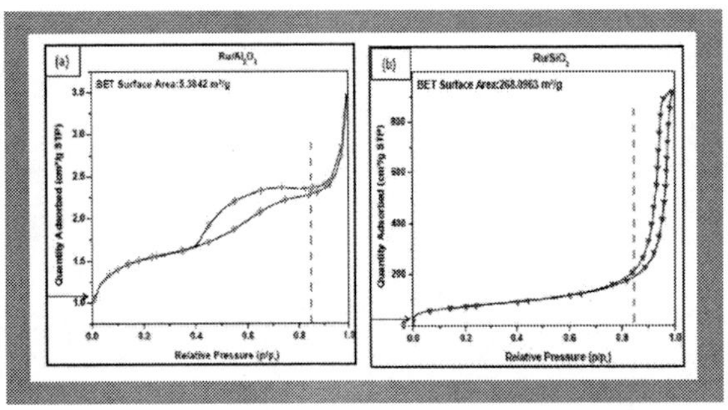

BET adsorption isotherms of supported ruthenium based catalyst

- BJH method was applied to the desorption branch of the isotherms to determine the pore size distribution and pore volume of the catalysts.
- BET analysis showed that Ru catalyst isotherms exhibits reversible Type II isotherm (mesoporous)

Physicochemical properties of Ruthenium supported catalysts

Catalyst type	S_{BET} (m^2/g)	Pore Volume, VP (cm^3/g)	Average Pore Diameter, D_{BJH} (nm)
Ru/Al_2O_3	5.3842	0.0042	6.4026
Ru/SiO_2	268.0963	1.4242	20.5468

PETRONAS

Open

XRD Analysis

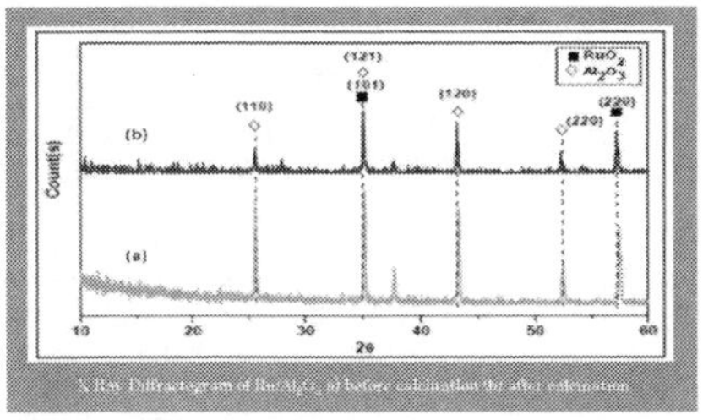

- The RuO_2 has a tetragonal crystal system whereby the prominent peaks were noticeable at 35.2° (101) and 57.9° (220).
- The Al_2O_3 has a rhombohedral crystal system with prominent peaks at 25.5° (110), 35.2° (121), 43.4° (120) and 52.6° (220).
- Both metal oxides peak overlaps at 35.2° making it harder to differentiate. However, crystallinity seems to be reduced when heated suggesting nanostructuring might have occurred.

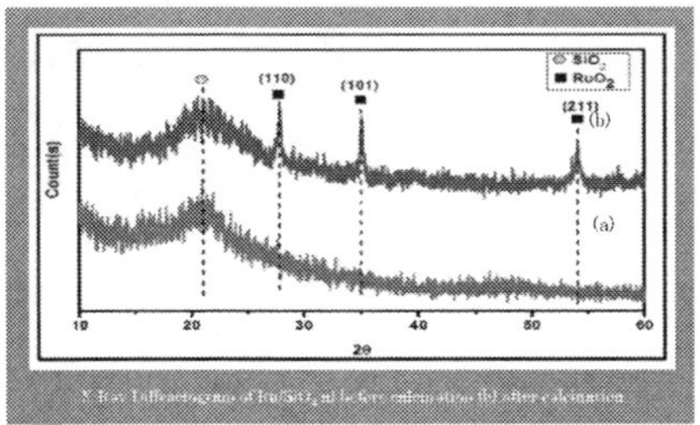

- After calcination at 350°C, RuO_2 peaks at 27.8° (110), 35.0° (101) and 34.0° (211) were evident.
- The amorphous region of SiO_2 remains the same even at high temperature.

FESEM Analysis

FESEM images of Ru supported catalysts (a) Ru/Al_2O_3 (b) Ru/SiO_2

- The catalyst has a hexagonal shape belonging to Al_2O_3 support with small Ru particles well dispersed on the support.
- The amorphous region of SiO_2 remains the same even at high temperature.

CATALYST PERFORMANCE

PETRONAS

Open

Catalyst Performance Evaluation

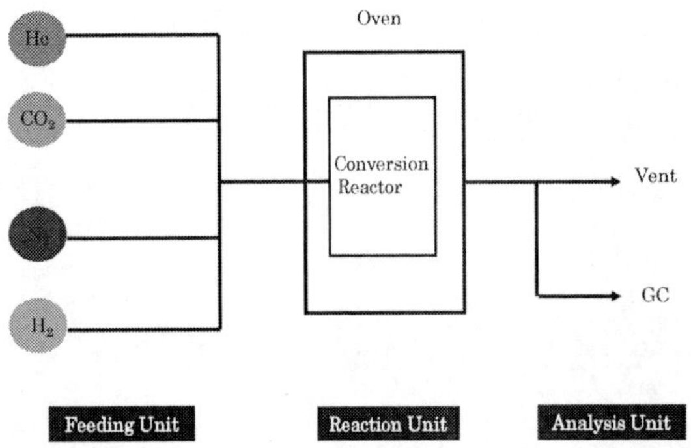

Variable	Gas System
Temperature (°C)	250 - 350
GHSV (h^{-1})	10,000 and 15,000
H_2:CO_2 ratio (mole)	4:1

PETRONAS

Open

Effect of Temperature on CO_2 Conversion and CH_4 Selectivity

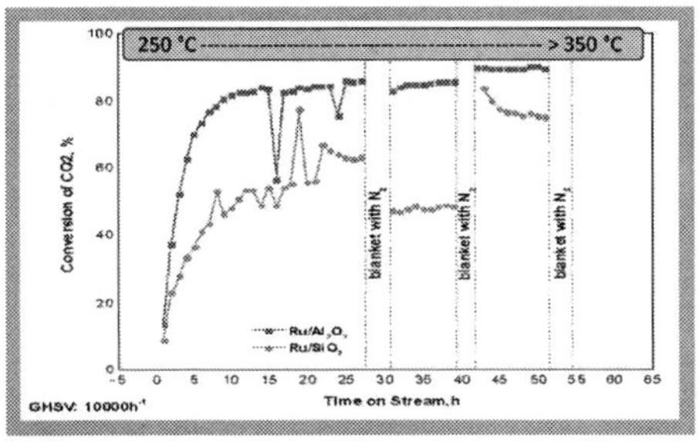

- Higher CO_2 conversion of **89.67%** achieved by Ru/Al_2O_3 at T = 310 °C.
- At lower T = 290 °C, Ru/Al_2O_3 achieved conversion by a factor of two as compared to Ru/SiO_2 at **82.80 %** and **47.03%** respectively.
- Ru/Al_2O_3 has maintained high CO_2 conversion as temperature varies in the range 250 °C – 350 °C with optimum value at 310 °C, with no sign of deactivation, maintaining its catalytic activity

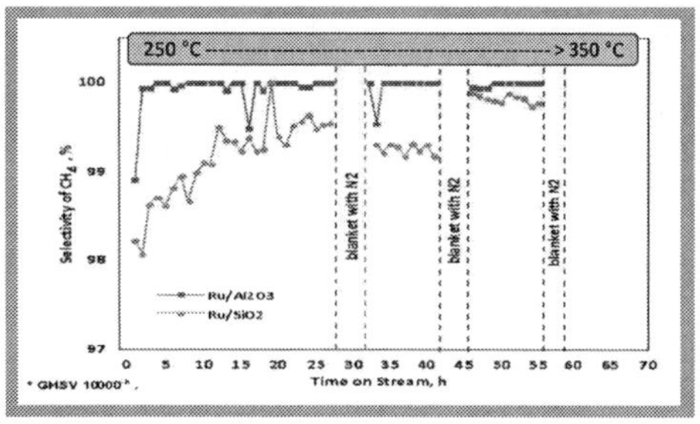

- The CH_4 selectivity for Ru/Al_2O_3 is comparable at all reaction temperatures with the highest selectivity at 100% at T = 310 °C.
- Nevertheless, as temperature increased, Ru/SiO_2 catalyst can reach the same performance as Ru/Al_2O_3 but suffer from catalyst deactivation.

PETRONAS

Open

Effect of GHSV at T = 310 °C on CO_2 Conversion

Catalyst ID	GHSV (h^{-1})	CO_2 Conversion(%)
Ru/Al_2O_3	10,000	89.67
Ru/SiO_2		75.43
Ru/Al_2O_3	15,000	91.60
Ru/SiO_2		80.92

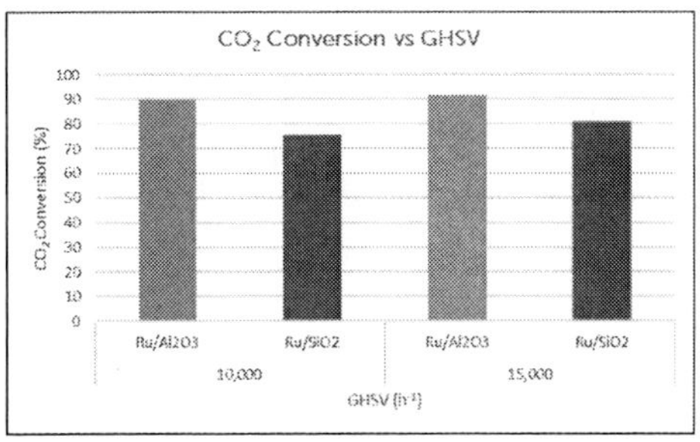

- Ru/Al_2O_3 obtained the highest CO_2 conversion at **89.67%** and **91.60%** at GHSV of 10000 h^{-1} and 15000 h^{-1} respectively.
- The bar chart indicates that working at 15000 h^{-1} seem optimal for both Ru/Al_2O_3 and Ru/SiO_2 catalysts.
- Interesting to note that:
  - High GHSV gave higher CO_2 conversion for both support systems, contrary to typical relationship which support conversion is inversely proportional to conversion.
  - This behavior implies a competitive reaction take place when reaction carried out at low GHSV.

PETRONAS

Open

CONCLUSION

PETRONAS

Open

Two catalyst precursors (Ru/Al_2O_3 and Ru/SiO_2) were synthesized, characterized and tested for performance via CO_2 Methanation.

Ru/Al_2O_3 exhibited highest CO_2 conversion and CH_4 selectivity at 91.60% and and 100% respectively with optimum reaction T = 310 °C and GHSV 15000 h^{-1}.

PETRONAS

Open

"When nothing is sure, everything is possible"

ACKNOWLEDGEMENT

The sincerest gratitude is expressed to PETRONAS, especially Gas Sustainability Technology department, Group Research & Technology, Project Delivery & Technology, for providing an opportunity with complete support to use PETRONAS data and study in order to complete this paper.

Lastly, many thanks are also expressed to family and colleagues, who provided many insightful comments and useful discussions for the completion of this paper.

PETRONAS

Open

HODLNG

JC FINIDORI, Founder

INTERNATIONAL GAS RESEARCH CONFERENCE 2020
WORKSHOP SESSION 22 - DIGITAL TECHNOLOGIES II

Oman, 25th of Feb. 2020

WHAT IS HODLNG?

- A Blockchain solution for liquefied natural gas (LNG); with smart contracts, we address anti-competitive clauses, triggering a fair profit-sharing mechanism. Our utility tokens generate additional revenue and contribute to social impact.

LNG MARKETS' ISSUES

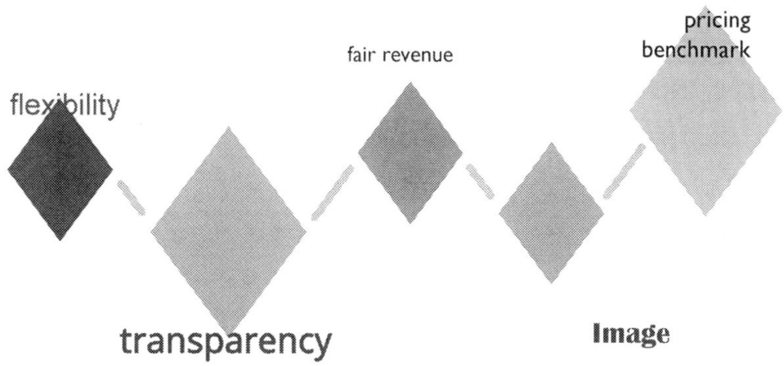

OUR SOLUTION

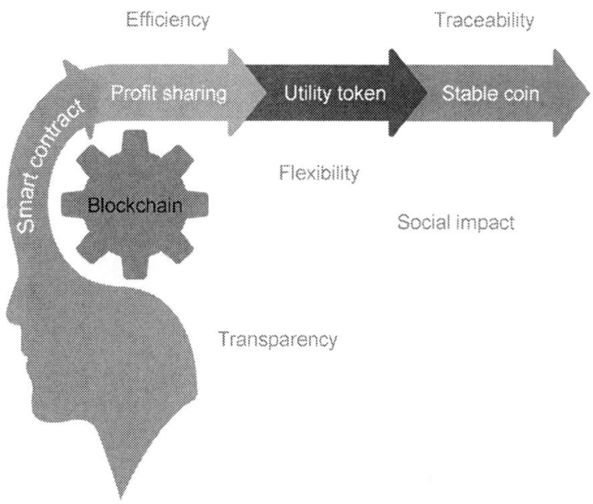

BLOCKCHAIN APPLICATIONS

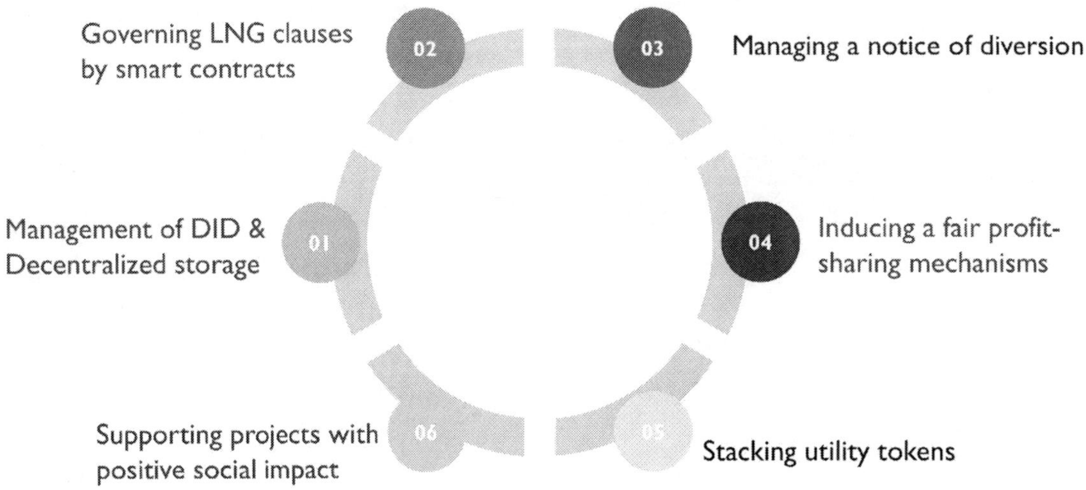

VALUE PROPOSITION

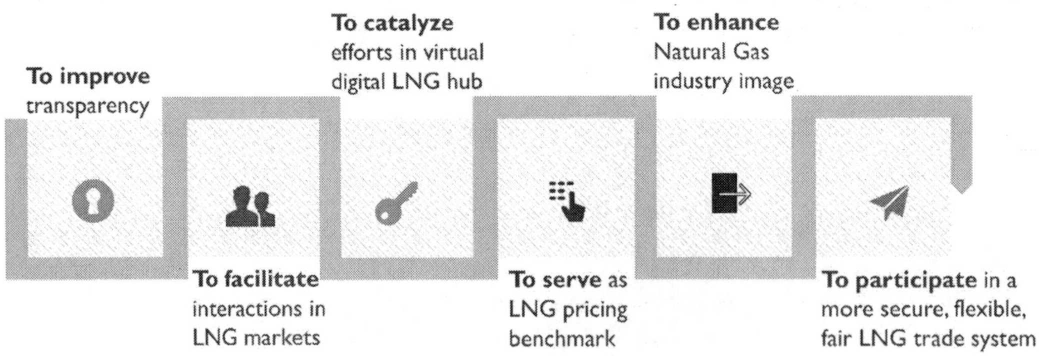

TERMINOLOGY

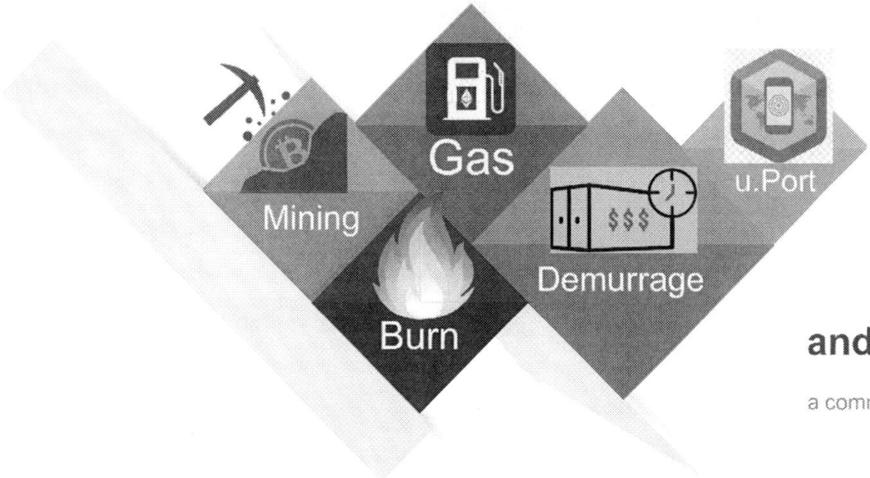

and, Ledger

a common language

SYSTEM ARCHITECTURE

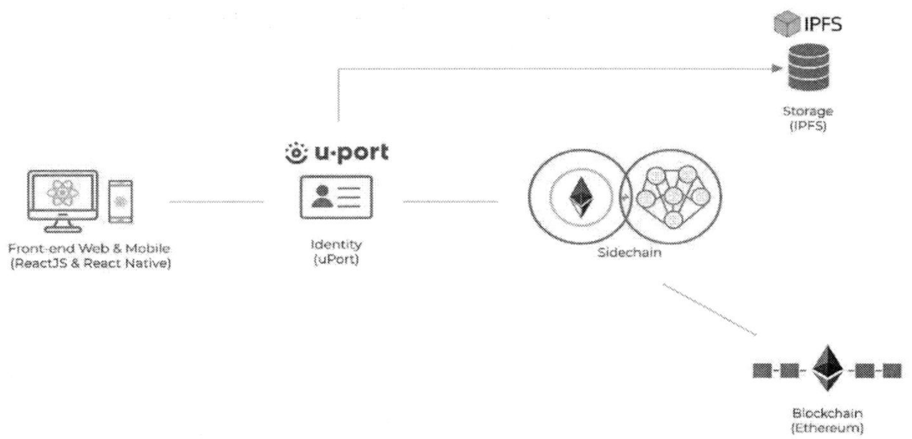

WORKFLOW

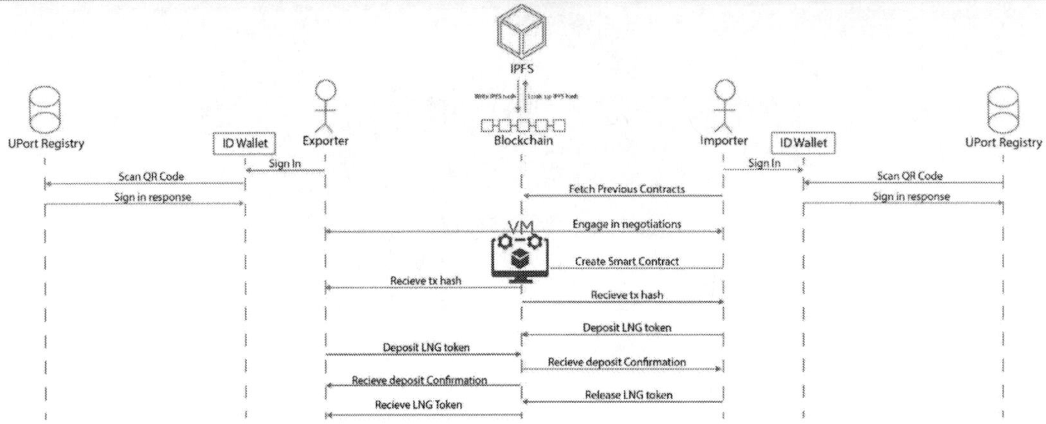

Interactive MOCK-UP

SEMANTIC OF THE SMART CONTRACTS

What is in a smart contract? There are two aspects:

- Contractual elements implemented: volume, price, cargos, destinations; and managing Notice of Diversion (in case of changes of destinations, or re-export).
  - The current paper contract: this should result in on transaction on the network, and decentralized storage; we need a smart contract that contains:
    - "deal = publishGasDeal(seller, buyer, conditions)";
  - The re-export:
    - "createOfferBasedOn(deal)", "createDemand,.."

UTILITY TOKEN & STABLE COIN

- Utility tokens:
  - an unit of gas; and an unit of valuation (exchange); and handling the scarcity
  - own and send any fungible and non-fungible tokens, as well as crypto
- Stable coins
  - Fiat-collaterized
  - Crypto-collateralised
  - Algorithmic Stable Coins
  - Asset-based

PRIORITIZATION OF EPICS

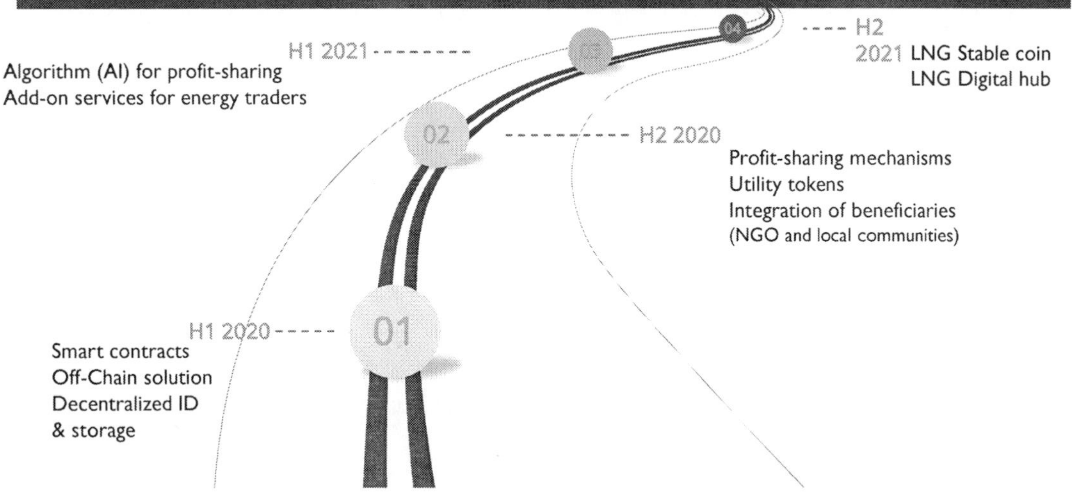

Block.IS

Project has received funding from the European Union Horizon 2020 research and innovation programme

@hodlngOfficial twitter

company/hodlng in

Hodlng F6S

THANK YOU FOR YOUR ATTENTION

JCFINIDORI@HODL.NG

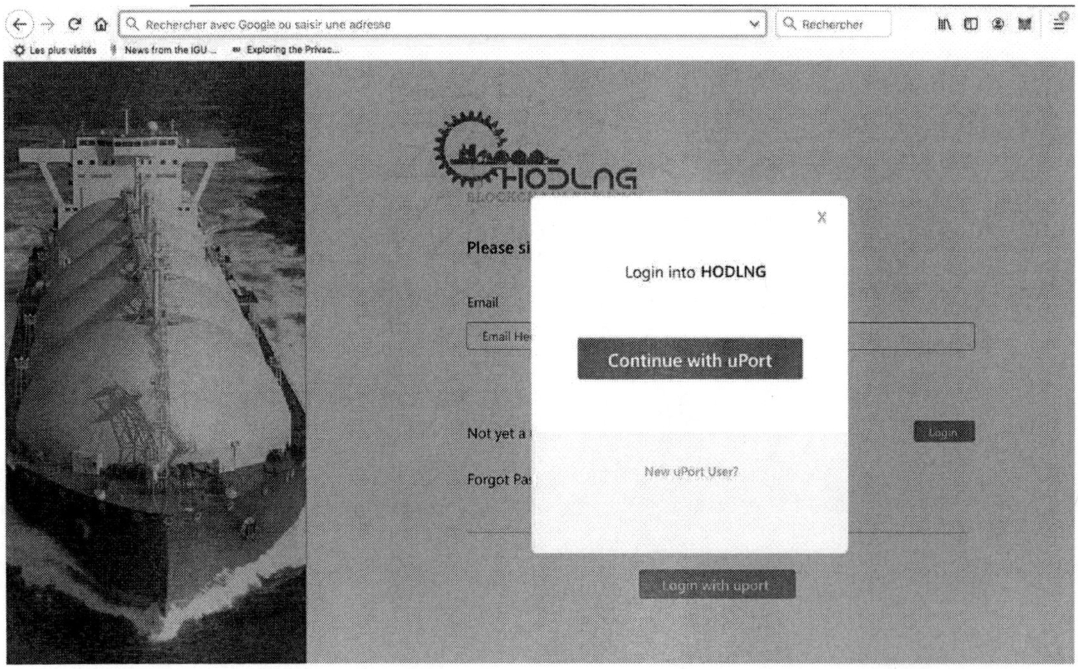

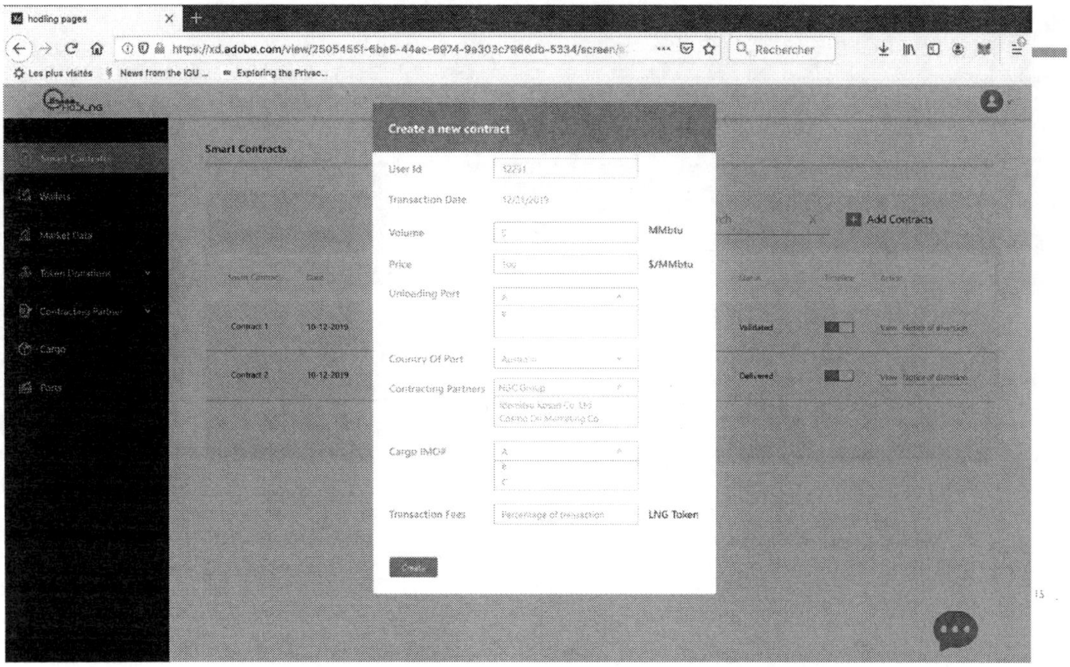
hodling pages
https://xd.adobe.com/view/2505455f-6be5-44ac-6974-9a303c7966db-5334/screen/
Rechercher
Les plus visités
News from the IGU ...
Exploring the Privac...
Smart Contracts
Create a new contract
User Id
Transaction Date
Volume
MMbtu
Price
$/MMbtu
Unloading Port
Country Of Port
Contracting Partners
Cargo IMO#
Transaction Fees
LNG Token
Add Contracts
Contract 1
Contract 2
Validated
Delivered

15

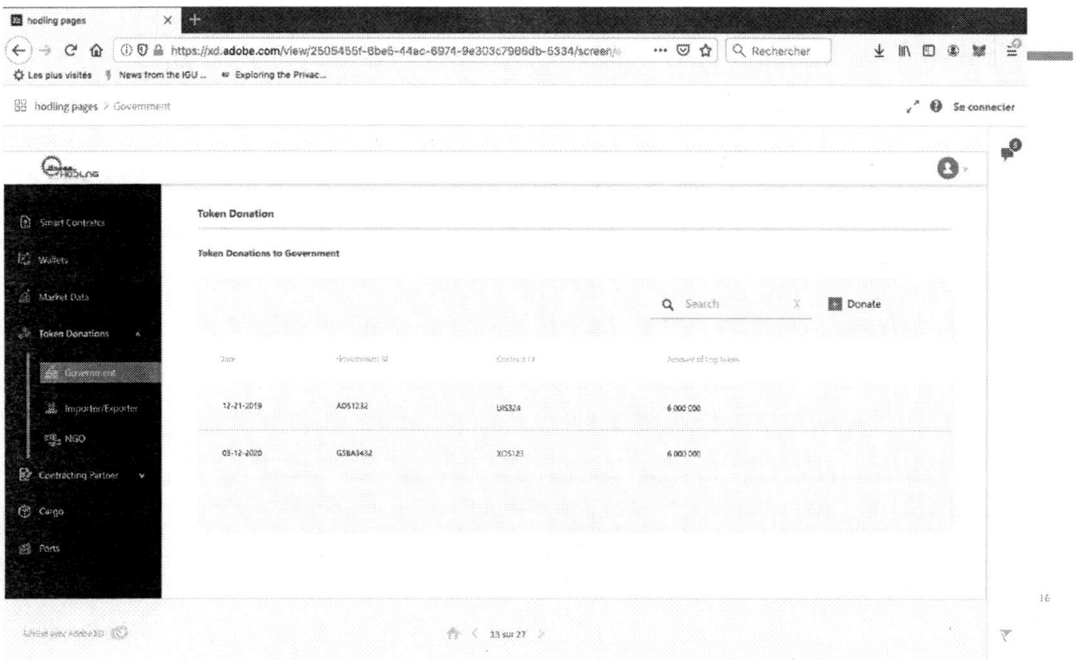
hodling pages
https://xd.adobe.com/view/2505455f-6be5-44ac-6974-9e303c7966db-5334/screen/
Rechercher
Les plus visités
News from the IGU ...
Exploring the Privac...
hodling pages
Se connecter
Smart Contracts
Wallets
Market Data
Token Donations
Importer/Exporter
NGO
Contracting Partner
Cargo
Ports
Token Donation
Token Donations to Government
Search
Donate
12-21-2019
ADS1232
UIS324
6 000 000
03-12-2020
GSBA3452
XDS123
6 000 000
33 sur 27

16

Definitions & cautionary note

The companies in which Royal Dutch Shell plc directly and indirectly owns investments are separate legal entities. In this **presentation** "Shell", "Shell group" and "Royal Dutch Shell" are sometimes used for convenience where references are made to Royal Dutch Shell plc and its subsidiaries in general. Likewise, the words "we", "us" and "our" are also used to refer to Royal Dutch Shell plc and subsidiaries in general or to those who work for them. These terms are also used where no useful purpose is served by identifying the particular entity or entities. "Subsidiaries", "Shell subsidiaries" and "Shell companies" as used in this **presentation** refer to entities over which Royal Dutch Shell plc either directly or indirectly has control. Entities and unincorporated arrangements over which Shell has joint control are generally referred to as "joint ventures" and "joint operations", respectively. Entities over which Shell has significant influence but neither control nor joint control are referred to as "associates". The term "Shell interest" is used for convenience to indicate the direct and/or indirect ownership interest held by Shell in an entity or unincorporated joint arrangement, after exclusion of all third-party interest.

This **presentation** contains forward-looking statements (within the meaning of the U.S. Private Securities Litigation Reform Act of 1995) concerning the financial condition, results of operations and businesses of Royal Dutch Shell. All statements other than statements of historical fact are, or may be deemed to be, forward-looking statements. Forward-looking statements are statements of future expectations that are based on management's current expectations and assumptions and involve known and unknown risks and uncertainties that could cause actual results, performance or events to differ materially from those expressed or implied in these statements. Forward-looking statements include, among other things, statements concerning the potential exposure of Royal Dutch Shell to market risks and statements expressing management's expectations, beliefs, estimates, forecasts, projections and assumptions. These forward-looking statements are identified by their use of terms and phrases such as "aim", "ambition', "anticipate", "believe", "could", "estimate", "expect", "goals", "intend", "may", "objectives", "outlook", "plan", "probably", "project", "risks", "schedule", "seek", "should", "target", "will" and similar terms and phrases. There are a number of factors that could affect the future operations of Royal Dutch Shell and could cause those results to differ materially from those expressed in the forward-looking statements included in this **presentation**, including (without limitation): (a) price fluctuations in crude oil and natural gas; (b) changes in demand for Shell's products; (c) currency fluctuations; (d) drilling and production results; (e) reserves estimates; (f) loss of market share and industry competition; (g) environmental and physical risks; (h) risks associated with the identification of suitable potential acquisition properties and targets, and successful negotiation and completion of such transactions; (i) the risk of doing business in developing countries and countries subject to international sanctions; (j) legislative, fiscal and regulatory developments including regulatory measures addressing climate change; (k) economic and financial market conditions in various countries and regions; (l) political risks, including the risks of expropriation and renegotiation of the terms of contracts with governmental entities, delays or advancements in the approval of projects and delays in the reimbursement for shared costs; and (m) changes in trading conditions. No assurance is provided that future dividend payments will match or exceed previous dividend payments. All forward-looking statements contained in this **presentation** are expressly qualified in their entirety by the cautionary statements contained or referred to in this section. Readers should not place undue reliance on forward-looking statements. Additional risk factors that may affect future results are contained in Royal Dutch Shell's 20-F for the year ended December 31, 2017 (available at and). These risk factors also expressly qualify all forward looking statements contained in this **presentation** and should be considered by the reader. Each forward-looking statement speaks only as of the date of this **presentation, September 18, 2019**. Neither Royal Dutch Shell plc nor any of its subsidiaries undertake any obligation to publicly update or revise any forward-looking statement as a result of new information, future events or other information. In light of these risks, results could differ materially from those stated, implied or inferred from the forward-looking statements contained in this **presentation**.

We may have used certain terms, such as resources, in this **presentation** that United States Securities and Exchange Commission (SEC) strictly prohibits us from including in our filings with the SEC. U.S. Investors are urged to consider closely the disclosure in our Form 20-F, File No 1-32575, available on the SEC website www.sec.gov.

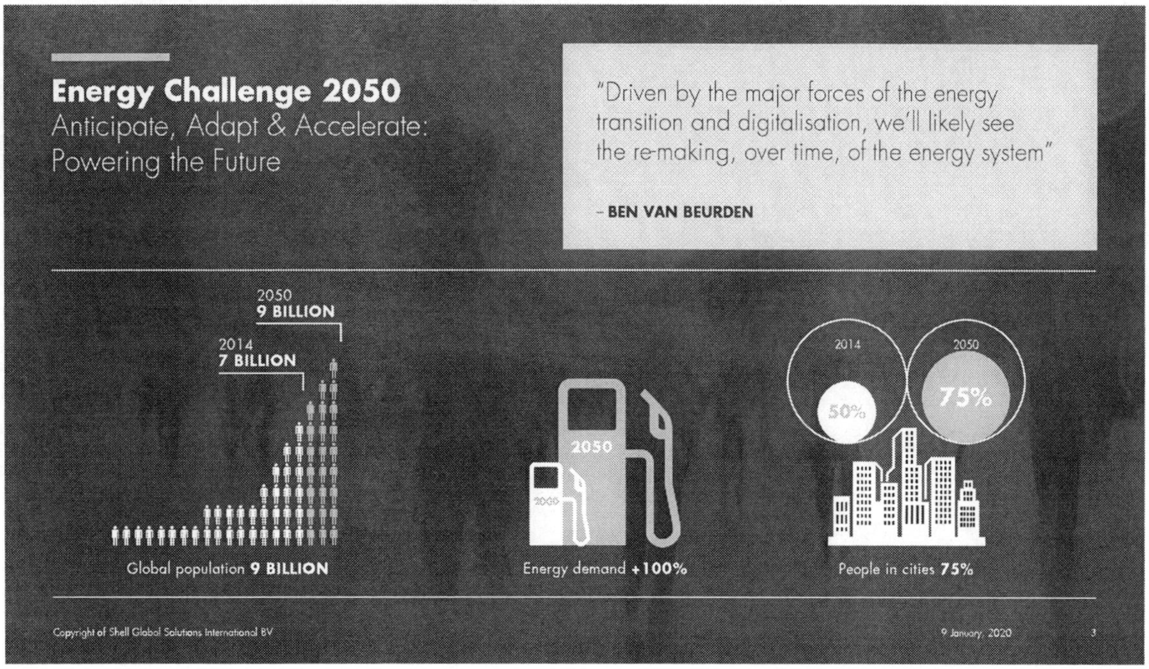

Energy Challenge 2050
Anticipate, Adapt & Accelerate:
Powering the Future
"Driven by the major forces of the energy transition and digitalisation, we'll likely see the re-making, over time, of the energy system"
– BEN VAN BEURDEN
2050
9 BILLION
2014
7 BILLION
2014
2050
50%
75%
2050
Global population 9 BILLION
Energy demand +100%
People in cities 75%
Copyright of Shell Global Solutions International BV
9 January, 2020
3

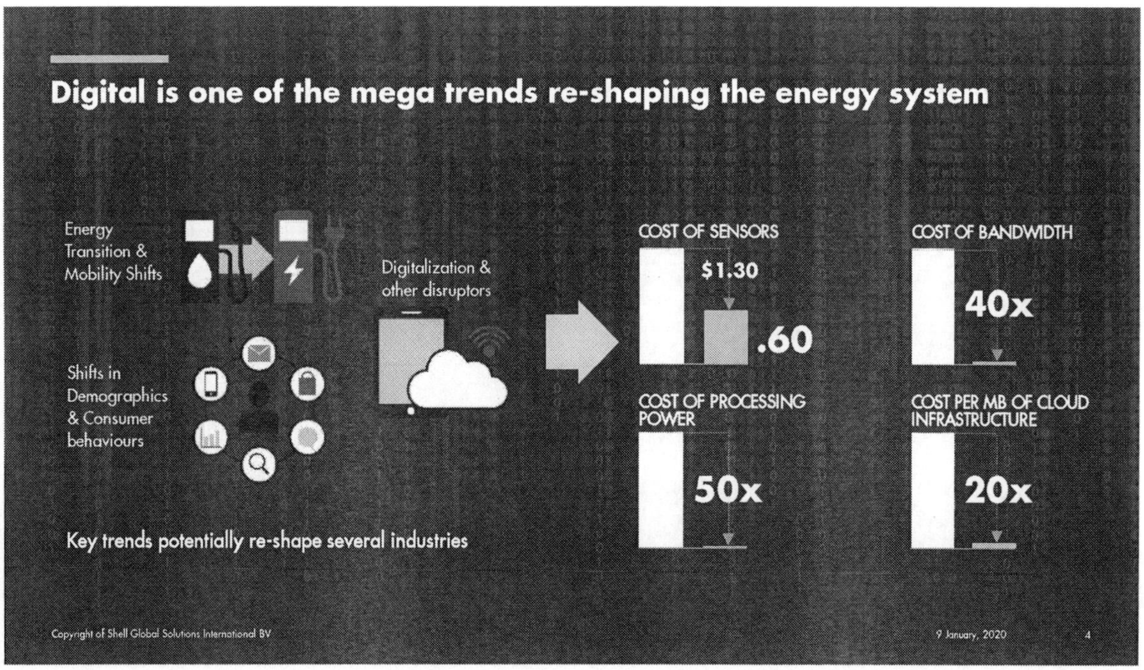

Digital is one of the mega trends re-shaping the energy system
Energy Transition & Mobility Shifts
Shifts in Demographics & Consumer behaviours
Digitalization & other disruptors
COST OF SENSORS
$1.30
.60
COST OF BANDWIDTH
40x
COST OF PROCESSING POWER
50x
COST PER MB OF CLOUD INFRASTRUCTURE
20x
Key trends potentially re-shape several industries
Copyright of Shell Global Solutions International BV
9 January, 2020
4

Shell's Digital Strategy – accelerating Digitalisation at scale

A coherent approach across Shell to realise and accelerate value through digital led by business and supported by Digital COE

FIVE DIGITAL DESIGN PRINCIPLES

- CUSTOMER/USER IS CENTRAL
- DATA IS AN ASSET
- BUSINESSES OWN DIGITAL
- BUILD IN-HOUSE CAPABILITY
- ACT OUR WAY INTO THE FUTURE

UNDERPINNING CRITICAL SUCCESS FACTORS

- CAPABILITIES
- OPERATING MODEL AND WAYS OF WORKING
- LEADERSHIP, MIND-SETS AND BEHAVIOUR

Digitalisation in Shell

Technology = enabler

Differentiators:

1. Data – insights – action conversion rate
2. Speed & scale of replication
3. Cultural transformation

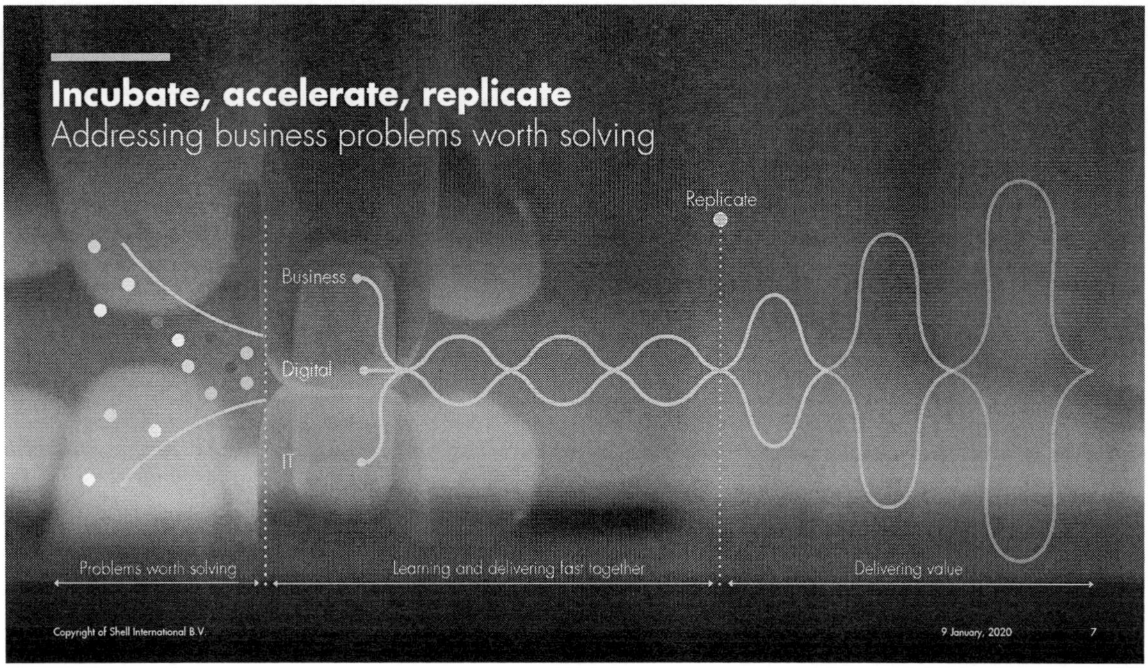
Incubate, accelerate, replicate
Addressing business problems worth solving
Replicate
Business
Digital
IT
Problems worth solving
Learning and delivering fast together
Delivering value
Copyright of Shell International B.V.
9 January, 2020
7

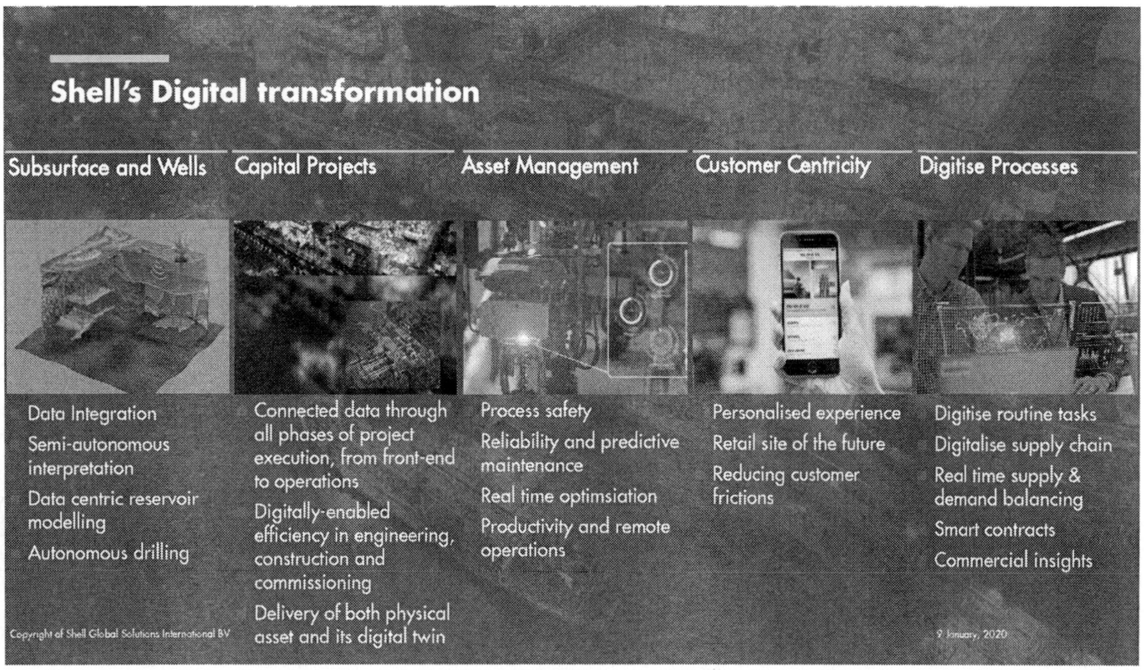
Shell's Digital transformation
Subsurface and Wells
Capital Projects
Asset Management
Customer Centricity
Digitise Processes
Data Integration
Semi-autonomous interpretation
Data centric reservoir modelling
Autonomous drilling
Connected data through all phases of project execution, from front-end to operations
Digitally-enabled efficiency in engineering, construction and commissioning
Delivery of both physical asset and its digital twin
Process safety
Reliability and predictive maintenance
Real time optimsiation
Productivity and remote operations
Personalised experience
Retail site of the future
Reducing customer frictions
Digitise routine tasks
Digitalise supply chain
Real time supply & demand balancing
Smart contracts
Commercial insights
Copyright of Shell Global Solutions International BV
9 January, 2020

Improving productivity
Wearable tech
& mobile connectivity
Copyright of Shell International B.V.
9 January, 2020
9

Pre-empting failure
Advanced analytics & machine learning
Compressor
Steady State
Outliers
Copyright of Shell International B.V.
9 January, 2020
10

Asset and Project of the future
Being built today
Pennsylvania Chemicals Project
Improving Construction Productivity through Digital
PENNSYLVANIA CHEMICALS
Play
NYHAMNA, NORWAY
Copyright of Shell International B.V.
9 January, 2020
11

The human factor
Developing in-house digital literacy
and problem-solving capability
Copyright of Shell International B.V.
9 January, 2020
12

Accelerating Shell's Digital Journey
Technology is the table-stake, insights and action are the differentiators
We focus on 'business problems worth solving'… and develop solutions to capture value
We quickly incubate, accelerate and replicate digital solutions across our assets
Digitalisation is about rewiring brains and adopting agile ways of working
Mastering digital operations is fundamental to our license to operate
Copyright of Shell International B.V.
9 January, 2020
13

NORDION ENERGI

Energy Determination when Renewable Gases are Injected into City Grids

Dr. Stefan Rickelt, SmartSim GmbH
Ylva Nordlund, Nordion Energi AB

SmartSim

Situation in Sweden & Europe

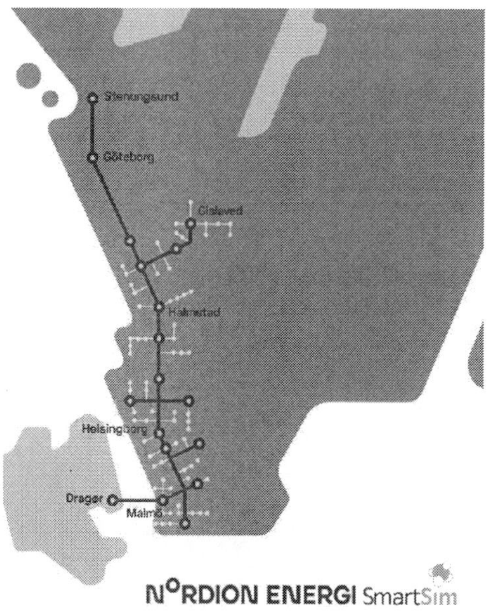

- Nordion Energi owns the TSO grid and the largest DSO grid
- Gas in Sweden 15 TWh (565 TWh)
- European Green Deal and GHG emission targets
- Integration of biomethane and hydrogen

→New demands on grid operators

- Biomethane upgrade

- Project: Green Compass in 2018
  - Geographic-, gas quality-, flow- and consumption pattern data
  - Green Compass + Smart Meters = Smart Sim in city grid

NORDION ENERGI SmartSim

Malmö grid - facts

SmartSim
NORDION ENERGI

- 340,000 inhabitants
- 1 biomethane and 3 TSO injections
- 16, 4 and 0.1 bar
- 519 km
- 79 reduction stations
- 12,000 measured customers
- 9,000 unmeasured customers (Stove)
- Pipes -1983 – today, Steel, PE and relined cast iron
- Redundant grid
- Gas flow in both directions

3

Malmö grid - challenges

SmartSim
NORDION ENERGI

- Limited control over grid-losses
  - Flow directions
  - Volume to unmeasured costumers
- Assets in need of modernisation and digitalisation
- Demand to accept renewable gases in the infrastructure
- Field tests showed that the grid needs full redundancy

→Impossible to ensure correct billing without propane upgrade

- Adjustments in the Malmö grid
- SmartSim

→ We could achieve our target even in a complex city grid as Malmö

4

Green Compass + Smart meters = SmartSim in a city grid

SmartSim
NORDION ENERGI

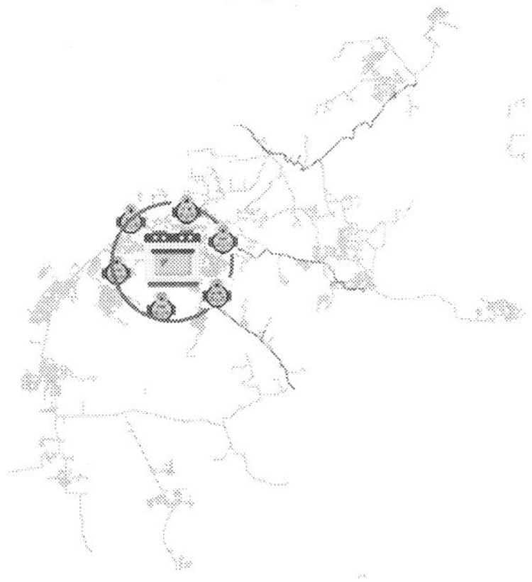

- Smart Meters for 12,000 customers
- Ultrasonic metering

Results

- Multiple injection points with different CVs
- Detailed network-specific forecasts
- A tool to plan additions to the grid
- Simulate new production and customer facilities
- At incidents, ensure flow directions

5

Renewable gases

SmartSim
NORDION ENERGI

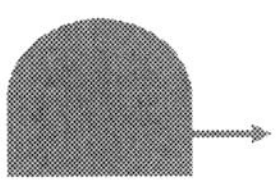

Biomethane

GCV:
~ 10.6 - 11.0 kWh/m³

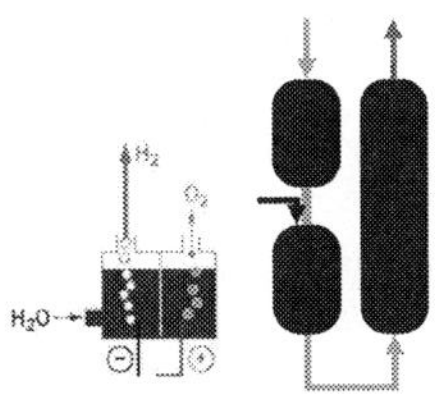

Hydrogen
(mixed with natural gas)
GCV:
3.5 kWh/m³ (100 % H_2)
~ 10.2 kWh/m³ (20% H_2 / 80 % natural gas)

Syngases, (bio-)
synthetic methane
GCV = ?

- Prerequisite: renewable gases / or the injected mix shall be interchangeable

→ Calorific Value (CV) differences of 6 - 15 % compared to typical natural gases

6

Energy determination

SmartSim
NORDION ENERGI

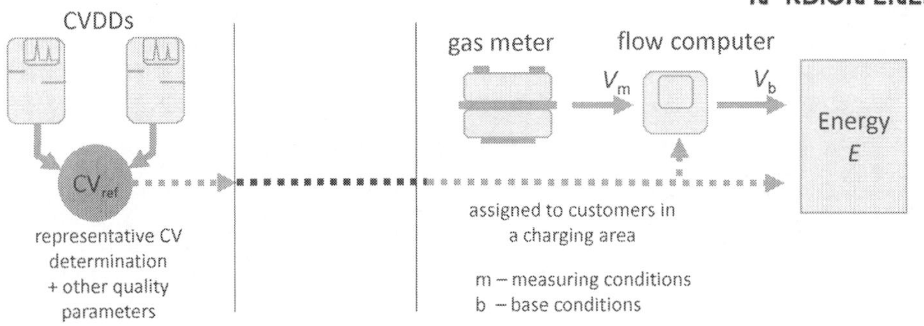

- Energy is determined from volume and energy content (CV) at base conditions
- Volume is measured locally, not possible for CV
- CV is usually determined at designated locations in the grid e.g. entry points
- A representative CV is determined and assigned to customers in a CV district

7

Representative CV and CV districts

SmartSim
NORDION ENERGI

How are representative GCVs determined?

- This depends on differences in gas qualities, grid configuration and volumes injected
- Two examples with real data showing the different gas flows

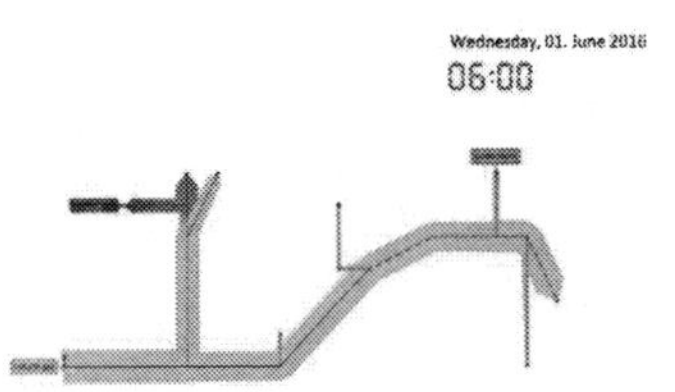

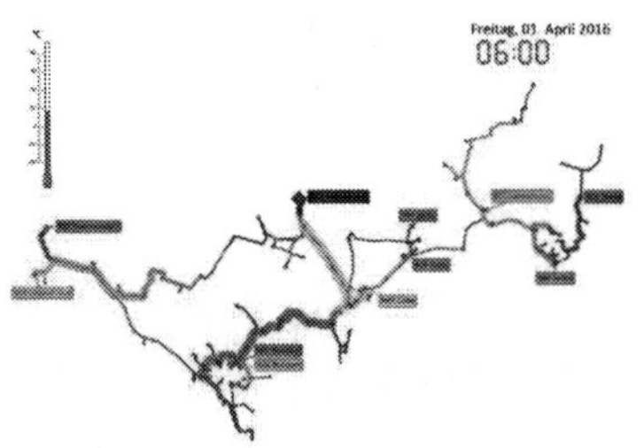

→ Difficult to ensure stable local gas mixtures in city grids
→ Gas quality tracking with SmartSim the most flexible solution

8

How SmartSim Works

SmartSim
NORDION ENERGI

Input Properties
- Grid Topology
- Volume flows at entry and exit points
- Gas quality at entry points (CV)

Problem in many distribution grids
- Offtakes are not always measured

Solution
- Determine volume flows at exit points based on standard load profiles in connection with a special correction algorithm

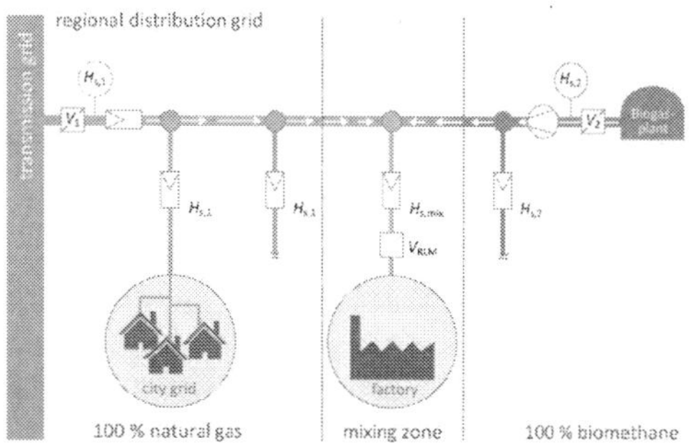

Calculation
- Grid simulation (pressures, flow velocities at all grid nodes)
- Tracking of "gas packages" with a new developed algorithm

9

Reference Projects

SmartSim
NORDION ENERGI

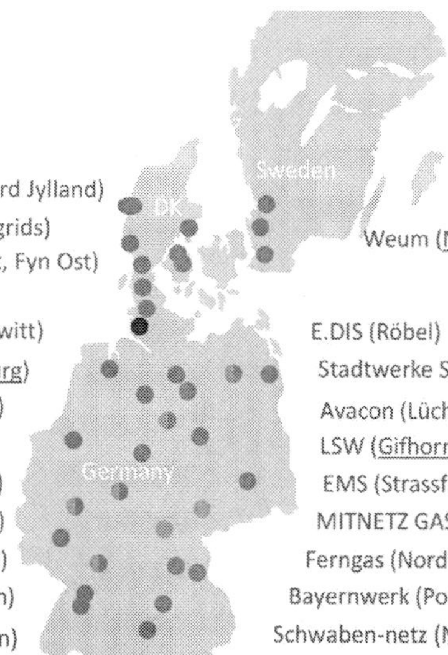

HMN Naturgas (Ringkøbing, Nord Jylland)
Dansk Gas Distribution (Vejen + 5 more grids)
NGF Nature Energy (Fyn West, Fyn Ost)
Hansewerk (Brunsbüttel, Handewitt)
EWE Netz GmbH (Oldenburg)
Innogy (Dorsten)
EnergieNetz Mitte (Kassel)
Oberhessengas Netz (Lich)
Energienetze Mittelrhein (Koblenz)
CREOS (Eifel)
MVV (Sinsheim)
Netze Südwest (Nordbaden, Oberschwaben)

Weum (Malmö, Laholm, Bjuv)

E.DIS (Röbel)
Stadtwerke Schwedt (Schwedt)
Avacon (Lüchow, Messenkamp, Gardelegen)
LSW (Gifhorn)
EMS (Strassfurt)
MITNETZ GAS (Hauptnetz)
Ferngas (Nordbayern, Thüringen)
Bayernwerk (Postbauer-Heng, Schwandorf)
Schwaben-netz (Nordschwaben, Schongau)

● Biomethane injection ● Different natural gases ● Power to Gas (H_2) Implementation Phase

10

Reference projects (TSOs)

SmartSim
NORDION ENERGI

● Biomethane injection ● Different natural gases ● Power to Gas (H_2) Implementation Phase

11

Results from Gas Quality Tracking with SmartSim

SmartSim
NORDION ENERGI

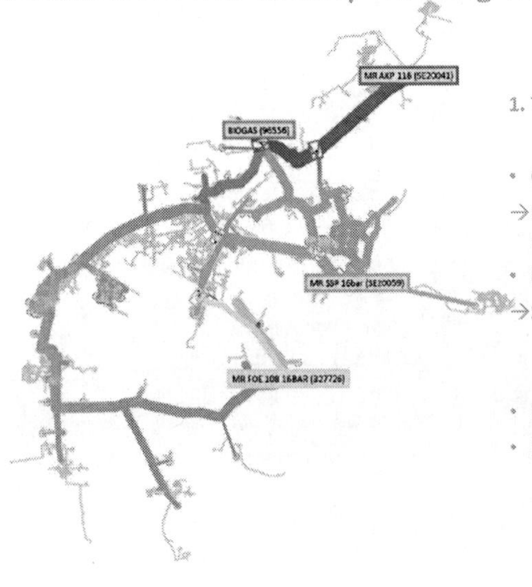

1. Topology simplification

- Grid contains several low-pressure downstream grids
→Not part of the simulation but represented by an exit node
- Many customers are directly connected to 4 bar grid
→Simplification by aggregation of exit points

- Improves simulation efficiency
- Does not reduce accuracy

12

Results from Gas Quality Tracking with SmartSim

SmartSim
NORDION ENERGI

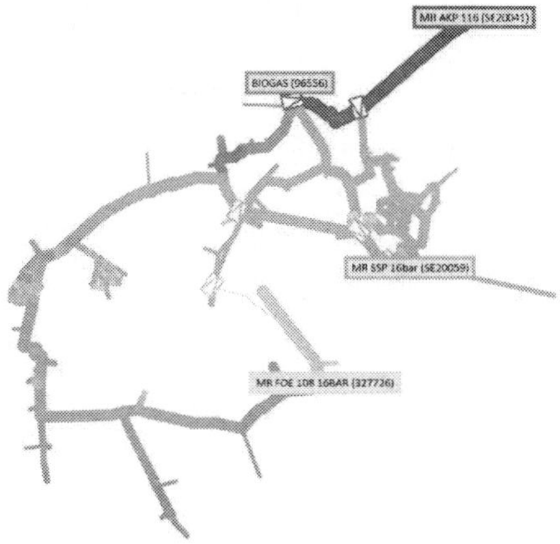

Group customers to CV districts

nartSim

- Performs flow simulation and tracks the CV for all nodes
- Calculates one representative CV for each CV district from node CVs
- Check ensures if customers in a CV district receive the same gas mixture
- Representative CV used for billing of all customers in one CV district

Reduces complexity for billing

13

Next steps & summary

SmartSim
NORDION ENERGI

- First project phase has shown that SmartSim is able to track the gas quality in the Malmö grid
- Next steps are validation according to EN ISO 15112:2018 Natural Gas - Energy Determination
  - Determination of uncertainty of the system grid Malmö/SmartSim
  - Experimental validation/field test
  - Regular operation of SmartSim planned during 2020

City grids

- Often have large capacities to receive renewable gases
- Have a key role in integration of renewable & decarbonized gases and reduce carbon footprint

SmartSim

- Gives grid operators a flexible solution to handle varying gas qualities
- Safeguards correct billing also in complex gas grids

14

Thanks for your attention!

Ylva Nordlund
ylva.nordlund@nordionenergi.se
+46 730 220 071

Dr. Stefan Rickelt
stefan.rickelt@smartsim.energy
+49 178 854 4585

SmartSim

Backup

SmartSim

16

Validation with mobile GC

SmartSim

- Validation for each system SmartSim/gas grid
- Numerous field tests over the last years

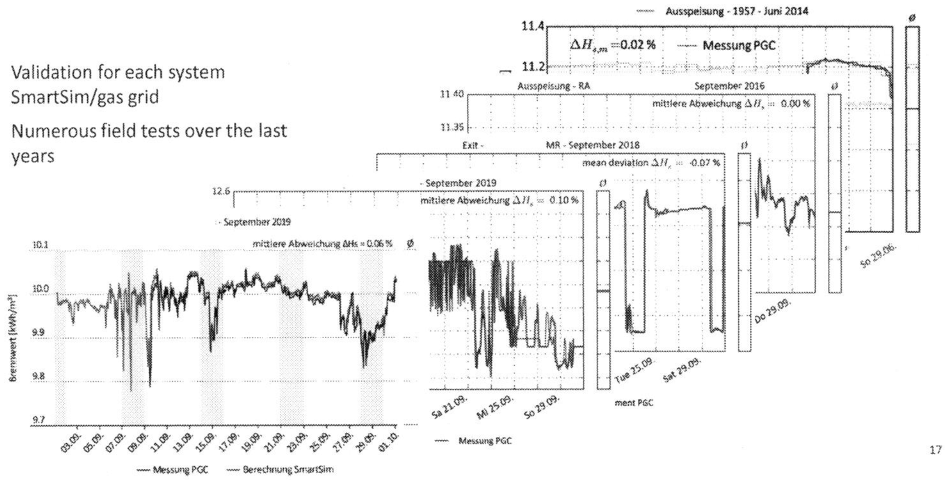

17

Comparison between Calculated Uncertainties with GC Measurements from Different Field Tests

SmartSim

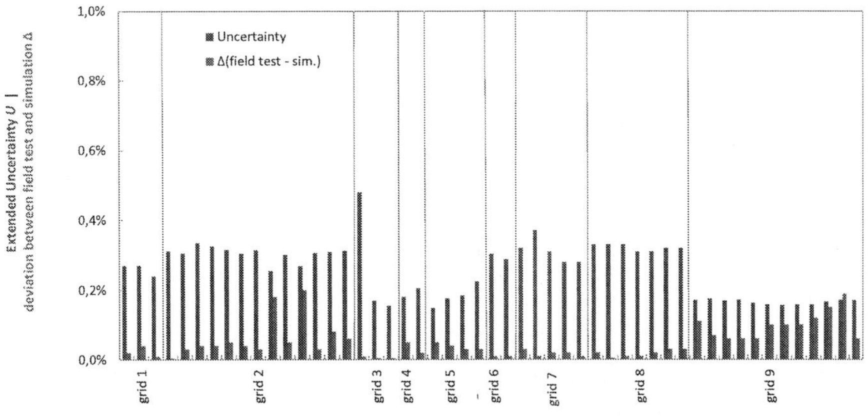

18

Impact of new gases on gas infrastructure

Snam pilot project: hydrogen injection in the gas grid

International Gas Union research conference
Muscat, 24 February 2020

Snam role in the gas value chain

An integrated natural gas infrastructure operator

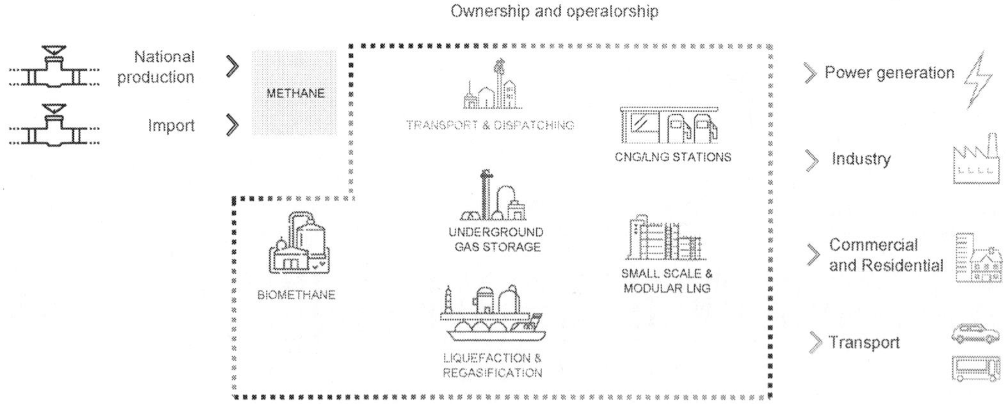

2

Snam transmission facilities...

Main figures

- Over 32,625 km of pipelines
- 11 compressor stations
- 3 LNG plants

✓ 72 BCM of gas demand in 2018
✓ ~170 clients

Market Shares

- 94% in gas transportation
- 96% in gas storage
- 3 out of 3 regasification terminals in Italy

3

... and Underground Gas storage

Storage in Italy

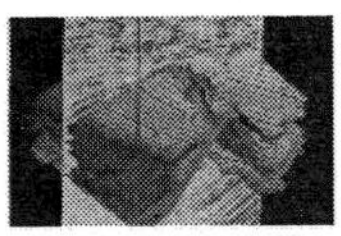

9 depleted gas fields
13 reservoirs
~430 wells

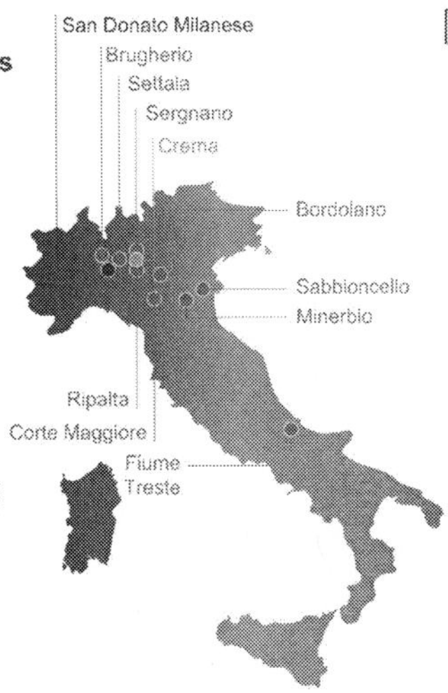

Main figures (2018)

Headquarters
San Donato Milanese (registered office)
Crema (operating office)

Storage concessions
9, of which 5 in Lombardy, 3 in Emilia-Romagna and 1 in Abruzzo

Clients
~ 90

Natural gas storage capacity
16.9 bcm

UGS volumes injected/withdrawn
~21 bcm

4

The role of hydrogen in the decarbonization roadmap

snam

Carbon free

- Can be produced with **zero carbon footprint**

Energy carrier

- Energy **transport** and **storage** with a **lower cost** with respect to electricity
- **Valorization** of the **existing infrastructure**

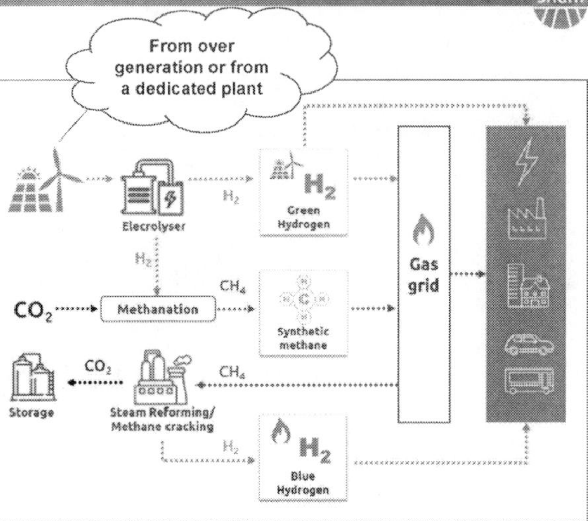

Fuel

- **Zero Emission fuel**
- **Decarbonization of the industrial processes**

  Considering the steelworks, the emission of ~ 10 mlnt/year of CO2 can be avoided, which are almost equal to the 20% of the total industrial emissions in Italy

5

Snam Hydrogen stategy: 3 Pillars

snam

Snam as Enabler

1 Readiness of the Asset

- **Asses** the **compatibility** of the existing infrastructure with respect to hydrogen transport
  - Storage
  - Piping
  - Metering
  - Compression stations

- **Ongoing activities:**
  - Injection campaign of **H2NG blend at 5% and 10 % vol. H2**
  - **Storage** analysis
  - Asses the blending effects on **turbo-compressors**
  - Asses the blending effects on **boilers**

2 Value chain development

- Promote the **technology development** along the entire value chain

- **Analysis of the innovative technologies:**
  - **Power to gas**
  - **Hydrogen mobility**

3 System design

- Definition of the **hydrogen integration roadmap** in the existing energy system

- Taking part to working groups focused on the **regulatory framework evolution**

6

Snam Approach

1 Pipeline 1st test — *April 2019*

- Mixture of H2NG (H2 – 5% volume)
- End Users:
  - Pasta Factory
  - Water Bottling

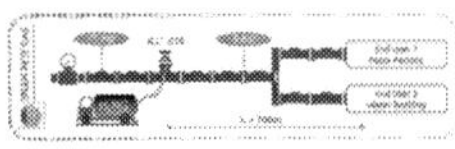

2 Pipeline 2nd test — *December 2019/ January 2020*

- Mixture of H2NG (H2 – 10 % volume)
- End Users:
  - Pasta Factory
  - Water Bottling

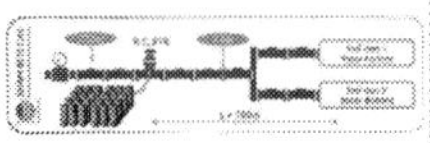

Completed

3 Compressor station gas turbines

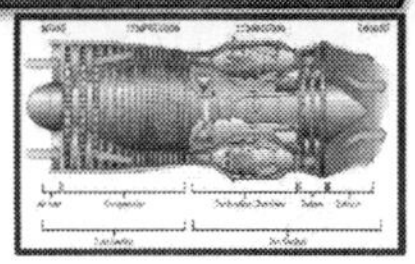

- Collaboration with **Snam gas turbine suppliers**

4 End Users with production processes sensitive to gas

- Collaboration with gas turbine suppliers of **Snam industrial clients**

In progress

7

H2 injection in the natural gas grid: Technical overview

In 2018 Snam started to approach the H2 market, and identified the main **technical constraints related to the hydrogen injection** in the natural gas grid, on the basis of bibliographic studies and analysis:

Thresholds of H2 injection in Natural Gas grid due to "technological" issues

	Pipeline Network	Storages (Porous Rocks)	CNG Mobility – Steel tanks
Hydrogen limits in H2NG blends:	5%	1%	2%
	• Gas Turbines behaviour • Gas chromatographs (0%)	Uncertainty on: • Elastomeric seal behavior • Phenomenon of hydrogen diffusion • Microbiological reactions	• Regulatory constraints (ISO 11439)

An **first injection campaign of a H2NG mixture** in the gas grid has been put in place, to **verify the compatibility of the current infrastructure** with respect to this blend

8

Readiness of the asset - 1° Injection campaign

Injection campaign of **a hydrogen-natural gas blend (H2NG)** in a portion of the Snam grid

CONSTRAINTS
:

CNG refuelling stations

Underground storage in porous rocks

Gas turbines of compressor stations

End Users with production processes sensitive to gas quality

Mixture features (Natural gas at 5% of H2)

- **Gas Quality parameters** set by the Ministerial Decree 18 May 2018 for the natural gas transport: **respected**;
- Materials HE (**Hydrogen Embrittlement): checked**;
- **AT**mosphères ed **Ex**plosives zones identification: **checked**.

Public bodies Involved

- **National and Local Fire Departments**

9

Readiness of the asset – 1° Injection campaign

Injection campaign of **a hydrogen-natural gas blend (H2NG)** in a portion of the Snam grid

Key features

Injection campaign of a H_2NG blend with H2 at 5% in volume in a portion of the Snam gas transport grid, in order to verify the readiness of the existing asset with respect to the transport of the considered blends.

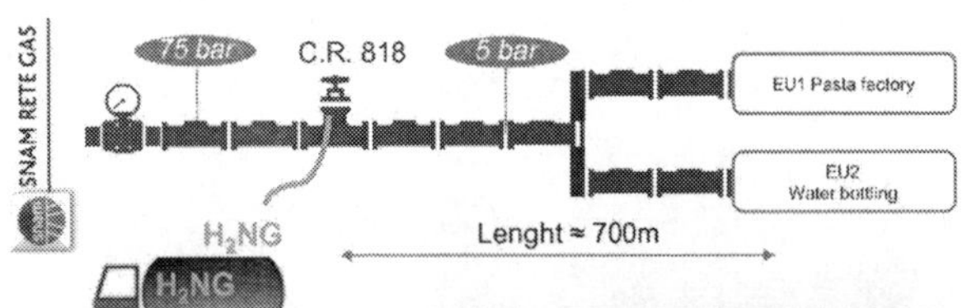

- The gas blend has been supplied with a trailer tank, with a capacity of 5.000Nm3, filled with the H2NG blend with H2 at 5% in volume, at a pressure of 200 bar.
- The injection campaign lasted one month

1° Injection campaign results overview

snam

Phase 1 –Snam plant only

- Correct operation of the **boiler located in the gas pre-heating section** of the pressure reduction plant
- Correct operation of the **pressure reduction plant components**

Phase 2 – Snam plant and end-users facilities(started on the 1° April 2019)

- **Correct operation** of the **facilities used** in the **industrial processes of the two end-users**
- Correct operation of the **hydrogen analyzer** installed in the Snam plant
- Correct operation of the **boiler** located in the **gas pre-heating section** and of the **gas pressure reducers** located in the plant

Focus in the next slides:
Exhaust gas composition of the gas pre-heating boiler

11

1° Injection campaign results: boiler exhaust gas analysis (1/2)

snam

Exhaust gas analysis of the boiler located in the gas pre-heating section

Key features

- **Measurements have been performed both before and after the H2 addiction** in the NG flow (with multiple repetitions for each test).
- **Measurement instrument:** gas analyser for combustion and exhaust gas composition analysis
- **No modifications of the boiler settings**, between the 100% NG case and H2NG blend at 5% vol. H2
- **Test analysis jointly developed with Politecnico di Torino**
- **The following exhaust gas components have been measured:** NOx , CO and CO2

Measurement with H2 addiction (5%)

- **H2 addiction in a NG flow brings the following effects on the exhaust gas composition of the boiler:**
  - **NOx**: H2 has multiple effects on the combustion parameters affecting the NOx production. Overall, a slight NOx increase has been measured, mainly due to the increase of the flame temperature with H2 addiction
  - **CO**: a slight increase has been registered, due to the higher combustion reactivity of H2 with respect to NG, which reduces the amount of air available for the complete combustion of the hydrocarbons . This effect is not confirmed with 10% of H2. CO decreases because the reduction of carbon concentration in the mixture cancels the previous effect.
  - **CO2**: a slight CO2 reduction has been registered, due to the reduced carbon concentration in the mixture with respect to NG only

12

Readiness of the asset – 2° Injection campaign

snam

Injection campaign of a H2NG with a double H2 content equal to 10% in volume

Key features

- Same grid and industrial end-users of the first injection campaign
- **Test protocol jointly defined with Politecnico di Torino**
- **The H2NG mixture has been supplied with bundles of cylinders,** due to the fact that the actual technical regulation does not allow the transport of a H2NG mixture at 10% H2 in cylinder trailers for NG. As a consequence, the volumes transported have been reduced with respect to the first campaign
- **Pilot duration reduced** with respect to the first campaign, due to a **more complex logistic of the mixture transport** with the bundles of cylinders(~ two weeks). Started in December '19

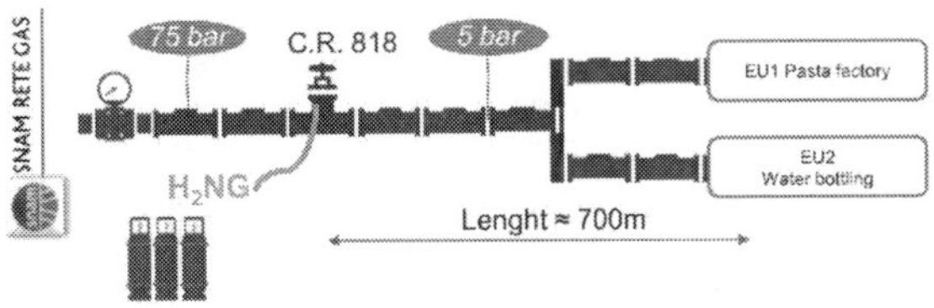

Hydrogen R&D Activities

snam

Supply	Monitoring the evolution of the production technologies (Electrolyzers, Methanation processes, Methane Cracking and Steam Methane Reforming)
Transport	Increase of the H2 content in the H2NG blend for the injection campaign, collaboration with a University for the results analysis Contacts with other European TSO's for joint initiatives/R&D projects
Storage	Analysis of projects to asses the H2 impact in storage fields
End-users	Impact analysis of the H2NG blend on turbo-compressors Impact analysis of the H2NG blends on CNG vehicle's engines and fuel tanks Decarbonization of the industrial end-users with high C02 emissions Impact analysis of the H2NG blend on boilers

14

Thank you for the attention

Head of Technical Hydrogen
Dina Lanzi
Dina.lanzi@snam.it

15

Artificial Intelligence
Maintenance Tab
Present by : Jasim AL Mahyijari
Senior technical development officer
Oman LNG L.L.C.
The Best At What We Do

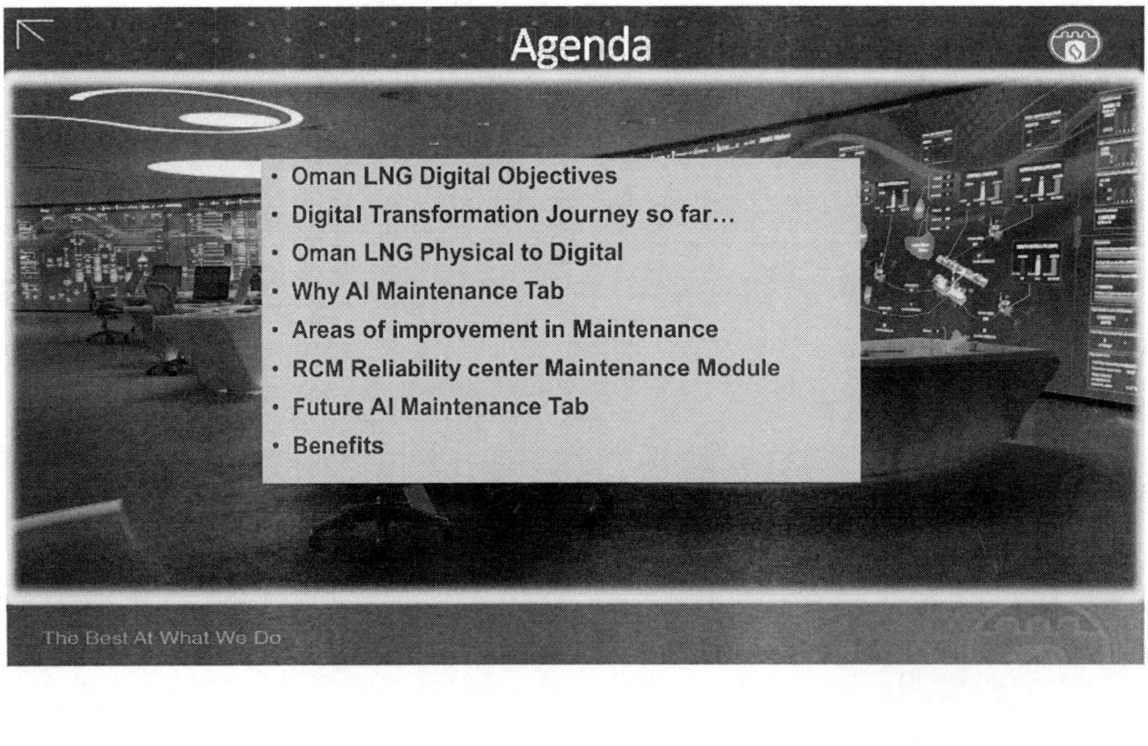
Agenda
• Oman LNG Digital Objectives
• Digital Transformation Journey so far...
• Oman LNG Physical to Digital
• Why AI Maintenance Tab
• Areas of improvement in Maintenance
• RCM Reliability center Maintenance Module
• Future AI Maintenance Tab
• Benefits
The Best At What We Do

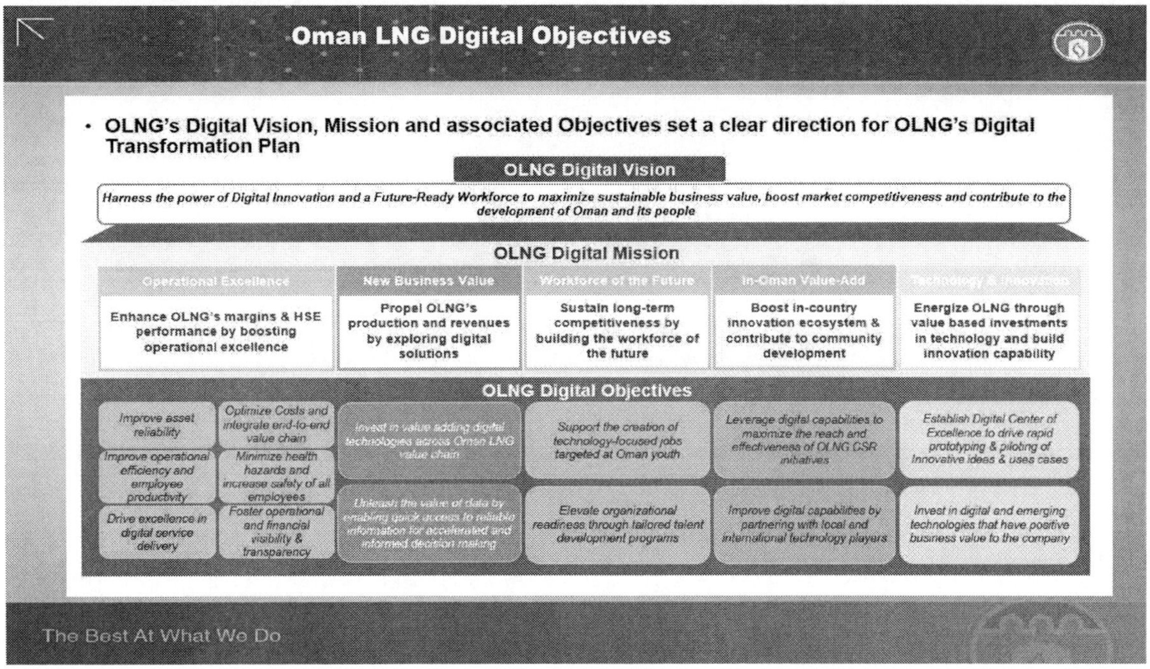

Oman LNG Digital Objectives
• OLNG's Digital Vision, Mission and associated Objectives set a clear direction for OLNG's Digital Transformation Plan
OLNG Digital Vision
Harness the power of Digital Innovation and a Future-Ready Workforce to maximize sustainable business value, boost market competitiveness and contribute to the development of Oman and its people
OLNG Digital Mission
Operational Excellence
Enhance OLNG's margins & HSE performance by boosting operational excellence
New Business Value
Propel OLNG's production and revenues by exploring digital solutions
Workforce of the Future
Sustain long-term competitiveness by building the workforce of the future
In-Oman Value-Add
Boost in-country innovation ecosystem & contribute to community development
Technology & Innovation
Energize OLNG through value based investments in technology and build innovation capability
OLNG Digital Objectives
Improve asset reliability
Optimize Costs and integrate end-to-end value chain
Improve operational efficiency and employee productivity
Minimize health hazards and increase safety of all employees
Drive excellence in digital service delivery
Foster operational and financial visibility & transparency
Invest in value adding digital technologies across Oman LNG value chain
Unleash the value of data by enabling quick access to reliable information for accelerated and informed decision making
Support the creation of technology-focused jobs targeted at Oman youth
Elevate organizational readiness through tailored talent development programs
Leverage digital capabilities to maximize the reach and effectiveness of OLNG CSR initiatives
Improve digital capabilities by partnering with local and international technology players
Establish Digital Center of Excellence to drive rapid prototyping & piloting of Innovative ideas & uses cases
Invest in digital and emerging technologies that have positive business value to the company
The Best At What We Do

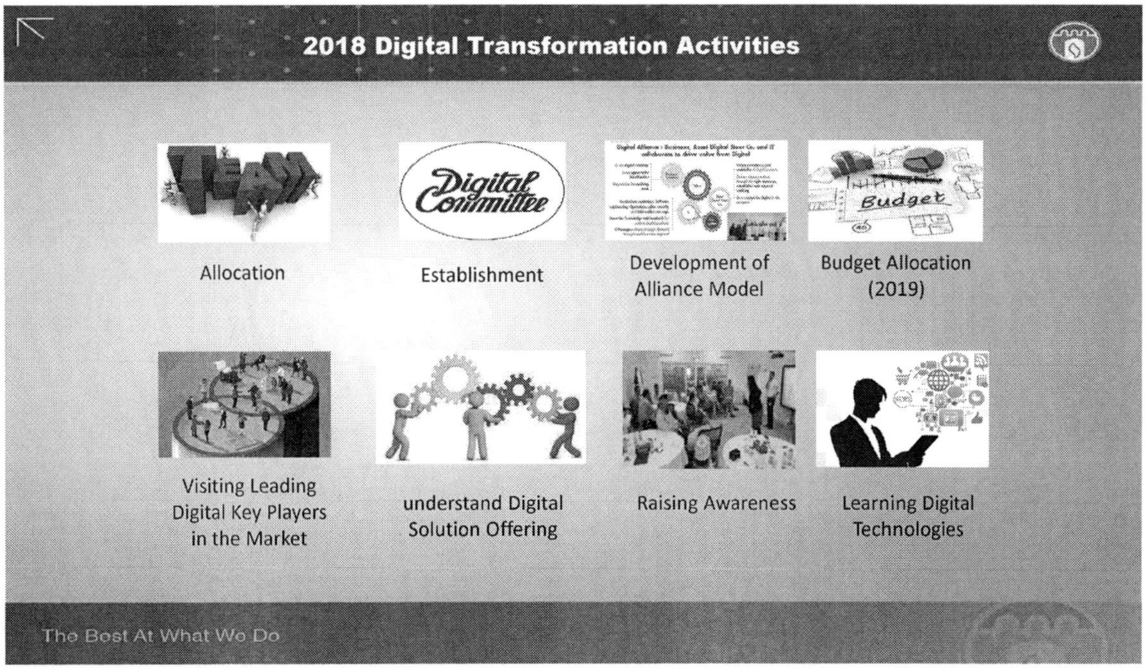

2018 Digital Transformation Activities
TEAM
Allocation
Digital Committee
Establishment
Development of Alliance Model
Budget
Budget Allocation (2019)
Visiting Leading Digital Key Players in the Market
understand Digital Solution Offering
Raising Awareness
Learning Digital Technologies
The Best At What We Do

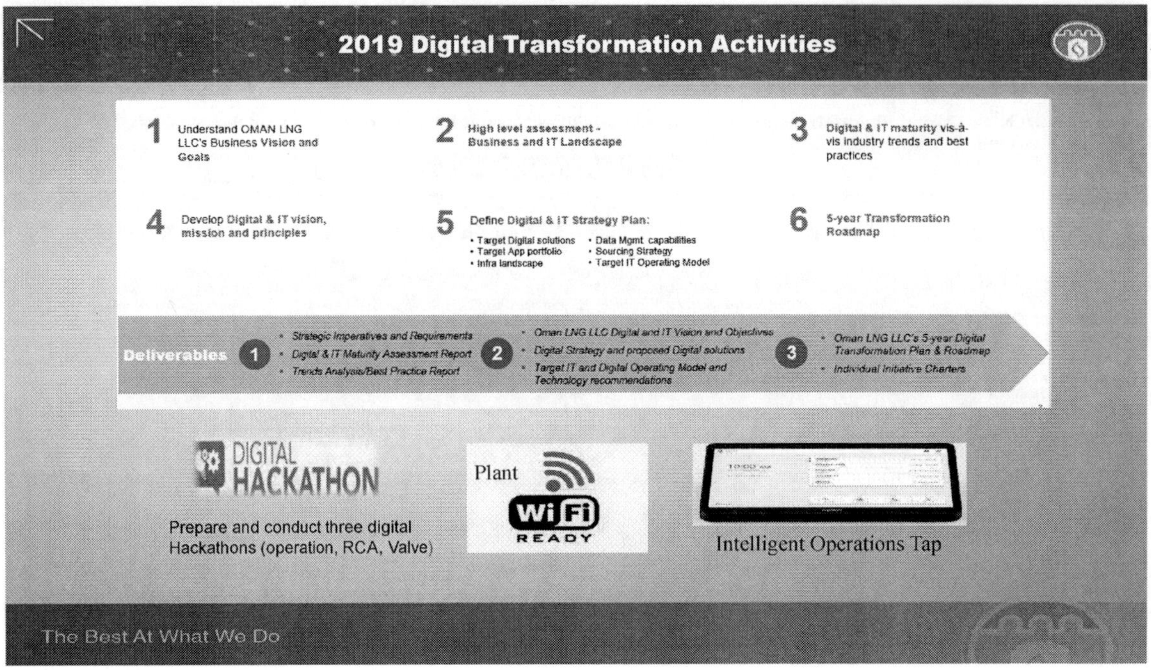
2019 Digital Transformation Activities
1 Understand OMAN LNG LLC's Business Vision and Goals
2 High level assessment - Business and IT Landscape
3 Digital & IT maturity vis-à-vis industry trends and best practices
4 Develop Digital & IT vision, mission and principles
5 Define Digital & IT Strategy Plan:
• Target Digital solutions • Target App portfolio • Infra landscape • Data Mgmt. capabilities • Sourcing Strategy • Target IT Operating Model
6 5-year Transformation Roadmap
Deliverables
1 • Strategic Imperatives and Requirements • Digital & IT Maturity Assessment Report • Trends Analysis/Best Practice Report
2 • Oman LNG LLC Digital and IT Vision and Objectives • Digital Strategy and proposed Digital solutions • Target IT and Digital Operating Model and Technology recommendations
3 • Oman LNG LLC's 5-year Digital Transformation Plan & Roadmap • Individual Initiative Charters
DIGITAL HACKATHON
Prepare and conduct three digital Hackathons (operation, RCA, Valve)
Plant
WiFi READY
Intelligent Operations Tap
The Best At What We Do

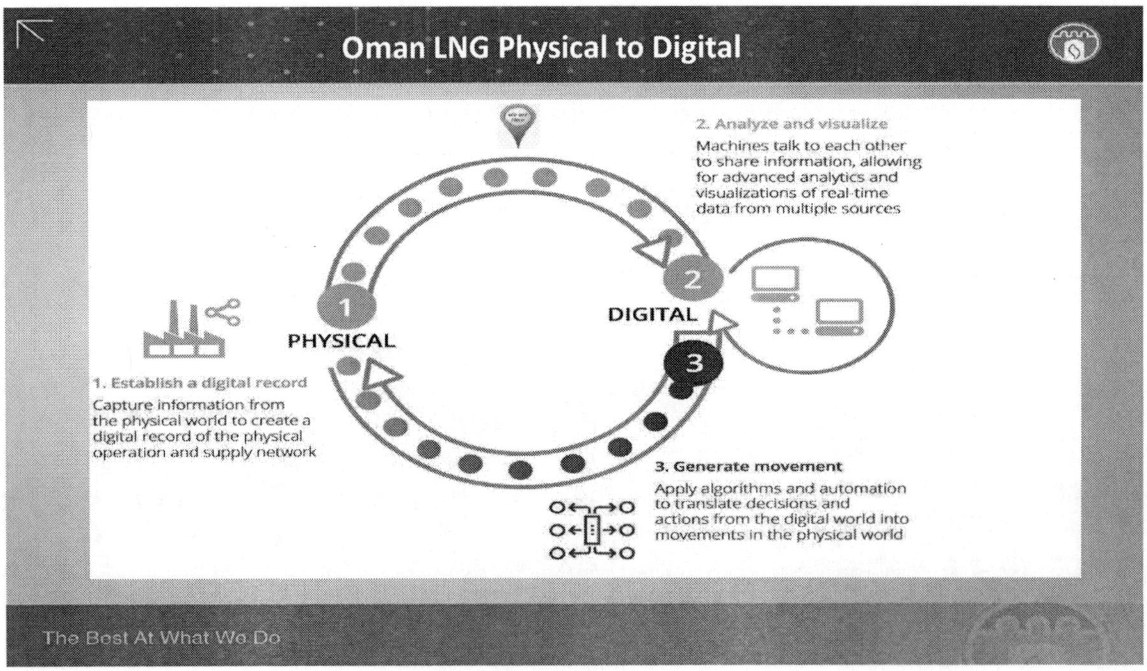
Oman LNG Physical to Digital
2. Analyze and visualize
Machines talk to each other to share information, allowing for advanced analytics and visualizations of real-time data from multiple sources
PHYSICAL
DIGITAL
1. Establish a digital record
Capture information from the physical world to create a digital record of the physical operation and supply network
3. Generate movement
Apply algorithms and automation to translate decisions and actions from the digital world into movements in the physical world
The Best At What We Do

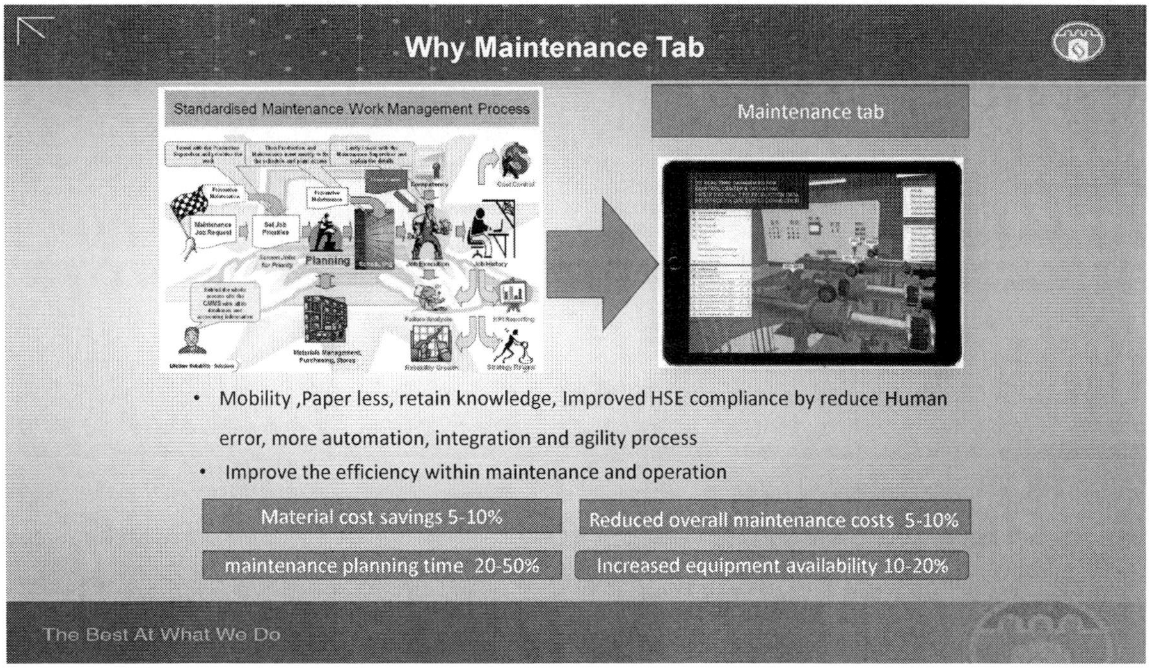
Why Maintenance Tab
Standardised Maintenance Work Management Process
Maintenance tab
• Mobility ,Paper less, retain knowledge, Improved HSE compliance by reduce Human error, more automation, integration and agility process
• Improve the efficiency within maintenance and operation
Material cost savings 5-10%
Reduced overall maintenance costs 5-10%
maintenance planning time 20-50%
Increased equipment availability 10-20%
The Best At What We Do

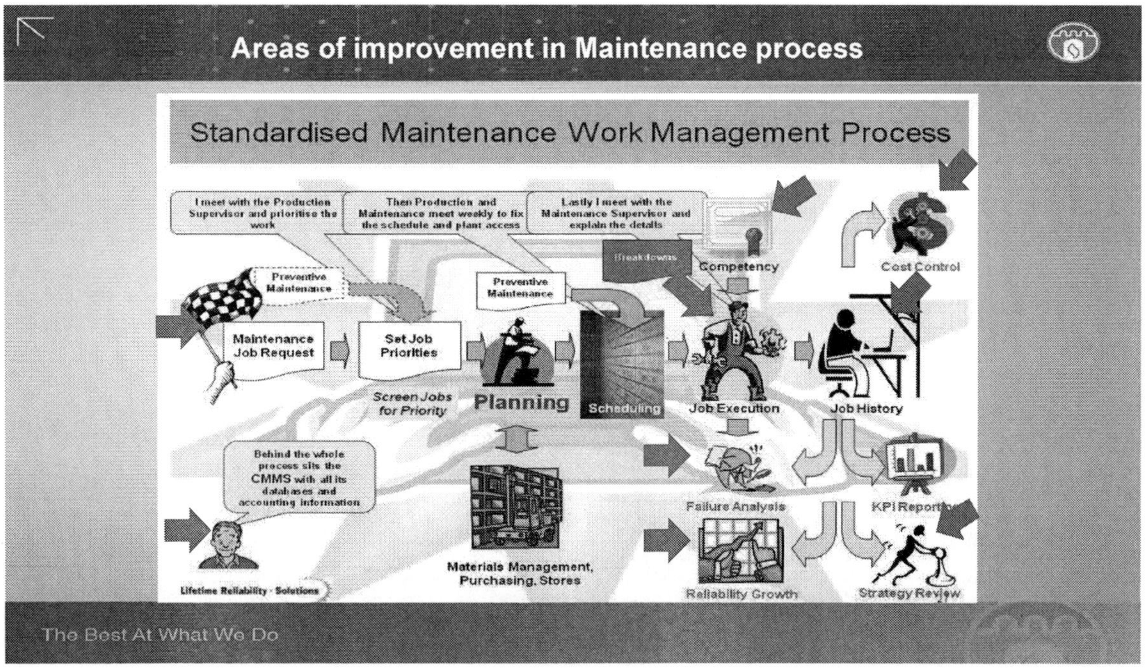
Areas of improvement in Maintenance process
Standardised Maintenance Work Management Process
I meet with the Production Supervisor and prioritise the work
Then Production and Maintenance meet weekly to fix the schedule and plant access
Lastly I meet with the Maintenance Supervisor and explain the details
Breakdowns
Competency
Cost Control
Preventive Maintenance
Preventive Maintenance
Maintenance Job Request
Set Job Priorities
Screen Jobs for Priority
Planning
Scheduling
Job Execution
Job History
Behind the whole process sits the CMMS with all its databases and accounting information
Failure Analysis
KPI Reporting
Materials Management, Purchasing, Stores
Reliability Growth
Strategy Review
Lifetime Reliability · Solutions
The Best At What We Do

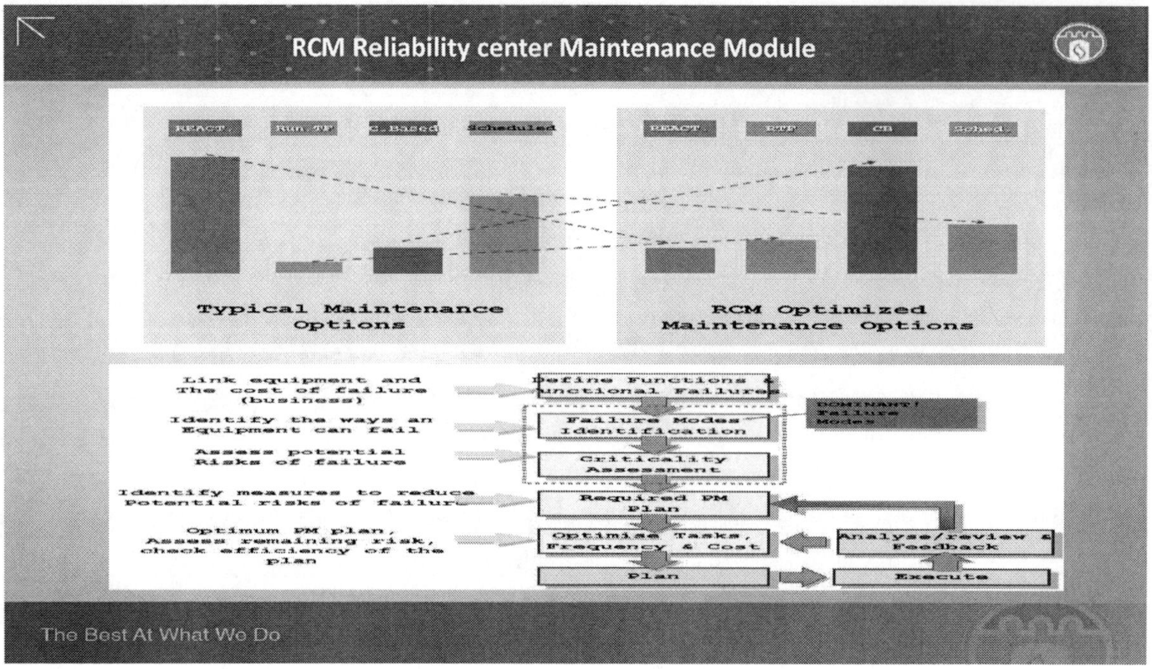

RCM Reliability center Maintenance Module
REACT.
Run.TF
C.Based
Scheduled
REACT
RTF
CB
Sched.
Typical Maintenance Options
RCM Optimized Maintenance Options
Link equipment and The cost of failure (business)
Identify the ways an Equipment can fail
Assess potential Risks of failure
Identify measures to reduce Potential risks of failure
Optimum PM plan, Assess remaining risk, check efficiency of the plan
Define Functions & Functional Failures
Failure Modes Identification
DOMINANT! Failure Modes
Criticality Assessment
Required PM Plan
Optimise Tasks, Frequency & Cost
Analyse/review & Feedback
Plan
Execute
The Best At What We Do

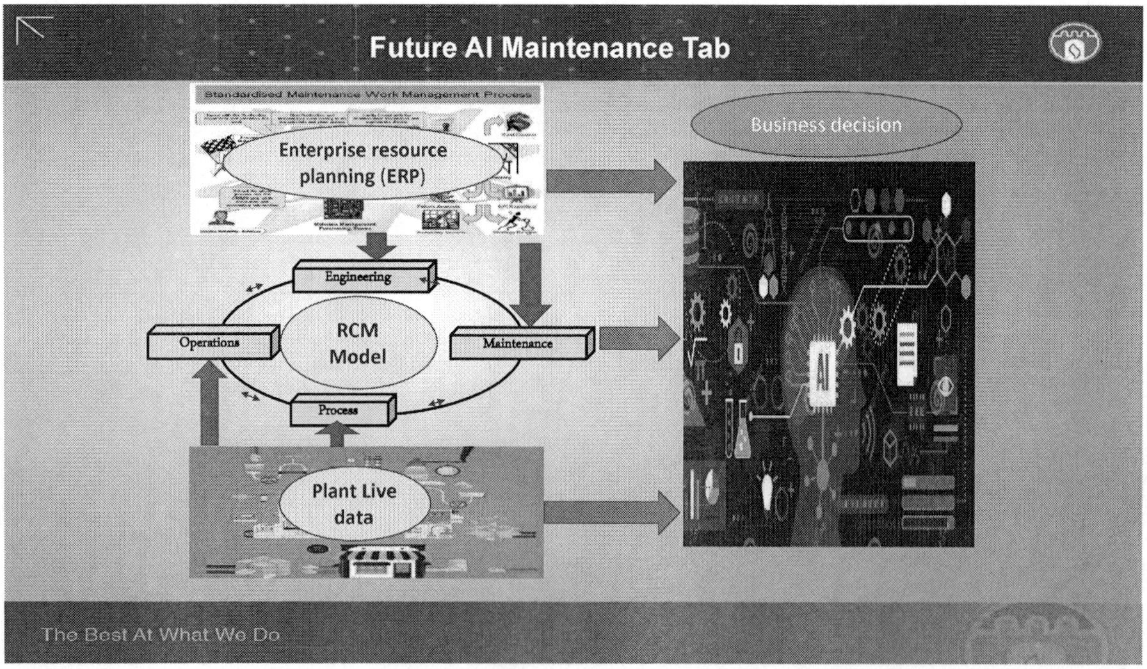

Future AI Maintenance Tab
Standardised Maintenance Work Management Process
Enterprise resource planning (ERP)
Business decision
Engineering
Operations
RCM Model
Maintenance
Process
Plant Live data
AI
The Best At What We Do

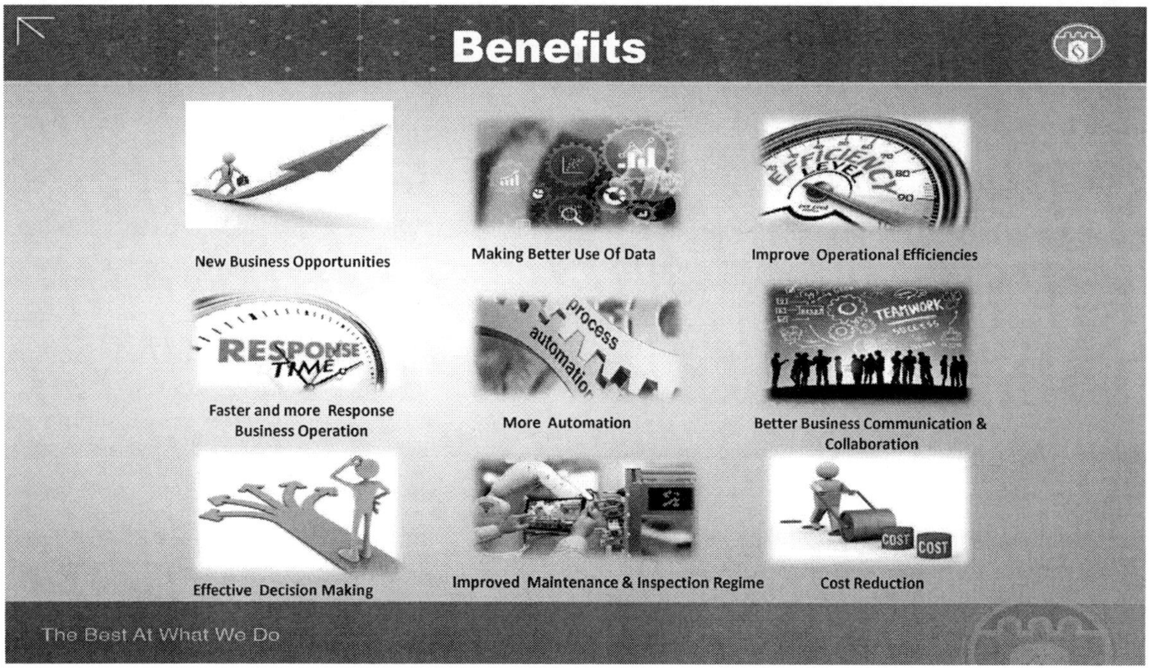
Benefits
New Business Opportunities
Making Better Use Of Data
Improve Operational Efficiencies
Faster and more Response Business Operation
More Automation
Better Business Communication & Collaboration
Effective Decision Making
Improved Maintenance & Inspection Regime
Cost Reduction
The Best At What We Do

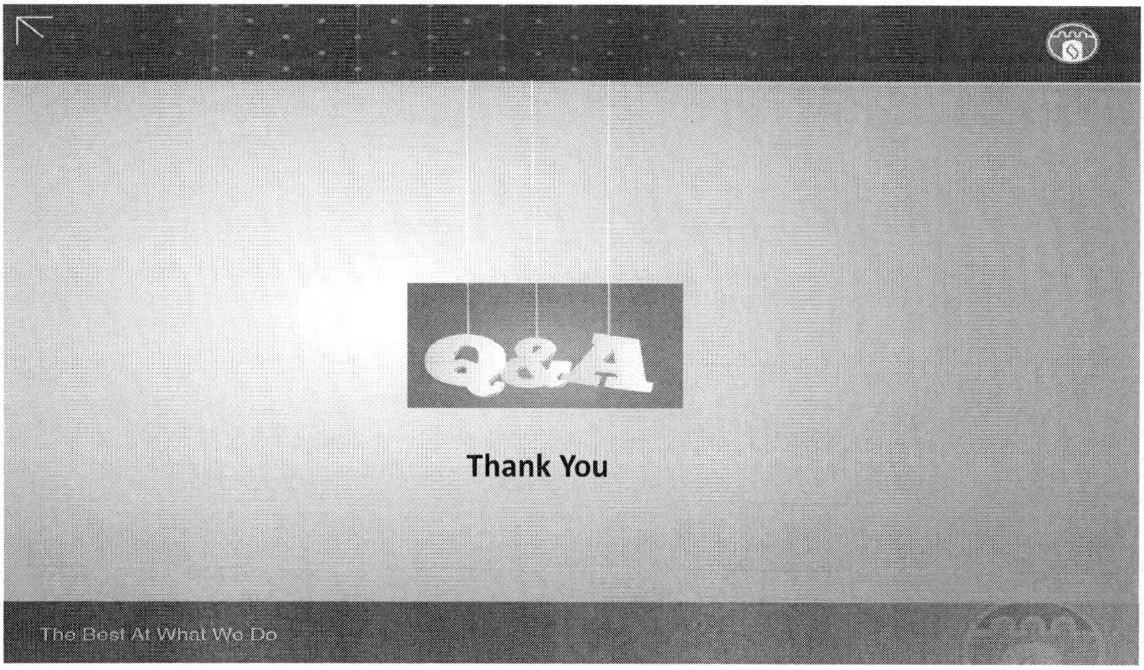
Q&A
Thank You
The Best At What We Do

العمانية للغاز الطبيعي المسال
Oman LNG

Variable cycling for molecular sieve dehydration

International Gas Union Research Conference 2020
Oman convention & exhibition center

Presenter:
Hamed Al Ismaili
Process technologies

Thursday, March 05, 2020

The best at what we do.

Introduction > Process overview > Fixed cycle > Variable cycle > Fuel gas savings > Lifetime extension > Conclusion > Q&A

Dehydration

- Absorption by glycol
- Refrigeration with hydrate inhibitors
- Solid bed adsorption

1. Footnote

2

Thursday, March 05, 2020

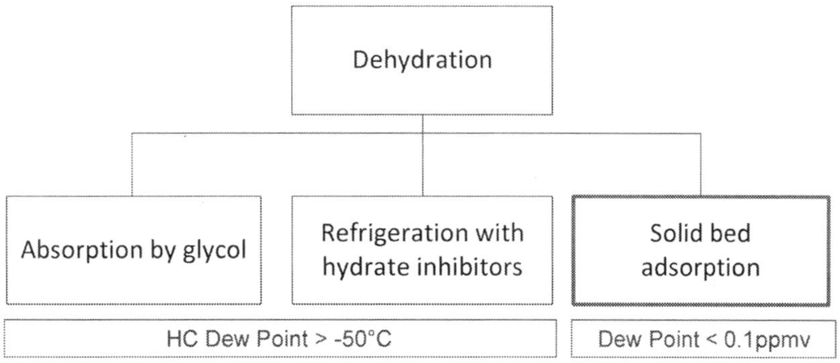

1. Footnote

Key definitions

- Adsorbent, Adsorbate
- Adsorption cycle, regeneration cycle
- Fixed cycle, Variable cycle.

1. Footnote

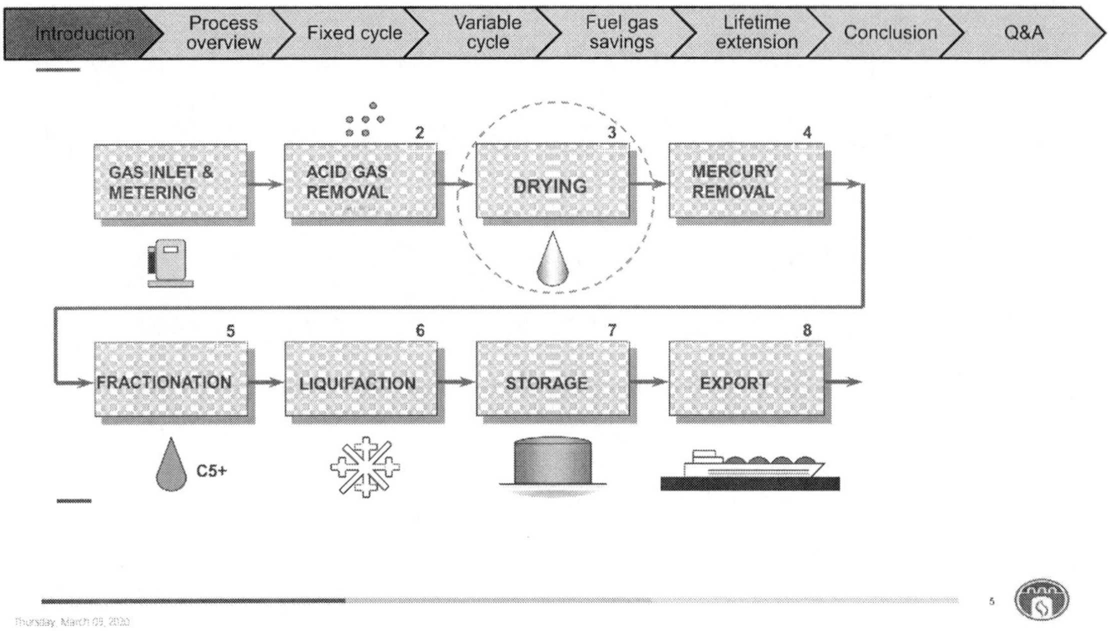
Introduction
Process overview
Fixed cycle
Variable cycle
Fuel gas savings
Lifetime extension
Conclusion
Q&A
GAS INLET & METERING
2
ACID GAS REMOVAL
3
DRYING
4
MERCURY REMOVAL
5
FRACTIONATION
6
LIQUIFACTION
7
STORAGE
8
EXPORT
C5+
5
Thursday, March 05, 2020

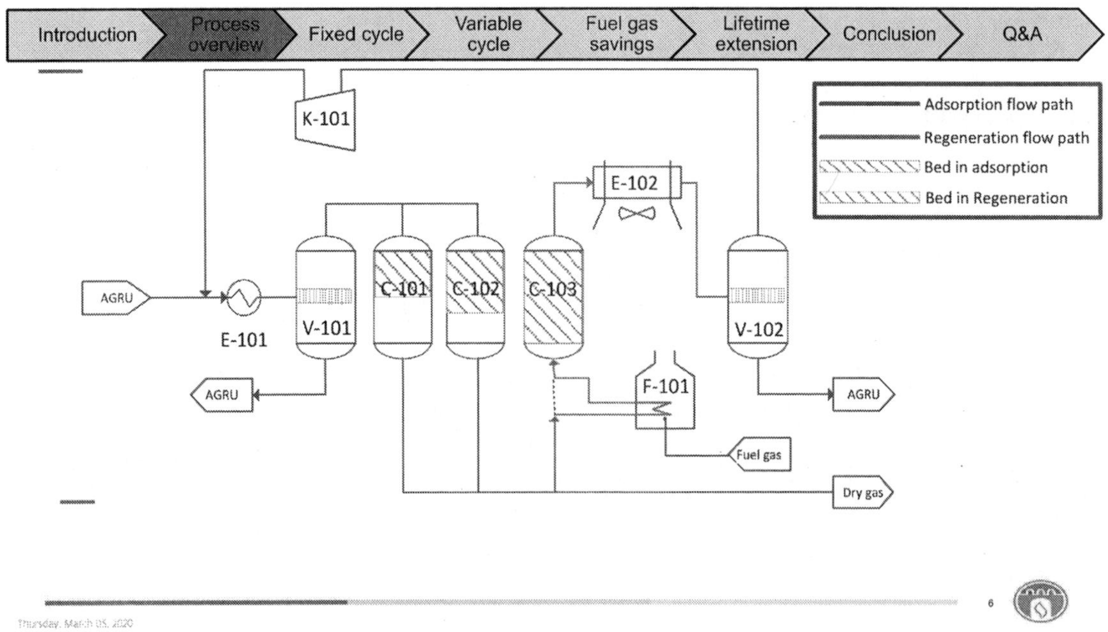
Introduction
Process overview
Fixed cycle
Variable cycle
Fuel gas savings
Lifetime extension
Conclusion
Q&A
Adsorption flow path
Regeneration flow path
Bed in adsorption
Bed in Regeneration
K-101
E-102
AGRU
E-101
V-101
C-101
C-102
C-103
V-102
AGRU
F-101
AGRU
Fuel gas
Dry gas
6
Thursday, March 05, 2020

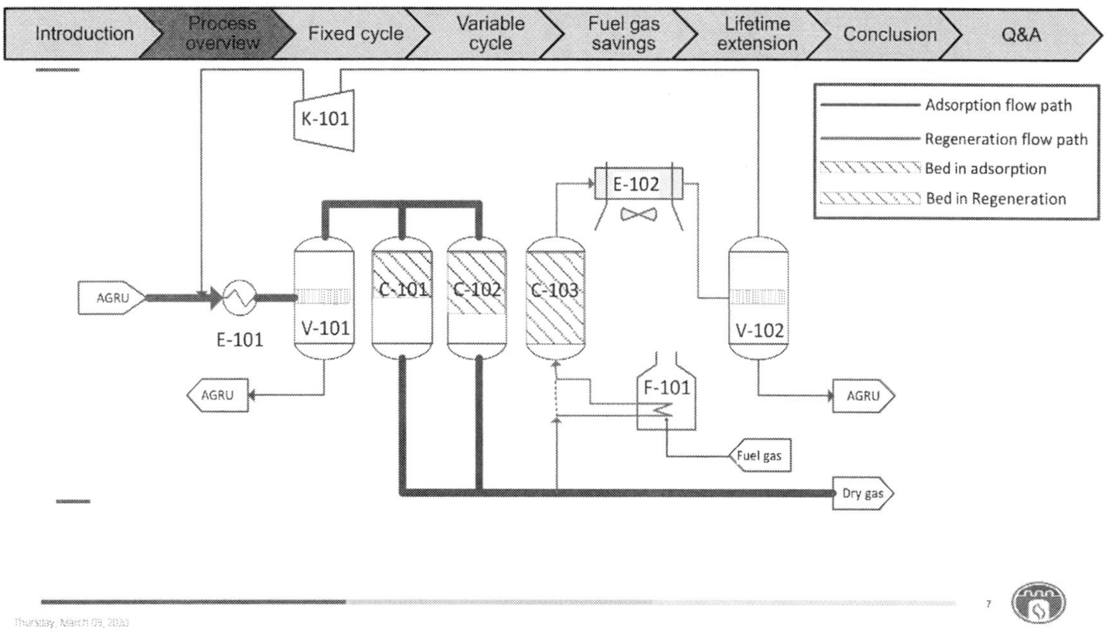
Introduction
Process overview
Fixed cycle
Variable cycle
Fuel gas savings
Lifetime extension
Conclusion
Q&A
Adsorption flow path
Regeneration flow path
Bed in adsorption
Bed in Regeneration
K-101
E-102
C-101
C-102
C-103
V-101
V-102
E-101
F-101
AGRU
AGRU
AGRU
Fuel gas
Dry gas
7

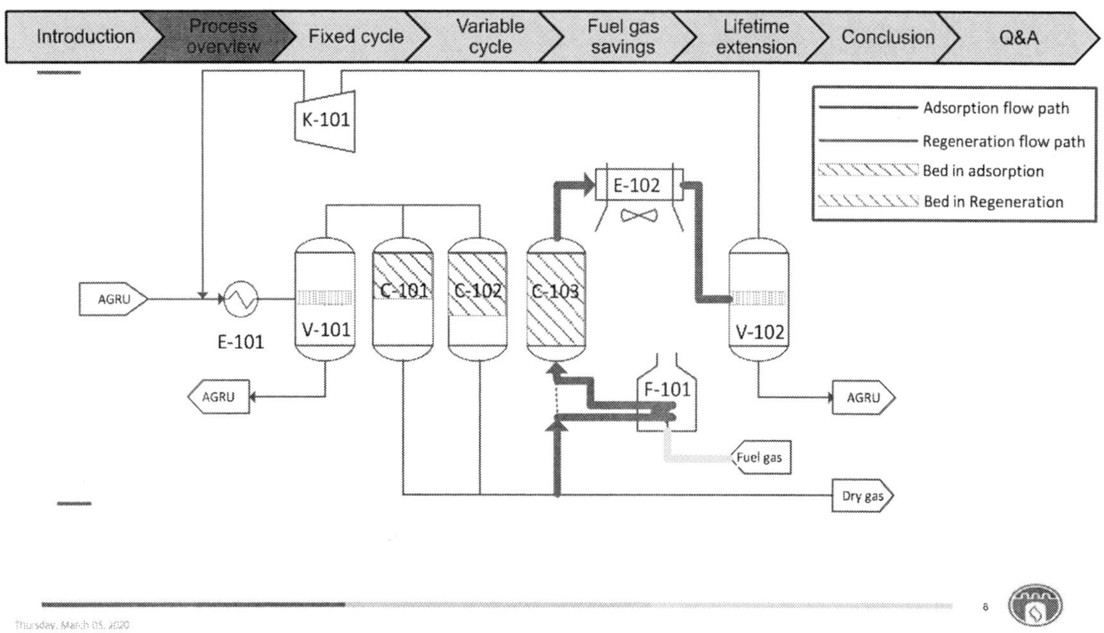
Introduction
Process overview
Fixed cycle
Variable cycle
Fuel gas savings
Lifetime extension
Conclusion
Q&A
Adsorption flow path
Regeneration flow path
Bed in adsorption
Bed in Regeneration
K-101
E-102
C-101
C-102
C-103
V-101
V-102
E-101
F-101
AGRU
AGRU
AGRU
Fuel gas
Dry gas
8
Thursday, March 05, 2020

Introduction | Process overview | Fixed cycle | Variable cycle | Fuel gas savings | Lifetime extension | Conclusion | Q&A

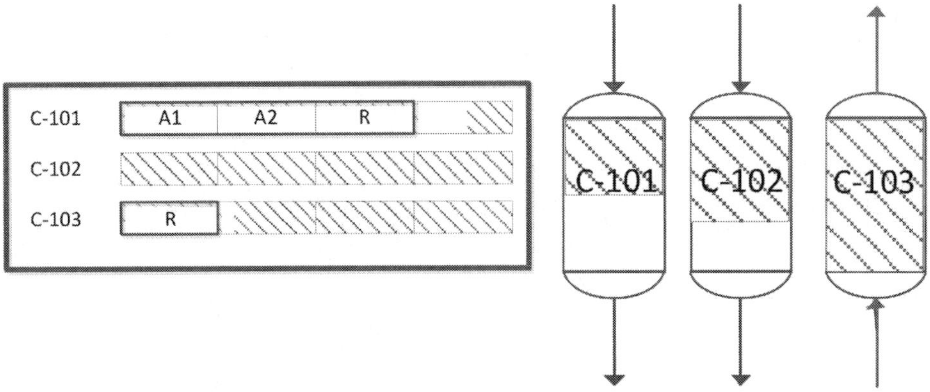

Introduction | Process overview | Fixed cycle | Variable cycle | Fuel gas savings | Lifetime extension | Conclusion | Q&A

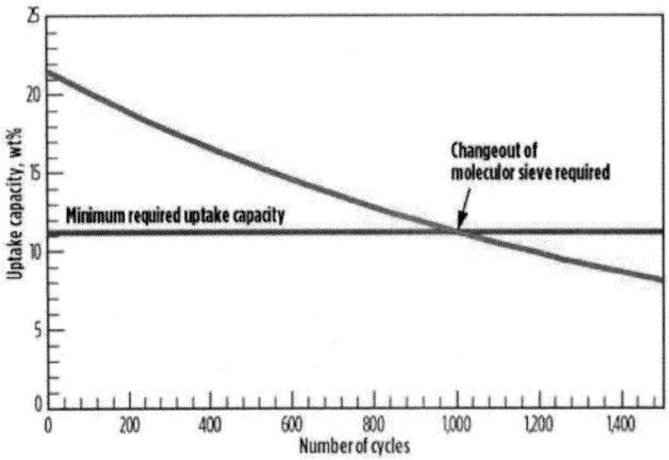

Herold, R., Mokhatab S. *Optimal design and operation of molecular sieve gas dehydration units—Part 1*, Amsterdam, The Netherlands, 2017, http://gasprocessingnews.com

Introduction | Process overview | Fixed cycle | Variable cycle | Fuel gas savings | Lifetime extension | Conclusion | Q&A

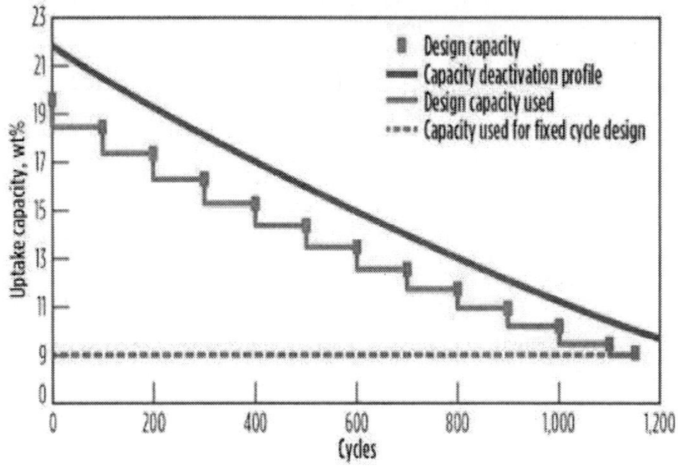

Herold, R., Mokhatab S. *Optimal design and operation of molecular sieve gas dehydration units—Part 1*, Amsterdam, The Netherlands, 2017, http://gasprocessingnews.com

11

Thursday, March 05, 2020

Introduction | Process overview | Fixed cycle | Variable cycle | Fuel gas savings | Lifetime extension | Conclusion | Q&A

The benefits and challenges

Benefits of adoption

1- Lifetime extension of the molecular sieves
2- Fuel gas savings & GHG emissions reduction

Challenges

1- Increased water load on the re-compressor knockout vessel
2- Increased likelihood of desorbed water condensation

12

Thursday, March 05, 2020

Introduction | Process overview | Fixed cycle | Variable cycle | Fuel gas savings | Lifetime extension | Conclusion | Q&A

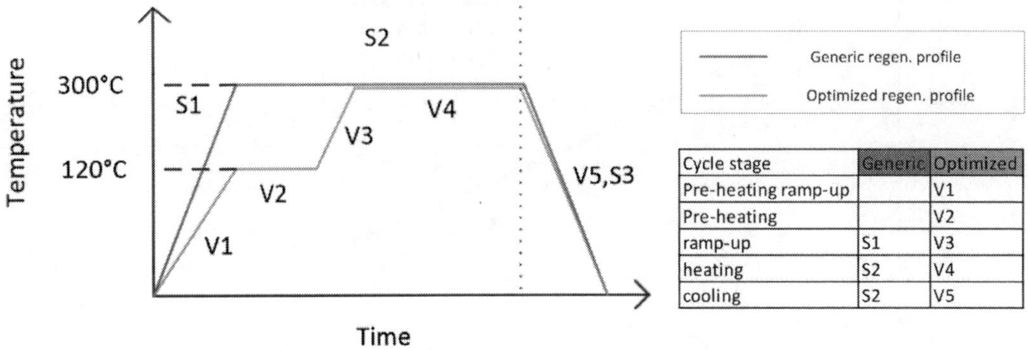

Cycle stage	Generic	Optimized
Pre-heating ramp-up		V1
Pre-heating		V2
ramp-up	S1	V3
heating	S2	V4
cooling	S2	V5

13

Thursday, March 05, 2020

Introduction | Process overview | Fixed cycle | Variable cycle | Fuel gas savings | Lifetime extension | Conclusion | Q&A

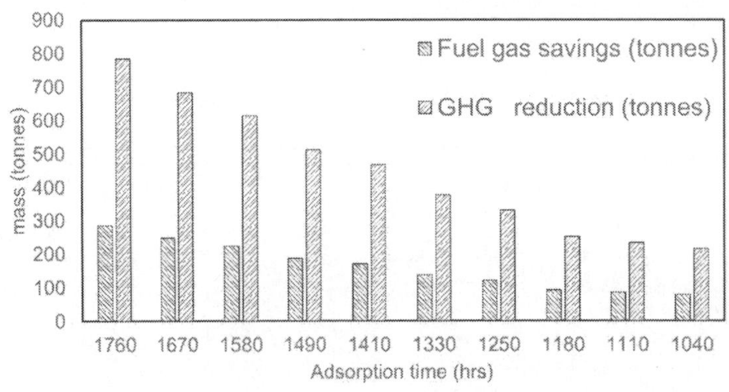

14

Thursday, March 05, 2020

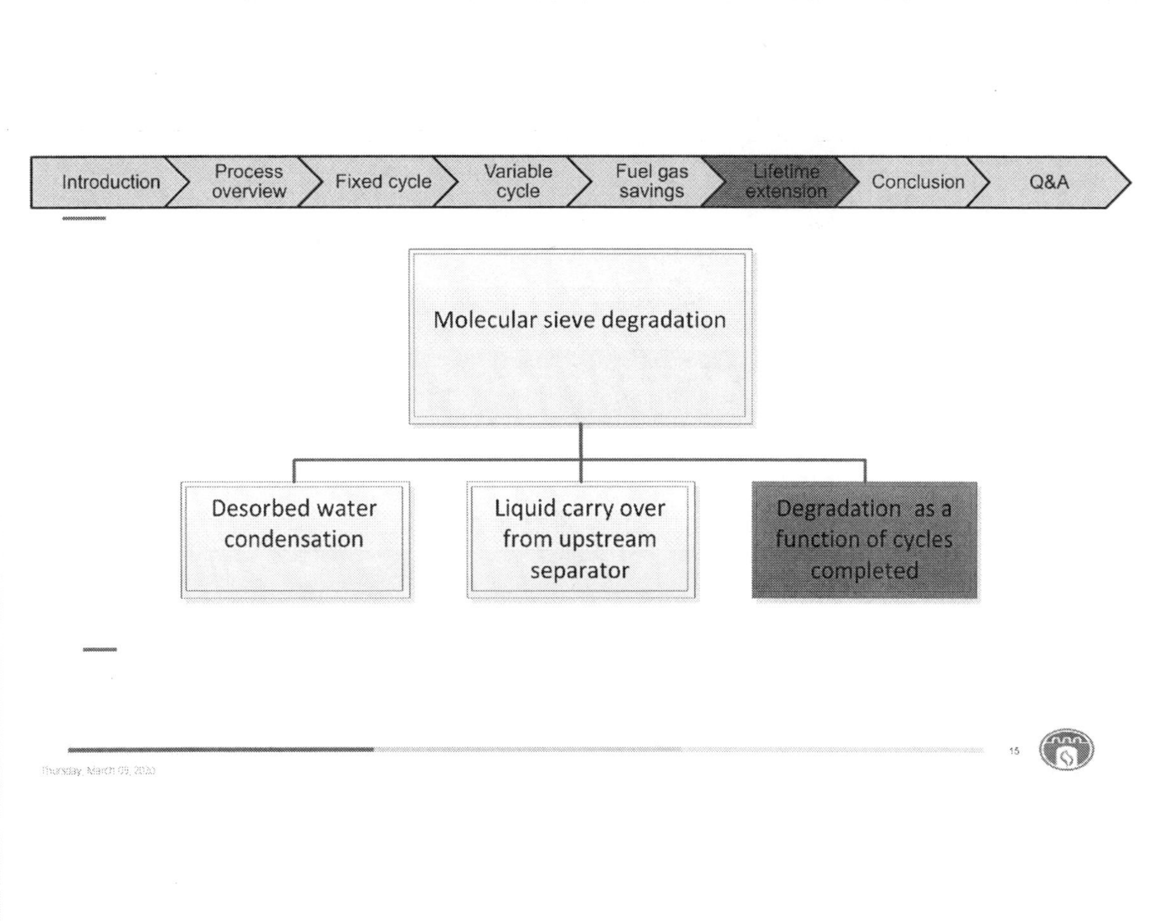
Introduction
Process overview
Fixed cycle
Variable cycle
Fuel gas savings
Lifetime extension
Conclusion
Q&A
Molecular sieve degradation
Desorbed water condensation
Liquid carry over from upstream separator
Degradation as a function of cycles completed
15
Thursday, March 05, 2020

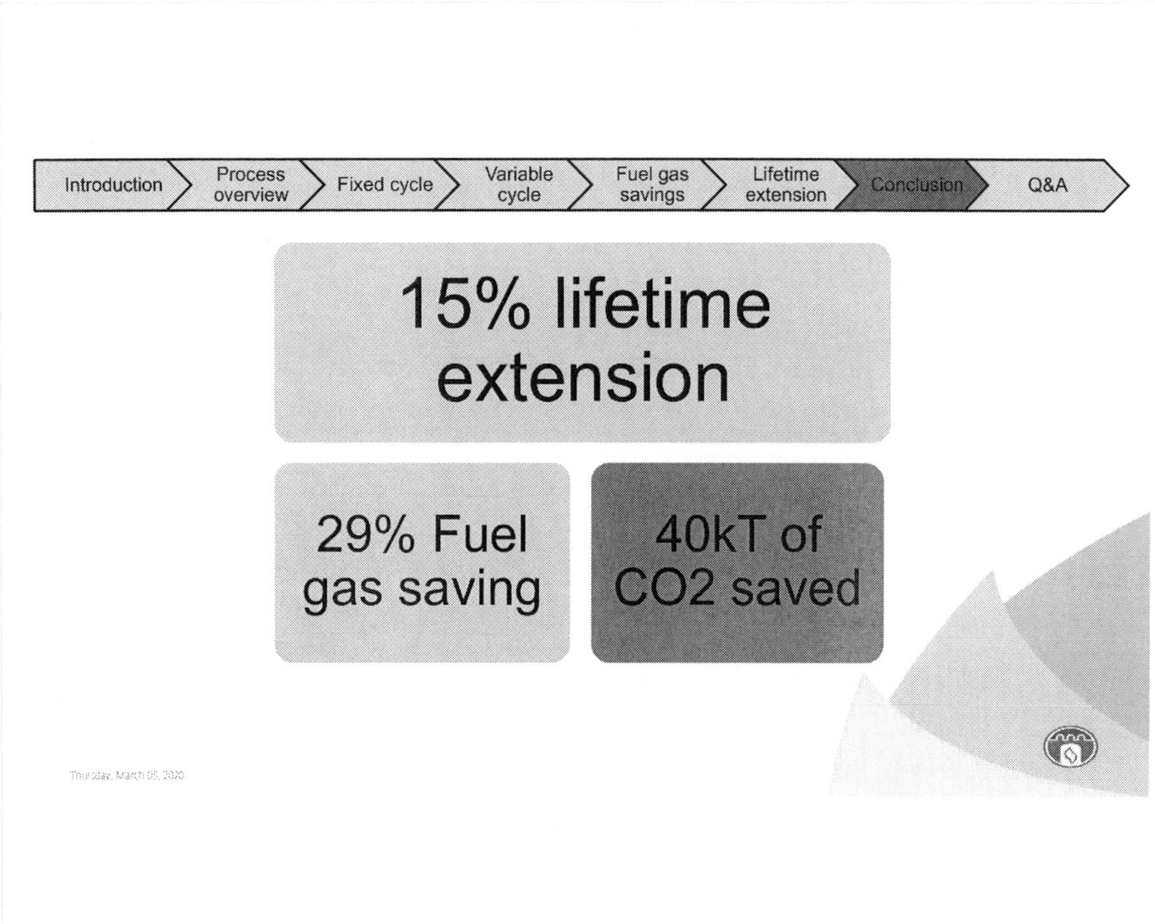
Introduction
Process overview
Fixed cycle
Variable cycle
Fuel gas savings
Lifetime extension
Conclusion
Q&A
15% lifetime extension
29% Fuel gas saving
40kT of CO2 saved
Thursday, March 05, 2020

Introduction | Process overview | Fixed cycle | Variable cycle | Fuel gas savings | Lifetime extension | Conclusion | Q&A

Thank you

Q&A

Thursday, March 05, 2020

Advancements in LNG Technology
ENGIE Lab CRIGEN
F. LEGRAND

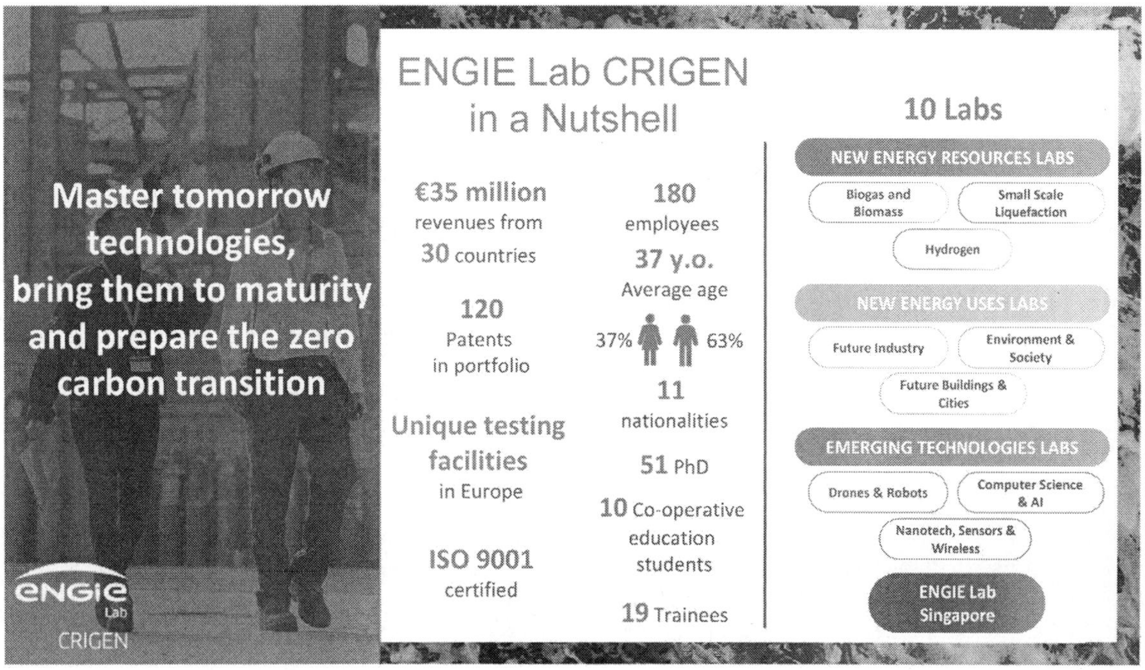
Master tomorrow technologies, bring them to maturity and prepare the zero carbon transition
ENGIE Lab CRIGEN
ENGIE Lab CRIGEN in a Nutshell
€35 million
revenues from
30 countries
120
Patents
in portfolio
Unique testing facilities
in Europe
ISO 9001
certified
180
employees
37 y.o.
Average age
37%
63%
11
nationalities
51 PhD
10 Co-operative education students
19 Trainees
10 Labs
NEW ENERGY RESOURCES LABS
Biogas and Biomass
Small Scale Liquefaction
Hydrogen
NEW ENERGY USES LABS
Future Industry
Environment & Society
Future Buildings & Cities
EMERGING TECHNOLOGIES LABS
Drones & Robots
Computer Science & AI
Nanotech, Sensors & Wireless
ENGIE Lab Singapore

ENGIE Lab LNG
Key figure

3 main fields of expertise

LNG behavior

Liquefaction technologies

Solutions for transport and distribution

A unique test facility

A partnership ecosystem

WÄRTSILÄ ARMINES MAN TOTAL

28 Patents

LNG behavior : 7

Liquefaction : 15

Transport and distribution : 6

Internationally recognized Software

LNG behavior :
Bunkering Software, LNGAUGE, LNG MASTER®, CompoCARGO®

Transport and distribution :
Optiretail, Stage

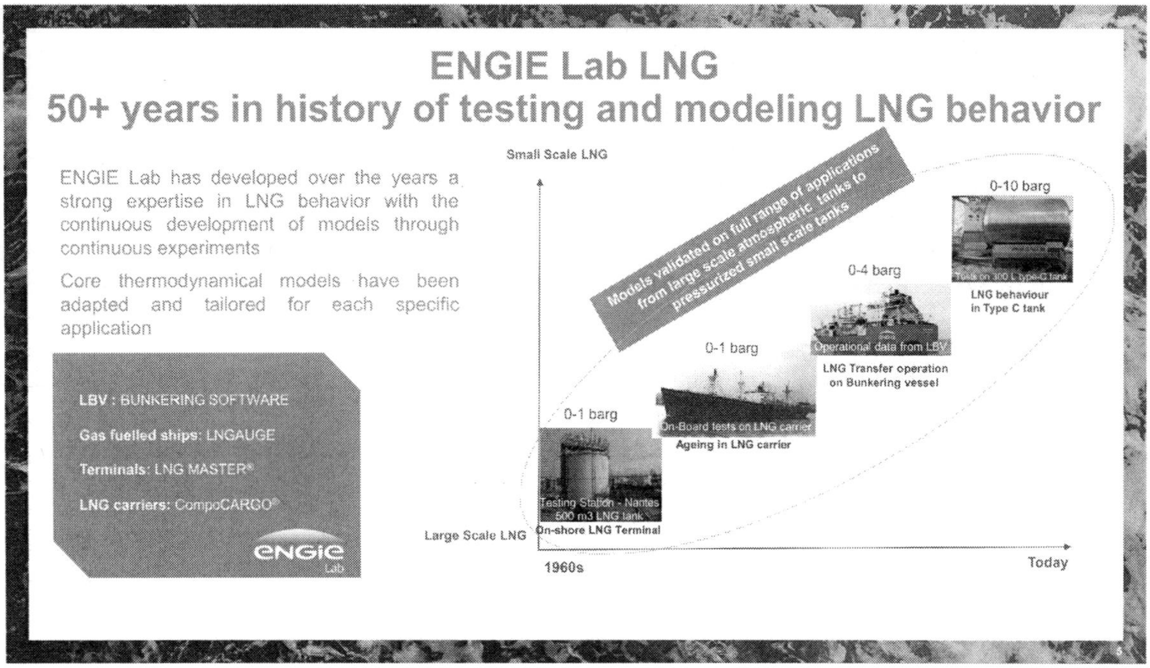
ENGIE Lab LNG
50+ years in history of testing and modeling LNG behavior
ENGIE Lab has developed over the years a strong expertise in LNG behavior with the continuous development of models through continuous experiments
Core thermodynamical models have been adapted and tailored for each specific application
LBV : BUNKERING SOFTWARE
Gas fuelled ships: LNGAUGE
Terminals: LNG MASTER®
LNG carriers: CompoCARGO®
Small Scale LNG
Large Scale LNG
1960s
Today
Models validated on full range of applications from large scale atmospheric tanks to pressurized small scale tanks
0-1 barg
Testing Station - Nantes 500 m3 LNG tank
On-shore LNG Terminal
0-1 barg
On-Board tests on LNG carrier
Ageing in LNG carrier
0-4 barg
Operational data from LBV
LNG Transfer operation on Bunkering vessel
0-10 barg
LNG behaviour in Type C tank

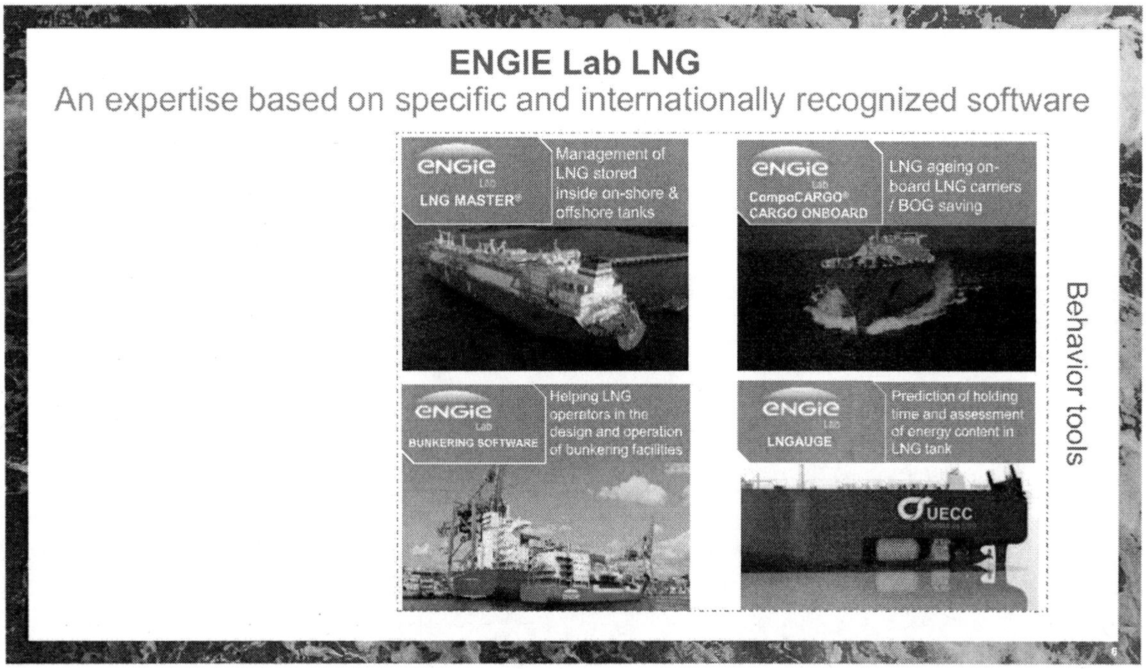
ENGIE Lab LNG
An expertise based on specific and internationally recognized software
LNG MASTER®
Management of LNG stored inside on-shore & offshore tanks
CompoCARGO® CARGO ONBOARD
LNG ageing on-board LNG carriers / BOG saving
BUNKERING SOFTWARE
Helping LNG operators in the design and operation of bunkering facilities
LNGAUGE
Prediction of holding time and assessment of energy content in LNG tank
UECC
Behavior tools

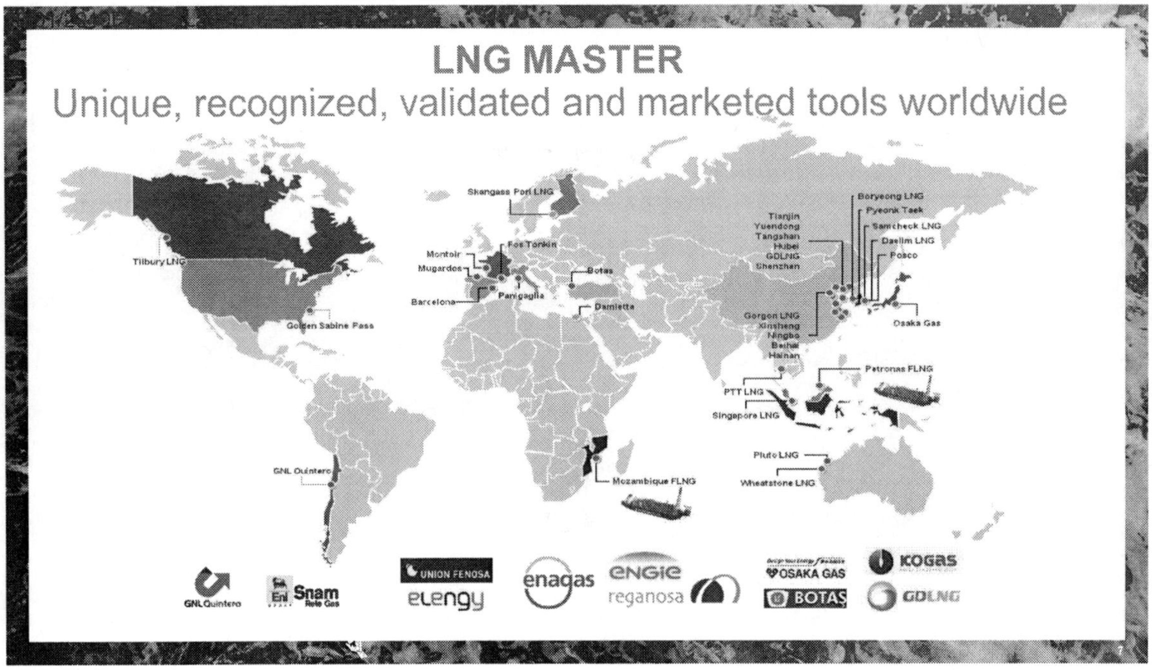
LNG MASTER
Unique, recognized, validated and marketed tools worldwide
Skangass Port LNG
Tilbury LNG
Golden Sabine Pass
Montoir
Mugardos
Fos Tonkin
Barcelona
Panigaglia
Botas
Damietta
GNL Quintero
Mozambique FLNG
Tianjin
Yuendong
Tangshan
Hubei
GDLNG
Shenzhen
Boryeong LNG
Pyeonk Taek
Samcheck LNG
Daelim LNG
Posco
Osaka Gas
Gorgon LNG
Xinsheng
Ningbo
Beihai
Hainan
Petronas FLNG
PTT LNG
Singapore LNG
Pluto LNG
Wheatstone LNG
7

Solutions for transport and distribution

Optiretail: digital innovation for downstream LNG

- ✓ **Design the optimal LNG supply** chain from sourcing to delivery by minimizing the total cost of the project (CAPEX and OPEX).
- ✓ **Applications for all types of SSLNG projects** (LNG-to-Power, LNG-as-fuel for road and marine transportation, LNG-for-industrials)
- ✓ Optiretail combines deep LNG and applied mathematics expertise

Optiretail is LNG Innovation award winner of the 27th World Gas Conference.

27th WORLD GAS CONFERENCE
WASHINGTON DC
JUNE 25-29 2018
IGU

9

Optiretail: digital innovation for downstream LNG

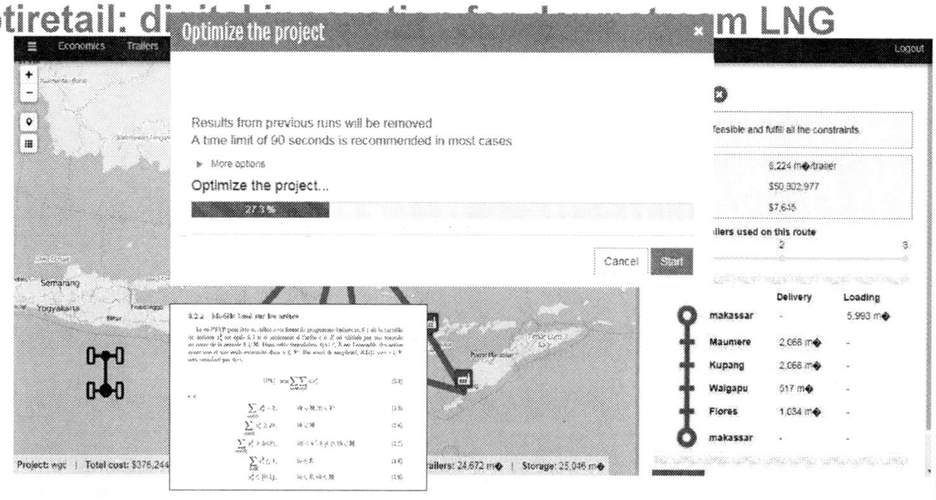

LocalSolver

10

Liquefaction solutions
11

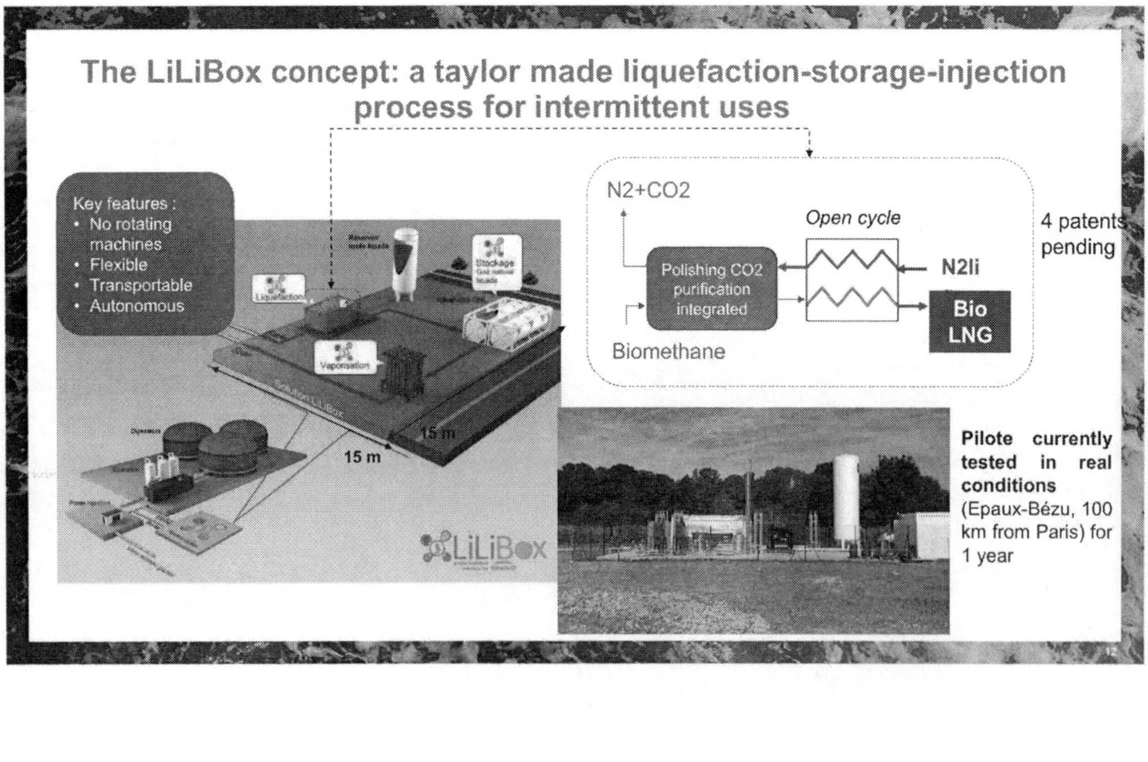
The LiLiBox concept: a taylor made liquefaction-storage-injection process for intermittent uses
Key features :
• No rotating machines
• Flexible
• Transportable
• Autonomous
Liquefaction
Stockage
Vaporisation
15 m
15 m
LiLiBox
N2+CO2
Open cycle
Polishing CO2 purification integrated
N2li
Bio LNG
Biomethane
4 patents pending
Pilote currently tested in real conditions (Epaux-Bézu, 100 km from Paris) for 1 year
12

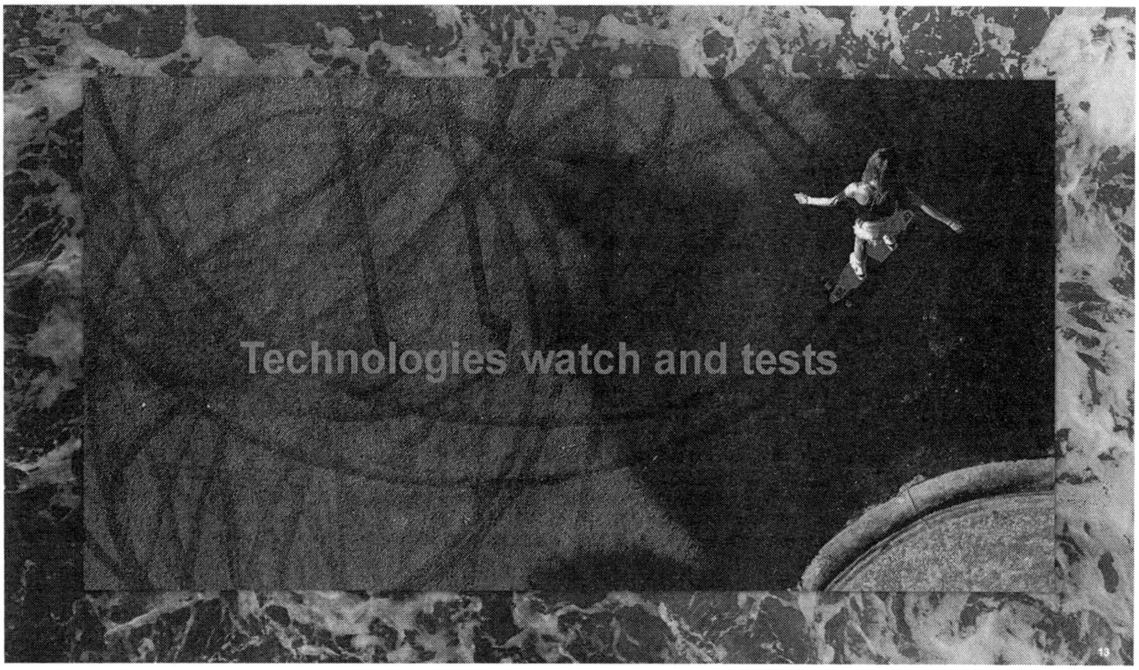
Technologies watch and tests
13

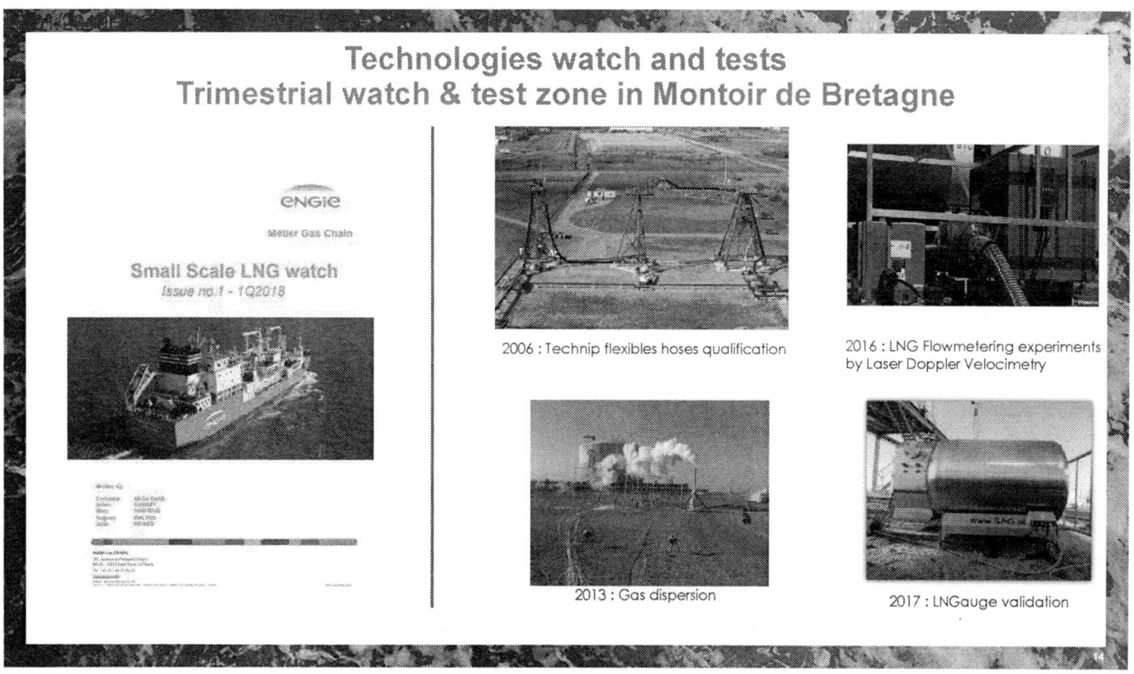
Technologies watch and tests
Trimestrial watch & test zone in Montoir de Bretagne
ENGIE
Metier Gas Chain
Small Scale LNG watch
Issue no.1 - 1Q2018
2006 : Technip flexibles hoses qualification
2016 : LNG Flowmetering experiments by Laser Doppler Velocimetry
2013 : Gas dispersion
2017 : LNGauge validation
14

ENGIE
Lab
CRIGEN

IGRC 2020 MUSCAT

TOHO GAS DENSO

Development of Solid Oxide Fuel Cell System for Commercial Use at TOHO GAS and DENSO CORPORATION

Yuichi Sasamata , Staff Engineer
Fuel Cell, Cogeneration & Air-conditioning System Technology Group
Technical Research Institute, TOHO GAS CO., LTD.

1

Main Topics

1. Company Profile of TOHO GAS
2. Feature of SOFCs
3. Appearance & Feature of 5kW SOFC system
4. Performance test results of 5kW SOFC power generation module

2

Company Profile of TOHO GAS

Chubu area (Nagoya City)

City Gas Provider

TOHO GAS

Automotive industry

TOYOTA

DENSO

AISIN

Ceramic industry

NTK TECHNICAL CERAMICS

NGK

Noritake

3

TOHO GAS's Business of Related to Fuel Cell

1. Sales & maintenance service of fuel cell mchp for residential, "ENE-FARM"
2. Operation of H_2 refueling station for Fuel Cell Vehicle(FCV)
3. Joint development of a 5kW class Solid Oxide Fuel Cell(SOFC) system for commercial use with DENSO

1. ENE-FARM(0.7kW)

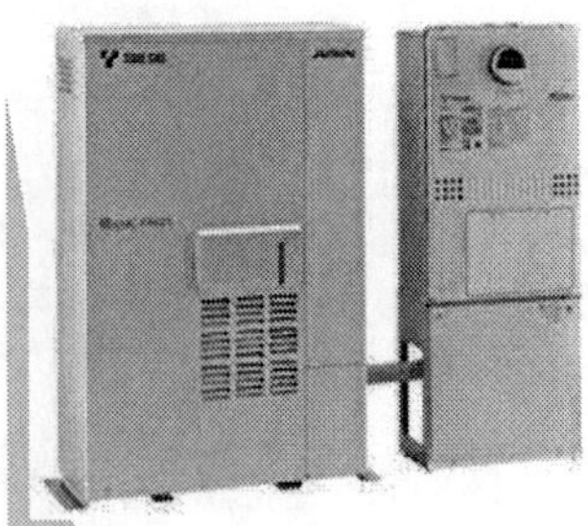

2. H_2 refueling station

3. 5kW class SOFC system

4

About SOFC

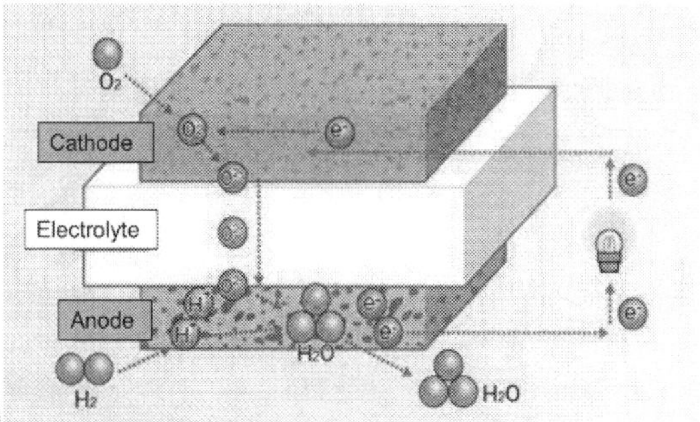

Operating principle of SOFCs

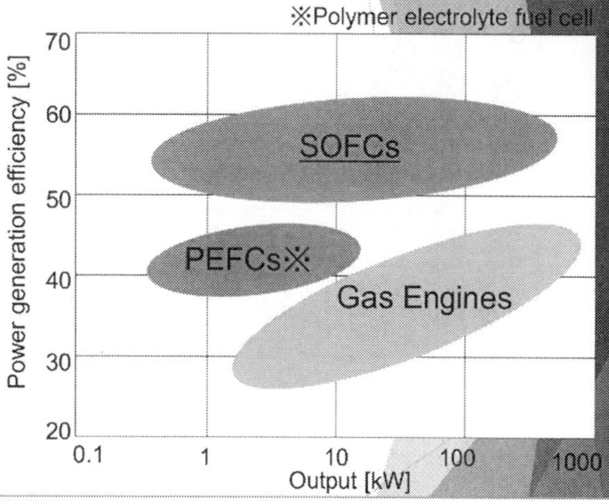

Power generation efficiency of Fuel cells and Gas Engines

SOFCs' efficiency are higher than PEFCs' and Gas Engines'

SOFC Systems in Japan

Company	AISIN	Kyocera	Miura Industries	DENSO	Hitachi Zosen
Output	0.7kW	3kW	4.2kW	4.5kW	20kW
Power generation efficiency(LHV)	52%	52%	50%	60% or more	52%
System total efficiency(LHV)	87%	90%	90%	(under consideration)	90%
Appearance	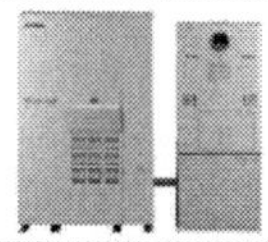	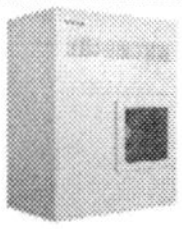			

Company	Fuji Electronics	Mitsubishi Hitachi Power Systems
Output	50kW	220kW
Power generation efficiency(LHV)	55%	55%
System total efficiency(LHV)	85%	73% (hot water) 65% (steam)
Appearance	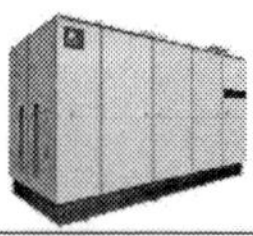	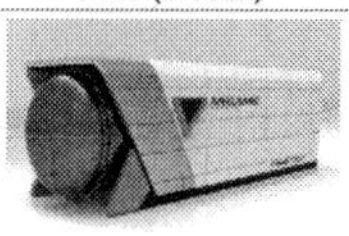

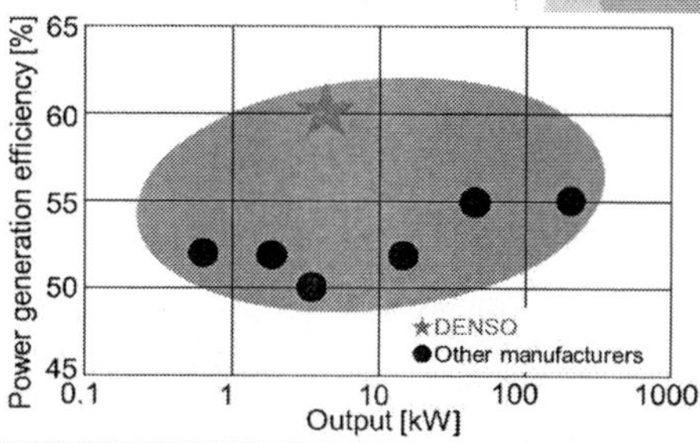

Power generation efficiency of SOFC systems

Overview of the 5kW Class SOFC System

5kW class SOFC System

Conceptual image of SOFC System

Target Specification of the 5kW Class SOFC System

Power Output	4.5kW
Power generating efficiency	AC 60% or more
Gas	City Gas / LPG

7

Benefits of System Introduction

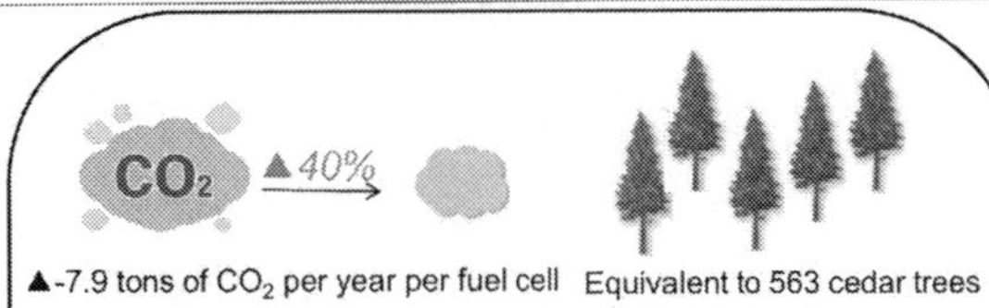

Reduction volume of CO_2 emission

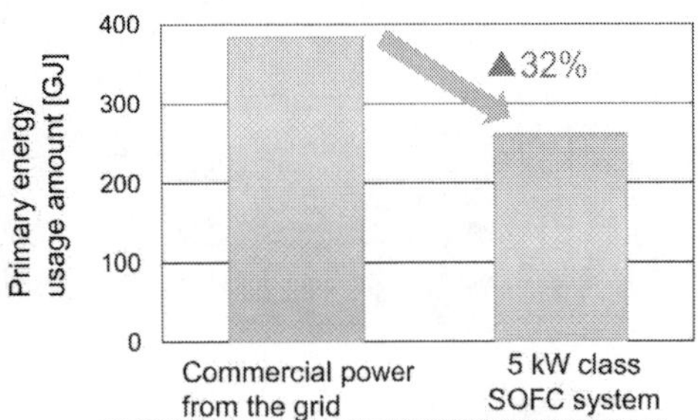

Reduction volume of Primary energy

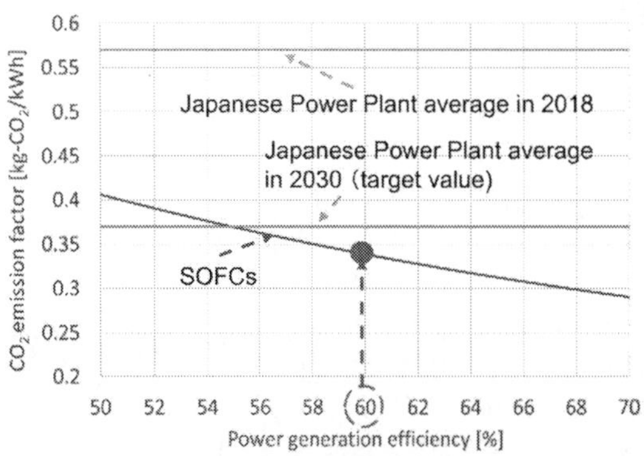

Compilation of CO_2 emission factor

5kW class SOFC System can reduce CO_2 emissions and its emission factor is lower than the average 2030 target of average emissions from the grid.

8

Concept of Power Generation Module

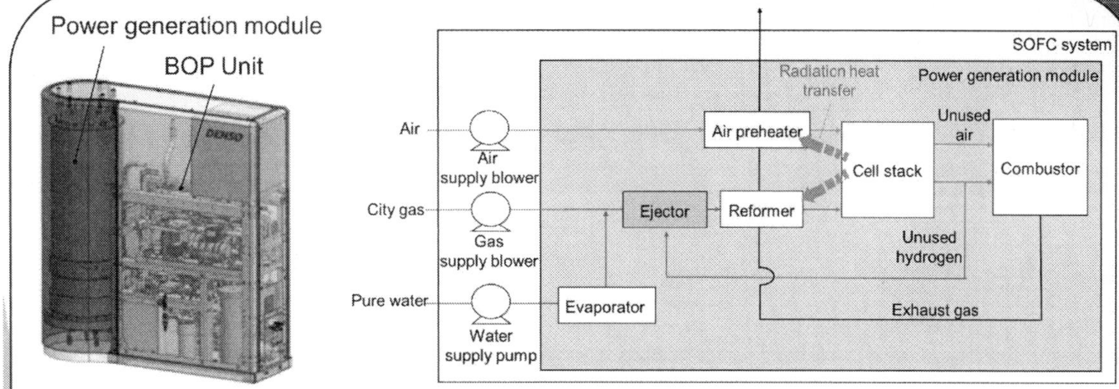

Schematic of 5kW Class SOFC System

Flow diagram of Power generation module

To increase power generation efficiency we use radiation heat transfer from cell stack, and an ejector.

9

Design of Power Generation Module

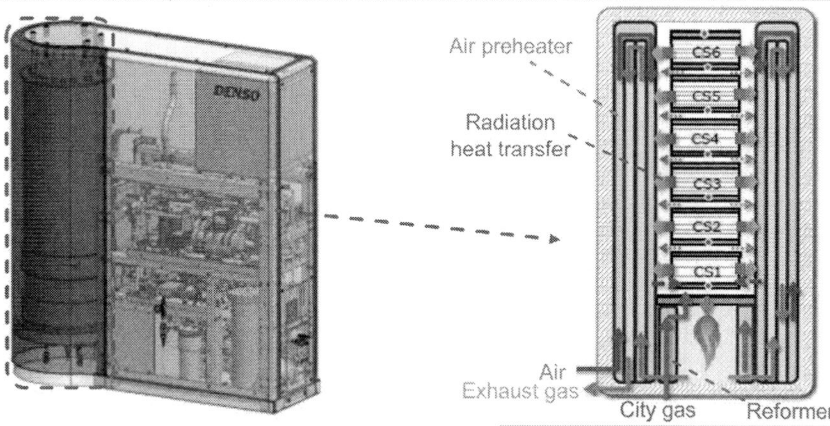

Schematic Power Generation Module

Using radiation heat transfer form cell stack. → The amount of air flow required for temperature control is reduced and the blower power can be reduced.

10

Fuel Recycling by Using Ejector

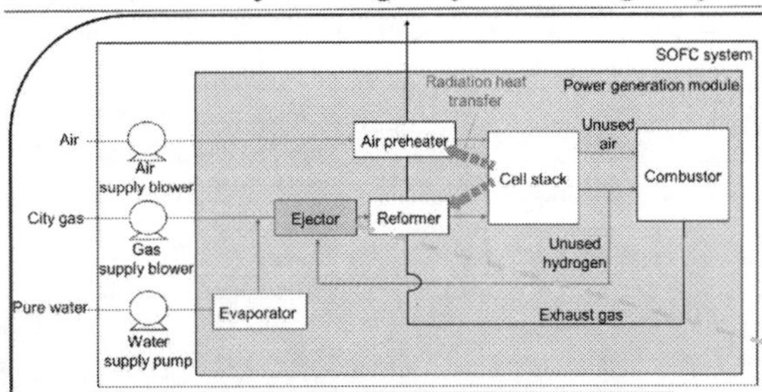

Flow diagram of Power generation module

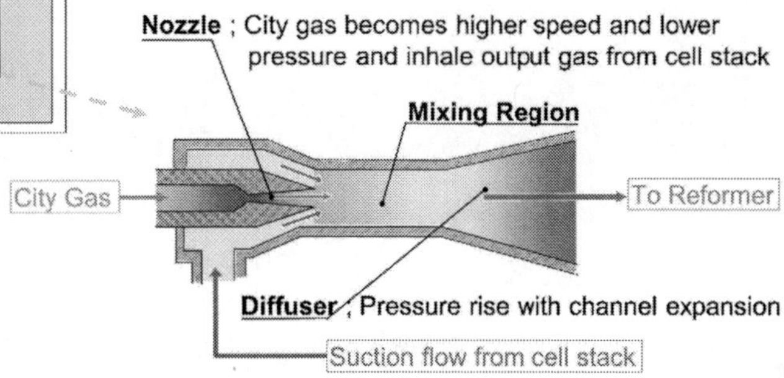

Schematic of ejector

Recycling output gas from the cell stack increases fuel utilization and improves system efficiency.

11

Power Generation Module Performance Test

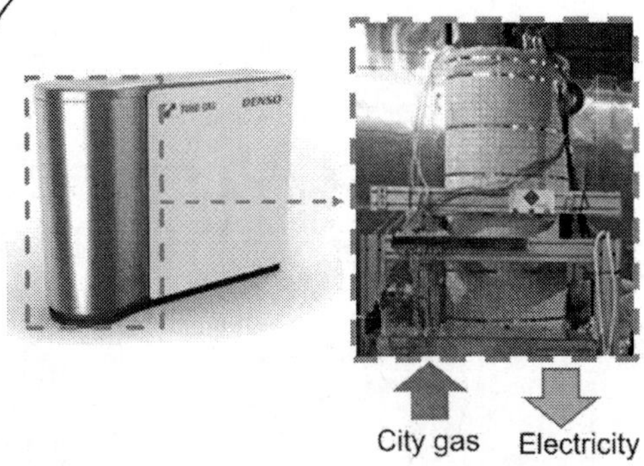

Image of performance test

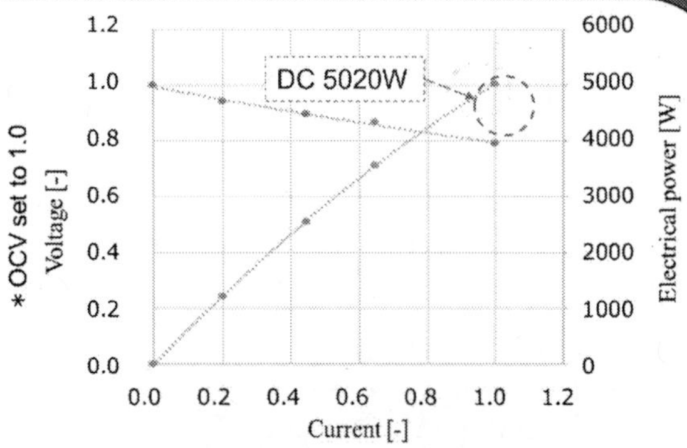

＊Rated current set to 1.0

Result of performance test

DC output : 5020W
DC power generation efficiency : 67.4%

AC output : 4520W
AC power generation efficiency : 60.7%

Assuming the balance of plant components loss is 10%

12

Power Generation Efficiency under Partial Load Conditions

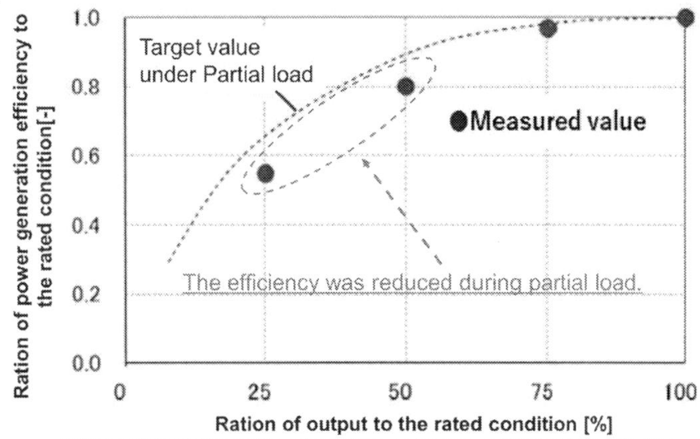

Ration of power generation efficiency to the rated condition

We need to optimize the specification of our SOFC system for maximum user benefit.

13

Concluding remarks

1. With the aim of contributing to the realization of a low-carbon society, TOHO GAS and DENSO have been developing the 5kW class SOFC system.

2. The energy-saving and CO_2 reduction benefits of introducing the 5kW class SOFC system in commercial facilities have been confirmed.

3. Achievement of 4.5kW output power generation and AC60.7% equivalent power generating efficiency.

14

Acknowledgements

This paper includes results of the "Technology Development for Promoting SOFC Commercialization" project organized by NEDO. We extend our gratitude to NEDO.

NEDO New Energy and Industrial Technology Development Organization

This paper was written in collaboration with Takahiro Tamura who belongs to NEGP Specific Development Dept., DENSO CORPORATION, Japan.

15

Thank you

16

Workshop Session 15

IGRC 2020 MUSCAT

24-28 FEB | WWW.IGRC2020.COM

Progress of SOFC Residential CHP system, ENE-FARM type S in Osaka Gas

T. Nakao, S. Inoue, M. Suzuki, S. Iwata

Residential Energy System Development Dept.,
Osaka Gas Co.,Ltd. , Japan

HOSTED BY Oman LNG L.L.C. | PRESENTED BY IGU | UNDER THE PATRONAGE OF

IGRC 2020 MUSCAT

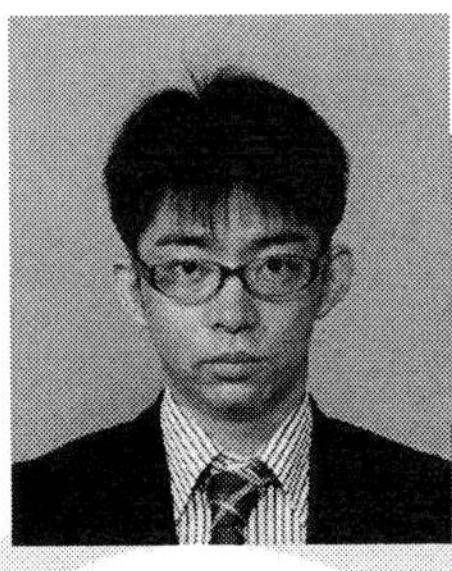

Takayuki Nakao
Assistant Manager (Osaka Gas)
Japan

ENE-FARM type S

Contents

1. Market of ENE-FARM type-S in Japan

2. Specifications of the residential SOFC CHP system
   - 2-1. 1st generation model (2012~)
   - 2-2. 2nd generation model (2016~)

3. Market Experience of Osaka Gas

4. Conclusion

About Osaka Gas

Becoming an innovative energy & service company that continues to be the first choice of customers

International Energy Business
(1) International upstream and trading
(2) Electric power generation
(3) International middle and downstream

Domestic Energy Business
(4) Electricity sales : 0.5M customers
(5) Gas distribution : 6.1M customers @ Nov 2017
(6) Gas manufacturing and sales
(7) LPG・LNG・SNG sales

Life and Business solutions
(8) Materials
(9) IT
(10) Real Estate
(11) Life solution service and others

R&D
(12) R&D

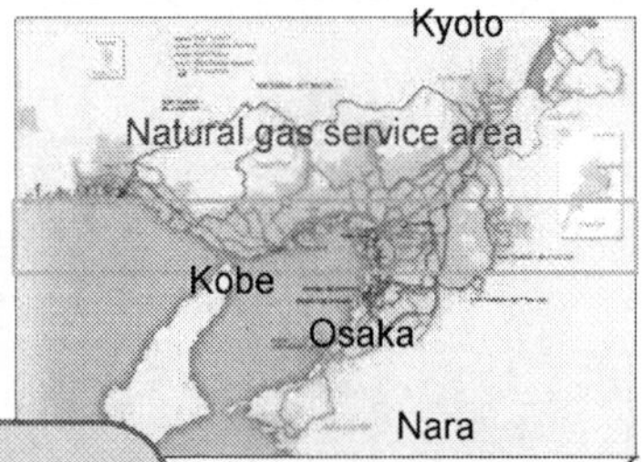

Develop and distribute "ENE-FARM" as a solution of residential energy issues

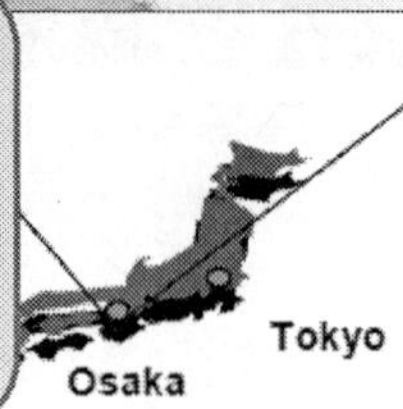

Status of Market Penetration

In March 2019, ENE-FARM have been installed *100,000* units in Osaka Gas area, meanwhile 330,000 in whole Japan. Further efforts are needed to realize self-sustaining growth by 2020.

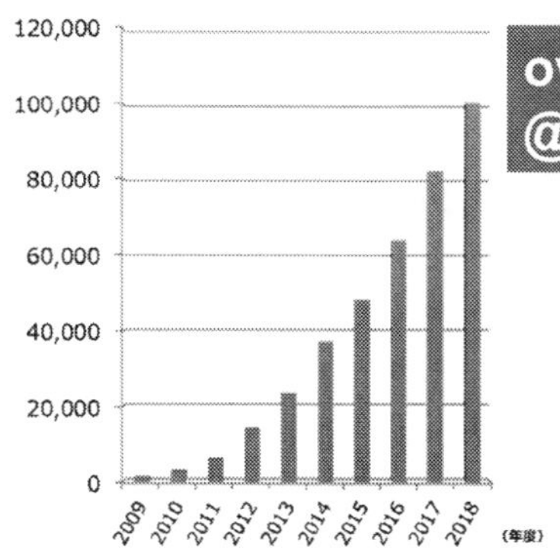

over 100,000 units @2019.3

Residential FC CHP system in Osaka Gas

Osaka Gas began commercializing residential FC CHP "ENE-FARM", PEFC type in 2009, and SOFC type in 2012 through joint development with manufacture.

PEFC *"ENE-FARM"*

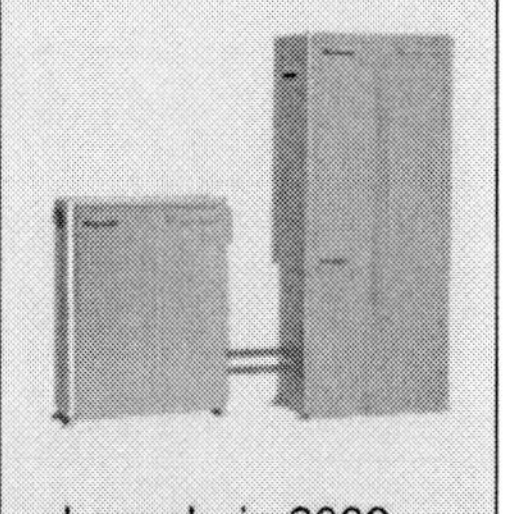

Launch in 2009

SOFC
"ENE-FARM type S"

Launch in 2012
Model change in 2016

→ High Power efficiency

Current Efforts

-Changes in power generation efficiency-

2003 ∽ 2009 — 2012 — 2014 — 2016 — 2018

10th anniversary in 2019

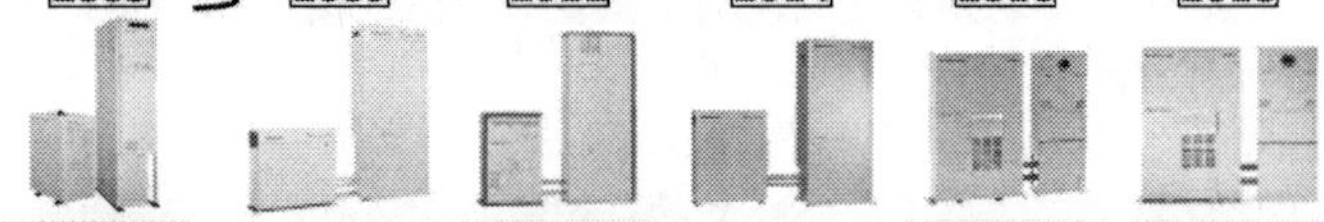

Eco Will	PEFC	SOFC(1G)	PEFC	SOFC(2G)	SOFC(2.5G)
20%	35%	46.5%	39%	52%	53.5%

net AC, LHV

-Expansion of added value and service)-

Dealing with blackout :2013~	Electricity purchase :2016~	IoT:2016~

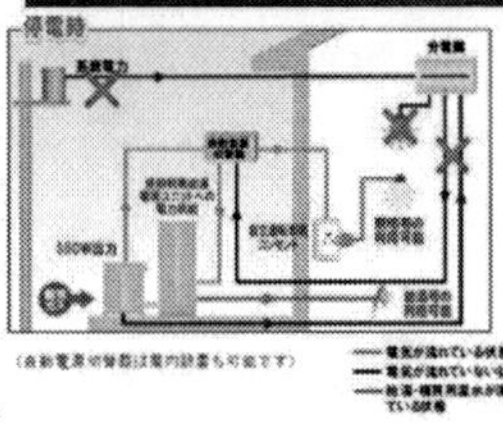

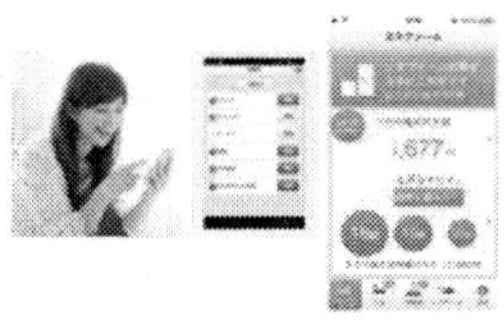

History of the SOFC development

2004 Joint development with Kyocera

2005 The first field test in an actual house in our experimental housing complex (NEXT21) in Japan

2007 Participation in "Demonstrative Research on Solid Oxide Fuel Cells" project conducted by NEDO / NEF

Osaka Gas installed over 120 units in residential houses.

2009 Reinforcement of joint development framework

KYOCERA OSAKA GAS TOYOTA AISIN

2012 Sales launch 1st generation model

2013 Stand alone model release

2014 Cost reduction / Internet connection

2016 Sales launch 2nd generation model

2018 *Further improve power generation efficiency*

Next Generation SOFC

Cell Stack of SOFC

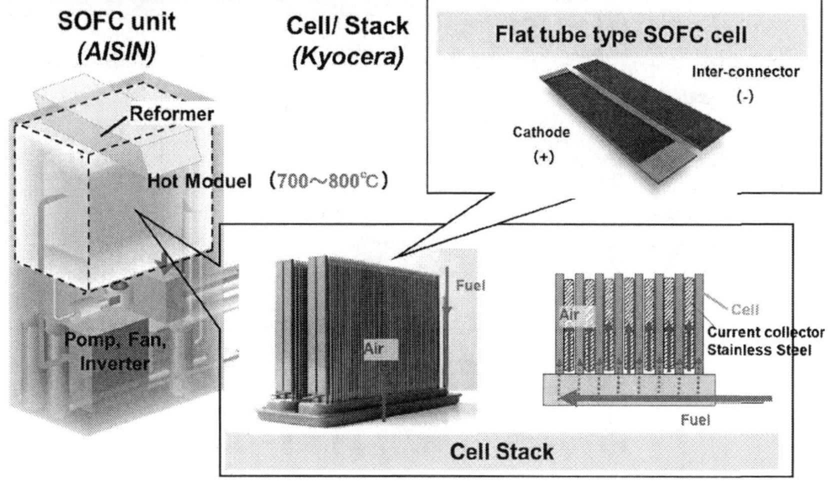

1st generation model (2012~)

Power generation efficiency was 46.5%.
It was placed in a newly built house market.

Power Generation unit

Dimensions (HxWxD)	935 x 600 x 335 mm
Weight	94 kg
Output	50~700W
Electric efficiency (rated)	46.5%(LHV)
Heat recovery efficiency	43.5%(LHV)

Manufacture: Aisin Seiki

Release Date; April 1, 2012.

Full maintenance support: 10years

Durability study and evaluation

Durability was crucial issue for commercialization.
For reliability and durability of the SOFC, Osaka Gas has conducted research and evaluation.

Stack components

SOFC Stack

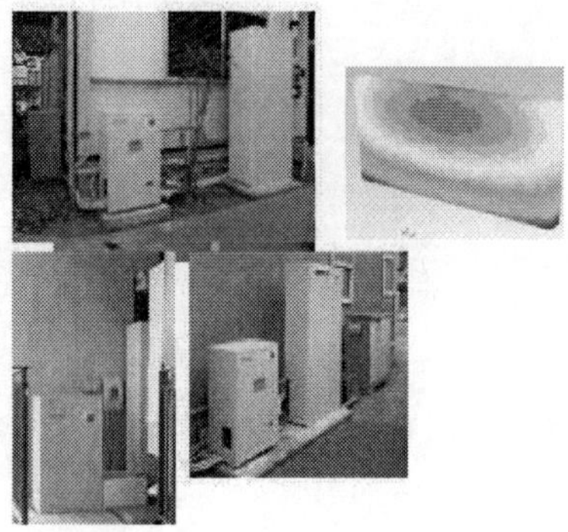

Power generation unit

Durability results in the field tests

In the "Demonstrative Research on SOFC project" (2007-2010 NEDO and NEF), Osaka Gas has set up a 121 units total, some of which continue to still operating.

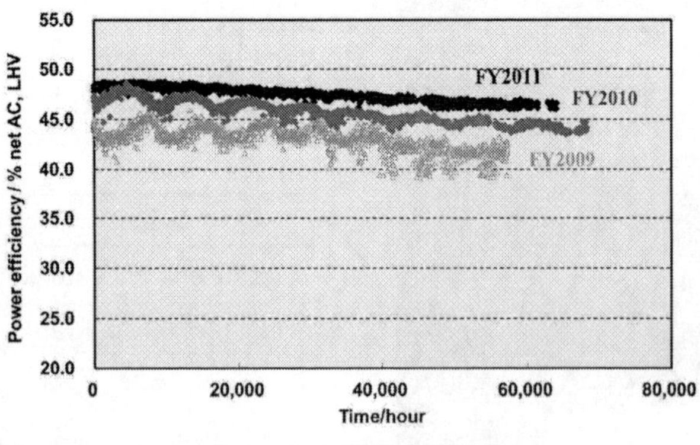

These are the results of the testing power generation unit in 2009 ~2011.
These are not exactly the same as the commercial unit.

Durability improvement

For durability, developed the current collector coating technology by Electrodeposition painting method.

Electrodeposition painting
- Already mass produced for automobiles and cameras etc.
- Uniform and flawless coating due to insulated coat.

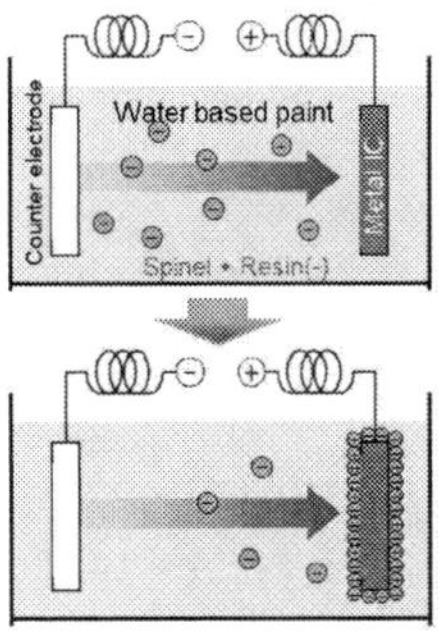

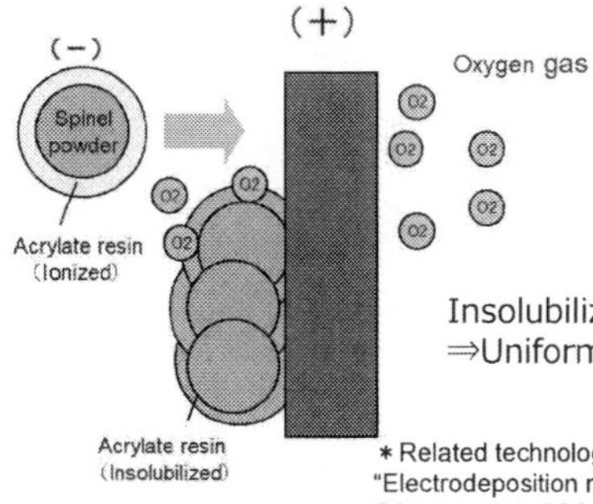

Insolubilized acrylate resin is insulated
⇒Uniform and flawless coating

* Related technology
"Electrodeposition method for SOFC interconnector coating"
S.Inoue *et.al.* ECS Transactions, 57 (1) 2217-2223 (2013)

Experience of 1st generation

Manufactures and Osaka Gas have conducted the following continuous efforts;

Productivity improvement
Quality improvement
Installation and maintainability improvement
Sales promotion and planning

,and more Product Specification improvement

2013 model
Adding a self-sustained operation function during power grid failure

2014 model
Price reduction

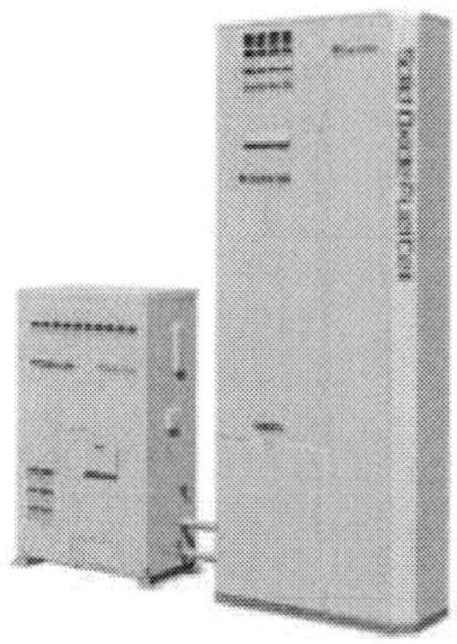

2nd generation model

Release Date : April 2, 2018

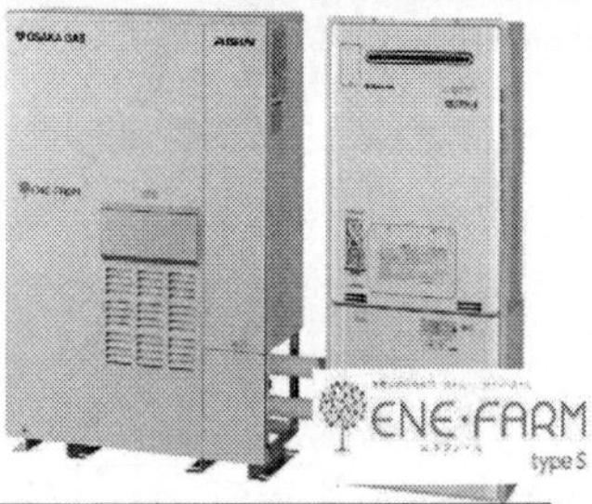

η_e	53.5 % (LHV)
η_H	87%
Power output	50W~700W
Installation space	1.4m^2
Hot water storage	28 L
weight	FC unit 100kg Boiler 33.5kg

1 High power efficiency

46.5%⇒ 52.0% ⇒ 53.5%(net AC,LHV)

2 Compact residential CHP system

3 Expand the application market

4 Cost Down

5 Increasing customer convenience by IoT

6 Surplus power purchase by Osaka Gas

7 High economic efficiency and environmental performance

Specification of ENE-FARM type S

		NEW model	2014 model	2012 model
Release Date		2016/4/1	2014/4/1	2012/4/1
Spec.	Power output	50W~700W	50W~700W	←
		During black out: 700W	During black out: 350W	-
	Power efficiency	**52.0% (LHV)**	**46.5% (LHV)**	←
	Total efficiency	87.0% (LHV)	90.0% (LHV)	←
	Hot water strage tank	28L	90L	←
	Hot water temperature	70℃	←	←
Size	Power Genration Unit	W780×H1195×D330(mm) (incl. Hot water tank)	W600×H935×D335(mm)	←
	Hot water unit	W480×H750×D240(mm)	W740×H1,760×D310(mm) (incl. Hot water tank)	←
Weight	Power Genration Unit	100kg	96kg	94kg
	Hot water unit	38kg	91kg	94kg
Full maintenance priod		10years	←	←
Suggested Price (non-tax price)	SET	1,785,000yen	2,150,000yen	2,620,000yen
	Power Generation Unit	1,425,000yen		

https://www.osakagas.co.jp/company/press/pr_2016/1234984_27712.html

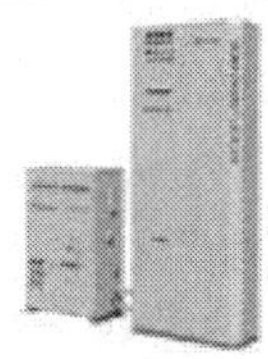
2014 model

2016 model

Installation / Market

IGRC 2020 MUSCAT

Residential Market

Additional installation

Pre-installed boiler is available.

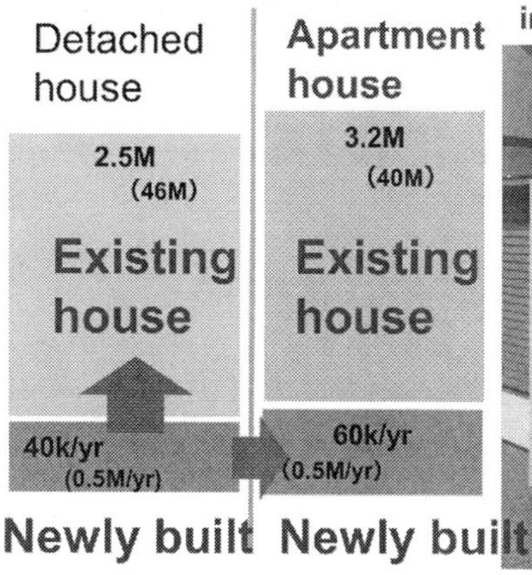

Osaka Gas area (Japan)

Apartment house in multi housing complex

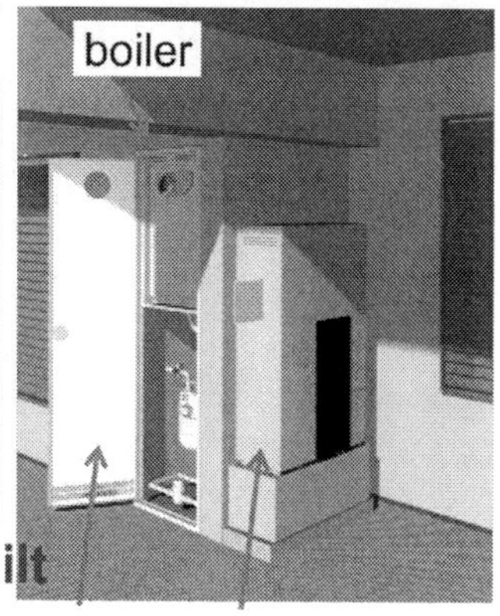

Pipe shaft SOFC unit

Concept of 2nd model

IGRC 2020 MUSCAT

The reductions of cost and installation space are critical for further market expansion. We set up the new model concept.

1st generation model (2012∼)

46.5 % LHV

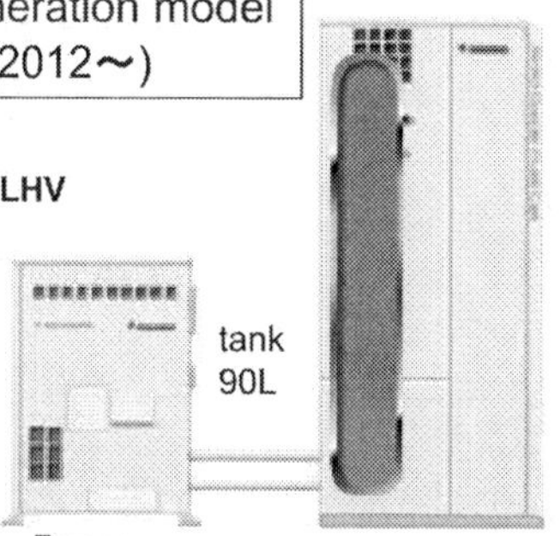

Power generation unit

Storage tank and boiler Unit

2nd generation model (2016∼)

52.0% LHV

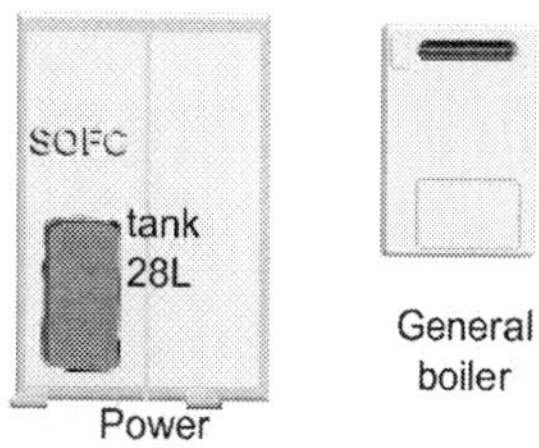

Power generation unit

General boiler

Value added Service with IoT

IGRC 2020 MUSCAT

ENE-FARM connects to OG server through wireless LAN.

- Operation data are used for QC and maintenance
- Osaka Gas provides mobile apps. , remote monitoring, etc.

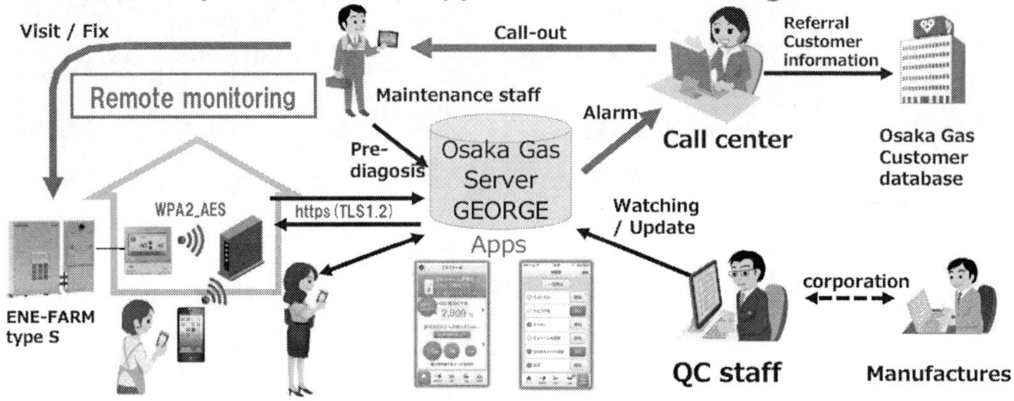

ENE-FARM with the power bussines

IGRC 2020 MUSCAT

On April 2016, Osaka Gas started to purchase the surplus power of ENE-FARM to coincide with full liberalization in the electricity market.

Osaka Gas's power source : 1.87BW

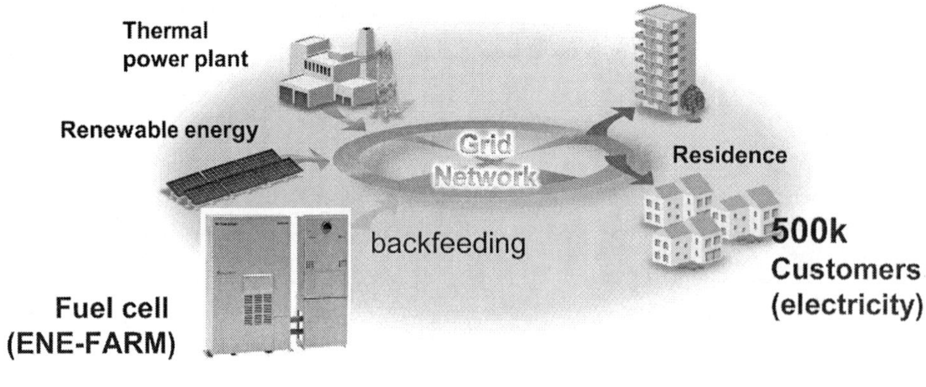

Merit of Back feeding

Expect to improve the power generation efficiency in addition to surplus power because ENE-FARM always generates at full rated power.

< (1) Profit of surplus power >

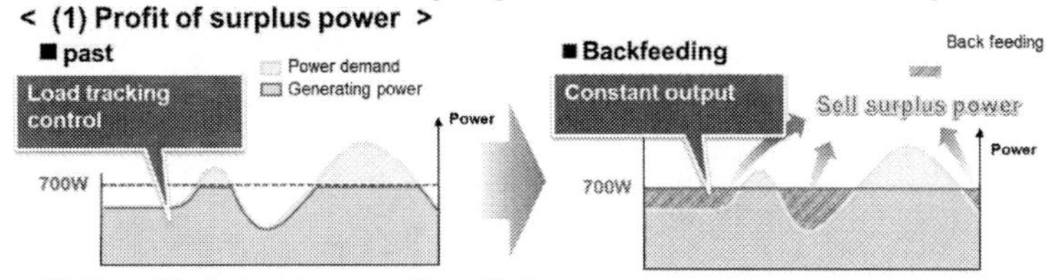

< (2) Benefit of virtual generating efficiency >

■ Generating efficiency

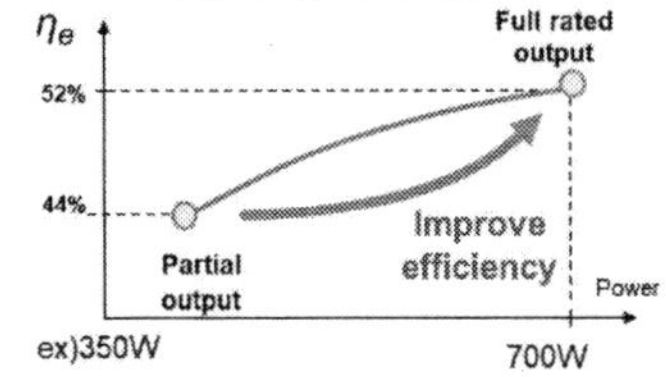

Difference Full or Partial :

$$\frac{\Delta \text{ power output}}{\Delta \text{ gas input}} \sim 60\% \text{ approx.}$$

virtual generating efficiency **60%** approx.

High Efficiency Control Mode(2018~)

Back feeding at full rated power operation makes improving generating efficiency possible.

Generating efficiency

52% ➡ **53.5%**

Load tracking control

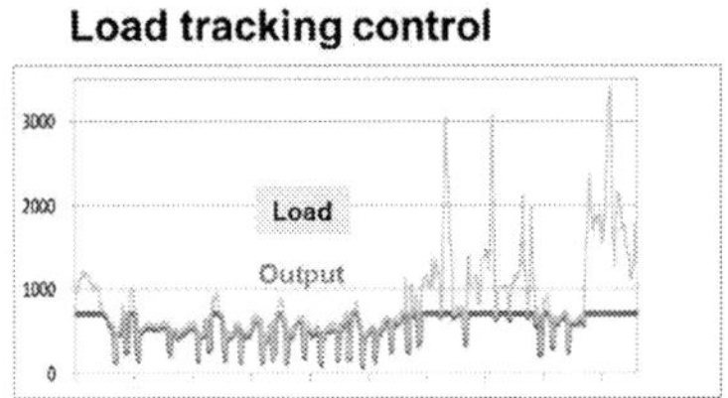

Past case:

Extra Gas input was needed to prevent a shortage of fuel

Back feeding with constant output

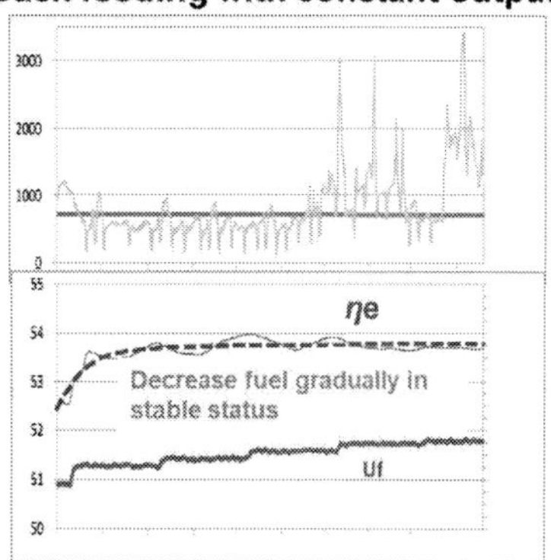

Status of Market Penetration

ENE FARM type S performed at about 2,000 homes though a power failure occurred mainly in Osaka due to Typhoon.

Stand alone outlet
(only blackout)

Eco-will(Stand alone model) at employees' home

Even during a power outage, the bath and hot water supply functions were available as usual.

Summary

- A current model of compact, high power efficiency of residential SOFC system was launched in April 2018. It has achieved cost reduction and performance improvement at the same time.

- We believe that the steady progress towards full-scale dissemination stage.

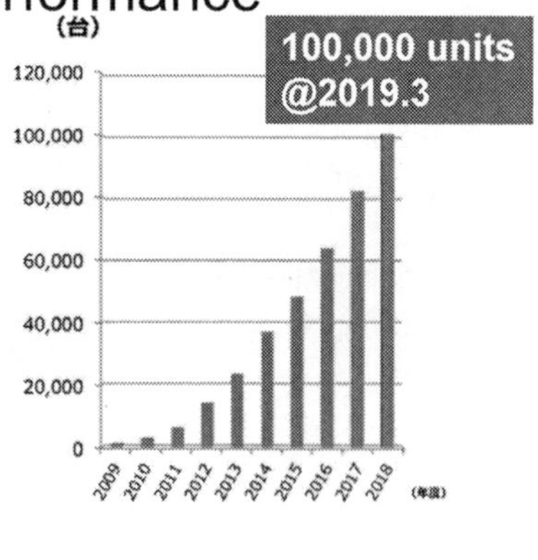

For future growth

Osaka Gas aims to early realize "7-8 year payback" and to contribute to the full and self-sustaining growth of ENE-FARM type S.

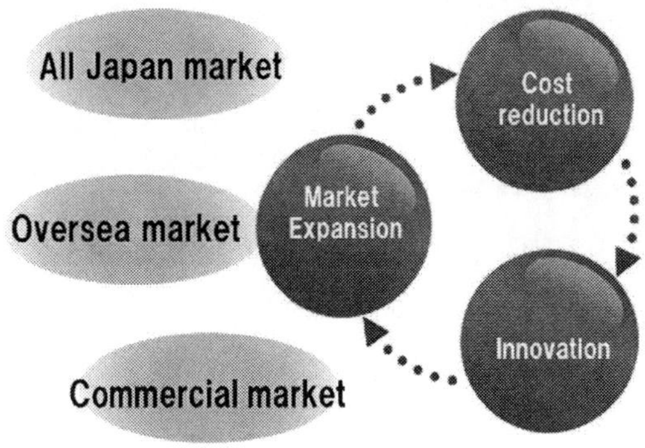

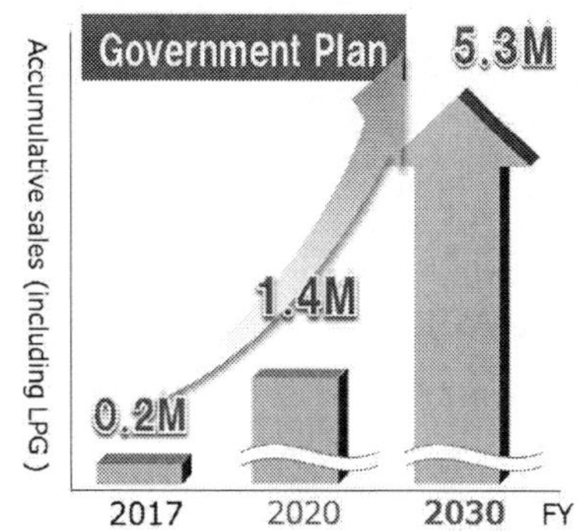

Acknowledgements

This paper includes the result of "Demonstrative Research on Solid Oxide Fuel Cells Project" conducted by NEDO/NEF and also includes the interim result with governmental program related to local production local consumption of renewable energy.

We wish to express our appreciation to people involved in these projects.

IGRC 2020
MUSCAT
THANK YOU
FOR YOUR ATTENTION

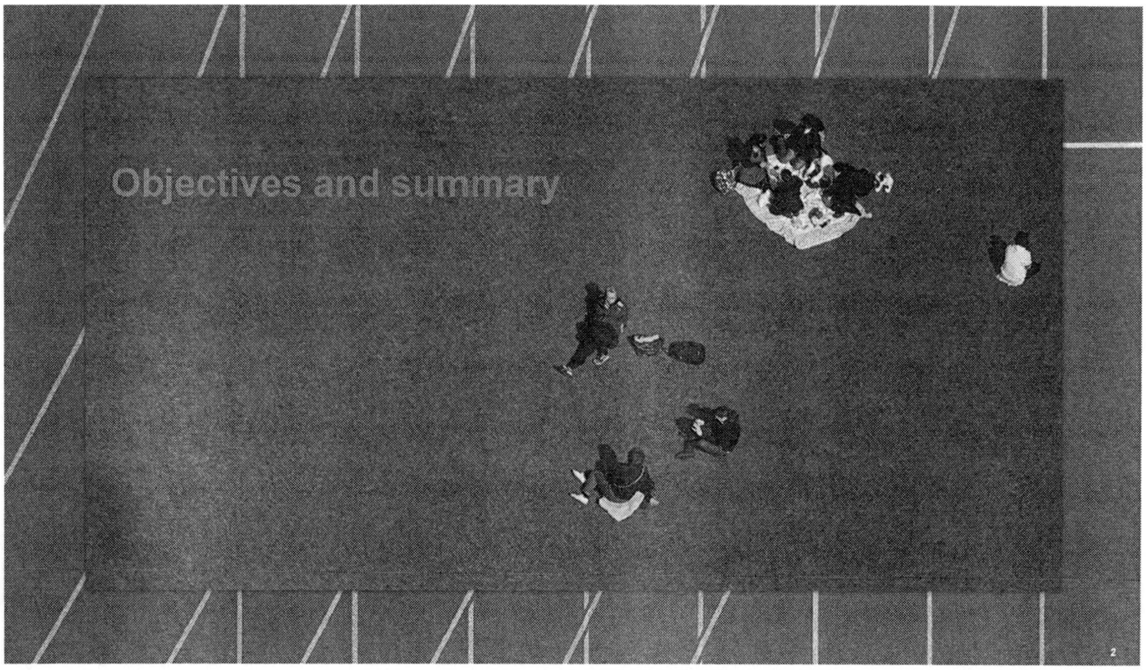
Objectives and summary
2

Context and objectives:
Micro-CHP in France, a quite complex situation

From a comfortable situation with feed-in tariff …

- Sizing based on maximization of electricity production
- CHP as a base during all winter

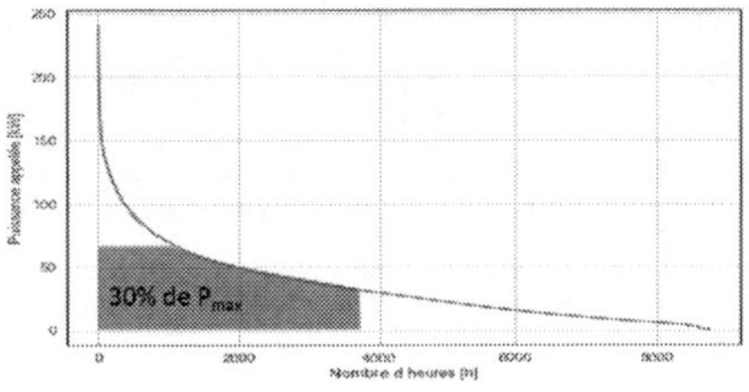

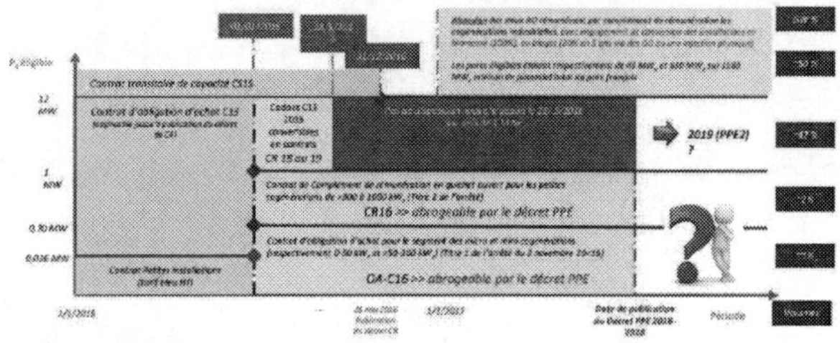

- But the context evolves with the end of the State subventions on fossil fuels

Context and objectives:
Micro-CHP in France, a quite complex situation

… to a market based on self-consumption

- CHP needs to modulate while remaining the base thermal producer
- Their sizing must be adapted to each building to optimize electricity self-consumption

→ Each installation is basically a new use case in order to optimize the installation (and payback)

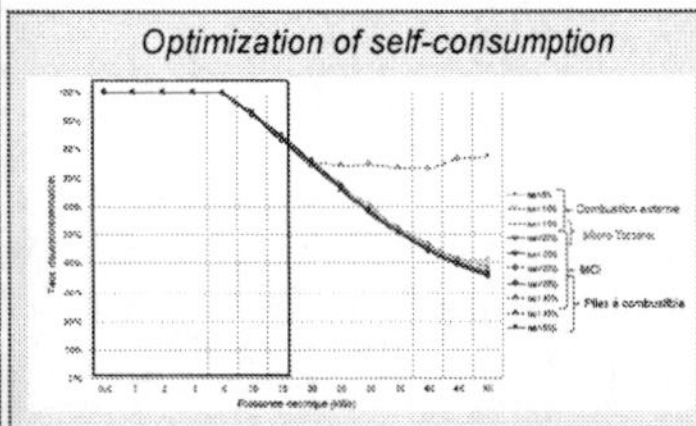

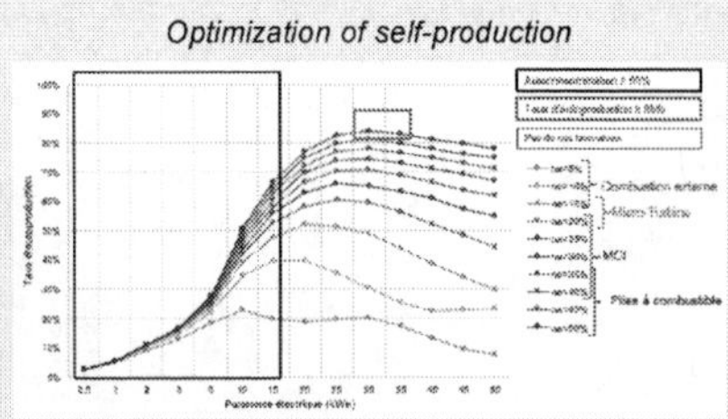

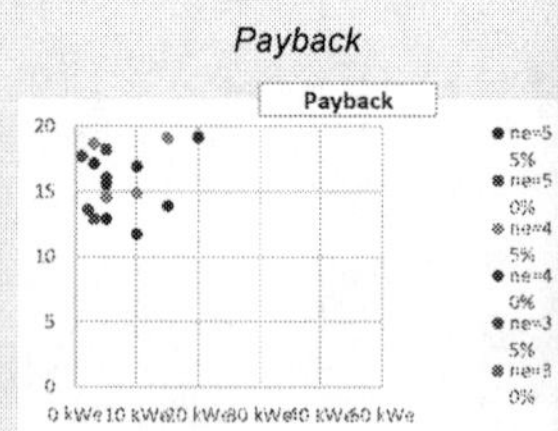

Example of a parametric study on theroretical mCHP sizings

Parametric studies allow us to have a good vision on sizing but they don't take into account the real issues of these systems: the hydraulic circuit and their installation

Context and objectives:
Micro-CHP in France, a quite complex situation

The good operation of the CHP is in the core of a correct behavior of the installation

Objectives of the presentation : evaluate, quantify & give insights

→ Proper sizing micro & min-CHP systems

→ Optimize its integration in hydraulic equipments

→ Enhance economic performances

Chosen Method : Modeling and simulation of real Buildings & equipments on DYMOLA platform

Summary

1. Methodology
2. Office Building
3. Multi-family Building
4. Conclusion & Recommandations

ENGIE Lab CRIGEN - Optimization of micro-CHP installation for renovation of commercial buildings and dwellings in France

5

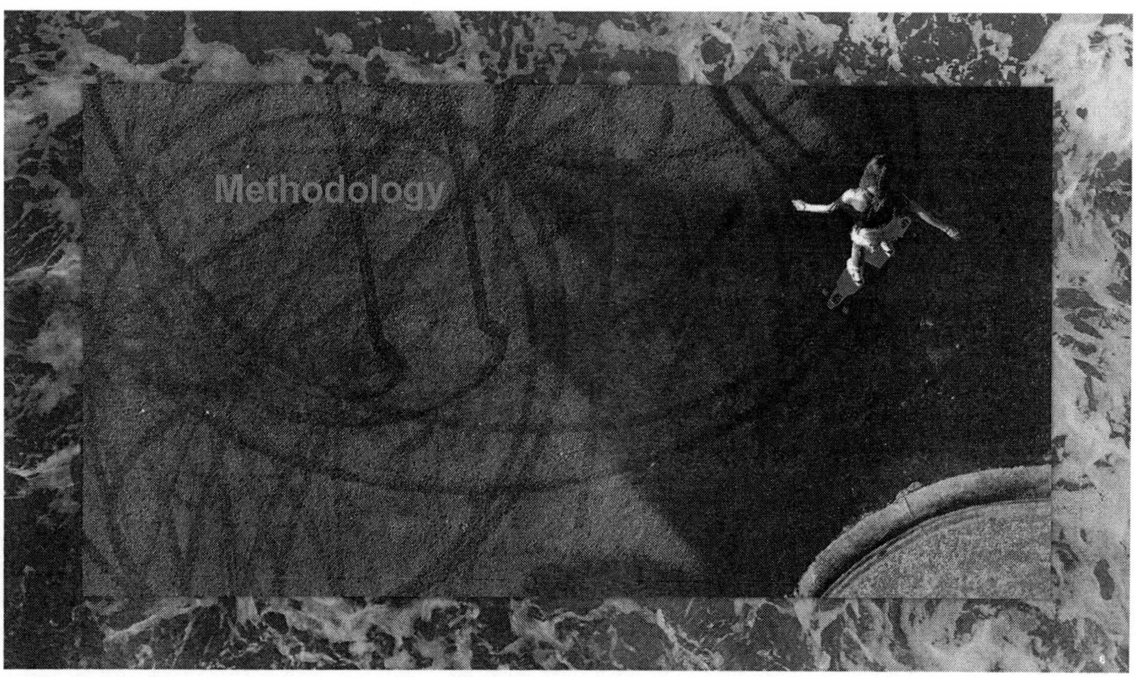

Methodology
A study based on manufacturer insight and dedicated tools

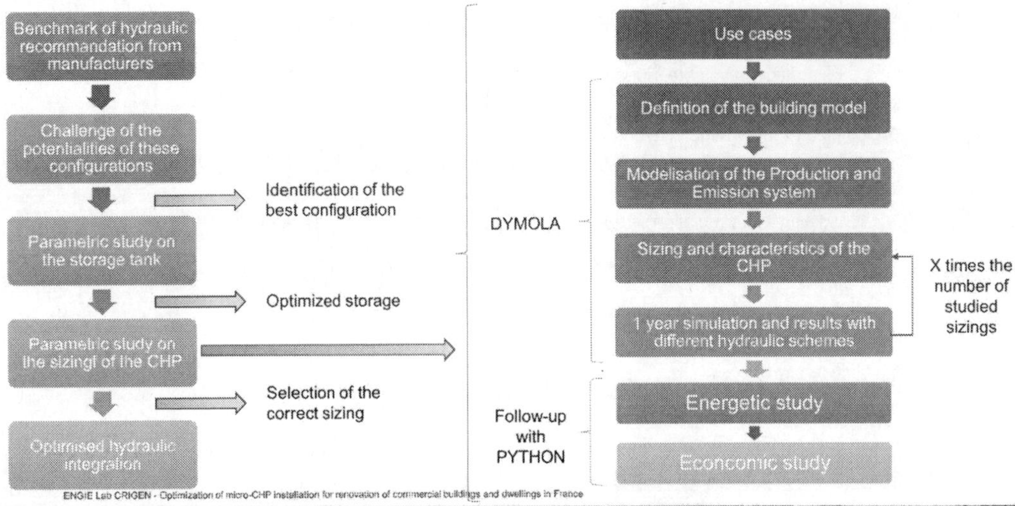

ENGIE Lab CRIGEN - Optimization of micro-CHP installation for renovation of commercial buildings and dwellings in France

7

Methodology
Models

Dynamic Models on Dymola

- Building: multizone approach
- Networks, including thermal losses
- Generation (mCHP, boilers)
- Detailed setiings of the installation (PID, hysteresis, dead-zones)
- Hydraulic behavior of storages
- Electricilty consumption load (coming from field-tests)

Real dynamic of heat production and use

Criterias:

- Variable time step (lower than 1min, chosen by the solver)
- Post treatement under PYTHON

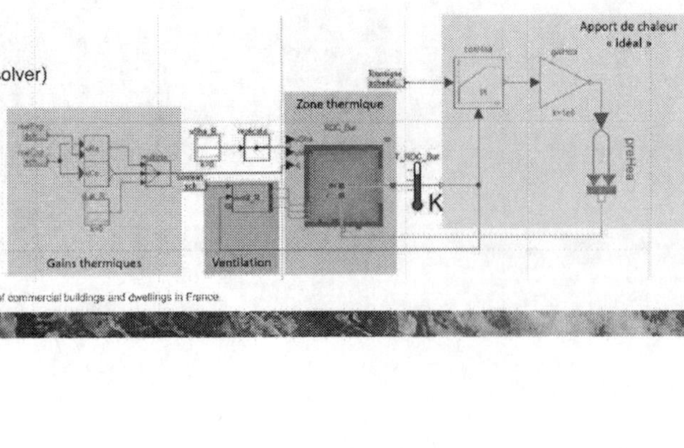

ENGIE Lab CRIGEN - Optimization of micro-CHP installation for renovation of commercial buildings and dwellings in France

8

Methodology
Hydraulic configurations (1/2)

Surprisingly, very few recommendation from mCHP manufacturers, they are mainly assemblers and not specialists of HVAC systems. We identified 8 different configurations from bibliography and real installations.

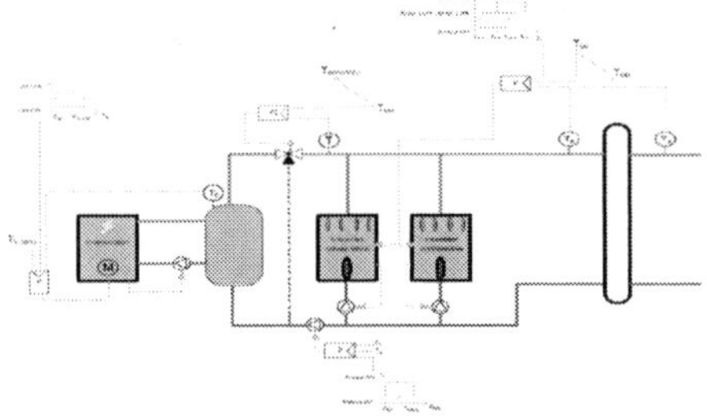

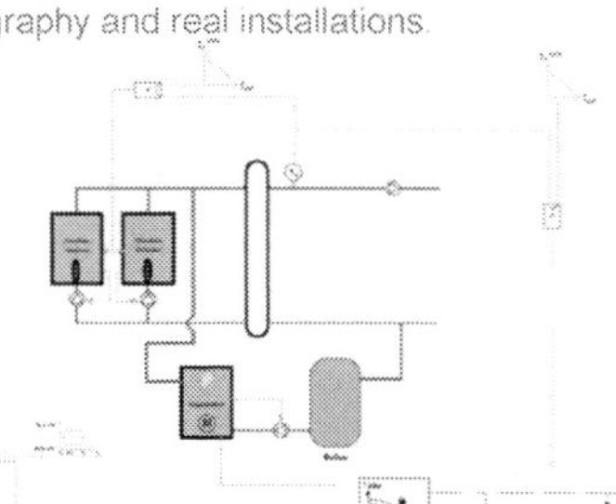

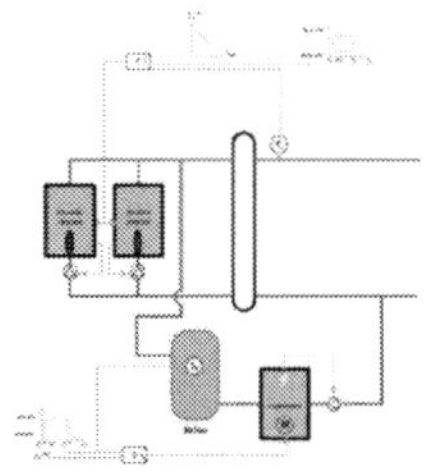

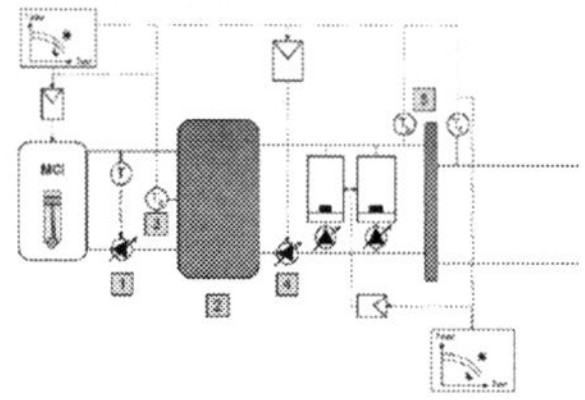

Main differences in terms of hydraulic circuit
- CHP and boilers in parallel/series
- Position of the storage or decoupling bottle
- Oversized storage volume

Methodology
Hydraulic configurations (2/2)

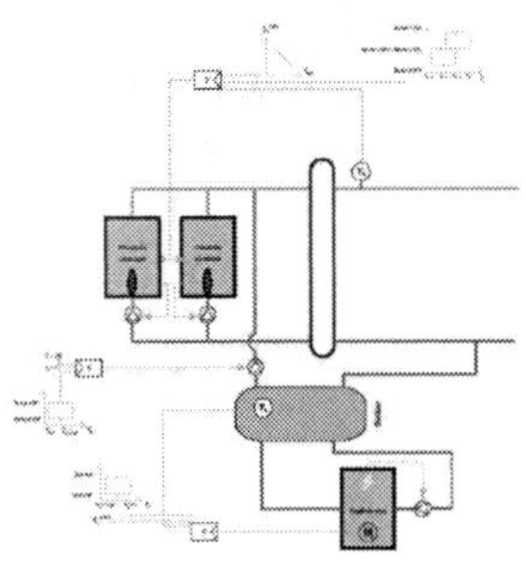

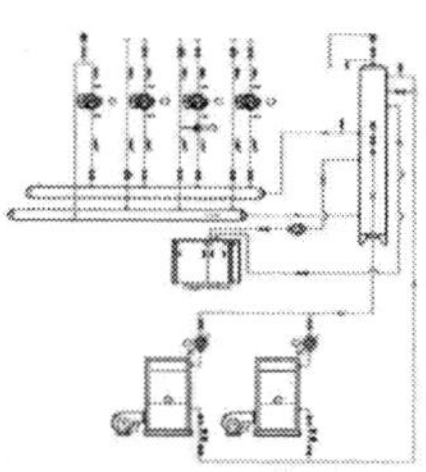

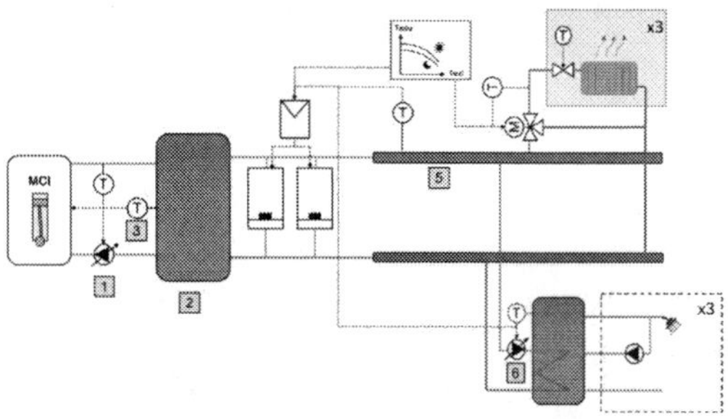

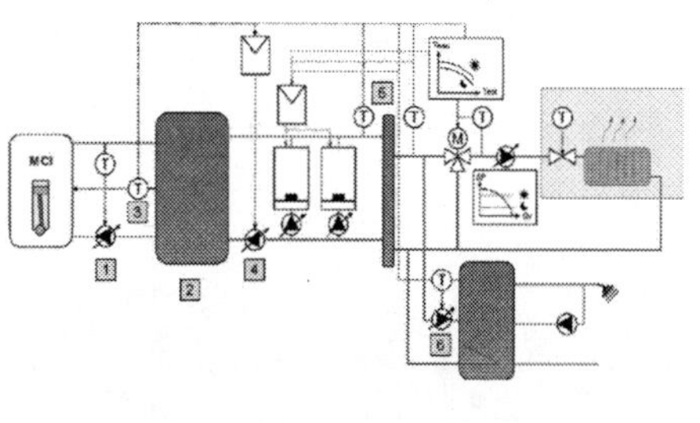

Main differences in terms of control
- Positions of temperature probes
- Control of CHP module
- Modulation on thermal load or electricity load

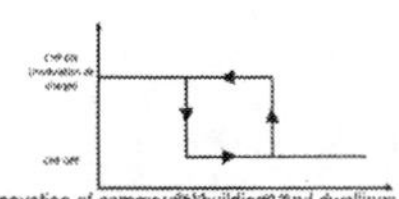

Methodology
Economical approach

Actualized global cost

$$CGA = I_0 + (D_1 - R_1) + \ldots + (D_n - R_n) + \ldots + (D_N - R_N)$$

With

- I0: Initial investment
- Dn: Actualized costs on year n
- Rn: Actualized revenues on year n

Inflation rate: 3%
Actualization rate: 0%

8 scenarios on economic optimization

	Collective Self-consumption (building including appartments)		Sel-consumption on Commun parts (boiler room, lights...)	
CHP operation	All year	Winter only	All year	Winter only
Revenues option 1	Feed-in tarif		Feed-in tarif	
Revenues option 2	Self consumption and feed-in tariff		Self consumption and feed-in tariff	

ENGIE Lab CRIGEN - Optimization of micro-CHP installation for renovation of commercial buildings and dwellings in France

11

Office building
Hypothesis

Building characteristics	New building (RT 2012)
Surface	3500 m² (4 floors)
Climatic zone	Paris (2015 climate reference)
Thermal annual demand	28 kWh/m²
Control of heating and ventilation	intermittent
Number of thermal zones	4
Terminal control	Thermostatic valves

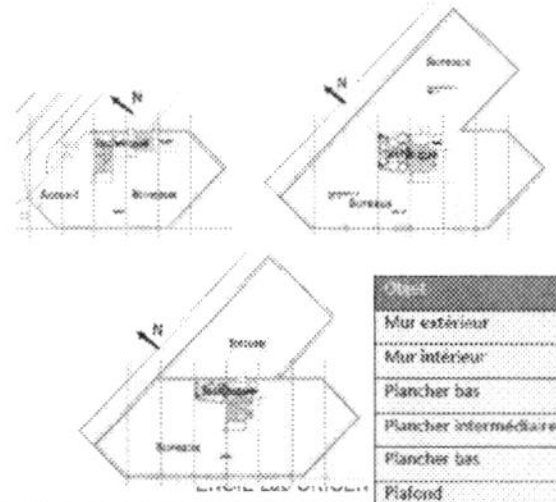

Objet	Matériaux (ext -> int)	Epaisseur [cm]	U [W/m²K]
Mur extérieur	Béton/Polystyrène	20/17	0.183
Mur intérieur	Brique	12	7.41
Plancher bas	Béton/Polyst./Polyréthane	20/197/8	0.10
Plancher intermédiaire	Béton/Isolation/Béton	20/15/20	0.19
Plancher bas	Béton/Polyréthane	20/14	0.14
Plafond	Béton/Polyréthane	17/23	0.09
Fenêtre	Vitre/Air/Vitre	0.4/1.5/0.4	1.4

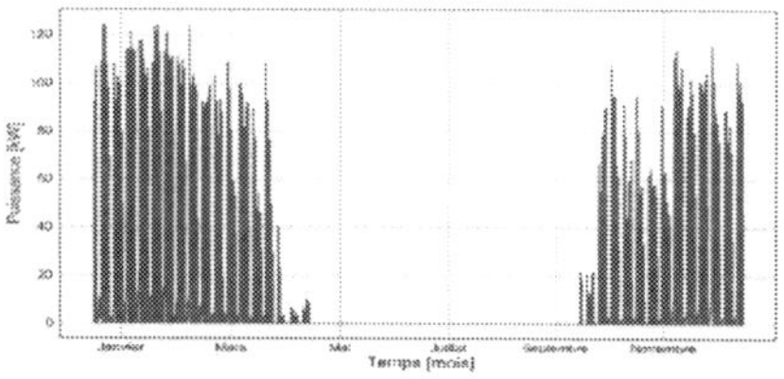

Office building
Understanding the models: Influence of hydraulic parameters

The models are quite complex and allow a total control of the parameters defining the boiler room setteings on primary and secondary circuits

- The 1st step of the project consisted in making sure that we have a correct behavior compared to real life installations (REX based on field-tests)
- Then, a large amount of time was dedicated in « playing » with all parameters to understand completely the in and out of each hydraulic configuration

Illustration: impact of the water flow on the secondary circuit on CHP/boiler loads and storage temperature

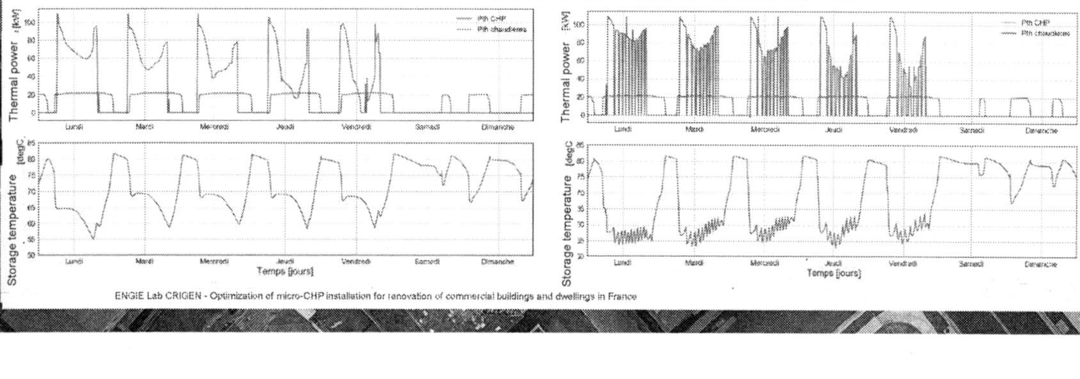

Office building

Understanding the models: optimization of settings

The optimization of each configuration beforehand is necessary to ensure the best efficiency

Comparison of thermal production per configuration /setting

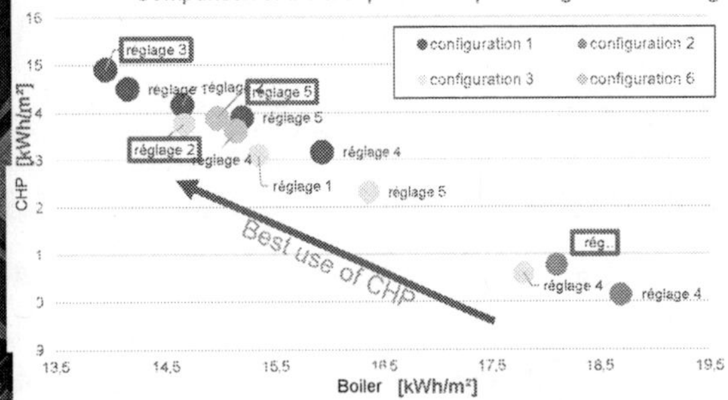

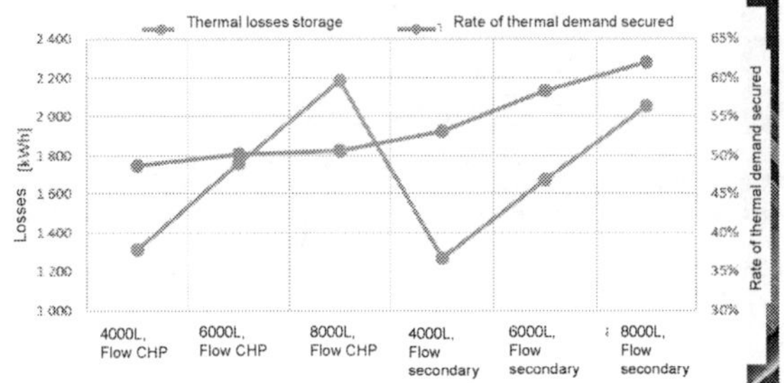

réglage 1	réglage 2	réglage 3	réglage 4	réglage 5
ΔTc=15 K, ΔTs = 10 K Ballon HX	ΔTc=20 K, ΔTs = 10 K Ballon HX	ΔTc=15 K, ΔTs = 15 K Ballon HX	ΔTc=15 K, ΔTs = 15 K Ballon normal	ΔTc=20 K, ΔTs = 15 K Ballon normal

*ΔTc : ΔT boiler, ΔTs: ΔT secondary

Office building

Results: best hydraulic circuit

Hydraulic configuration N°4 is selected

- Optimize the thermal energy produced by the CHP
- Best electrical efficiency
- Good balance in terms of condensing rate

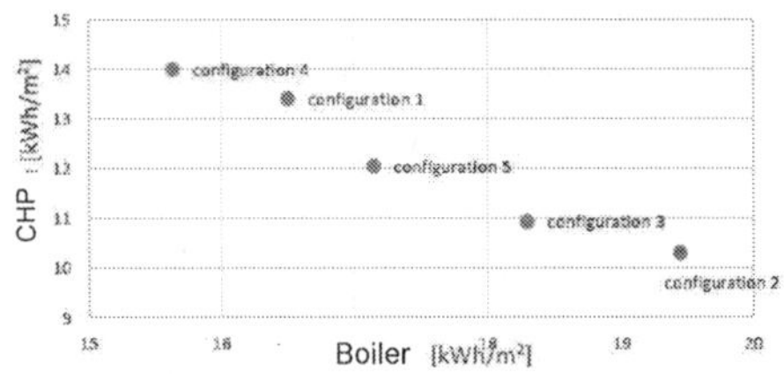

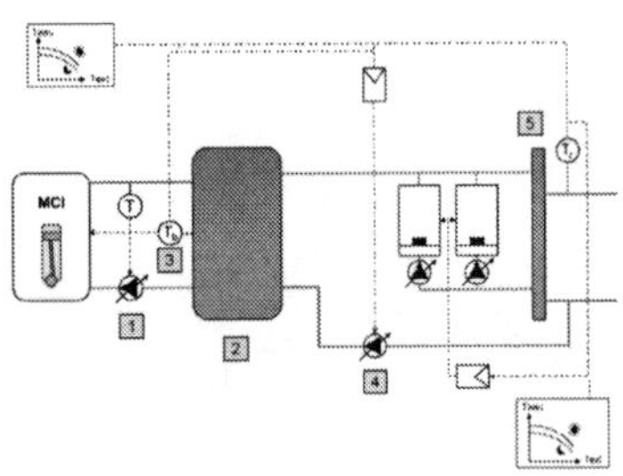

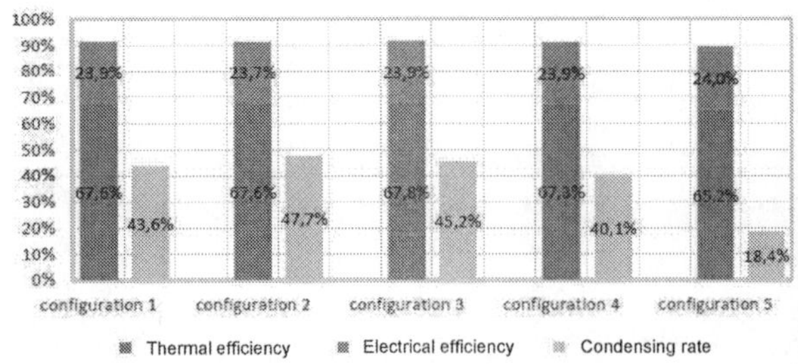

Office building

Results: best sizing of the mCHP

Best CHP power is 22kWe

- The highest the power ,the lowest the thermal efficiency
- However electrical efficiency increses
- The last criteria is economic, the 22kW presents the best payback

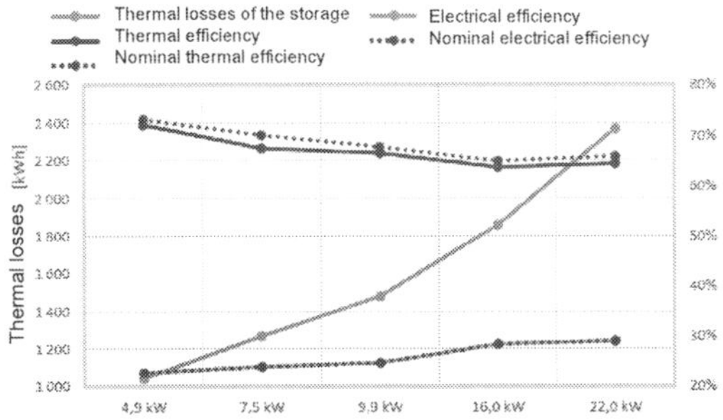

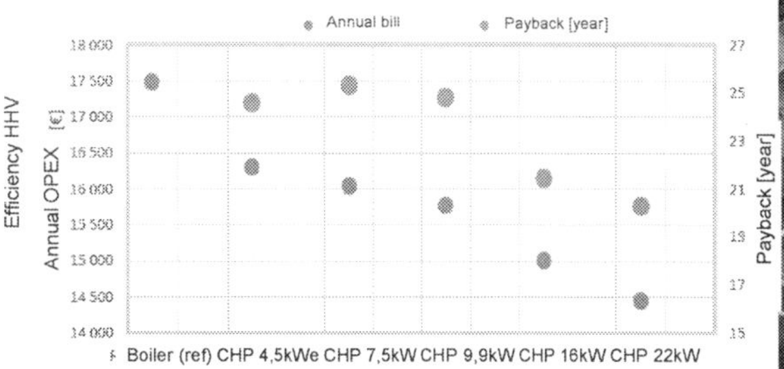

ENGIE Lab CRIGEN - Optimization of micro-CHP installation for renovation of commercial buildings and dwellings in France

17

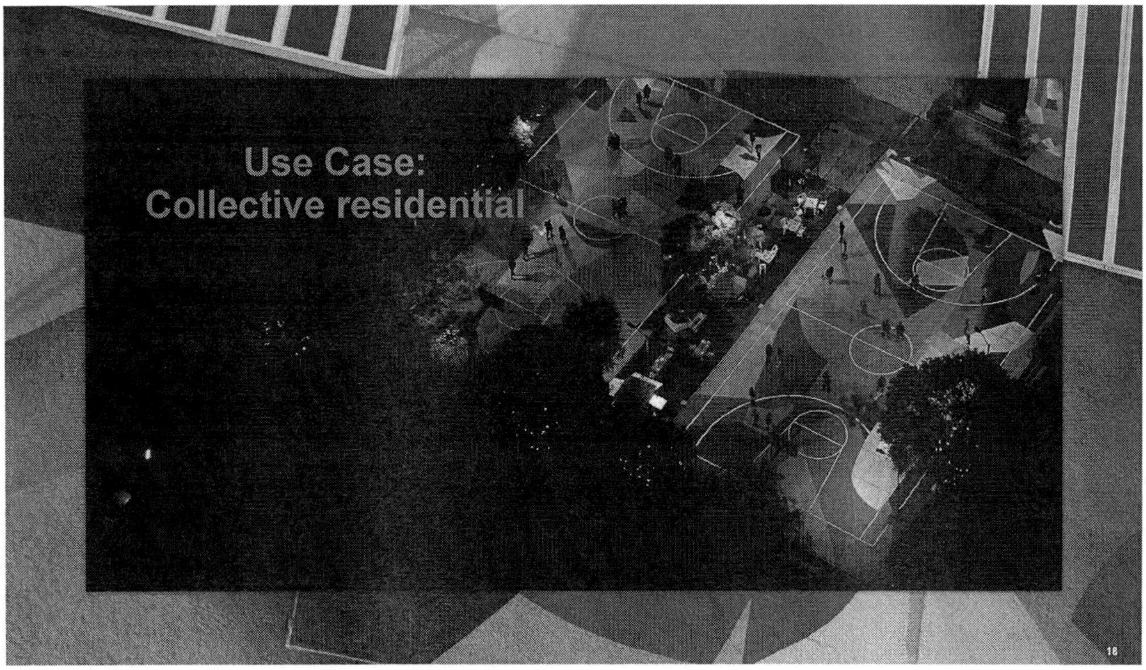

Collective residential building

Hypothesis

Building characteristics	Existing building (RT 1988)
Surface	5 520 m²
Number of appartments	3 x 28 *(84)*
Climatic zone	Paris (2015 climate reference)
Thermal annual demand	43,5 kWh/m²
DHW annual demand	222 MWh

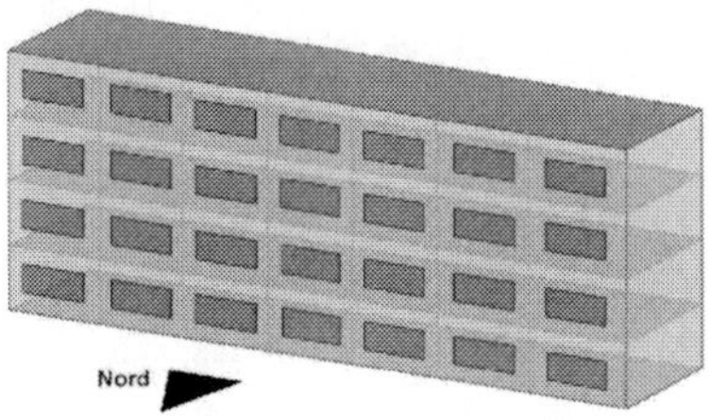

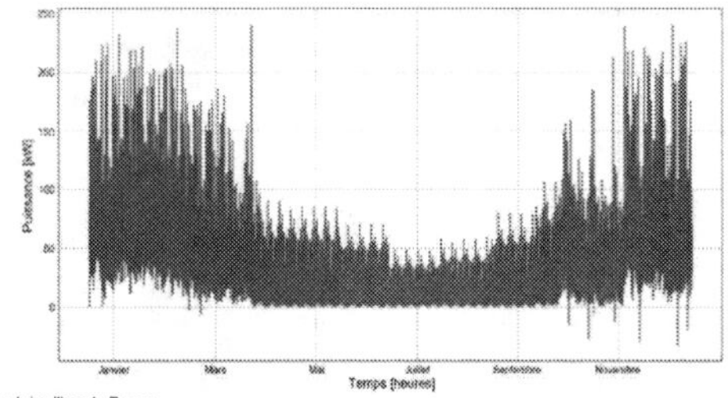

Collective residential building

Results: parametric study

Studies on new control rules in order to maximize self-consumption in summer, « hybrid » control, a mix between

- Control on electricity load
- Control on thermal load
- Anti short cycles

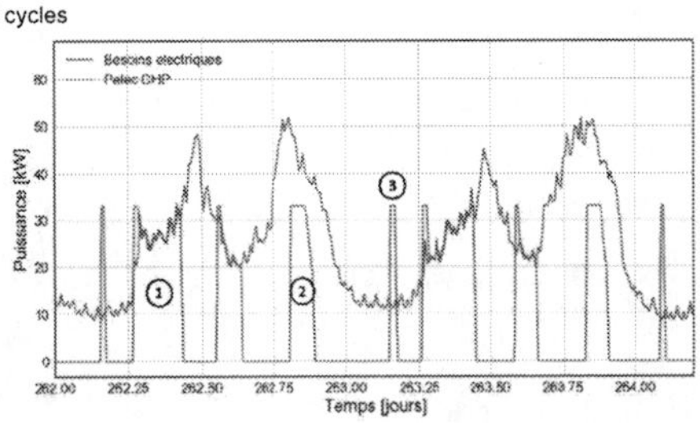

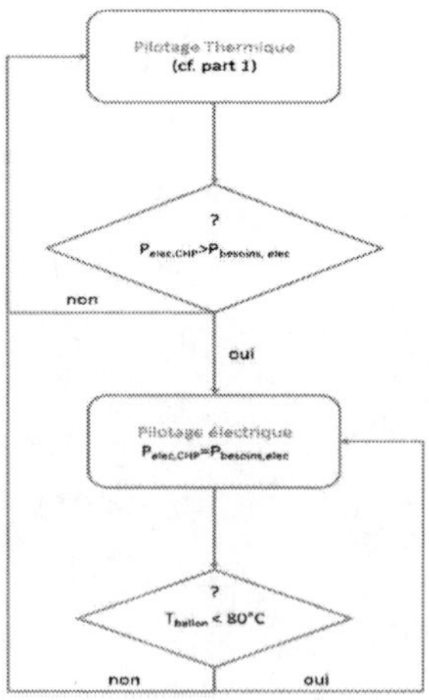

Conclusions

- 5% of economic gain on the electricity bill
- Global primary efficiency gain is lower (1%) because less electricity is produced)
- 1% gain on global costs on 15 years, hybrid control is advised but with marginal economic interest

Collective residential building

Results: identification of the best configuration

All configurations give more or less the same results (except config. 5)

Configuration 1 is chosen for its slightly better efficiencies.

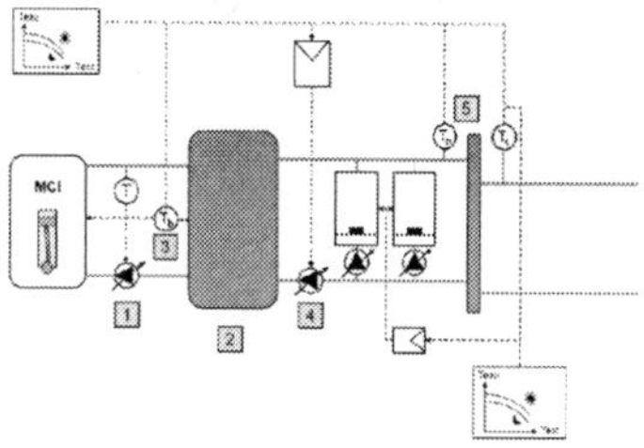

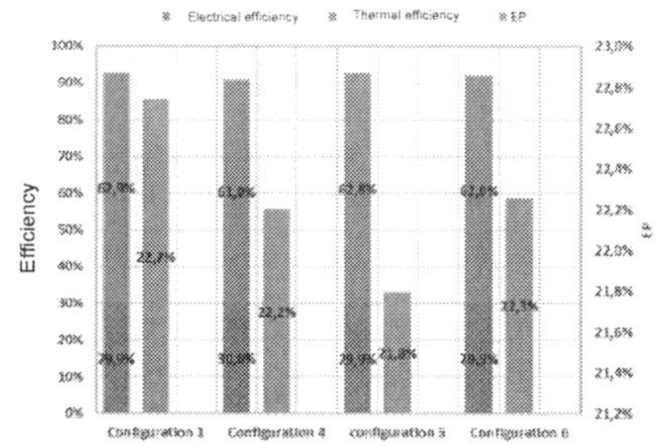

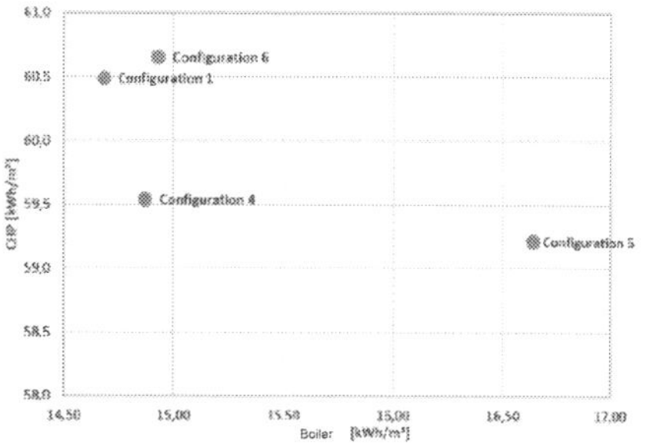

ENGIE Lab CRIGEN - Optimization of micro-CHP installation for renovation of commercial buildings and dwellings in France

21

Collective residential building

Results: best sizing of the CHP

The size of the mCHP is traditionnaly set at 30% of the max thermal load

The study focused on the impact of CHP sizing

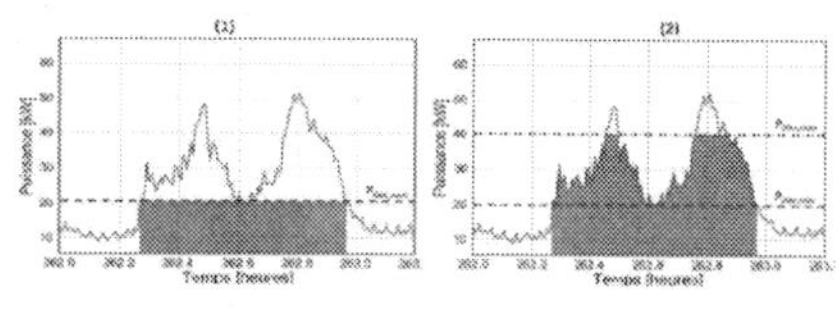

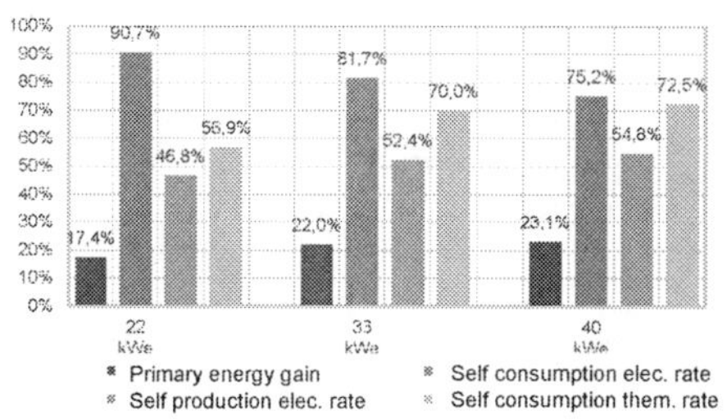

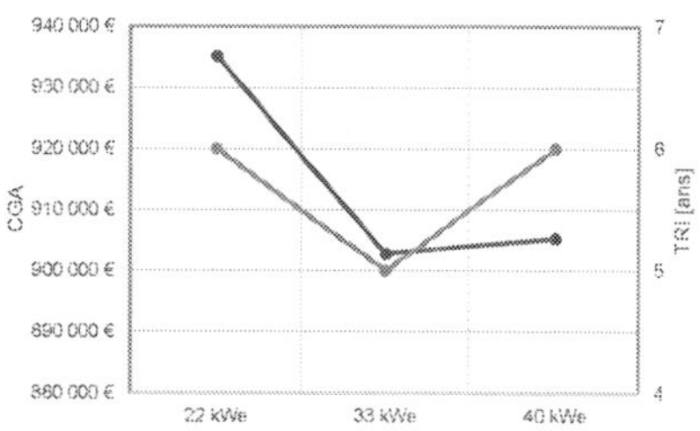

- The highest the capacity, the higher the part of thermal power produced by the CHP, and also the self production of electricity
- Economically, the 33kWe mCHP is the best compromise between self-consumption and self productoin with payack lower than 5 years

ENGIE Lab CRIGEN - Optimization of micro-CHP installation for renovation of commercial buildings and dwellings in France

22

Collective residential building
Results: Cost optimization

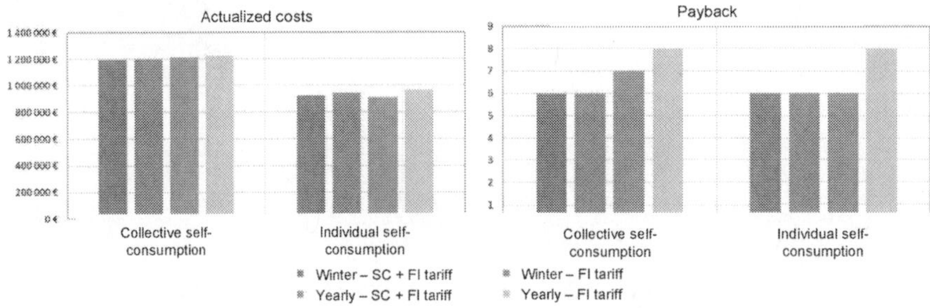

Main results

- Not interesting to have the CHP operationnal in summer with Feed-In tariffs (low cost of produced elecicity in summer)
- Collective self-consumption allows a similar payback but the interest is to integrate the appartment owners in a global scheme, that responsabilizes them and offer them an active part in the collectivity
- However, the optimization of self-consumption is of crcucial importance, especially in summer

ENGIE Lab CRIGEN - Optimization of micro-CHP installation for renovation of commercial buildings and dwellings in France

23

Conclusions

Micro-CHP in France is still a quite complex situation, but modelling tools can help to the optimization of the behavior of the installations

- They provide an ability to test the response to changes of sizings and parameters
- They also give us decision tools that allow for quick identification of the most cost effective solutions while giving possibility of services
  - Optimization of hydraulic circuits to ensure base operation of the CHP
  - Control over the number of cycles of the CHP and therefore its maintenance

Now, we all now that modelling tools remain tools, and that experimental assessment is imperative to give a direction agreed on by the sector. → next step Harware in the Loop approach

Conclusions

ENGIE LAB CRIGEN's HIL bench will be used to validate

- The models of operation of the CHP
- The hydraulic circuits themselves

} Optimization of installations based on real behaviors

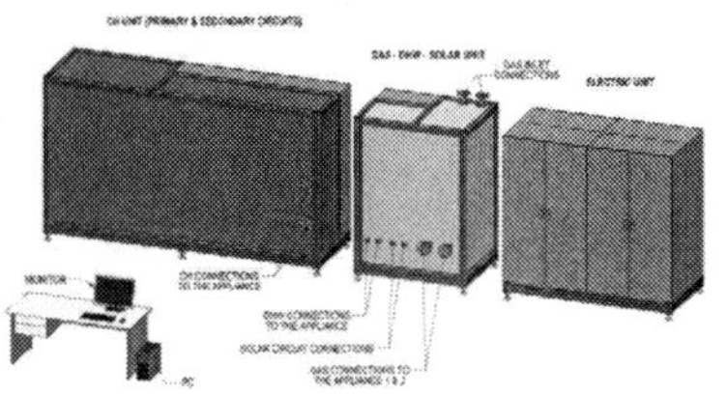

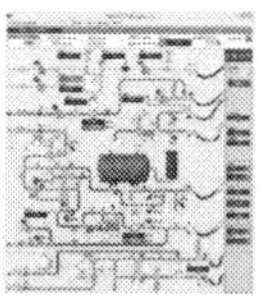

AUTHOR INDEX

Akimoto, K. 371
Al-Azri, N. 94
Al-Jahwari, F. 314
Al-Kindi, M. 94
Al-Mohannadi, H. 124
Ballestas, D. 278
Beerthuis, T. 428
Benhamidcha, A. 155
Berahim, N. 408
Bode, C. 243
Buchet, P. 250
Caumette, G. 89
Devaraj, D. 400
Discantiny, I. 219
Donnellan, P. 400
Fernandes, E. 108
Finidori, J. 420
Gersen, S. 34
Hagyard, P. 50
Hanashi, K. 177
Hardy, M. 271
Inoue, S. 483
Ismaili, H. 458
Ito, K. 233
Iwata, S. 483
Khalaf, S. 350
Knijp, J. 288
Lanzi, D. 444
Lawati, H. 165
Legrand, F. 44, 467
Louvat, A. 278
Lovat, A. 262
Mahrouqi, A. 304
Mahyijari, J. 452
Millin, P. 497
Mohammadzaheri, M. 358
Mori, T. 6
Murshid, G. 204
Nadhairi, I. 1
Nagata, T. 371
Nakao, T. 483
Noh, M. 408
Nordlund, Y. 435
Pervez, T. 314
Qamar, Z. 314
Quintino, F. 108
Raisi, F. 84
Rickelt, S. 435
Riyami, S. 103
Sammarraie, L. 366
Sano, F. 371
Sasamata, Y. 475
Schmitz, G. 243
Schweitzer, J. 147
Seman, A. 408
Senkel, A. 243
Stopa, J. 73
Suzuki, M. 483
Syron, E. 400
Tamimi, Y. 172
Tastard, C. 278
Tawk, R. 59
Tholen, H. 227
Torun, S. 131
Venkatesan, K. 428
Yanaka, M. 188
Yasuda, K. 342
Yousef, H. 117